Encyclopedia of FLIGHT

Volume 1

Accident investigation - Guernica, Spain, bombing

Indexes

Edited by
Tracy Irons-Georges

Consulting Editor
James F. Marchman III
Department of Aerospace and Ocean Engineering
Virginia Polytechnic Institute and State University

Project Editor
Heather Stratton Williams

SALEM PRESS, INC.
Pasadena, California Hackensack, New Jersey

Managing Editor: Christina J. Moose
Developmental Editor: Tracy Irons-Georges
Project Editor: Heather Stratton Williams
Copy Editor: Leslie Ellen Jones
Assistant Editor: Andrea E. Mitchell
Acquisitions Editor: Mark Rehn
Photograph Editor: Philip Bader
Research Supervisor: Jeffry Jensen
Research Assistant: Jeff Stephens
Production Editor: Cynthia Beres
Page Design: James Hutson
Layout: Eddie Murillo
Cover Design: Moritz Design, Los Angeles, Calif.

∞ The paper used in these volumes conforms to the American National Standard for Permanence of Paper for Printed Library Materials, Z39.48-1992 (R1997).

ISBN 1-58765-046-0 (set : alk. paper)
ISBN 1-58765-047-9 (v. 1 : alk. paper)

First Printing

PRINTED IN THE UNITED STATES OF AMERICA

Contents

Publisher's Note

Many books on aviation and other aspects of flight are written either at a very basic level, designed for a juvenile audience, or at a technical level, intended for members of the aviation industry. The *Encyclopedia of Flight* bridges the gap between theoretical concepts and practical applications, between scientific information and historical issues. A unique addition to the market, this illustrated, three-volume work provides information about animal and human-made flight in a way that is accessible to high school and undergraduate students, general readers, and aviation enthusiasts. It examines a wide range of topics, from birds and balloons to jets and spacecraft.

The broad scope of the *Encyclopedia of Flight* allows readers to gain a rich understanding of the more than two-hundred-year-old history of human flight and the natural factors and aerodynamic principles that led to its development. The encyclopedia addresses all categories of flying craft, both civil and military, including kites, gliders, dirigibles, biplanes, hovercraft, rockets, missiles, and satellites. In addition, entries can be found on numerous types of aircraft, from Sopwith Camels and B-52's to stealth fighters and the space shuttle, as well as on several famous aircraft from history, such as the *Spirit of St. Louis*, the *Spruce Goose*, and the *Hindenburg*. Also discussed are the principles of aerodynamics and design. Some entries focus on flying insect and animal species. Articles such as "Avionics," "Ailerons and flaps," "Guidance systems," "Instrumentation," and "Turboprops" cover the mechanical and technical aspects of flying. Readers can learn about the features of the 707, DC, and MD plane families.

The encyclopedia also analyzes many procedural and social issues pertaining to the contemporary airline industry, from ticketing and airport security to air rage and hijacking, and it offers profiles of major aircraft manufacturers and air carriers, both past and present. Biographical entries highlight many pioneers in aviation and aeronautics, such as Sally Ride, Alberto Santos-Dumont, Konstantin Tsiolkovsky, the Wright brothers, and Chuck Yeager. The use of flight in various world conflicts is examined, from pigeons and balloons in the Franco-Prussian War (1870-1871) to Patriot missiles in the Gulf War (1991). Many topics in the history of spaceflight are covered, such as "Astronauts and cosmonauts," "Apollo Program," and "Orbiting." It should be noted, however, that travel in space is not the primary focus of this encyclopedia; more extensive information can be found in our three-volume *USA in Space* (2d ed., 2001).

The ability of human beings to take to the skies in flight had become routine, even mundane, by the start of the twenty-first century. For many people, that attitude changed on September 11, 2001, when hijacked commercial jetliners became weapons of mass destruction at the World Trade Center in New York City and the Pentagon in Washington, D.C. The *Encyclopedia of Flight* was in its final stages of editorial work, and many articles written in prior months—such as "Terrorism," "Hijacking," and "Airport Security"—were quickly updated to reflect the new aviation landscape and to speculate on the future finances and regulations of this worldwide industry.

The 321 alphabetically arranged entries in the *Encyclopedia of Flight* range from one to six pages in length. Each article begins with ready-reference top matter that defines the topic and outlines its significance. Dates and alternate names are given where applicable. The main text was written with the needs of students and general readers in mind; for example, articles such as "Aerodynamics," "Mach number," and "Wind shear" present clear discussions of these topics, explaining any terms or references that may be unfamiliar. Every essay ends with an annotated bibliography that directs readers to further sources of information. Cross-references at the end of the article direct readers to related topics, forging connections between events, inventions, and biographies. Distributed throughout the *Encyclopedia of Flight* are valuable photographs, line drawings, charts, tables, and maps, as well as historical time lines describing developments in the history of flight as well as significant disasters. Boxed sidebars discuss important concepts, key developments, or examples to expand on topics mentioned in the text.

The *Encyclopedia of Flight* contains a number of useful tools to aid readers in locating entries in their areas of interest. An Alphabetical Index of Entries and a Categorized Index of Entries appear at the end of each volume. In addition, volume 3 concludes with a Glossary of relevant terms, a general Bibliography, a directory of helpful Web Sites arranged by category, an annotated list of flight-related Organizations and Agencies, contact information for major Flight Schools and Training Centers in North America and for notable aviation Museums of North America, annotated lists of International Airports and of major Air Carriers throughout the world, an extensive list of various Airplane Types grouped chronologically, a Time Line of important events in the history of flight, a list of Air Disasters and Notable Crashes, and a full Subject Index.

Reference works such as the *Encyclopedia of Flight* would not be possible without the help of many experts in the field. More than one hundred contributors, including pilots, professionals, and academicians, have lent knowledge and insight to this project. Their names and affiliations are listed in the pages that follow. We are particularly grateful to the project's consulting editor, James F. Marchman III, Professor of Aerospace and Ocean Engineering at Virginia Tech in Blacksburg, Virginia, whose hand helped guide the project at every step.

Contributors

Richard Adler
University of Michigan-Dearborn

Robert L. Ballantyne
Reading Area Community College

R. Kurt Barnhart
Indiana State University

Maryanne Barsotti
Independent Scholar

Wendy S. Beckman
Parks College of Engineering and Aviation

Raymond D. Benge, Jr.
Tarrant County College

Alvin K. Benson
Brigham Young University

Kenneth H. Brown
Northwestern Oklahoma State University

Douglas Campbell
Independent Scholar

Roger V. Carlson
Jet Propulsion Laboratory

Willie Jane Cave-Dunkel
Southern Illinois University

Frederick B. Chary
Indiana University Northwest

Monish R. Chatterjee
Binghamton University, SUNY

Joseph F. Clark III
Embry-Riddle Aeronautical University

Douglas Clouatre
Kennesaw State University

Veronica T. Cote
Bridgewater State College

John A. Cramer
Oglethorpe University

Scott R. Dahlke
United States Air Force Academy

Ursula Malluvius Davidson
Aviation Education Services, Inc.

Bruce J. DeHart
University of North Carolina at Pembroke

James S. Douglas
Douglas Aircraft Company of Ohio

Ellen Elghobashi
Independent Scholar

Said Elghobashi
University of California, Irvine

James C. Elliott
Independent Scholar

Victoria Erhart
Catholic University of America

Ronald J. Ferrara
Middle Tennessee State University

Alexandra Ferry
Independent Scholar

David G. Fisher
Lycoming College

Richard D. Fitzgerald
Onondaga Community College

Triantafyllos G. Flouris
Auburn University

George J. Flynn
SUNY-Plattsburgh

David E. Fogleman
Southern University at Shreveport

Alan S. Frazier
Glendale Community College

C. George Fry
Winebrenner Theological Seminary

K. Fred Gillum
Colby College

Richard E. Givan
Eastern Kentucky University

Daniel G. Graetzer
Independent Scholar

Oliver Griffin
Weber State University

Pamela M. Gross
Adams State College

Robert Harrison
University of Arkansas Community College at Batesville

Paul A. Heckert
Western Carolina University

Paul Hodge
University of Washington

William H. Hoffman
Independent Scholar

Niles R. Holt
Illinois State University

Willem J. Homan
Western Michigan University

Cass D. Howell
Embry-Riddle Aeronautical University

W. N. Hubin
Kent State University

Thomas Inman
Pennsylvania College of Technology

Tracy Irons-Georges
Independent Scholar

Jamey D. Jacob
University of Kentucky

Lance Janda
Cameron University

Bruce E. Johansen
University of Nebraska at Omaha

John C. Johnson
Embry-Riddle Aeronautical University

Leslie Ellen Jones
Independent Scholar

Richard C. Jones
Texas Woman's University

Maureen Kamph
Independent Scholar

Lori Kaye
Independent Scholar

Narayanan M. Komerath
Georgia Institute of Technology

Kenneth M. Krongos
Independent Scholar

Donald L. Kunz
Old Dominion University

Josué Njock Libii
Purdue University-Fort Wayne

M. A. K. Lodhi
Texas Tech University

John L. Loth
West Virginia University

Matthew G. McCoy
Arizona State University

Dana P. McDermott
Independent Scholar

Nancy Farm Mannikko
Independent Scholar

James F. Marchman III
Virginia Polytechnic Institute and State University

Carl Henry Marcoux
University of California, Riverside

Robert Maxant
Independent Scholar

Mark Miller
Independent Scholar

Randall L. Milstein
Oregon State University

Walter Nelson
RAND Corporation

Eugene E. Niemi, Jr.
University of Massachusetts, Lowell

Cynthia Clark Northrup
University of Texas at Arlington

Jim Oppermann
Ohio State University

Jani Macari Pallis
Cislunar Aerospace, Inc.

Robert J. Paradowski
Rochester Institute of Technology

Alan Prescott Peterson
Gordon College

John R. Phillips
Purdue University-Calumet

George R. Plitnik
Frostburg State University

Aaron D. Purcell
University of Tennessee, Knoxville

Stephen M. Quilty
Bowling Green State University

Steven J. Ramold
Doane College

P. S. Ramsey
Independent Scholar

John David Rausch, Jr.
West Texas A&M University

Frank J. Regan
Independent Scholar

Kevin B. Reid
Henderson Community College

R. Smith Reynolds
Embry-Riddle Aeronautical University

Dawna L. Rhoades
Embry-Riddle Aeronautical University

Charles W. Rogers
Southwestern Oklahoma State University

David M. Rooney
Hofstra University

William B. Rourke
Metropolitan State College

Alison Rowley
Duke University

Frank A. Salamone
Iona College

Mary Fackler Schiavo
Ohio State University

R. Baird Shuman
University of Illinois at Urbana-Champaign

Sanford S. Singer
University of Dayton

Billy R. Smith, Jr.
United States Naval Academy

Larry Smolucha
Benedictine University

Polly D. Steenhagen
Delaware State University

Barry M. Stentiford
Grambling State University

Robert J. Stewart
California Maritime Academy

Sue Tarjan
Independent Scholar

Gregory S. Taylor
Grambling State University

Leslie V. Tischauser
Prairie State College

Lance Wayne Traub
Texas A&M University

Mary Ann Turney
Arizona State University

Robert J. Wells
Society for Technical Communication

Hugh Wheeler
Independent Scholar

Robert Whipple, Jr.
Creighton University

David R. Wilkerson
Oklahoma State University

Heather Stratton Williams
Independent Scholar

Seth B. Young
Embry-Riddle Aeronautical University

Introduction

Few words better capture the dreams, desires, fantasies, and fears of humankind than "flight." The English language is filled with its imagery. People have flights of fancy, let their imaginations soar, aspire to greater heights, fly into the face of adversity, rise to the occasion. They sometimes refer to their apprehensions about life as "fear of flying." To fly is to go fast, to excel, to be excited, to live life to its fullest. Yet, except for a tiny fraction of the span of human existence, flight could only represent a dream.

Images of flight have been an important part of humankind's experience for thousands of years. Ancient myths and legends are pervaded with fantasies of flight. The earliest known civilizations used everything from cave paintings to stone carvings on temple walls to record visions of gods and animals and even humans endowed with wings. There are legends of flying carpets, flying broomsticks, flaming flying chariots, lions and horses with wings, ancient rulers carried aloft by giant birds, and numerous mortals who fell to earth while attempting to soar on birdlike wings. Sacred writings of most religions tell of gods and their messengers who migrate through the skies to communicate with humankind. To fly is to emulate the gods, to dare to enter their heavenly domain. Flight was, and still is, a kind of magical fantasy, capable of transforming the toils, frustrations, and limitations of day-to-day existence into the world of dreams.

Children often dream of flying with Peter Pan to far-off lands and adventures. They may drape towels over their shoulders and run down the nearest hill, hoping that their next leap into the air will send them soaring like their favorite comic book hero. Children and adults may share similar fantasies as they read of fictional wizard Harry Potter swooping above a playing field on his broomstick. Adults gaze at travel brochure illustrations of jet planes gliding through the clouds, imagining that they are carried off to dream vacations in distant, exotic locales.

Some people take their fascination with flight a step further than mere imagination. History is filled with those who dared to take that step, one which for thousands of years only led to disaster. The first were those who were convinced that they could take to the air like birds, if only they could build their own set of wings. Although these early "tower jumpers" ranged from commoners to kings and lived in ancient lands in all parts of the world, their story was always the same—gravity proved too strong a bond between person and planet. Today people can build their own wings and hang glide with relative ease, but it took centuries of ill-fated experiments before this became possible.

Most who dreamed of flight spent hours watching the motions of birds and, like early engineering geniuses such as Leonardo da Vinci, thought the key was in devising the means of using a human's limited muscle power to flap a set of strong yet lightweight wings. This method worked well for birds, so it seemed that it should be the best way for humans to take to the skies. There were, however, fundamental flaws in this approach, one being that a human's power output-to-weight ratio was no match for that of a bird. Another was that a bird's wings do not merely flap up and down; they rotate and change shape and angle as they flap. There is no record that Leonardo da Vinci ever attempted to fly the machine in his drawings, but many others apparently tried to emulate birds. From Icarus in ancient Greece to King Kai Koos of Persia to King Bladud (father of King Lear) of medieval England to the hapless but determined experimenters seen in old-time newsreels, only broken wings and bodies were left in their wakes.

Finally, in the mid-1800's, experimenters and scientists such as Sir George Cayley of England realized that if people were ever to fly, they must separate the two tasks of propulsion and lift and must learn to emulate the actions of a bird's tail in providing balance in flight. Meanwhile, in the previous century, the Montgolfier brothers in France had successfully taken to the air in balloons, carried aloft by captured hot air from open fires and propelled, sometimes unintentionally, by the winds.

It was balloons filled with hot air and later with hydrogen and helium, rather than birdlike wings, that first took human beings into the sky. Early balloons were often tethered to the ground for safety and used primarily to provide their passengers never-before-seen views of the surrounding landscape or of their enemies in wartime. By the end of the nineteenth century, these structures had evolved into helium- or hydrogen-filled bags that carried boatlike craft equipped with crude propellers and rudders which, on rare windless days, could transport one or two passengers for a few miles in a selected direction. Flight, of a sort, had finally entered the realm of the possible for humankind.

Meanwhile, the attempt to fly using wings continued, coming closer with each new trial success. Gliding flight was achieved by experimenters such as Otto Lilienthal in Germany. He learned through trial and error the proper way to suspend himself beneath batlike wings and to shift his weight from side to side and forward to aft to maintain equilibrium as he soared a few feet above his artificial hillside. In the United States, Octave Chanute corresponded with Lilienthal and others while conducting experiments of his own. His publications on flight inspired others in the United States, including Samuel Pierpont Langley, head of the Smithsonian Institution, and Orville and Wilbur Wright in Ohio to their own efforts to turn their dreams to reality.

Adding primitive propellers powered by steam or early internal combustion engines to gliders of widely varying sizes allowed inventors and adventurers of all type to experience everything from utter disaster to the thrill of a short powered "hop" into the air. Others around the world claimed success in momentarily getting a heavier-than-air machine off the ground with a person on board. Langley sent his powered but pilotless aerodrome on a successful flight down the Potomac River near Washington, D.C.

Virtually alone among all their peers, the Wright brothers learned that successful flight required successful control, and their patient experimentation and analysis culminated in their first flight in December, 1903, on the sands of Kitty Hawk, North Carolina. In the same month, Langley's second attempt to launch his scaled-up and piloted aerodrome from a houseboat in the Potomac River ended in failure, with his fragile craft lying broken in the water and the newspaper headlines bemoaning the waste of government funds on his fantasy.

In the years since that day at Kitty Hawk, flight has grown far beyond the wildest dreams of those early "aeronauts." Not even the Wrights envisioned their invention being able to carry large numbers of passengers or tons of cargo over thousands of miles. Any mention of flight faster than the speed of sound was thought of as lunacy for at least the first third of the twentieth century. Children now play with flying model aircraft and rockets that are far more sophisticated than the vehicles imagined by their great-great-grandparents. Today's programmable, rechargeable, battery-powered, radio-controlled craft were not predicted by Jules Verne in his novels of the time.

Verne did foresee space travel, however, although his imaginary flight to the Moon was launched from a gun instead of a solid- or liquid-fueled rocket booster. The rocket has gone from its roots as a Chinese invention used for celebration and warfare to become the basis for sending people and scientific probes into ever-expanding reaches of space. Proponents of spaceflight such as Robert H. Goddard experimented with multistage, liquid-fueled rockets that would take their payloads far above the realm of winged flight. After World War II, U.S. and Soviet scientists built upon German weapons technology to design rockets capable of reaching orbit.

The 1957 launch of the Soviet Sputnik satellite shocked the United States out of its scientific complacency, and since that time, space exploration and travel has moved from science fiction beyond the solar system into deep space. In 2001, Dennis Tito, the world's first paying "space tourist," launched on a Russian rocket for a stay in the International Space Station. In 1969, the world watched with anxiety and elation as men from Earth left their footprints in the loose soil of the Moon. Today's dreamers focus on Mars, the planet that has fascinated stargazers and readers of science fiction for centuries. Will scientists find water there or dig up evidence of long-extinct microbes which populated a once-living planet?

Spaceflight continues to capture the imagination of young and old alike, as did the adventures of daring aviators such as Charles A. Lindbergh in the 1920's. For many of today's young people, the start of their path into space will begin at a local airport just as it did for those of Lindbergh's time who looked to the skies and envisioned their futures in Ford Tri-Motors and DC-3's. Today's local airports may bear little resemblance to the dirt- and grass-covered airstrips of the 1920's and 1930's, but they remain, for many, the point of departure from forces of gravity to the skies and beyond. Instead of the open-cockpit, fabric-covered, and wire-braced biplanes of an earlier era, today's student pilots make their first flights in instrument-filled cabins of Cessna and Piper aircraft. They will be guided by a GPS signal or a VOR needle instead of railroads and highways below. Soon NASA's Highway-in-the-Sky-based Small Airplane Transportation System will allow pilots to navigate from home to destination virtually automatically.

If one lingers long enough at any small airport, one will be fascinated to watch student pilots, young and old, anxiously anticipating their next trip into the sky with a flight instructor. Even though many hours of flying are required before new pilots pass the flight test for a license, there is little that compares to the excitement of the student pilot's first solo flight. Students may feel like they have been doing "touch-and-goes" forever, learning to handle the small aircraft in the most demanding parts of any flight, takeoffs and landings. There is a building passion finally to test one's new knowledge alone in the cockpit, to fly solo. The student's training has included slow flight and stalls, and the instructor is relatively confident that the student can handle the plane alone. After a couple of routine trial takeoffs and landings, the plane pulls to the side of the runway and the teacher steps out, telling the student to go it alone.

As a nervous instructor looks on, the student, with a combined feeling of absolute elation and rising fear, taxis the plane out to the end of the runway and begins accelerating toward takeoff. It takes a minute to recover from the shock felt as the small plane lifts off the ground much sooner than it had in previous flights, now relieved of one-half its normal payload. Right away it is time to think about landing as this first solo is a mere loop around the field in the airport traffic pattern, and the lone pilot begins to wonder how different this landing will be from the scores of touchdowns made previously with a heavier plane. The plane turns onto the final approach and descends, ever so slowly because of its light weight. Sweat breaks out as the student pilot begins to think the wheels will never reach the ground as they glide farther along the runway than ever before.

Finally, the wheels touch and an excited student pilot fights to control both built-up emotions and crosswinds while taxiing proudly back to the hangar and to the waiting, and now much less nervous, teacher. Then, a time-honored ritual is carried out by the flight instructor, who cuts off a portion of the student's shirt back and, with an appropriate blend of seriousness and celebration, writes the student's name and the date of the flight on the cloth and pins it to the bulletin board in the pilot's lounge. The student's "tail feathers" have been cut, both literally and symbolically. A new pilot-to-be is ready to take to the sky, free and alone. Although many hours of further instruction remain, both student and teacher are now confident that the license to fly is within reach. A pilot's future awaits, perhaps as an airline or fighter pilot or even as an astronaut. For most, a pilot's license will lead to a life-long hobby that will add depth and dimension to life.

Other seekers of flight will head for the mountains and the seaside dunes, finding their ultimate adventure in hang gliding, a modern-day return to the experiments of Lilienthal, Chanute, and the Wright brothers. Some will add a small motor to such a lightweight craft and enter the world of ultralight aviation. Others will opt for the new sport aviation class of aircraft, which will offer much of the improved safety of conventional small airplanes at lower cost and with less regulation.

Others who may never pilot their own craft will play a role in the world of flight by becoming aerospace engineers, airport managers, mechanics, technicians, or one of the hundreds of other professionals who are needed to design new airplanes and spacecraft and to keep them flying. It takes thousands of engineers to design and certify a new airliner and years of research and testing to develop new space probes. Highly trained electronic and computer technicians and engine mechanics are needed to keep these planes and spaceships flying. It takes thousands of people with all manner of educational backgrounds to keep the world of aviation operating.

For most, however, dreams of flight will be realized through the hustle and bustle of airport hubs in business or vacation travel, an environment in which the dream can easily become a nightmare in the crowded skies of the twenty-first century. In contrast to every jaded and seemingly bored business traveler on today's packed airliners, however, there is one excited five-year-old or equally enthusiastic eighty-year-old who looks out the airplane window with eyes full of wonder as the engines roar and the ground disappears beneath them.

It is for those people whose interest in flight ranges from idle curiosity to jubilant enthusiasm that these volumes are intended. The *Encyclopedia of Flight* should provide a handy first reference for those wanting to know a bit more about aircraft and spacecraft, how they are designed and built, and how they operate. In addition, the *Encyclopedia of Flight* covers a wide range of related topics, both historical and technical. It is designed to give authoritative definitions, explanations, biographical sketches, and general information about hundreds of flight-related topics. Every entry is written by an identified expert in the field and has been edited to ensure that it will be understood by the nonengineer or nonscientist. Each entry includes references recommended by its author that can provide the interested reader much more detail and depth about the chosen subject.

Flight has been the dream of humankind for all of recorded history. For a fortunate few generations, flight has become a reality, but human curiosity and excitement about flight, whether to Walt Disney World or to Mars, is as great as ever. Whether one's interest is in how an airplane flies, in how a rocket engine works, in the history of flight, or in any one of the scores of aviation and space-related topics, the *Encyclopedia of Flight* should prove an outstanding resource, either giving the quick answer being sought or providing a starting place for an adventure of discovery.

James F. Marchman III
Department of Aerospace and Ocean Engineering
Virginia Polytechnic Institute and State University

ENCYCLOPEDIA OF
FLIGHT

A

Accident investigation

Definition: The examination into the causal factors of an aircraft mishap or incident.

Significance: The investigation of aircraft accidents is important to the aviation industry for many reasons. The study of accident factors helps airlines determine accountability, educate inexperienced pilots, and prevent future accidents.

Perceptions and Realities

Those of the flying public who are not airline pilots, and even some pilots who fly as passengers, are sometimes nervous while doing so. Although airline accidents occur infrequently, those that do occur can be catastrophic events involving a great loss of life. Media coverage of airline accidents is usually extensive, fueling the uneasy feelings many people have about airline travel. However, the periods after aviation accidents are often the safest times in which to fly. Further, contemporary aviation accident rates are very low, in relation to other types of accidents, such as automobile accidents.

Because aviation always involves the risk of an accident, accident investigation is an important element of aviation education. By studying the accidents of other pilots, less experienced aviators can avoid making similar mistakes.

Accident Patterns

A commonly noted pattern in aviation accidents is that there is rarely only a single reason for the accident. Experienced pilots refer to the events leading up to an accident as an error chain. Individual links of the chain, when combined, cause an accident to happen. For instance, bad weather alone might not cause an accident, but bad weather combined with darkness and the fact that the pilot became lost might. The error chain is weather, darkness, and becoming lost. The elimination of any one of these factors may prevent the mishap. In other words, if one link of the chain were broken, the accident would not happen.

The key to accident investigation, then, is to determine the error chain leading up to the event. Rectifying the situation regarding an accident under investigation is impossible, because the accident has already happened. However, accident investigators can study each link of the chain and prepare documents to report their findings. In turn, other flight crews can study the reports and avoid the same fate.

The vast majority of aviation accidents result from human-factor errors. All those who have roles in launching an aircraft, including pilots, mechanics, air traffic controllers, cabin attendants, and baggage handlers, can make mistakes that may cause an airplane accident.

Accident Rates

In the study of aviation accident investigation, an important statistic is the number of accidents compared to the hours flown. The number of accidents per 100,000 hours determines a figure known as the accident rate. With the exception of the mid-1990's, the accident rate has been declining since 1982. The number of accidents has declined, as has the number of fatalities. The Federal Aviation Administration (FAA) and others in the industry attribute better pilot education and technological advances for improvement of aviation safety statistics.

Causal Factors

According to statistics compiled by the industry, approximately 70 percent of air carrier accidents are the results of flight crew error. Maintenance error constitutes another 5 percent, while air traffic control or other airport issues account for about 4 percent. Human error is responsible for a total of almost 80 percent of the commercial airline accidents worldwide. Of the remaining 20 percent, mechanical failures make up 11 percent, weather factors account for about 4.5 percent, with the remainder categorized as miscellaneous or other.

Pilots and first officers are responsible for most of the human error accidents. Reasons for the flight crew's mistakes are many, including loss of situational awareness, flight crew fatigue, and training and operational issues.

Accidents can occur during all phases of flight, from pushback to arrival at the destination gate. However, the majority of accidents, almost 56 percent, happen during the approach and landing phase of the flight.

The reasons for approach and landing accidents vary. If a flight has been a particularly long one, crew fatigue can play a significantly greater role at the end of a flight than at the start. Being tired or fatigued can impair a crew's decision-making process. Poor destination weather combined with a tired flight crew could be a recipe for disaster.

Post-Accident Sequence of Events

During an accident investigation, there are many simultaneous issues requiring attention. The first and most important consideration is to assist the injured. Medical personnel are needed immediately to administer to those on site, and provisions are needed to transport patients to the nearest medical facility as quickly as possible. The rescuers also need to determine where the injured were sitting in the aircraft and where they ended up after the accident. The next task is dealing with the survivors of anyone killed in the accident; if there is even one fatality, the loss touches many people.

After the first officials arrive on the scene, their first order of business is to secure the area. Another important task is to observe evidence that is transient in nature. For instance, a popular twin-engine aircraft seemed to be crashing for no reason. It took four such crashes of a similar nature before investigators arrived at the wreckage quickly enough to determine that ice forming on the aircrafts' horizontal stabilizers had caused the accidents. After the previous accidents, the ice had melted before anyone could see or record its presence.

After first seeing to the injured, personnel guarding the accident scene have several responsibilities. They must make certain the wreckage is not disturbed, because if someone moves the wreckage, aviation accident investigators will no longer be able to see the parts of the aircraft as they came to rest after the accident. Consequently, investigators will lose many clues that may help them determine a probable cause of the accident. Another essential task is to determine whether hazardous materials were being transported and are present at the scene. If so, personnel must take measures to protect everyone on scene from the dangers of the hazardous materials.

It is important for the accident investigators to photograph the scene. Photographs of the wreckage can preserve the visual evidence of the accident for later analysis and should include all aspects of the accident: from individual parts of the aircraft to any fatalities where they lie. Marks caused by the craft's hitting the ground and aerial photos are important to show the path of the aircraft as it crash-landed.

In addition to photographs of the crash scene, the investigators should also draw sketches and maps of their observations. By using the photographs, sketches, and maps, they will be able to create a diagram of the last moments of the flight. This diagram of the flight's last moments is essential to understanding the decisions made by the crew.

Guarding the wreckage is an important responsibility for those tasked with the job. Guards should be somewhat familiar with aviation. They must protect the property, the wreckage, and the crash site from being disturbed. They have the difficult duty of making sure that people do not wander through the area. They also collect the names, addresses, and telephone numbers of anyone who may have witnessed the crash. If an accident occurs in a remote location and all aboard sustain fatal injuries, eyewitness statements may be nonexistent. Anyone with knowledge of the accident must be located because witnesses can be very important in helping to determine the cause of an aviation accident.

The best witnesses to an aviation accident are not other pilots, or those involved in the industry. In fact, they are not even adults. Children often provide the most accurate and unbiased statements about aviation accidents. Adults often tend to put their own spin on an accident. Pilots who witness accidents may inject far too much opinion into their account of what happened. Children very simply report what they see.

National Transportation Safety Board

Once notification of a major aircraft mishap reaches the authorities, the National Transportation Safety Board (NTSB) launches a go team. The team originates from Washington, D.C., where the members of the NTSB rotate the duty of being on the team. While the go team is en route, it is the job of the local authorities on scene to organize the agencies to start the rescue or recovery procedures.

The accident investigator is indeed a detective. Typically, it is very difficult to determine exactly what caused an aircraft accident. The investigator's first order of business is to sort through the pieces of wreckage, cataloging the more and less obvious clues leading to the most probable cause of the mishap. In many cases, it is much like looking for the proverbial needle in a haystack.

Once the team is on site, it will survey the wreckage to determine where the aircraft initially struck the ground. Damaged shrubbery and trees may mark the path of the airplane into the crash scene. The investigators will note the general condition of the wreckage while trying to account for all the parts. If parts or components of the aircraft are missing from the accident scene, this may be indicative of an in-flight structural failure. If the empennage, or tail, or the wing of the aircraft came off the airplane at a high altitude, that could well be the cause of the accident. Those parts, once separated from the main aircraft, will fall to the

After an airplane crash, investigators arrive at the scene to inspect the debris. Comparisons of where different parts originated on the craft and where they landed on the crash site are often key to discovering the cause of the accident. (AP/Wide World Photos)

ground sometimes miles away from the main site of the wreckage.

The investigators begin with the physical aspects of the accident. They collect parts, examine each one, and map its final placement against the original installation on the aircraft. They also have to be certain that the aircraft is complete; parts of the aircraft missing from the wreckage site suggest an in-flight breakup in which other essential components of the airplane have landed elsewhere.

From the physical evidence, the investigators can then determine the aircraft's approximate angle and speed of impact. They can also determine whether the engines were working at the time of impact. They map and photograph the wreckage to preserve as many of the clues as possible. After this work is complete, they will begin their true detective work.

This detective work begins with a review of the pilot's qualifications and an examination of the pilot's training and certification documents and medical records. The investigators conduct interviews with the pilot's friends, peers, and relatives. They review autopsy results. They create a pathological history for a seventy-two-hour time period leading up to the accident and conduct weather data analysis, among other things. The investigators check into the pilot's physical and psychological makeup and try to determine the pilot's state of mind at the time of the accident. They question the pilot's aeronautical decision-making abilities, look into recent history of the pilot's judgment, and even evaluate the pilot's training and experience.

Investigators also try to determine whether the weather was a factor in the accident, relying upon official weather reports and forecasts. They look for indications of low visibility, turbulence, extreme wind shear, or heavy rains that may been contributors.

Reconstruction

Finally, investigators examine the aircraft wreckage to determine whether mechanical malfunctions may have caused the crash. This is one of the more intense segments of the investigation. The aircraft will undergo reconstruction in a secure hangar or other facility. Plans of the aircraft are helpful in determining that all parts of the aircraft have been retrieved from the wreckage site and adjacent areas.

The aircraft's reconstruction helps the examiners find signs of structural failure. The key to determining structural failure involves asking whether engine failure caused the accident or whether it caused the breakup of other parts of the airplane. Investigators try to figure out where such a breakup first occurred. This is the most intriguing part of accident investigation, and it may go well beyond the expertise of the investigators. On many occasions, expert witnesses, such as metallurgists, are necessary to assist in finding the answers.

Every aspect of the aircraft is under dissection during the accident examination. Disassembly of each component and system of the aircraft will follow for investigation of any possible failures. Examination of the flight controls may reveal a frayed cable or a broken bearing; a control pushrod may have become bent, allowing aerodynamic flutter to start. That aerodynamic flutter may have caused an actual structural failure of the elevator or rudder, hastening the accident.

Investigators also check switch positions at the time of the crash. They are especially interested in the positions of the switches and controls and the relative positions of the associated components. These indicate whether the pilot may have done something improper to cause the accident, such as raising the flaps at the wrong time or unintentionally dumping fuel, resulting in fuel exhaustion. A component failure may be indicated if a switch was properly set and the component discovered is not positioned per the switch selection.

Investigators are also intensely interested in the instrument readings at the time of impact. After removal from the crash site, each instrument is sent to an appropriate laboratory for intense post-accident analysis. At the time of impact, each needle on the face of an instrument leaves impact marks that allow the technicians to determine exactly what measurement the instrument was indicating at the time of the crash. With this technique, investigators can determine the speed of the aircraft. They can also make corroborations between engine operations and other instrument indications that may help in explaining the accident.

The investigators have a high interest in the navigational instruments and radios, along with everything else in the cockpit. This is particularly true if the accident happened in poor weather. By careful analysis of the frequencies selected, the switch positions of the units, and the readings at impact, examiners can determine if a navigational error factored into the cause of the accident. All other systems in the airplane, such as the generators or alternators, the vacuum systems, pneumatics, and hydraulics, are also scrutinized during the investigation.

Flight and Cockpit Recorders

Flight recorders and cockpit voice recorders, carried in the tail of an aircraft, are important elements in accident investigation. They provide critical clues in solving the mysteries associated with many of the world's air disasters and are invaluable in helping to prevent future accidents. Although they are known as black boxes, they are actually painted bright orange to aid in their recovery following an accident.

Aircraft flight recorders record many different operating conditions of a flight and provide information that may be difficult or impossible to obtain by any other means.

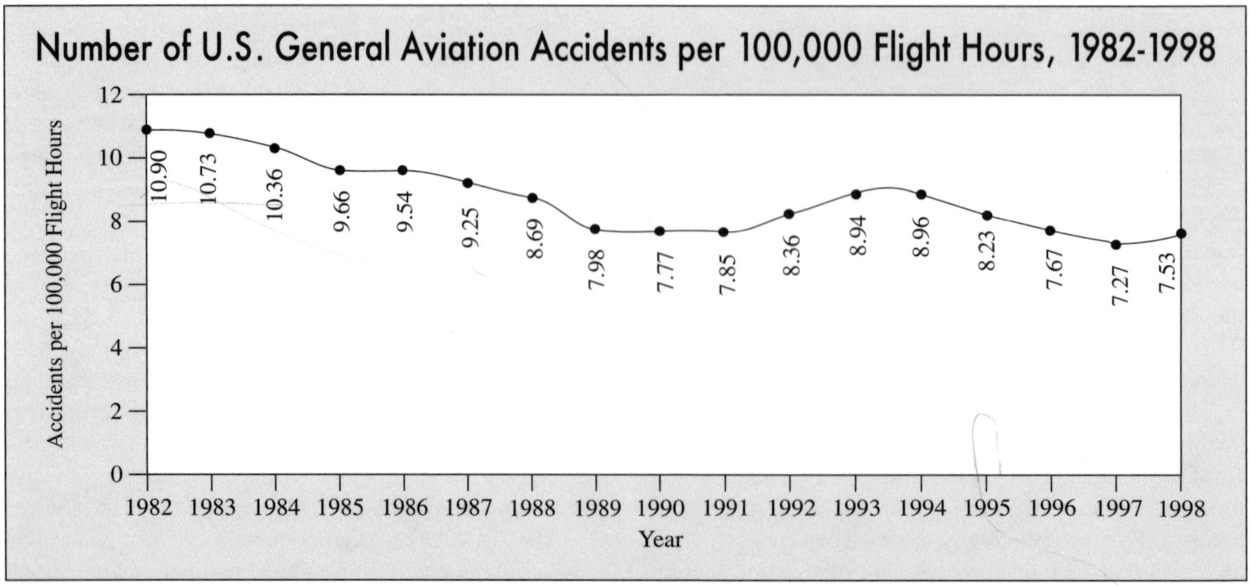

Source: National Transportation Safety Board, 2000.

Cockpit voice recorders record the flight crew's voices, as well as other sounds within the cockpit, including communications with air traffic control, automated radio weather briefings, and conversation between the pilots and ground or cabin crew. Sounds of interest to an investigation board, including engine noise, stall warnings, landing gear extension and retraction, and any clicking or popping noises, are also typically recorded. Based on these sounds, important flight parameters, such as speed, system failures, and the timing of certain events can often be determined.

In the event of an accident, an investigation committee creates a written transcript of the cockpit recorder tape. Local standard times associated with the accident sequence are determined for every event on the transcript. This transcript contains all the pertinent portions of the cockpit recording. Due to the highly sensitive nature of the verbal communications inside the cockpit, a high degree of security is provided for the cockpit recorder tape and its transcript. The timing of release and the content of the written transcript are strictly regulated.

Autopsies

The examiners find the pilot's autopsy results particularly important in determining whether pilot incapacitation caused the accident. The pilot may have experienced a heart attack, stroke, or some other medical factor that caused incapacitation. The pilot may have passed out due to hypoxia or carbon monoxide poisoning. The results of the autopsy enable the investigators to assign probable cause to the pilot or rule out incapacitation.

The airplane also undergoes a mechanical autopsy of sorts. After an aviation accident, the maintenance records and logbooks of the accident aircraft are collected. NTSB examiners examine the records in search of evidence of possible material defects or mechanical malfunction. The inspectors may determine that there were metallurgical or manufacturing defects in the history of the aircraft. They may uncover improper maintenance procedures, such as that a life-limited part has been allowed to exceed its time in service. These are only a few of the possible explanations for the accident.

These observations and examinations comprise the bulk of the investigators' work. The compilation of information on the accident and the examination of evidence may take months, or even years. Many people in the agency are involved in the search for answers about the cause of the accident. As the work is completed, many more people await the results, some patiently, others less so.

The meticulous work of the accident investigators takes time, however. Although this challenging work is sometimes tedious and demanding, it must always be thorough. Its reward is the promise of the prevention of future accidents.

Joseph F. Clark III

Bibliography

Ellis, Glenn. *Air Crash Investigation of General Aviation Aircraft: With Emphasis on the Crash Scene Aspects of the Investigation.* Greybull, Wyo.: Capstan, 1984. An explanation of the investigative process followed by the Federal Aviation Administration and National Transportation Safety Board regarding the general aviation industry.

Hawkins, Frank H. *Human Factors in Flight.* 2d ed. Brookfield, Vt.: Ashgate, 1987. The definitive volume on the study of humans as they relate to aviation and aviation accidents and required reading for anyone considering a position with the NTSB.

Panas, John. *Aircraft Mishap Photography: Documenting the Evidence.* Ames: Iowa State University Press, 1996. An excellent guide to aircraft accident investigation. Although the emphasis is on photography, the text remains an excellent source of information regarding aviation accidents and inquest.

Wells, Alexander T. *Commercial Aviation Safety.* 2d ed. New York: McGraw-Hill, 1997. A reference work that covers all aspects of aviation safety, including the management of human error, aircraft safety technology, the nature of accidents, and flight standards and rule making.

See also: Flight recorder; National Transportation Safety Board; Runway collisions; Safety issues

Advanced propulsion

Definition: Any means for launching or propelling spacecraft beyond the use of traditional chemical rocket engines.

Significance: If humanity is ever to explore the solar system and access resources beyond Earth, cheaper, faster, and more efficient propulsion systems must be developed.

The Fundamentals

If the space shuttle's external fuel tank were placed on the pedestal of the Statue of Liberty, it would stand just taller than Lady Liberty's torch. At launch, the mass of the space shuttle is about 2,040 metric tons (4.5 million pounds), but it can deliver only 6.5 percent of that mass to low-Earth orbit, and that costs $20,000 per kilogram ($9,100 per pound). For comparison, in 2001, gold sold for about $9,000 per kilogram. To achieve a stable low-Earth orbit, the payload must simultaneously be lifted about 300 kilometers above Earth's surface and accelerated to a horizontal speed of nearly 8 kilometers per second (about Mach 23, or twenty-three times the speed of sound).

At 111 meters (364 feet), the Saturn V rocket that took the Apollo astronauts to the Moon stood over twice as tall as the space shuttle. In ascending to low-Earth orbit, the Saturn's first three stages burned for a total of 11.5 minutes, using 75 terajoules (75×10^{12} joules) of chemical energy. That was about 1.5 percent of all of the energy in the world produced from fossil fuel during those 11.5 minutes. Only 6 percent of that energy went into lifting and accelerating the Apollo payload into orbit, while most of the remaining 94 percent was expended on lifting and accelerating the fuel used on the way up.

There are many plans for more efficient spacecraft. The Venture Star project featured the more efficient aerospike rocket engine. Solar sails, plasma bubbles, gravitational slingshots, the Pegasus spacecraft, and laser- or microwave-launched spacecraft are schemes that leave all or part of the main power source behind and thereby reduce the spacecraft's mass. Scramjets gather part of their fuel, the oxidizer, in flight, and the hypothetical Bussard interstellar ramjet would gather fusion fuel in flight. Ion drives, plasma drives, and drives using nuclear fission and fusion are all schemes to increase the exhaust velocity of the propellant.

Aerospike Engines, Scramjets, and Pegasus

Worldwide, there are a number of projects under way to develop a fully reusable launch vehicle that will orbit payloads for one-tenth the cost of the space shuttle. The X-33 "Venture Star" is a sleek, wedge-shaped craft designed to take off vertically like a rocket and glide to a landing like an airplane. It pioneered the use of lightweight graphite composites in its structure and fuel tanks, and its efficient lifting body shape allowed it to fly with only stubby wings for stabilizers. Although many important technological advances were achieved, development problems led to the withdrawal of support by the National Aeronautics and Space Administration (NASA) in March, 2001, but work continued on the X-33's Boeing Rocketdyne XRS-2200 aerospike engines.

A conventional rocket engine has a combustion chamber that opens into a bell-shaped nozzle. Fuel and oxidizer are mixed and burned in this chamber, and high-speed combustion products escape through the nozzle. For greater efficiency, the pressure of the exhaust plume should match the surrounding air pressure. The aerospike nozzle is V-shaped and is turned inside out: Fuel and oxidizer are mixed in ten combustion chambers (five on each side of the V), and the exhaust plume sprays down the outside of the V. Since the outside of the exhaust plume is open to the atmosphere, it automatically blooms outward until it matches the ambient pressure, while the inside of the plume pushes against the V and provides thrust.

A scramjet is a supersonic combustion ramjet. Scramjets can be more efficient than rockets because they use oxygen from the air and must carry only the oxygen that they will use in space. A scramjet engine has no moving parts; it uses its supersonic speed (about Mach 10) and internal shape to compress air coming into its engine instead of using the rotating compressor of a normal jet engine. Hydrogen fuel is injected into the airstream in the engine, and the hot combustion gases (mostly water vapor) escape from the rear of the engine to provide thrust. A scramjet must be launched at supersonic speed before it can fly. In June, 2001, a B-52 aircraft lifted an X-43A scramjet mounted on a Pegasus-based rocket 6,000 meters (20,000 feet) into the atmosphere and launched the combination. The rocket was to accelerate to the scramjet's cruising speed and then release it. Unfortunately, a structural failure occurred shortly after rocket ignition and the mission was terminated.

Several nations are working on scramjets. In August, 2001, India announced plans to develop the Avatar, a 25-metric-ton craft believed to be able to cheaply carry a 1-metric-ton payload into a 100-kilometer-high orbit. The Avatar will take off and land like a conventional airliner using a combination of turbofan, ramjet, and scramjet engines fueled with hydrogen. A unique feature is that it is to cruise at Mach 8 for an hour at an altitude of 10 kilometers while it takes in and liquefies 21 metric tons of oxygen before it uses a hydrogen-and-oxygen-fueled rocket to push into space.

The Pegasus rocket has placed dozens of satellites into orbit and is the most successful small commercial launch vehicle in the world. The "Stargazer" Lockheed L-1101 aircraft carries Pegasus to a launch point 12 kilometers high, above the densest and most turbulent part of the atmosphere. The three-stage rocket is then released and ignited. It can carry a 450-kilogram payload into low-Earth orbit.

Solar Sails and Plasma Bubbles

The surface of the Sun is a fearsome place—a seething, turbulent ocean of blinding incandescent gases, incessantly rocked by sonic booms as gigantic gouts of matter race upward through the photosphere. The flood of energy from the Sun tears particles from its outermost part, the solar corona, and constantly drives this sun-stuff into space. This is the solar wind: electrons along with ionized hydrogen and helium atoms streaming outward at an average speed of 400 kilometers per second.

Earth's magnetosphere is the region surrounding the planet that is dominated by its own magnetic field, not the Sun's. Geophysicist Robert Winglee of the University of Washington realized that the solar wind pushing against Earth's magnetosphere pushes Earth away from the Sun, except that Earth is far to heavy for this to produce any measurable effect. However, Winglee proposed that if a light spacecraft could generate a large magnetic field, the solar wind would propel the spacecraft. He calls this hypothesis Mini-Magnetospheric Plasma Propulsion, or M2P2.

Winglee and his colleagues suggest that a 200-kilogram spacecraft (including 50 kilograms of helium) might be built around an electromagnet coil powered by solar cells. Winglee's group demonstrated that injecting ionized helium into a coil's magnetic field forces the field to expand like a bubble, becoming a mini-magnetosphere. They calculate that in space this magnetosphere would be 15 to 20 kilometers in radius, and with the solar wind pushing on it, the craft should reach speeds of 50 to 80 kilometers per second after three months. This is ten times faster than the speeds previously reached by chemical rockets. Such a craft could reach Saturn in six months instead of the seven years required for the Cassini mission.

The mini-magnetosphere is not really spherical. Its shape depends upon its interaction with the solar wind and on the parameters of the coil. To oversimplify, a mini-magnetosphere may be pictured as a flat sheet of paper orbiting the Sun. If the sheet is tilted so that its leading edge is closer to the Sun than its trailing edge, solar wind particles bouncing off of it will push it forward in its orbit and make it go faster and spiral outward from the Sun. Conversely, if the trailing edge is closer to the Sun, solar wind particles will slow it in its orbit, and the Sun's gravity will pull it inward. Since the magnetosphere is practically without mass, it should be easy to maneuver it by simply rotating the field-generating coil.

The concept of propelling a craft with solar sails is similar to M2P2, but these sails are propelled by sunlight, not the solar wind. At the orbit of Earth, the pressure of sunlight is about 9 newtons (2 pounds) per square kilometer. An 820-meter square-rigged sail, named "the clipper" by its designers, is expected to have a mass of about 2,000 kilograms. Carrying a 2,000-kilogram payload, it could travel from Earth to Mars or to the outer planets in about the same time, or less, than a chemical rocket would require. Once solar sail technology is achieved, its use would be cheaper than the use of chemical rockets because it requires much less mass to be lofted into Earth orbit for each mission. Solar sails are also reusable. They can be returned to Earth orbit, but their sunward speed is limited by the relatively weak pull of solar gravity. Energy Science Laboratories of San Diego, California, has developed a novel sail fabric, a very porous mesh of carbon fibers. They have demonstrated that the fabric is light enough to be pushed by laser light, and that it can withstand temperatures of 2,500 degrees Celsius. This is important because, someday, solar sails may be given a push by shining a battery of high-intensity lasers on them.

Electric Propulsion

The thrust produced by a rocket depends upon how much reaction mass is ejected per second, and how fast it is ejected. Chemical rockets can deliver a large amount of thrust because they can push out a great deal of mass per second, but ejection speed is limited by the amount of energy released by the chemical reaction of fuel and oxidizer. Electric propulsion engines typically deliver a small thrust with high efficiency since they can handle only a small amount of mass per second, but they can eject it at very high velocities. With a few exceptions, electric propulsion has been commonly employed only in the thrusters used by satellites for station keeping (staying where they are supposed to be).

Resistojets use electric resistance to heat propellent gases and thereby increase their ejection speed. They have operated with ammonia, biowastes, hydrazine, and hydrogen. Arcjets ignite an electric arc in the propellant flowing through a rocket nozzle. Arcjets are twice as fuel efficient as chemical thrusters, but ion engines are more efficient yet. The 480-kilogram spacecraft Deep Space 1 (DS-1) is propelled by an ion engine and powered by 2,400 watts from solar arrays. Launched on October 24, 1998, its mission was to test twelve new technologies, including the ion drive and a relatively autonomous navigation system. While DS-1 came within 26 kilometers of asteroid Braille on July 28, 1999, problems kept it from obtaining any closeup images. Its extended mission was to fly through the coma (head) of comet Borrelly on September 22, 2001, when it sampled the materials of the coma and photographed the comet's nucleus.

DS-1's ion engine uses xenon, a gas 4.5 times heavier than air, for a propellent. Xenon in the engine chamber is bombarded by electrons that ionize the xenon. The rear of the chamber is fitted with two wire mesh screens; the first is positively charged, while the second is negatively charged with up to 1,280 volts. Positive xenon atoms passing through the first screen are accelerated to 28 kilometers per second by the voltage on the second screen. The ejection of these ions into space propels the craft forward. Electrons sprayed into the exhaust stream keep the craft from building up a static charge. Although the engine exerts no more force than the weight of a sheet or two of paper, its 82 kilograms of xenon is enough for 6,000 hours of operation and can increase DS-1's speed by 4 kilometers per second. It is ten times more efficient than a chemical engine with the same weight of fuel.

Nuclear Power

The great attraction for using nuclear power in space is that nuclear reactions pack millions of times more energy than chemical reactions. While the United States placed a single nuclear reactor in space in 1965, the former Soviet Union has used small nuclear reactors to provide electrical power on dozens of satellites. Both nations have used radioisotope thermoelectric generators (RTGs) that convert the heat from radioactive decay directly into electricity, but neither nation has used nuclear power for propulsion. Since they have no moving parts and are well constructed, RTGs are considered to be relatively safe, but they are not very efficient. However, using electricity from an RTG to power an ion engine in the regions beyond Mars, where solar power is weak, is an attractive possibility.

The Nuclear Engine for Rocket Vehicle Application (NERVA) was almost ready for flight testing when the project was canceled in 1972. Under development for a manned mission to Mars, the NERVA engine heated hydrogen by passing it through the reactor core and then expelled it from a rocket nozzle. Uranium carbide fuel elements were coated with carbon and niobium to protect them from corrosion by the hydrogen propellent. The Mars craft would be assembled in Earth orbit and, using nuclear engines, it could travel to Mars, stay for two months, and return to Earth in the space of about one year. A program to develop a nuclear engine code-named Timberwind began in the 1980's and continues under the Space Nuclear Thermal Propulsion (SNTP) Program. Fluidized bed reactors and other advanced reactors that can operate at higher temperatures are being studied since they should be more efficient than the NERVA engine.

The most audacious nuclear engine is the nuclear pulse rocket that was the basis of the ORION project, which ended in 1965. The mass of the ORION vehicle was a grandiose 585 metric tons. The rear of the vehicle was connected by shock absorbers to a massive pusher plate. Every few seconds, a small fission bomb with a ten-ton yield was to be dropped out the back end and exploded about 100 meters behind ORION, so that the blast wave would drive ORION forward. About 2,000 bombs would be required for a 250-day round trip to Mars. To prove the concept, a small prototype was successfully launched from the ground with tiny chemical bombs, but international treaties now prohibit nuclear explosions in space, and therefore the ORION project is unlikely to be revived.

None of the proposed nuclear engines are very efficient at converting nuclear energy into a means of propulsion, but they are still attractive because of the large amount of energy in nuclear fuel. If the rare artificial element americium-242m could be produced in significant quantities, a much more efficient engine might be constructed. The key is that a thin film of americium-242m can sustain a chain reaction. High-energy fission fragments escape from a thin film and can be directed by magnets out the rear of the craft to provide propulsion. A spacecraft with such an engine might travel to Mars in two weeks instead of the eight to ten months required by chemical rockets.

Tethers and Bolos

Tethers up to 20 kilometers long have already been tested in space. A tether is a cable that can be unreeled from an orbiting craft such as the space shuttle. A mass on the far end of the tether will help keep it stable. The tether may be deployed upward by letting centrifugal force carry it farther from Earth, or it may be deployed downward by letting Earth's gravity carry it down. If the tether includes a conducting cable, it can be used to convert a satellite's momentum into electrical energy, since a conductor moving in Earth's magnetic field will act like a generator. To keep a current flowing in the cable, electron guns will expel electrons into space and prevent the buildup of a static charge. If used long enough, this system will bring down a satellite from low-Earth orbit, and thereby save the roughly 20 percent of rocket fuel that is reserved to deorbit spent satellites. If solar cells are used to produce a current in the tether, the generator becomes a motor, and the spacecraft's orbit will be raised. Because of the air resistance that exists in low-Earth orbit, the International Space Station needs a boost from time to time. If it were boosted with tethers powered by the station's solar panels, up to two billion dollars in fuel costs might be saved over ten years.

A bolo consists of two masses connected by a tether and set spinning. The end masses are equipped with grapples and thrusters to adjust position. Long tethers will probably be Hoytethers, a loosely woven Kevlar web. Their open structure makes Hoytethers less likely to be severed by meteoroids. If a bolo station (at the center of mass of the bolo) is in low-Earth orbit and therefore has a speed of 7.7 kilometers per second and the rotating tether's tip speed is 2.4 kilometers per second with respect to the station, the bolo's rotation direction is such that the tip speed subtracts from the orbital speed for the tip closest to Earth. A spacecraft launched from Earth need only be traveling at 5.3 kilometers per second when it rendezvouses with, and is seized by, the lower grapple. If the bolo is much more massive than the spacecraft, the spacecraft will be lifted and accelerated by the tether so that the spacecraft is traveling at 10.1 kilometers per second when the tip is farthest from Earth.

At the appropriate time, the bolo will release the spacecraft to travel to its next destination, perhaps a second bolo in geosynchronous orbit, which in turn might pass it on to a bolo in lunar orbit, which might set it on the Moon. The great efficiency of such a system is that it minimizes the fuel that must be lifted and accelerated from Earth. However, the bolos will slow down or fall into lower orbits as they give energy to the spacecraft. The bolo in low-Earth orbit could be boosted by using solar panels and a conducting tether. Other bolos might be boosted with solar-powered ion engines. Only steering energy would be required if the amount of mass going from the Moon to low-Earth orbit were the same as that going from low-Earth orbit to the Moon. (The falling mass would provide the energy to lift the rising mass.) A nearly constant flow of traffic would be required to make a bolo system cost-effective.

Charles W. Rogers

Bibliography

Forward, Robert L. *Indistinguishable from Magic*. Riverdale, N.Y.: Baen, 1995. A speculative look at spacecraft propulsion from tethers to antimatter. Written for general audiences, it contains an interesting mix of engineering and wishful thinking. Each chapter is followed by a short story based on the concepts discussed.

Mauldin, John H. *Prospects for Interstellar Travel*. San Diego, Calif.: Univelt, 1992. A book written at a popular level, published by the American Astronautical Society. It covers propulsion by solar sails, nuclear fission and fusion, ion engines, and mass drivers, along with many other topics.

Miller, Ron. *The Dream Machines*. Malabar, Fla.: Krieger, 1993. A fascinating chronological collection of pictures, drawings, historical notes, and descriptions of most of the spaceships ever built or seriously dreamed about.

Wright, Jerome L. *Space Sailing*. Philadelphia: Gordon & Beach Science, 1992. An excellent discussion of solar sails, including their types, construction techniques, handling, possible use of beamed power, and missions.

See also: Advanced Space Transportation Program; Jet Propulsion Laboratory; National Aeronautics and Space Administration; Propulsion; Rocket propulsion; Rockets; Saturn rockets; Space shuttle; X planes

Advanced Space Transportation Program

Date: Established on August 5, 1994

Definition: A collection of research projects of the National Aeronautics and Space Administration (NASA) designed to improve space transportation beyond technologies existing at the beginning of the twenty-first century.

Significance: The goal of the Advanced Space Transportation Program (ASTP) is to develop new technologies to make space travel safer and more economical in the future than it is currently. In order to accomplish this task, the ASTP seeks to develop new methods of propulsion and new spacecraft designs.

Background

During the early days of space exploration, many of the top rocket designers in NASA thought that the best way to fly into space was with a self-contained spacecraft, reusable in much the same way that an aircraft is reusable after each flight. However, the technology to develop such a reusable spacecraft did not exist in the 1960's. In order to compete with the Soviet Union to develop a crewed space program, NASA decided to adapt existing missile technology as boosters for crewed spacecraft. Thus, the early Mercury and Gemini Programs used Redstone, Atlas, and Titan missiles as boosters. The difficulty with these boosters was that they had a tendency to fail in flight. Special care, therefore, was given to the individual boosters used in the crewed space program. The Saturn rockets, developed for the Apollo Program, were designed from the beginning as boosters for crewed spaceflight. The Saturn rockets, however, were essentially very large versions of the type of rocket used on earlier flights. Though not a single Saturn rocket ever failed catastrophically in flight, it was not considered to be much safer than earlier rockets. The designs of the Saturn rockets were an upgrade of existing technologies. Upgrading existing technologies was a way of achieving a lunar vehicle in the shortest time possible, with a goal of beating the Soviet Union to the Moon. Though mishaps occurred during the Apollo missions, none resulted in loss of life while in flight due to rocket failure.

In addition to the safety issues surrounding early rocket designs, another difficulty was that the spacecraft and rockets could be used only one time. A great deal of effort and expense went into the construction of the rockets, but they were used for only a few seconds to launch the spacecraft from the Earth. The spacecraft itself was a small capsule that was designed to be used for only one flight. With each space mission requiring its own rocket booster and spacecraft, space travel proved to be extremely expensive. As early as the 1970's, before the Apollo missions had even finished, NASA was investigating the possibility of a new spacecraft that would be safer and cheaper to operate. Such a spacecraft would be reusable numerous times. Due to budget considerations, the resulting spacecraft, the space shuttle, was a compromise solution. Only part of the spacecraft was reusable. Furthermore, the space shuttle was launched into space using modifications to existing technologies used to launch some uncrewed spacecraft. The space shuttle ultimately proved to be not nearly as reliable and safe as had been hoped. Following the solid rocket failure and the resulting explosion that destroyed the space shuttle *Challenger* on January 28, 1986, NASA implemented new stringent safety measures that added to the cost of space shuttle missions. Even with these new safety measures, it was recognized that there remained a significant chance for additional catastrophic launch vehicle failures that could result in loss of the spacecraft and crew.

One of the goals of the space shuttle program had been to provide an inexpensive and safe transportation system into low-Earth orbit (LEO) in order to promote and assist future space activities, such as launching and repairing satellites or constructing and servicing LEO space stations. The space shuttle has achieved a tremendous success record, and has made important strides toward opening LEO to greater development. However, by the late 1980's, NASA had come to realize that the space shuttle was far too expensive and unreliable for the needs of the foreseeable future. Ideas began to emerge for a replacement to the space shuttle as the primary workhorse of the crewed space program.

Reusable Launch Vehicles (RLVs) are part of NASA's Advanced Space Transportation Program, intended to lower the cost and efficiency of the United States' space program. (NASA)

Origins

The early work on a replacement for the space shuttle tended to be unfocused and was conducted by different departments and divisions within NASA. The National Space Transportation Policy of 1994 finally formalized the goal to develop new space transportation systems. This policy statement divided the responsibility of developing new space transportation systems into two categories: new reliable, expendable launch vehicles and new reusable launch vehicles. Research toward developing new expendable launch vehicles was to be done by the Department of Defense. Research on the development of a second-generation reusable launch vehicle was to be NASA's responsibility. Soon afterward, NASA organized the Advanced Space Transportation Program (ASTP) to oversee the development of new reusable spacecraft. The ASTP is headquartered at the Marshall Space Flight Center in Huntsville, Alabama. Research related to the ASTP, however, is conducted at nearly all NASA centers and in many university and aerospace industry laboratories.

The primary goals of the ASTP are to reduce the costs

of launching payloads into LEO from more than $10,000 per pound in the year 2000 to about $1,000 per pound by approximately the year 2010, and as little as about $100 per pound by 2025. The ultimate goal is to reduce payload launch costs to only about $10 per pound by 2040. Such inexpensive launch technology would permit a great expansion of space-related activities by both government agencies and private enterprises. In addition to reducing costs, the ASTP seeks to increase safety. Upgrades and improvements to the space shuttle are expected to increase safety margins by a factor of ten by the year 2010, when new space transportation systems may become available. Within twenty-five years, NASA hopes to make space travel one hundred times safer, and within forty years, to make space travel nearly one thousand times safer. In order to achieve the increased cost-effectiveness and safety, NASA cannot simply rely on improvements to the space shuttle. New technologies and spacecraft must be developed. This was the responsibility of the ASTP.

Integrated Space Transportation Plan

The National Space Transportation Policy of 1996 reinforced the policy statements of 1994. The ASTP was enlarged and expanded and became the Integrated Space Transportation Plan (ISTP). The ISTP goals and timetables were the same as the ASTP's for reducing costs and increasing safety for space travel. The first steps toward those goals were to upgrade and improve the space shuttle fleet, NASA's first generation of reusable launch vehicles. The ISTP would also upgrade technologies to develop a new reusable vehicle to service the International Space Station (ISS). This new vehicle is to be a second-generation reusable launch vehicle operating similarly to the space shuttle, but much more efficient and advanced. The second-generation vehicle is expected to achieve the cost and safety goals for 2010. To achieve the remaining goals, however, requires the development of entirely new technologies and systems, not just revisions and upgrades of existing technologies and systems. Thus, the ASTP was incorporated within the ISTP to develop the third- and fourth-generation reusable launch vehicles needed for the future.

One of the ASTP's areas of research is in new rocket systems designed to operate with more efficient engines or using new propellants. Another consideration is the possibility of using radically new types of propulsion, such as nuclear or solar power. Further considerations include using external propulsion, in which the rocket is pushed into space by some force outside it, such as magnetically levitated craft, beamed energy systems, or tethers. In addition to researching new propulsion systems, the ASTP also is investigating new spacecraft designs capable of higher-speed atmospheric flight, and self-diagnostic systems capable of detecting failures before they occur. Most ASTP projects are still in the planning stages, and a majority may well never be constructed as new ideas emerge. Because the technologies envisioned with the ASTP do not yet exist, it is difficult to say exactly what form they will take.

New Rocket Designs

New rocket designs range from more efficient uses of existing technology to radical new technologies. An example of an engine based on existing technology is the Fastrac engine, designed with a small number of readily available existing parts. The engine is reliable and inexpensive to build; it may ultimately cost as little as 10 percent of the construction costs of current rocket engines. A new rocket engine technology is the pulse detonation rocket engine. This engine operates in pulses rather than continuously, as with most rocket engines. Propellant is injected into the reaction chamber and detonated with a spark plug, creating a short burst of thrust. Such an engine can be made to be very powerful and efficient.

In addition to new designs for rocket engines, ASTP researches new propellants. Currently, launch boosters require separate fuel and oxidizers. A monopropellant, a propellant that is self-oxidizing, would require less storage space in a rocket and fewer propellant tanks, thus saving rocket weight and permitting heavier payloads to be lifted for the same cost.

Another strategy being considered for more cost-effective rockets is an air-breathing rocket. Current rockets carry both fuel and oxidizer as propellant. If a rocket could take oxygen from the atmosphere as it flies, in much the same way that a jet engine does, then there would be no need to carry oxidizers and the savings in weight could be used to carry additional payload. Such an engine could conceivably permit a spacecraft to take off and land in a manner similar to that of a commercial jet aircraft, and may be completely reusable, reducing spacecraft cost.

New Spacecraft Designs

In order to utilize some of the new propulsion technologies, new airframe designs are needed to permit hypersonic flight many times the speed of sound. Furthermore, lighter-weight yet strong airframes would permit heavier payloads to be carried, reducing the per-pound cost of launches. Spacecraft design goals include improvements to permit a single spacecraft to operate for up to one thousand missions, a tenfold increase beyond the operational lifetime of the space shuttle. Furthermore, spacecraft servicing is expected to be simplified, so that a spacecraft may be serviced, much like a transoceanic airliner, between missions by only a few dozen personnel in less than a day. These innovations could reduce the cost of putting a payload into LEO to only a few hundred dollars per pound within two decades. To facilitate quick and effective maintenance, an integrated vehicle health management system is envisioned. In such a system, all parts of the spacecraft would have sensors that would be linked to a central computer that would monitor vehicle health and performance. The goal of an integrated vehicle health management system would be to detect weaknesses and defective parts before they become problems. This would make maintenance both easier and much less expensive.

New Propulsion Systems

In order to achieve the goals of the ASTP, entirely new propulsion systems may be required. Areas of research include both active and passive propulsion. Passive propulsion systems include magnetically levitated vehicles that would accelerate up to 600 miles per hour prior to takeoff along a magnetic track in much the same way as a magnetically levitated train. This initial velocity would mean that less rocket propulsion would be needed to achieve orbit. Focused-beam energy systems, such as high-powered lasers or microwave transmitters, could also be used to push objects into space. In such a system, fuel would not be needed and the spacecraft could be designed to carry primarily a payload.

Many payloads need to be deployed beyond LEO. Once in space, miles-long tethers can be deployed to utilize the Earth's magnetic field to change spacecraft orbits without the need of rockets. A tether system was tested with limited success from the space shuttle in 1996. Away from Earth, solar electric ion propulsion may be used for uncrewed interplanetary trips. Solar cells can create electricity from sunlight, and this electricity can be used to accelerate ions from the engine to propel a spacecraft. NASA tested such an ion-propelled spacecraft, called Deep Space 1, in 1998. Other proposals include the use of nuclear reactors to power spacecraft. The nuclear reactor could be used to power an ion engine. If a suitable fusion reactor could be developed, then it may even be possible for a spacecraft to scoop hydrogen atoms from space as it travels to use as nuclear fuel, thus minimizing the need to carry vast amounts of hydrogen with it.

It is hoped that the ASTP will yield new technologies such as these or even ones not yet considered. These new technologies will gradually displace the current generation of chemical-powered rockets used for space travel, thus yielding a safer and more cost-effective space transportation system.

Raymond D. Benge, Jr.

Bibliography

McCurdy, Howard E. *Space and the American Imagination*. Washington, D.C.: Smithsonian Institution Press, 1997. A history of space exploration containing some speculation as to future space needs for LEO travel.

Marshall Space Flight Center. *Advanced Space Transportation Program: Paving the Highway to Space*. Huntsville, Ala.: Author, 1999. A fact sheet summarizing the areas of research in the ASTP.

National Aeronautics and Space Administration. *Introduction to NASA's Integrated Space Transportation Plan and Space Launch Initiative*. Washington, D.C.: Government Printing Office, 2001. A thorough synopsis of the ISTP, including the ASTP.

Office of Aero-Space Technology. *Advanced Space Transportation Program R&T Base Program Plan*. Washington, D.C.: Government Printing Office, 1999. Extensive description of the organization and areas of study of the ASTP.

See also: Advanced propulsion; Apollo Program; Gemini Program; Mercury project; National Aeronautics and Space Administration; Rocket propulsion; Rockets; Saturn rocket; Space shuttle; Spaceflight

Aer Lingus

Also known as: Aer Lingus Teoranta

Date: Founded and incorporated on May 22, 1936

Definition: Government-owned international airline of the Republic of Ireland. Aer Lingus is a member of the oneworld Alliance.

Significance: Aer Lingus is the national flag carrier of the Republic of Ireland. The state-owned company was originally made up of two separate airlines: Aer Lingus Teoranta and Aerlinte Eireann Teoranta. These separate entities merged into one corporate structure during the 1970's.

Origins

The name Aer Lingus originated from the Gaelic term *aer loingeas*, which means "air fleet." The first Aer Lingus flight took place on May 27, 1936, when a De Havilland 84 took off on a flight to Bristol, England, from Dublin's old Baldonnel airfield. Beginning in 1946, operations were divided between Aer Lingus Teoranta for regional and European activities and an associate company, Aerlinte Eireann, for international flights.

Routes and Expansion

In 1938, Aer Lingus relocated to Dublin's new airport at Collinstown. The company grew steadily, and flight operations continued during World War II, with scheduled flights between Dublin and Liverpool and the Isle of Man. After the war, Aer Lingus's expansion accelerated, and service was initiated to London and Paris. Transatlantic flights to New York were attempted in 1947, but the Irish government soon decided to cancel these flights due to the high costs of operation. In the 1950's, Aer Lingus expanded its network to various European destinations. In 1958, the flights to New York resumed and by the early 1960's, the company started service to Boston and Canada.

During the late 1970's and 1980's, Aer Lingus found itself in the middle of a crisis. To remain competitive in the changing marketplace, the company was forced to initiate a major reorganization, to abandon numerous overseas destinations, to reduce its air fleet, and to close several of its international sales offices. However, the airline continued to grow regionally. In 1984, a subsidiary, Aer Lingus Commuter, was created to serve the shorter routes. In the 1990's, Aer Lingus reestablished itself as a leaner and more competitive carrier. Several of the routes that had previously been eliminated were reinstated, and flights to Chicago, Newark, Los Angeles, and several continental destinations were added to the airline's schedule.

Fleet and Safety

The first Aer Lingus aircraft was a six-seat De Havilland 84 Dragon biplane, which entered service in May, 1936. Operating initially with different versions of the DH-84 and DH-89, the airline added two Lockheed 14 and two Douglas DC-3 aircraft to its fleet in 1939. However, after acquiring seven Vickers Vikings in 1947 for continental services, the company decided to make the switch to more reliable DC-3 aircraft.

In 1954, Aer Lingus took delivery of its first jet-prop aircraft, the Vickers Viscount 707. Fokker F-27's were added in 1958. During the same period, Aer Lingus was leasing Super Constellations for the service to New York, but these aircraft were soon replaced when company management decided to purchase its first real jets: the B-720. Later that decade, B-707's were introduced on the transatlantic routes and BAC 1-11's on the continental routes. The company continued to change its fleet mix to become more efficient. In the 1970's, Aer Lingus introduced the B-747 to its fleet and purchased several B-737 aircraft. The following decade, Aer Lingus Commuter entered service with Irish-made Shorts 330/360 aircraft. However, by the late 1990's, the company was operating British Aerospace BA-146's and Fokker 50's on many of its routes. Also, several older, midrange aircraft were replaced with more modern Airbus A321 equipment. In 1995, Aer Lingus retired all its B-747's and decided to lease Airbus A330 aircraft to serve the transatlantic routes. By 2000, the Aer Lingus fleet, serving both Europe and the United States, consisted of about forty aircraft, among which were fifteen European Airbuses, thirteen B-737's, and twelve regional jets. Overall fleet modernization continued well into the new millennium, as new continental destinations were added to the flight schedule.

Early hull losses at Aer Lingus were significant. Between 1947 and 1952, the carrier lost two DC-3 aircraft in crashes. Twenty-three people died in a 1952 accident at Gwynynt Lake, England. Three more airplanes, Vickers Viscount aircraft, were destroyed during the period from 1967 to 1968. Sixty-one people died in the accident that occurred in Wexford Harbor, Ireland, in 1968. After the 1968 disaster, however, Aer Lingus's safety record improved significantly. For the period from 1969 to 2000, no mishaps were reported.

Company Strategy and Alliances

From its inception, Aer Lingus was managed as two separate companies. In the mid-1970's, Aer Lingus Teoranta and Aerlinte Eireann completely merged all operations under the umbrella of Aer Lingus, and the parent company adopted its distinctive shamrock logo. During that same

Events in Aer Lingus History

1936: Aer Lingus Teoranta is established by the Irish government.
1947: Aer Lingus's associate company, Aerlinte Eireann, plans international service.
1958: Aerlinte Eirann makes first transatlantic flight from Dublin to New York.
1984: Aer Lingus Commuter airline is established.
1987: An independent Irish carrier, Ryanair, creates low-fare competition for Aer Lingus.
1999: Aer Lingus adds service to Los Angeles.
2000: Aer Lingus joins the oneworld Alliance.

period, Aer Lingus, like many other airlines, decided to diversify. The company acquired a major stake in the Irish Intercontinental Hotels, invested in engineering firms and aircraft brokerage companies, and actively developed its own air-charter business.

During the challenging 1980's, Aer Lingus's company management scaled back its previous ventures, eliminating many to focus on the airline's core business. During this period, Ryanair, a new low-cost Irish carrier, became a major competitor for Aer Lingus on many of its United Kingdom and continental routes. Aer Lingus struggled through the early 1990's but got back on track when air traffic rose by 30 percent in the middle of the decade. However, to deal with the market pressures and the rapidly changing global airline environment, Aer Lingus management decided to pursue partnerships with other international carriers. In 2000, Aer Lingus joined the oneworld Alliance, made up of American Airlines, Qantas, British Airways, and many others, allowing the company to extend its global reach as an established international carrier.

Willem J. Homan

Bibliography

Donoghue, J. A. "Timely Turnaround." *Air Transport World*, September, 1997, 55-59. An article describing several business successes at Aer Lingus during the second half of the 1990's and praising new management for turning around the company's fortunes.

Hengi, B. I. *Airlines Worldwide*. 3d ed. Leicester, England: Midland, 2001. An excellent review of essential data of more than 350 airlines worldwide, with an overview of the different aircraft fleets.

Share, Bernard. *The Flight of the* Iolar: *The Aer Lingus Experience, 1936-1986*. Dublin: Gill and Macmillan, 1986. A historical overview, full of anecdotes and personal recollections, authored by one of the editors of the Aer Lingus company magazine.

See also: Air carriers; Airports

Aerobatics

Definition: Any aerial maneuver involving abrupt or extreme bank or pitch angles, unnecessary for normal flight.

Significance: Aerobatics are an integral part of military flight tactics, air show demonstrations, and sport flying. An aerobatic pilot's ability to retain spatial orientation and control an airplane in three dimensions provides an extra measure of safety in the event of an accidental upset.

Regulations

Most aerobatic flying is for pleasure, but regional and national contests are held every year, and a world championship contest is held every other year. Although there is no separate aerobatic rating, aerobatics can be safely learned only in an aircraft that is certified for the maneuvers and only under the tutelage of an experienced instructor.

Specifically, the U.S. Federal Aviation Regulations require approved parachutes when two or more occupants in an airplane intentionally exceed a bank of 60 degrees or a pitch angle of 30 degrees relative to the horizon. The basic aerobatic maneuvers are the slow roll, loop, spin, snap roll, aileron or barrel roll, and the wingover/hammerhead stall. Competition and air show figures combine these basic maneuvers into complex upright and inverted versions. In the absence of a special waiver and to protect passengers and the general populace, intentional aerobatic maneuvers must be performed away from crowded air space, above only sparsely populated areas, and at altitudes greater than 1,500 feet above the surface.

Aerobatic aircraft include some gliders and helicopters. Because aerobatics places extra structural and stability demands on an aircraft, only approved maneuvers may be performed in a particular aircraft. For aerobatic certification in the United States, an airplane must be capable of withstanding g-load factors from minus 3 to 6 without permanent deformation and loads of up to 50 percent greater (minus 4.5 to 9) without structural failure. The g-load factor, popularly known as the number of "g's," refers to the acceleration of gravity. Sitting still on Earth, one experiences an acceleration of 1 g, or a gravitational force of 1, the normal sensation of gravity. During periods of changing acceleration, such as a banking turn in an airplane, the so-called g-loading will change. Although the g-load factor in upright level flight is 1, it becomes minus 1 in inverted level flight. The best aerobatic aircraft, including those suitable for competition at the highest level, are stressed for load factors of 12 or more g's.

Aerobatics places extra physical demands on the pilot as well: loss of consciousness (positive g-load factors) or burst blood vessels (negative g-load factors) result from sustained high load factors. Military pilots have g-suits that help keep blood in their heads during positive load factors, whereas competition pilots use reclined seats and muscle tensing. A pilot's tolerance to g-loads increases with practice.

Slow Roll

The slow roll is the most basic roll maneuver and the hardest to learn. It must be mastered before solo aerobatic flight should be considered. In this maneuver, the aircraft is rolled about its longitudinal, or nose-to-tail, axis without altering the direction of flight.

Differential aileron deflection provides the torque that produces the roll. The other two controls, the elevator and the rudder, are used to keep the airplane from turning. When the roll is initiated, the opposing rudder must be used, and this reaches a maximum at about one-quarter, or about 90 degrees, through the roll. As the wings lose lift, the elevator must simultaneously be moved toward neutral. For the next quarter-roll, the rudder pressure is reduced, and forward elevator is added, as the wings are asked to generate negative lift. For the next 90 degrees of roll, rudder pressure in the direction of the roll is added and the elevator is gradually neutralized. In the last 90 degrees, elevator pressure is increased to the value before the roll was initiated, in level flight. The roll can be stopped at any point by neutralizing the ailerons; a momentary stop every 90 degrees, for example, yields a four-point roll.

The slow roll is difficult to learn because elevator and rudder inputs are constantly changing in a manner completely different from those of other maneuvers, because the forces on the pilot are so different and constantly changing, and because even a small error can place the aircraft in an inverted dive from which a safe recovery can be difficult. If the roll is initiated from level flight, the pilot senses an apparent weight that varies from normal to zero to upside down to zero to normal, corresponding to g-load factors varying from (at least) 1 to 0 to –1 to 0 to 1.

Attempts to teach oneself this maneuver will almost certainly cost a great deal of altitude and exceed design speeds and loads. Beginning pilots often fail to add enough opposite elevator as they near inverted flight, causing the nose to drop and allowing the speed to drop and then build very rapidly. At this point, pilots are disoriented and distracted by hanging from their shoulder harness and will relax the aileron pressure, causing the roll to stop. The natural and almost guaranteed reaction is then to pull back on the stick or wheel, attempting a recovery with a dangerous half-loop. A similar disastrous reaction can be expected from a nonaerobatic pilot when wingtip vortices or atmospheric turbulence flips the plane well past a 90-degree bank.

The slow roll has been mastered when the control inputs are instinctive, based on what the pilot wants the nose to do. Rudder pressure on one side always moves the nose in that direction, and forward movement of the stick or wheel always moves the nose away from the pilot. Once this concept is learned, slow rolls in any direction—straight up, straight down, or at an angle to the horizon—can be safely executed. However, the vertical, climbing roll is always a challenge, because it lacks a forward reference point and poses the danger of an inadvertent tail slide.

A slow roll is anything but slow in a modern, competition aerobatic airplane, in which roll rates of 720 degrees per second are not uncommon. The roll can be completed so rapidly that there is little time in which to encounter difficulties. Jet fighters can roll very rapidly without requiring rudder input.

Loop

A loop is one of the prettiest and most enjoyable aerobatic maneuvers, but skill is required to perform it safely and well. If the pull-up is made too abruptly, the aircraft can suffer either structural damage or a high-speed stall and will not complete the top of the loop. If the pull-up is too gradual, or if there is inadequate speed, the aircraft will run out of speed and fall inverted out of the maneuver.

A smooth but noncircular loop requires a g-load factor of 3 to 3.5, whereas a competition-quality circular loop may require a g-load factor of 6. Good aerobatic aircraft are fully symmetrical and can loop from level flight from either erect or inverted flight. A wingtip can be used for spatial reference during the second quarter of the loop, when the horizon will be hidden but, once over the top, the pilot will look overhead for the beautiful sight of the reappearing horizon. Competition-quality "square" loops can generate momentary g-load factors of 10 or more.

The first pilots to perform the loop, in 1913, were Petr Nesterov of Russia and Adolphe Pégoud of France. In 1928, Speed Holman of Minnesota broke the world's upright looping record by performing 1,433 consecutive loops in a five-hour period.

Spin

A spin's downward spiral makes it a crowd pleaser, although it is not a particularly pleasant maneuver for the occupants of the plane. A spin is normally initiated at a speed close to the stall speed with power off, neutral aileron, and full rudder and elevator deflection. After about one turn, the spin should stabilize in a nose-low position, and the airspeed should stabilize at a relatively slow speed, because both wings should be stalled, one more than the other, creating considerable drag. Recovery is usually effected with full opposite rudder to stop the rotation and then at least a relaxation, if not a reverse deflection, of the elevator control. Pulling out of the resulting dive generates a g-load factor of 2 or more. All aerobatic pilots must be very well versed in the spin characteristics of their aircraft, because any failed maneuver often degenerates into a spin.

In a true, stable spin, the spin can be continued as long as altitude remains and the airspeed does not increase. Utility aircraft certificated for spinning may appear to give a good spin entry, but the spin may become a diving spiral, increasing the speed. The same will happen in a good aerobatic airplane if the pilot does not hold full elevator and rudder deflection.

Heavy aircraft such as fighter aircraft may show wild gyrations upon spin entry and an oscillating pitch attitude once the spin is established. The World War II P-51D Mustang, for example, would oscillate from near-vertical to above the horizon and would lose about 1,000 feet per turn; spins were not to be performed below 12,000 feet.

Fully aerobatic aircraft can perform inverted spins as well as upright spins, but the aircraft recovers to inverted, stalled flight when the rotation is stopped, from which recovery to level flight should be made with a slow roll to minimize altitude loss. The rudder may suffer less blanking in inverted spins, allowing recovery to be faster. The inverted spin is much more disorienting than an upright spin and the pilot must concentrate on maintaining full elevator deflection, or the spin will transition to a diving spiral with rapidly increasing speed. If recovery from an upright spin is forced with down elevator and power, some aircraft will flick into an inverted spin.

If the aircraft is not certified for spins, or if the center of gravity is too far aft, the spin may be an unrecoverable flat spin with the nose on the horizon, yawing almost entirely rather than exhibiting nearly equal yaw and roll. Modern aerobatic aircraft with fully inverted fuel and oil systems, however, can force an upright or inverted spin to go flat with power and aileron deflection against the spin. These flat spins not only are recoverable but also form an important part of many air show routines.

Because it is such an important maneuver, the spin is the only aerobatic maneuver required of pilots seeking to become flight instructors. The requirement for parachutes is waived if an instructor is teaching an instructor-student. Considering that a low-altitude stall that degenerates into even an incipient spin remains a leading cause of fatal accidents, it would seem reasonable for more pilots to become familiar and comfortable with efficient recoveries from incipient spins, entered in the same fashion as accidental spins. Lieutenant Wilfred Parke of England is generally credited with first using what became the classic spin recovery method, in 1912.

Snap Roll

A snap roll, also known in England as a flick roll, uses the same control inputs as the spin, but in a snap roll, the controls are applied with power on and at speeds well above the unaccelerated stall speed. The resulting differential lift of the wings produces a rapid roll that can be very difficult to stop at a precise point. Good aerobatic aircraft can execute three or more consecutive snap rolls, both upright and inverted, before the axis of the roll changes excessively and the roll degenerates into a power spin. The load factor varies, as the square of the entry speed is divided by the unaccelerated stall speed, but a considerable twisting moment is also applied to the fuselage. This maneuver, among others, teaches the aerobatic pilot that an aircraft can exceed the critical angle of attack at any airspeed and at any angle relative to the horizon.

Aileron Roll

The most comfortable rolling maneuver is the aileron roll, also known as the barrel roll. It is performed through coordinated use of the ailerons and rudder, basically continuing a climbing steep turn to a 90-degree bank, letting the nose fall through the horizon with reducing elevator pressure as the roll continues to inverted flight and then recovering with increasing elevator pressure back to upright flight. The nose will trace out a sort of circle around a point on the horizon. The radius of the circle depends on the roll rate; if the roll is slow, the circle must be large and the top of the circle must be far above the horizon to keep the nose from dropping too low and building up a great deal of speed in the lower half of the maneuver. G-load factors of close to 1 throughout the maneuver are achievable. An expert pilot can perform this kind of roll in almost any airplane; in 1955, test pilot Tex Johnston barrel-rolled the prototype Boeing 707 airliner at a flight demonstration for potential customers.

Wingover and Hammerhead Stall

The hammerhead stall and the wingover are the most common turnaround maneuvers used by air show performers to maintain their presence in front of the audience. A wingover is a maneuver that changes the flight direction through 180 degrees with negligible net change in altitude. It is performed by simultaneously raising the nose and smoothly banking to a 90-degree bank angle as the flight direction changes by 90 degrees and then smoothly reducing the bank angle to 0 degrees in a descending turn to level flight in the opposite direction. Load factors should be in the range of 0 to 2 for a smoothly executed wingover, because there is no attempt to maintain level flight in the steeply banked turn.

In the hammerhead stall, known in England as the stall turn, the aircraft is pitched straight up with power on until it is pointing straight up. Shortly before the craft runs out of airspeed, full rudder is used to rotate the nose to the right or the left, and the rotation is stopped when the aircraft is heading straight down. Recovery may be to either upright or inverted level flight. Load factors need not exceed 2 or 3 if the initial entry and the pullout in recovering are smooth and to upright flight. The "stall" part of the maneuver's name is a misnomer, because the angle of attack is close to zero during the maneuver, and no stall buffet should be felt. An aircraft with a clockwise propeller rotation from the pilot's view will rotate best to the left. The greatest danger is waiting too long to use full rudder, allowing the aircraft to slide back on its tail, known as a tailslide, which could damage some of the control surfaces on otherwise aerobatic aircraft.

Advanced Aerobatic Maneuvers

The Immelmann turn, named after German World War I fighter pilot Max Immelmann, is a half-loop followed by a half-roll to upright flight. If the speed is low or the loop is stopped too abruptly, a sudden flick into an inverted spin is possible.

The Cuban Eight combines three-quarters of a loop, a roll to upright, another three-quarters of a loop, and a roll to upright again. From the ground it appears in the form of a horizontal eight.

The rolling turn, a very demanding maneuver to do well, combines a 360-degree turn with a roll, either to the inside or the outside of the turn. The square loop attempts to minimize the radius of the turns at the top and bottom of the loop and generates some of the highest momentary load factors.

The lomcovàk is a spectacular, twisting, tumbling maneuver invented by the Czech Ladislav Bezák in 1957. It is usually entered from a climbing, inverted snap roll and is commonly demonstrated at air shows.

Another spectacular maneuver is the torque roll, in which the airplane is rolled pointing straight up, and the roll is continued, with the help of engine torque, for a number of fuselage lengths in the ensuing tailslide.

Powerful aerobatic airplanes can generate enough fuselage lift and horizontal thrust component to maintain level flight in a 90-degree bank, known as knife-edge flight. Russian pilots have demonstrated the cobra maneuver, in which a jet fighter, flying in level flight, is abruptly pitched up through 90 degrees of rotation or more, recovering to level flight with relaxation of the stick.

Aresti Symbols

The distinguished Spanish aerobatic pilot Colonel José Luis de Aresti Aguirre developed a shorthand notation for aerobatic maneuvers. First published in 1961, Aresti symbols have become universally used to outline aerobatic routines for both contests and air shows. Each figure in Aresti's dictionary includes a difficulty, or "K," factor, by which, in contests, judges' scores—from 0 to 10—are multiplied.

W. N. Hubin

Bibliography

Carson, Annette. *Flight Fantastic: The Illustrated History of Aerobatics*. Newbury Park, Calif.: Haynes, 1986. A treasure trove of international aerobatics history, including the pilots, the planes, and the maneuvers, with many pictures and figures.

DeLacerda, Fred. *Surviving Spins*. Ames: Iowa State University Press, 1989. Factual and theoretical information and advice for pilots encountering inadvertent spins.

Kershner, William K. *The Basic Aerobatic Manual*. Ames: Iowa State University Press, 1987. An excellent basis for an introductory and basic aerobatic course in the Cessna 150 or 152 Aerobat.

Müller, Eric, and Annette Carson. *Flight Unlimited*. London: Eastern Press, 1983. A description of aerobatic flying, from basic to highly advanced maneuvers, by a highly experienced aerobatic champion and aerobatic instructor.

O'Dell, Bob. *Aerobatics Today*. New York: St. Martin's Press, 1984. Good information and advice for pilots seeking to enter aerobatic competition.

Szurovy, Geza, and Mike Goulian. *Basic Aerobatics*. Blue Ridge Summit, Pa.: Tab Books, 1994. Pilot and airplane preparation, background information, flying techniques for all the basic maneuvers, and advice for recreational or competition or air show aerobatic flying.

See also: Air shows; Airplanes; Barnstorming; Skywriting; Wing-walking

Aerodynamics

Definition: The study of airflow over bodies.

Significance: Knowledge of aerodynamics allows for the prediction of the forces and moments on airplanes. This allows the design of safe and efficient aircraft that can perform a large variety of tasks ranging from small radio-controlled craft to airliners and supersonic military airplanes.

Historical Aspects

In the late seventeenth century, English physicist Isaac Newton laid the foundations for not only modern mechanics and calculus but also fluid mechanics. Newton's analysis of fluid flow considered air to be composed of individual particles that struck a body's surface. This analysis was applied to determine the drag of an object in a moving fluid stream but gave poor results, because it did not account for the effect of the wing or body on the oncoming air. Interestingly, it later proved to be far more valuable in hypersonic flow analysis. Swiss mathematician Daniel Bernoulli and his father, Johann I, both published treatises in the 1740's that greatly clarified the understanding of the behavior of fluid flows. Eighteenth century Swiss mathematician Leonhard Euler noted the problems with Newton's model and proposed a more accurate formula for drag in 1755.

Subsequent aerodynamic theories developed in the 1800's and early 1900's were based on the works of Newton, Euler, and the Bernoullis. In 1894, British inventor Frederick William Lanchester developed a theory that could predict the aerodynamics of wings. However, Lanchester published this work many years later, in 1907. An acquaintance with Lanchester's theory might have saved considerable effort for Orville and Wilbur Wright, who first flew a heavier-than-air craft in 1903. Instead, the Wrights gained an understanding of aerodynamics through numerous wind-tunnel experiments conducted in their homebuilt wind tunnel. Subsequent advances in aerodynamics are associated with individuals, including Max Munk, Adolf Busemann, Ludwig Prandtl, and Robert Jones, who developed the principles of aerodynamic analysis.

Aerodynamic Flight Regimes

Fluids comprise both gases and liquids. A major difference between a fluid and a solid is that a fluid deforms readily. The major distinction between a gas and liquid is that a liquid is difficult to compress. The atmosphere is a gas composed of 78 percent nitrogen, 20.9 percent oxygen, 0.9 percent argon, 0.03 percent carbon dioxide, and in very small quantities, neon, helium, krypton, hydrogen, xenon, ozone, and radon, based on their volume. The study of the behavior of a body immersed in a moving liquid is called hydrodynamics; in a moving gas, gas dynamics; and in air, aerodynamics.

Aerodynamics may be categorized as either low- or high-speed, depending on where the fluid behavior changes. A common demarcation is subsonic and supersonic flow, where the latter has airspeeds greater then the speed of sound. Transonic flow, where both sub- and supersonic flow may exist, is also usually treated as a distinct regime. Increasing airspeed sees supersonic flow evolving into hypersonic flow at about five times the speed of sound. Difficulty in the analysis of airflow has additionally resulted in airflows being divided into viscous flows and inviscid flows, in which the latter are assumed to have no viscosity and are generally much simpler to analyze. The basic principles underlying aircraft flight are well described assuming inviscid flow.

Basics

The flow of air over a body is governed by the so-called continuity equation and the momentum equations. These equations state that mass can be neither created nor destroyed and that the sum of the forces experienced by a body equals its rate of change of momentum, or its quantity of motion. Analysis of these equations applied to various flight problems laid the foundations of aerodynamics. As air flows over an airplane, the plane causes the air to change its velocity, which also leads to changes in the static pressure distribution over the aircraft. The static pressure is the pressure that is felt when moving at the speed of the airstream. The static pressure distribution causes forces and moments, or torques, over the aircraft. The equation that relates velocity and static pressure is referred to as Bernoulli's principle.

Subsonic Airflow over Airfoils

The forces and moment that an aircraft experiences are affected by the air density, which in turn is affected by air pressure, temperature, and the amount of moisture in the air, as well as the speed and size of the airplane. As the aircraft flies through the air, it displaces air downward. By pushing the air down, the aircraft's wings experience a reaction force that tends to push the airplane up, creating lift. The lift is defined as being perpendicular to the oncoming airstream. One may imagine the lift of a wing flying along a wave of high pressure to be somewhat like a surfer on a surfboard riding an ocean wave.

In cross section, the wing of an airplane is composed of an airfoil profile. The shape of the airfoil profile's camber line, which is the line equidistant between the upper and lower surface of the airfoil, increases the lift generated at a given angle of attack if the airfoil has positive camber. Positive camber indicates that the leading edge and trailing edges, or the front and back of the airfoil, are curved down. If the airfoil has negative camber, the lift generated at a particular angle of attack is reduced compared to that of a flat airfoil with no camber. Consequently, positive camber or curvature of the camber line has the effect of increasing the lift by a constant amount for a given angle of attack, compared to a flat or symmetrical airfoil. The larger the camber of the airfoil, that is, its curvature, the greater the lift the airfoil will generate. This effect is most pronounced as the location of the maximum camber, or highest point of the airfoil, moves to the trailing edge, or back of the airfoil. The thickness of the airfoil, with reasonable accuracy, does not directly affect the lift the airfoil section generates, but it may affect the nature of the airfoil's stall.

The shape of the airfoil profile and its thickness distribution have a profound effect on the nature of the airfoil's stall. When an airfoil's angle of attack is greater than approximately 12 degrees, the majority of airfoils will stall. This condition is due to the air's viscosity and is caused by a disruption and separation of airflow over the airfoil's upper surface. Stall causes lift to decrease as the airfoil's angle of attack is increased. Flow separation also causes a large increase in drag, referred to as pressure drag. For very thin wings, or a flat plate, for example, the stall is quite moderate, and the loss of lift is gradual. For airfoils with a maximum thickness in the 9 percent range, where the maximum thickness of the airfoil divided by the length, or chord, of the airfoil is 0.09, the nature of the stall is quite sharp, and the loss of lift is dramatic. Thicker airfoil profiles, analogous to very thin airfoils, also have weak stalls, with a gradual loss of lift.

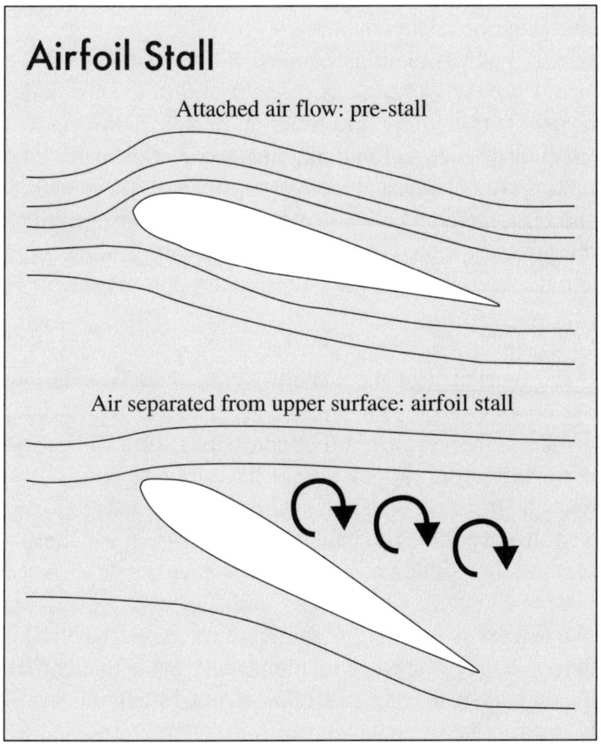

Numerous methods and devices have been developed to delay the stall of airfoils. These usually comprise a modification to the nose of the airfoil and typically involve pointing the nose down or extending it off the airfoil and rotating it down. These devices are referred to as leading-edge flaps, or slats. Some birds use a similar concept with a feather called the alula, which forms a slat and stops the bird's wing from stalling. For a given angle of attack below stall, these leading-edge devices generally do not much change the lift of the airfoil. However, they do extend the lift range of the airfoil and can increase it up to 10 degrees beyond the typical stall angle. These types of devices can be seen, and often heard extending or retracting, on airliners extending from the front of the wing at takeoff or landing.

On modern aircraft, all components are streamlined, that is, smoothly blended. The importance of streamlining became evident in the 1920's, when it was found that smoothly faired, or joined, bodies, such as aircraft wheels with aerodynamic fairing, had much lower drag than non-faired bodies. The fairing allowed the air flowing over the wheel to conform smoothly to the surface. Without the fairing, air would separate off the wheel and form large turbulent eddies, or swirling motions, in the wake behind the wheel, greatly increasing drag. The effect of streamlining is to reduce the tendency of the flow to separate off the surface. This separation is caused by the viscosity of the air.

Theoretically, at low speed in an inviscid airstream, an airfoil does not suffer any drag. This condition is known as d'Alembert's paradox, after eighteenth century French mathematician Jean le Rond d'Alembert, who calculated this apparent anomaly but was unable to explain it. The reason for the paradox was the exclusion of the effects of the air's viscosity in d'Alembert's calculations. Due to viscosity, airfoils experience a component of drag called skin friction drag. The skin friction is caused by the viscosity of the fluid layers near the airfoil surface. On the wing surface, the speed of the air is zero, a condition referred to as the no-slip condition. However, at some small distance above the airfoil surface, the airspeed reaches that which would occur if the flow were inviscid. The region between the surface and this point is referred to as the boundary layer. The nature and behavior of this boundary layer have a significant impact on the skin friction drag and stalling characteristics of the airfoil.

Airplane manufacturers quickly realized that sleek, streamlined aircraft flew most efficiently. This P-51 Mustang, photographed in 1942, illustrates the principles of aerodynamic design. (Library of Congress)

The boundary layer can either be laminar, turbulent, or transitional from laminar to turbulent. A laminar boundary layer is composed of air moving in orderly lines. A turbulent boundary layer has air moving close to the airfoil surface in swirling motions. A laminar boundary layer has far lower skin friction drag than the turbulent boundary layer; however, it is also more prone to separate from the airfoil surface. Thus, most airfoils have an initially laminar boundary layer that flows from the front of the airfoil back along the surface. At some point, the boundary layer transitions from laminar to turbulent and is typically turbulent from this point to the trailing edge of the airfoil. Boundary layer transition can be caused by disturbances of insects, ice crystals, high airspeeds, and roughness or imperfections on the airfoil surface. To improve performance at high angles of attack by keeping the boundary layer attached to the airfoil upper surface, an aircraft designer may choose to cause the boundary layer to transition from laminar to turbulent at some point on the airfoil. This may be achieved using small protuberances attached to the airfoil's surface.

Ultra Low-Speed Flight

Efficient flight at very low speeds, such as those of slow-moving birds and insects, presents unique complications. Typical airfoil shapes do not generate much lift at these low airspeeds. The boundary layer at these low speeds is normally always laminar, and so easily separates off the airfoil surface. When this occurs, the lift of the airfoil decreases significantly and its drag increases. Insects and small birds such as hummingbirds use complicated wing motions to create lift at their low airspeeds. These insects and birds develop both so-called steady and unsteady lift, the latter of which is caused by the acceleration of the wing and its carefully performed motion through the air.

Subsonic Flow over Wings

If the wingspan of the aircraft were infinitely long and the air were assumed to have no viscosity, the wing would theoretically generate a lift force and a moment but no drag. However, aircraft do not have infinite wings, and thus an aircraft in steady cruise experiences lift and drag, as well as a pitching moment, which tends to move the aircraft nose up or down, and possibly either a side force, or yawing moment, which tends to displace the nose from side to side, or a rolling moment, which causes the aircraft to roll about its fuselage such that one wing is higher than the other. The lift of an airfoil is reduced when the airfoil is incorporated into a wing of finite length. The shorter the wingspan is relative to the chord of the wing, the less lift is generated. The largest losses of lift are near the wingtips.

On a finite-length wing, air from the lower surface of the wing tries to curl up around the wingtip to the upper surface, causing the formation of two tornado-like structures, known as wingtip vortices, that trail from both wingtips backward. These vortices possess high rotational speeds and pose a significant threat to other aircraft that may fly through them. These vortices require delays between takeoffs and landings of aircraft using the same runways at airports, in order that vortices may have time to weaken.

The component of drag due to the aircraft having a finite-length wing is called vortex drag. Generally, a wing's vortex drag is independent of its span. Thus, wings with either a large or a short span will, to a first approximation, develop the same vortex drag. The larger-span wing will, however, generate far more lift, and thus the vortex drag will have a greater effect for short-span wings. To keep the amount of vortex drag low compared to the lift generated, a wing should have a large span. This is the reason why airliners have wings with large spans, and also the reason why gliders have narrow-chord, large-span wings.

Aircraft may have many different types of wing shapes that are dictated by the aircraft's function. A glider flies at low speed but needs to generate a large amount of lift with little drag. As a result, glider wings have large spans but small chords. An airliner needs to fly efficiently but is limited in its wingspan by airport considerations. A large wingspan results in a heavy wing, which is required to support the wing structure. As a result, airliner wings have a large span but not as large as their chord.

As aircraft fly faster and approach the speed of sound, the flow over the wing changes. Shock waves may appear on the wings, even though the aircraft is still flying subsonically. An airfoil accelerates the air flowing over its upper surface such that it may become locally supersonic. A shock wave is a very thin flow discontinuity that occurs in supersonic flow and causes the airflow through it to slow down significantly. Shock waves are accompanied by large increases in drag on the airplane and are thus undesirable. A way to delay the onset of shock waves on wings is to sweep the wings back, a commonly seen design on airliners, in which most wings have a sweep of about 20 to 30 degrees. This sweep effectively reduces the airspeed that causes the shock waves to form and so allows the plane to fly closer to the speed of sound, normally about 760 miles per hour at sea level. The speed of sound varies with the square root of the air's temperature.

Supersonic Aerodynamics

When the airspeed is greater than the speed of sound, the airflow is said to be supersonic. Aircraft that are designed to fly supersonically have distinctive design features. At supersonic speed, a new component of drag, called wave drag, appears in addition to the vortex, pressure, and skin friction drag. The wave drag is usually caused by the presence of shock waves on the wing or airfoil. This drag component is sensitive to the thickness of the wing and the lift that the wing is generating and increases with both. To keep wave drag as low as possible, supersonic airplanes may have very thin wings, such as those seen on fighter aircraft, highly swept wings, or a combination of both.

The wing on the Concorde is an excellent example of a supersonic wing design. A popular wing planform shape is the delta, or triangular, wing, upon which the Concorde's wing is based. The design requirements for efficient flight at supersonic speed and subsonic speed are contradictory. At low speeds, a large-span wing is desirable, whereas at high speeds, a highly swept wing is most effective. These requirements have led to the development of the so-called swing wing, seen on aircraft developed in the 1960's and 1970's, such as the European Panavia Tornado and the U.S. B-1 bomber. For low-speed flight, the wings sweep forward, whereas for high-speed flight, the wings sweep rearward. However, this design's prohibitive cost and structural weight have generally hindered its widespread use.

A problem with wings designed for supersonic flight is that, due to their large sweep and small wingspans, they are poor lift generators. That is, they do not develop a large amount of lift for a particular angle of attack, which can pose serious difficulties when these aircraft either take off and land at very high speeds requiring long runways. One way to alleviate this problem is by designing the highly swept wing to have a sharp nose or leading edge. This design causes the airflow over the wing to form two tornado-like vortices that lie above the wing. These vortices may be clearly seen in photographs of the Concorde taking off or landing on humid days. These vortices greatly increase the lift of the wing, but they also significantly increase drag.

Lance Wayne Traub

Bibliography

Anderson, J. D. *Fundamentals of Aerodynamics.* 3d ed. Boston: McGraw-Hill, 2001. An excellent, if mathematical, presentation of the foundations of aerodynamics. Subjects are treated in a thorough and logical manner.

Barnard, R. H., and D. R. Philpott. *Aircraft Flight: A Description of the Physical Principles of Aircraft Flight.* 2d ed. Harlow, Essex, England: Longman Scientific and Technical, 1994. A comprehensive and lucid explanation of the principles underlying airplane flight, in a nonmathematical formulation.
Kermode, A. C. *Flight Without Formulae.* 5th ed. Harlow, Essex, England: Longman Scientific and Technical, 1989. A clear and well-illustrated text that explains aircraft flight in a logical presentation.
Shevell, R. S. *Fundamental of Flight.* 2d ed. Englewood Cliffs, N.J.: Prentice Hall, 1989. A thorough introduction to both the aerodynamics and mechanics of airplane flight.
Wegener, P. P. *What Makes Airplanes Fly? History, Science, and Applications of Aerodynamics.* 2d ed. New York: Springer-Verlag, 1996. A somewhat technical presentation that traces the development of aircraft and aircraft technology.

See also: Aeronautical engineering; Airfoils; Airplanes; Birds; Concorde; Forces of flight; Heavier-than-air craft; Insects; Ludwig Prandtl; Supersonic aircraft; Wind tunnels; Wing designs; Wright brothers

Aeroflot

Date: Founded in 1932
Definition: The major airline company of the Russian Republic.
Significance: Aeroflot is the major airline of the Russian Republic. Before 1991, it was the national airline of the Soviet Union, but privatized after the breakup of the Soviet Union.

History

The Russian Federation of Soviet Socialist Republics formed its first airline, Dobrolet (an acronym for "Russian volunteer air fleet") in 1923, and two years later, Ukrainian Airways, or Ukvozdukhput, was started. In 1928, the Soviet government merged the two lines into a national airline under the name "Dobroflot" (also known as Dobrovolnii Flot, "volunteer fleet"). In 1932, during the period of the First Five-Year Plan, the government reorganized the company as the Main Civil Air Fleet Administration, or Aeroflot, and also established aircraft and aircraft engine plants. By 1935, Aeroflot had routes throughout the Soviet Union.

Aeroflot handled all civilian air service, including passenger and freight transport on international and domestic routes, and engaged in other air activities such as crop spraying, aerial surveying, air rescue, and medical transport. During the Soviet period, Aeroflot used Soviet-made aircraft exclusively, initially making Li-2's with twenty-four seats modeled on the Douglas DC-3. The company introduced the Ilushin Il-12 during World War II and the Il-14 in 1947. The Il-14's and Li-2's became the airline's standards. To meet the growing need for passenger service, in 1956 Aeroflot became the first airline in the world to introduce nonmilitary jet service with Tu-104 turbojets carrying one hundred passengers. From 1957 to 1959, the company added the turboprops Il-18, Tu-114, An-10, and An-24, and in 1965 the turboprop Antei (An-22), the world's largest transport plane. In 1968, it became one of the first companies to use supersonic aircraft, the Tupolev Tu-144, for civilian flights. The Tu-154, also introduced in 1968, carried 164 passengers at a speed of 1,000 kilometers per hour.

In 1991, Aeroflot had 5,400 planes, including more than 1,300 airliners, plus thousands of smaller craft, and carried 138 million passengers to over 3,500 cities in the Soviet Union as well as 100 countries on its international routes. Aeroflot was then the world's largest airline, accounting for 15 percent of all scheduled commercial civilian traffic. However, with the breakup of the Soviet Union in 1991, the company was greatly reduced. It remained the airline of the Russian Republic, changing its name to Aeroflot-Russian International Airlines, and invited private investors to participate along with the government in backing the company. The airline faced new problems associated with operating in free markets, such as competition from private companies and airlines of the other former Soviet republics not only in domestic travel but also on international routes. Furthermore, in 1999 the Russian government charged the company with embezzling funds through connections with Boris Berezovsky, a Russian oil and media magnate suspected of criminal activity. The line's general director, Valery Okulov, son-in-law of then Russian president Boris Yeltsin, continued in his position, however. Okulov fired ten top executives and discontinued routes which had been involved with the embezzlement. Yeltsin's and Okulov's opponents questioned the latter's leadership of the airline, but in 2000 the airline improved its image and performance.

Organization

In June, 1992, the Russian Federation reorganized Aeroflot into a public joint stock company, with the Russian govern-

Events in Aeroflot History

February, 1932: Aeroflot is established by the Soviet government as the nation's official airline.
1937: Russian pilots discover the shortest transpolar air route, flying from Russia over the Arctic Ocean to Vancouver, Canada.
September, 1956: Aeroflot inaugurates regular passenger turbojet service, two years ahead of British and American airlines.
December 31, 1968: The first supersonic airliner, the Tupolev Tu-144, developed by Aeroflot, makes its maiden voyage.
1977: Aeroflot begins offering supersonic passenger service, which is suspended the following year.
1998: Aeroflot Boeing 777's begin making regular flights between New York and Moscow.

ment owning 51 percent of the stock. The other shareholders included both Russian (chiefly the company's fourteen thousand employees) and foreign investors. In 1997, Okulov was appointed general director. The company had 151 representatives abroad and 3 in Russia: St. Petersburg, Novosibirsk, and Khabarovsk. In the post-Communist period, Aeroflot bought foreign airplanes as well as those made in Russia. At the end of the twentieth century, the fleet had well over one hundred planes, including Boeing 737's and 767's, Airbus A310-300's, and Ilushin 96-300's. Some of the Ilushins had American Pratt & Whitney engines. The company carried about four million passengers and 90 tons of cargo annually. In contrast, a competing airline such as Delta had over 550 planes and carried over ninety million passengers. After 1991, over three hundred spinoff lines called "babyflots" emerged, taking over much of Aeroflot's equipment and routes. At the beginning of the twenty-first century, Aeroflot flew to 150 cities in 93 countries. It still provided 70 percent of all Russian air travel.

Like other post-Communist Russian companies, Aeroflot presented a flashy exterior, with luxurious corporate offices in New York and London, attractive reports and brochures, a World Wide Web site, an offshore subsidiary in Switzerland (which was, however, connected to the Berezovsky scandal), but it still had problems stemming from Soviet inefficiency, and depended on the Russian government for subsidies and keeping monopolies on internal Russian routes. In 1997, American and European banks loaned the company $1.5 billion to buy planes. Although Aeroflot service on domestic flights was inferior, its international service improved and its prices were very competitive. It also retained smoking sections, unlike other airlines. Nevertheless, in 2000, travelers voted Aeroflot the world's worst airline.

Crashes

During the Soviet period, crashes were not reported and are difficult to track. In the post-Communist period, when news of air disasters was reported, the world became aware of the dismal record of Aeroflot. In 1994, there were a record number of domestic crashes and the United States embassy warned its personnel not to use the line for travel within Russia. In the same year, an Airbus A310 crashed on its way to Hong Kong, after the pilot allowed his teenage son to fly the plane. The company worked with the U.S. Federal Aviation Administration to improve its safety record, and the crash rate declined. Nonetheless, on August 29, 1996, an Aeroflot plane carrying 140 passengers crashed in Norway.

Frederick B. Chary

Bibliography

Davies, R. E. G. *Aeroflot, an Airline and its Aircraft: An Illustrated History of the World's Largest Airline.* Shrewsbury, U.K.: Airlife, 1992. A popular history for the general reader. Includes many illustrations and diagrams of the planes.

Duffy, Paul. "Fighting Back." *Flight International* 152, no. 4586 (August 6, 1997). An article in a professional magazine analyzing the problems of Aeroflot after the fall of Communism.

Macdonald, Hugh. *Aeroflot: Soviet Air Transport Since 1923.* London: Putnam, 1975. An older but excellent history.

See also: Accident investigation; Air carriers; Airbus; Airports; Boeing; Federal Aviation Administration; Safety issues; Supersonic aircraft; Andrei Nikolayevich Tupolev

Aeromexico

Definition: A major Mexican air carrier of passengers and cargo.

Significance: Aeromexico is the second largest air carrier in Mexico.

Background

The North American country of Mexico has a long and rich aviation tradition. Mexico's airlines are not as well known as those of many other countries because they have not developed the extensive international route networks

The Aeromexico Fleet

Aircraft	Number in Service	Length (meters)	Wingspan (meters)	Seats	Cruising Speed (kilometers per hour)
Metro 23-11	23	18.1	17.4	19	500
Saab 340-B	9	19.7	21.4	33	530
DC-9-30	17	36.4	28.5	97	820
MD-87	5	39.8	32.9	109	820
MD-82/83/88	34	45.1	32.9	142	820
Boeing 757-200	8	47.3	38.0	175/177/180	820
Boeing 767-200 ER	4	48.5	47.6	179/181	850
Boeing 767-300 ER	1	54.9	47.6	209	850

Source: Data taken from (www.aeromexico.com/ingles/acerca_am/flota/flota.htm), June 5, 2001.

for which carriers of the United States and Europe are famous. However, the early history of aviation in Mexico was as turbulent and exciting as that of any country in the world. In the first half of the twentieth century, over one hundred airlines started domestic service in Mexico. Many of these airlines later merged with one of Mexico's two national carriers, Mexicana and Aeromexico. These two carriers have been competitors almost since their beginnings. Each has strived to be the predominant domestic and international carrier of Mexico. Through a series of mergers, Aeromexico developed a strong domestic route structure linking Mexico City, the United States, and Canada to most of the tourist destinations of Mexico. Although its international network continues to be weaker than that of Mexicana, it has grown to serve destinations in the United States, Central America, South America, and Europe.

Early Years

Aeromexico, then called Aeronaves de Mexico, began as a small regional carrier serving the Pacific coast of Mexico in 1934. At that time, it operated flights between Mexico City and the newly developing tourist destination of Acapulco. It continued as a small, regional carrier until the U.S. air carrier Pan Am World Airways purchased 40 percent of its equity in 1940. With the new capital provided by Pan Am, Aeronaves de Mexico began acquiring a series of other small carriers along Mexico's Pacific seaboard. In 1952, the airline expanded into north central Mexico with the acquisition of Lineas Aereas Mineras, S.A. (LAMSA) from the U.S. carrier United Air Lines. The following year, Aeronaves de Mexico purchased Aerovias Reforma to further serve the Pacific coast. The 1957 opening of service to New York City heralded Aeronaves de Mexico's entry into the international air transport market. This same year they joined the International Air Transport Association (IATA), an organization of airlines affiliated with the United Nations and responsible for promoting safe and secure air travel throughout the world. IATA is the premier organization for coordinating airline policies and procedures and training airline personnel in all aspects of aviation.

Financial Troubles

Aeronaves de Mexico's expansion was temporarily halted when a strike in January, 1959, threatened the company's financial health. The Mexican government moved quickly to assume control of the company, taking official ownership in July of that year. The board of directors appointed by the Mexican government proceeded to upgrade Aeronaves de Mexico's fleet and merged it with Aerovias Guest, the first Mexican carrier to serve Europe. Aeronaves de Mexico continued to expand its domestic and international route structure throughout the 1960's. Its acquisition of Servicios Aereos Especiales (S.A.E.) in 1970 left Mexico for all intents and purposes with only two airlines, Aeronaves de Mexico and Mexicana.

Aeronaves de Mexico underwent a second financial crisis in the early 1970's. In an attempt to revitalize the airline, the aircraft color scheme was changed to red, the Aztec warrior tail design was modernized, and the company name was shortened to Aeromexico. The company's financial health improved following efforts to upgrade its fleet and enter new markets in North America opened up by the United States deregulation of its own airline industry in 1978. However, Aeromexico was forced to declare bankruptcy in 1988 due to economic uncertainty and overcapacity in the Mexican market. The company was reorganized under the name Aerovias de Mexico, retaining the Aeromexico name for marketing purposes. As part of the reorganization, Aeromexico laid off approximately ten thousand staff, hired industry outsiders to help them improve quality and financial performance, and strengthened their route structure. The company also purchased 47 percent of the shares of Aeroperu in 1992, allowing it to open a hub in Lima, Peru. This provided the first South American connecting point for Aeromexico, allowing them to tie together the Americas from Canada to Argentina.

The financial crisis that struck Mexico in 1994 brought Aeromexico and its competitor Mexicana to the brink of bankruptcy. In 1995, both companies were purchased by the Corporacion Internacional de Aviation (CINTRA), a consortium of banks. The two airlines now cooperate on ground handling, training, and computer reservations. These efforts have allowed the two carriers to improve service and lower costs. Although both companies are subsidiaries of CINTRA, they remain separate entities and continue to compete in many areas. Aeromexico continues to be the stronger domestic competitor and maintains a fleet of Boeing aircraft. Mexicana remains the dominant international competitor and has begun purchasing Airbus aircraft to serve its markets. Aeromexico has joined the SkyTeam Alliance composed of Delta, Air France, and Korean Air Lines. Mexicana is a member of the STAR alliance, whose chief members include United Air Lines, SAS, Lufthansa, Varig, Air Canada, and Singapore Air Lines.

With the growth of the North American free trade area, which is lifting trade restrictions between the United States, Canada, and Mexico, the prospect for further growth in the Mexican air transport market looks promising. In addition, growing trade between South America and the United States is providing Aeromexico with opportunities to link both areas. Despite several periods of financial crisis, Aeromexico has survived and looks forward to continued growth.

Dawna L. Rhoades

Bibliography

Davies, R. E. G. *Airlines of Latin America Since 1919.* Washington, D.C.: Smithsonian Institution Press, 1984. An extensive review of the history of aviation in Latin America. This book is filled with charts, tables, and graphs outlining the early development of Latin aviation.

Magnusson, M. *Latin Glory: Airlines of Latin America.* Osceola, Wis.: Motorbooks International, 1995. This small book contains a brief history of most of the Latin American carriers, as well as color pictures of each airline's livery.

Moody's Transportation Manual. New York: Mergent FIS, 2000. The Moody's series presents up-to-date financial information as well as a brief company history and listing of corporate offices and officers.

See also: Air carriers; Airports; Pan Am World Airways; United Air Lines

Aeronautical engineering

Definition: The study, design, and manufacture of aircraft and spacecraft.

Significance: Aeronautical engineering is responsible for the development of and advancements in aviation and spaceflight.

Engineering

In the first century of crewed flight, which began in December, 1903, the application of the new science of aerodynamics was translated into flying machines by people who understood engineering and problem solving. The industry that grew from this small beginning made amazing strides in the first century of air travel. It is an industry built around visionary engineers and pilots.

Aeronautical engineering had its true beginning before Orville and Wilbur Wright but the two brothers were the pioneers in the techniques, processes, and system testing that were at the heart of the engineering design and development of aircraft and spacecraft. The conceptualization of an aircraft begins with the identification of something useful to be accomplished by an air machine. The process begins with sketches of an air vehicle to fulfill the performance expectations for the aircraft. In the first two decades of aircraft design and operations, many concepts were proposed, but by the end of World War I, the basics of successful aircraft design were established. Future refinements would come through better tools, materials, and concepts. At the beginning of the second century of crewed flight, the process involves digitally created drawings that are sent to machines that make the basic parts, which are then assembled, tested, and prepared for flight test.

Twentieth century aircraft engineering refinements moved at a speed unseen in any previous period of the industrial world. The motivation and excitement of flying higher, faster, and with larger payloads seemed to drive innovation and to demand engineering solutions. By the end of World War II, the aviation industry was fully established as a significant contributor to the economic and military strength of the United States. European aerospace also produced leaders in this field. Companies were built on the talents of engineers and the skills of craftsman. Engineering disciplines expanded, and in the late 1950's, aeronautical engineering became aerospace engineering. In most aircraft manufacturing firms, the engineering department was second in size only to the production groups.

Typically, in the middle of the twentieth century, modern aerospace companies spread their products between commercial enterprises and government contracts. The bread-and-butter contracts came from the federal government until the end of the Cold War. Commercial applications of engineering ideas were spun off from aircraft and missiles that had been developed for the military. However, by 1990, the industry was in decline. Following the Gulf War in 1991, the downsizing of the air arms of the military accelerated. The demand for large numbers of new and different military aircraft came under such scrutiny that few of the new programs survived. On the commercial side of the industry, the engineering of new and better transports and aircraft destined for the air carrier markets stopped in favor of building on existing concepts to build bigger aircraft with bigger engines. Airspeed, comfort, and passenger loading ceased to be major requirements and took a back seat to economically viable air transport.

William H. Pickering, James A. van Allen, and Wernher von Braun, the engineers behind the Explorer 1, show off a model of the satellite in January, 1958, shortly before it was launched into space. (NASA)

Research and Development

There are three significant eras in the expansion of the aerospace industry. These coincide with technology improvements as well as political changes that affected the industry. The first period started with the Wright brothers' successful efforts at powered flight and ended with the advent of the jet engine. The next period began when jet engines were being put into all new aircraft designs, and this period saw rapid advances in aircraft performance. The last period began with the introduction of digital computer controls for the aircraft. This development made it possible to design and build incredibly safe and reliable aerospace systems.

Out of World War II came large bombers and cargo aircraft. When jet engines were added to these aircraft they held promise for faster and higher, hence more efficient and comfortable, air transportation for the public. The first such jet transport built for the British Overseas Airway Corporation (BOAC, which became British Airways) was the Comet. However, the understanding of structural issues arising from rapid changes in pressure on certain parts of the aircraft, along with manufacturing techniques from the 1940's, resulted in an unsafe aircraft. After two exploded in flight due to structural failure and one burst during ground pressure testing, the world of aeronautical engineering became aware of fatigue failures and the need to design fail-safe structures. At the time, the U.S. Air Force had Boeing designing and building a jet tanker using technology similar to that applied to its highly successful swept-wing B-47 jet bomber program. What came out of that work was the most successful aircraft transport design in history. The Boeing 707 model was the forerunner of all of today's large jet transports.

The Industry

After World War II, the growth in the aviation industry, both commercial and military, saw a proliferation of new prime contractors who were building and selling aircraft. A prime contractor was defined as the company that was responsible for the concept, design, development, and final introduction of the new aircraft into operational use. In short, a prime contractor was responsible for all aspects of the life cycle of the aircraft. The prime would have subcontractors, perhaps hundreds, with which it did business.

At the start of the 1970's, and at the height of the Vietnam War, there were many primes in the aerospace business. The biggest and most successful were Boeing Aircraft, Douglas Aircraft, McDonnell Aircraft, Lockheed Aircraft, Republic Aircraft, General Dynamics, Grumman Aircraft, North American Aviation (North American Rockwell), Northrop Aviation, LTV Aerospace (part of LTV, which used to be Chance Vought), Northrop Aircraft, Bell Airplane and Bell Helicopter, Sikorsky Helicopter, and a handful of general aviation companies, including Cessna, Beech, Piper, and others. At the end of the twentieth century there were only three major aerospace companies left, with all others being absorbed into the remaining companies or having gone out of business. Boeing took over McDonnell Douglas, which used to be McDonnell Aircraft and Douglas Aircraft. Lockheed Martin absorbed General Dynamics Aircraft Division and Martin Marietta. Northrop and Grumman joined, adding pieces of LTV and others. In addition, Raytheon Corporation, which was a small missiles and electronics outfit in the 1960's, took over Beech Aircraft and other subsidiary companies. Cessna and Piper nearly went out of business during the 1970's and 1980's, due to changes in liability laws. Chance Vought became Ling Temco Vought in the mid-1960's and changed its name to LTV Corporation in the 1970's. It was one of the first prime contractors that attempted product diversification, with markets in steel, appliances, missiles and aircraft; the corporation went bankrupt in 1986.

Future Developments

Compared to the days of the Wright *Flyer* and the Curtiss JN-4, aircraft which were very difficult to control and which carried very small payloads, the F-22 automated advanced fighter and the Boeing 777 automated, large twin-engine transport are engineering marvels. At the beginning of the twenty-first century, there are several different paths that may provide the next major step forward in aeronautical engineering.

In June, 1963, President John F. Kennedy, speaking at the commencement of the fifth class to graduate from the United States Air Force Academy, announced that the federal government would seek to develop the world's first supersonic passenger transport (SST). This never happened, for two reasons. The first was the economic issue. Such an aircraft, designed using late 1950's and early 1960's technology, would be very expensive. Airlines could not justify the costs to operate them. The second issue was environmental. Warnings and concerns about the pollution or damage to the upper atmosphere from turbojet engines and problems with sonic booms, which are caused by the shock waves from a supersonic aircraft, led to a premature end of the SST. Europe, in a cooperative move between British and French aircraft firms, did pursue a smaller version of the SST, called the Concorde. It operated successfully starting in January, 1976, although it was under a limitation forbidding it from flying supersonically over the United States. Technology improved during the twenty-two years the Concorde was operating, and by the late 1990's, the National Aeronautics and Space Administration (NASA) attempted to resurrect the SST concept. By then, the problems of jet exhaust and its impact upon the upper atmosphere had been nearly resolved. Ways to reduce the pressure from the sonic booms were being planned. The program ended in 1999 when, for the second time, the economic issues surrounding operational costs of a large SST overrode advances in the aerospace engineering field.

The next hope for large transport aircraft lies in engineering a craft that will cruise just under Mach 1. Most large aircraft can cruise efficiently at Mach .75 to .9 (the percent of the speed of sound) but if they could fly efficiently at 95 percent of the speed of sound this would mean a 5 to 20 percent increase in true airspeed (35 to 155 miles per hour). A speed increase of that magnitude would shorten the flight time from New York to Paris by approximately an hour and fifteen minutes. The potential savings in fuel, the increase in the number of aircraft that could fly the same route, and other factors make this an appealing possibility. It is not an easy engineering task, but then, most of the history of aviation has faced such challenging engineering tasks.

The ultimate flight would be one that takes the passenger into low-Earth orbit and flies across both continents and oceans. That aircraft will probably come about once the space program has fully established the safety and reliability of such travel. Aeronautical engineering and the companies that have come to the forefront in both engineering and applied sciences for aerospace purposes will be able to achieve these ideas.

R. Smith Reynolds

Bibliography

Anderson, David F., and Scott Eberhardt. *Understanding Flight*. New York: McGraw-Hill, 2000. A basic explanation of how flying works, including coverage of aircraft engineering and the principles underlying successful designs.

Heppenheimer, T. A. *A Brief History of Flight: From Balloons to Mach 3 and Beyond*. New York: John Wiley & Sons, 2000. Covers human flight and the inventions that made it possible, looking at social, political, and economic influences on engineering advances.

Launius, Roger D. *Innovation and the Development of Flight*. College Station: Texas A&M University Press, 1999. A collection of essays covering many aspects of aeronautical engineering, from the Wright brothers to current developments.

See also: Airplanes; Boeing; Concorde; Forces of flight; McDonnell Douglas; Manufacturers; Military flight; Spaceflight; Supersonic aircraft

Aerospace industry, U.S.

Definition: Manufacturers directly involved in the production of aircraft, engines, and ancillary products for use in aviation and space travel.

Significance: The aerospace industry became a critical part of the U.S. economy following World War II. The industry benefitted from the postwar emphasis on military and commercial aviation, as well as spaceflight.

The Aerospace Industry Through 1945

The United States' adventure in aviation went from its first flight in 1903 to flights to the Moon in 1969. Despite this impressive record of accomplishment, aircraft manufacturing proved to be a difficult business. Early companies, notably the Wright Company, founded by Wilbur and Orville Wright, and the Curtiss Aeroplane Company, established by Glenn H. Curtiss, sold a handful of planes to the military but did not find a lasting market for their aircraft. The federal government recognized the importance of aviation by establishing the National Advisory Committee for Aeronautics in 1915 but did little to help struggling manufacturers. Although World War I forced the United States to produce greater numbers of aircraft, most American pilots flew French planes during the conflict.

Although U.S. airplanes did not make an impact during the war, they did serve to train postwar aviation enthusiasts. These surplus planes undercut manufacturers to some degree, but the demand for increased performance gave companies an opportunity to introduce new designs. Despite widespread interest in aviation in the years leading up to World War II, the industry catered primarily to the military. Even companies, such as Boeing and Lockheed, that aggressively targeted the commercial market with private planes and airliners looked to the military for a significant proportion of their business. Other companies, notably Grumman and Douglas, dealt almost exclusively with the military.

World War II put a stop to commercial aviation plans and forced all manufacturers to focus on military aircraft. The leaders of the U.S. postwar industry clearly emerged during this period. Boeing, North American, Lockheed, Grumman, and Douglas all established themselves as mainstays in aerospace manufacturing. The war also necessitated enormous advances in technology. By the end of the war, jet fighters had taken to the air, heralding the future of the aviation industry. Finally, World War II established aviation as an indispensable component of both military and civilian life in the years to follow.

Postwar Industry Trends

The aerospace industry became increasingly important during the Cold War. The United States relied on technology to offset the numerical advantage of the Soviet Union. Many of the aircraft manufacturers that had done well during World War II remained at the forefront of the industry. These companies concentrated on four areas: military aircraft, missiles, rockets and space exploration vehicles, and commercial aircraft. The advent of space exploration prompted journalists to coin the term "aerospace" in the 1950's, reflecting the new era of U.S. aviation manufacturers. The aircraft, aerospace, and parts industry had become the largest U.S. employer by 1959, and cities connected to the industry, including Los Angeles, Seattle, and Phoenix, exploded in population.

Military Aircraft. Aerospace manufacturers worked hard to win lucrative government contracts following World War II. The United States demanded advanced fighters and bombers to meet the Soviet threat. While these contracts provided the backbone of the industry, they also placed extraordinary demands on the manufacturers. The new planes required expensive engines and complicated alloys, both of which added a great deal of expense to the planes. Construction of the aircraft usually necessitated new techniques and equipment. Government designs often included overly complicated ideas that added to the weight of the aircraft. The industry did not help matters by overpaying executives and using unnecessarily expensive components. These problems led the U.S. Congress to require new levels of bureaucracy and paperwork to control costs. Furthermore, Congress could decide at any point to cancel a project, leaving the contractors heavily in debt with no potential market.

Despite these problems, aerospace companies could not disregard the billions of dollars that the government contracts offered. The enormous sums granted to various manufacturers also allowed much more funding for research and development, accelerating advances in technology. The United States ended World War II somewhat behind Great Britain in jet engine construction, but by the mid-1950's, American manufacturers Pratt & Whitney and General Electric had become the leaders in jet technology. The increasing reliance on computers in the design stage led to continual improvements in microtechnology. Talented individuals such as Clarence "Kelly" Johnson at Lockheed and Ed Heinemann at Douglas created brilliant designs that exceeded government specifications and kept costs down.

The biggest problems aerospace manufacturers faced after World War II were not the technical demands of the new aircraft. Given enough time and money, men such as Johnson and Heinemann could overcome those obstacles. Unfortunately, the political demands of the defense issue often took precedence. Companies simply could not afford to spend several years and millions of dollars to develop an aircraft that would not enter service. Consequently, manufacturers went to great lengths to make their projects successful. Lockheed received a considerable amount of bad publicity in the 1970's when investigations revealed that the company had relied heavily on bribery to ensure foreign contracts for its F-104 fighter during the preceding decade. Northrop also suffered for its use of bribery in the Middle East in an effort to find a market for its F-S fighter. Even companies that avoided politics could not disregard the new era in the industry. In the late 1960's, Grumman expanded its facilities to begin manufacturing the Gulfstream II corporate jet. The company, which had always eschewed marketing, placed its new facility in Savannah, Georgia, which was represented by an important member of the House Armed Services Committee and the home state of another influential member of the Senate Armed Services Committee.

American defense cutbacks forced manufacturers to consider other markets. In the mid-1970's, General Dynamics designed the single-engine F-16 fighter. The lower cost associated with using only one engine made the plane attractive to European nations with limited budgets. General Dynamics did have to allow European countries to manufacture some of the planes, but the consortium reduced costs for all companies and promoted sales around the world. Difficulties in controlling costs finally forced U.S. competitors to begin working together as well. Northrop, with little experience in carrier aircraft, had to turn to McDonnell Douglas for help with a new carrier-based fighter. The result, the F-18 Hornet, became a great success. Not only did the Navy and Marine Corps adopt the plane, but its low cost ensured brisk sales to air forces around the world. The F-18 program convinced manufacturers that collaboration had become necessary to control spiraling costs.

The early 1980's saw a resurgence in Cold War tensions. President Jimmy Carter reinstated previously canceled programs such as the MX Peacekeeper intercontinental ballistic missile (ICBM) and the B-1 bomber. The new U.S. military buildup offered greater opportunities for military manufacturers, but these advantages were offset by the fact that the government demanded small numbers of extremely complex aircraft. This trend accelerated after the end of the Cold War, as the United States slashed its defense budget even further. The Air Force could not afford advanced programs such as the F-22 fighter and B-2 bomber, the Navy canceled its search for a new attack plane after well-publicized cost overruns, and crashes of new aircraft eroded public confidence, leaving manufacturers to fight over a shrinking sector.

Missiles. As military aircraft contracts forced manufacturers into hard-fought competition, America's missile program gradually came to represent a larger share of the industry's production. Between 1956 and 1961, airframe companies increased the percentage of missiles within their military business from 5 percent to 44 percent. In missile technology, many of the same manufacturers that dominated aircraft production also took a leading role in missile development, but companies such as TRW and Morton Thiokol made significant contributions to the industry.

The United States saw missiles as an important part of the nation's Cold War arsenal. The government took great pains to secure the services of Germany's leading missile designers at the end of World War II, but the growing Soviet threat made the development of ballistic missiles a high priority. These weapons, like the aircraft and space vehicles of the Cold War, proved much too expensive for individual companies. Missile projects required subcontracting and cooperation between manufacturers. By 1960, U.S. ballistic missile projects included two thousand contractors and forty thousand employees.

In the late 1950's, the United States' first intermediate range ballistic missiles (IRBM) entered service. Thor, produced by Douglas, and Jupiter, produced by Chrysler, went into installations in Britain, Italy, and Turkey. The United

States soon succeeded in fielding ICBMs, which could be launched from the United States and attack targets within the Soviet Union. The first two ICBM programs were Atlas and Titan. These programs used separate contractors for each major system in order to facilitate competition and force companies to deliver their products on time.

The Air Force did not like the complicated Atlas and Titan missiles and granted a contract to Boeing to manufacture the Minuteman, which entered service in 1962. The Minuteman program did not use separate contractors for each system, but allowed Boeing to subcontract the component manufacturing. Morton Thiokol, Aerojet-General, Hercules Incorporated, North American, Sylvania, Avco, and General Electric all supplied systems for the Minuteman, which were then assembled by Boeing. This approach proved much more effective, and Boeing produced more than one thousand Minutemen, making the missile the foundation of the U.S. ICBM arsenal, even after the MX Peacekeeper missile entered service in the 1980's.

Space Exploration. American interest in rocket technology before World War II scarcely existed. Robert H. Goddard conducted pioneering research in the field, but few people gave his theories much notice. During World War II, tactical rockets for battlefield use proved their effectiveness and teams at American universities and corporations began work on the weapons. The success of these weapons combined with the German breakthroughs in ballistic missile technology ensured that rockets would be a key component of national defense.

A logical outgrowth of work on ballistic missiles was the idea of space travel. Goddard had theorized about using rockets to reach the Moon, and the conquest of space quickly became an important Cold War achievement. The Soviet Union's successful launch of Sputnik in October, 1957, revealed that the U.S. space program lagged behind its rival. In response, the United States took a number of drastic steps to improve the nation's position in the space race. Schools instituted new curriculums with heavy emphasis on math and science, while the government combined military and civilian rocket research and in 1958 created a new agency, the National Aeronautics and Space Administration (NASA), to replace the National Advisory Committee for Aeronautics (NACA).

The Soviet lead in the space race allowed it to put the first human in space in 1961, but the United States soon made up the gap. The focused space program administered by NASA stressed corporate cooperation rather than competition. The tremendous cost of developing space vehicles prevented any one company from dominating the field. Instead of using one contractor, NASA used components from a wide variety of manufacturers to create finished products. Companies that failed to meet NASA's specifications and deadlines risked losing contracts after having spent millions of dollars on research and development. Grumman, General Electric, and North American all revamped their manufacturing and management techniques after aggressive analysis from NASA.

The Apollo Moon-landing program illustrated NASA's approach. No individual company could develop the equipment necessary for such a task. The agency used a variety of contractors to produce a handful of rockets and spacecraft. The Saturn rockets that carried the crews to the moon were a result of components produced by companies including Chrysler, Boeing, and North American. The Saturn V rocket stood 363 feet high and had a diameter of 33 feet, dwarfing any rocket the United States had yet produced. The huge size required companies to invest in new jigs and welding fixtures, new techniques in fabrication, and static test stands that were far larger than any in existence. The research and development and production costs of the Saturn rockets alone totaled $9.3 billion. Grumman, the main contractor for the Lunar Module, also faced tremendous challenges and suffered through numerous delays and cost overruns before delivering the finished product. The companies involved often complained about NASA's unrealistic expectations, but the two sides generally found mutually agreeable solutions and manufacturers often found ways to streamline their manufacturing processes.

Following the conclusion of the Apollo Program in 1972, American interest in space exploration waned. NASA conducted Skylab missions and a joint mission with the Soviet Union in 1975, but these offered little financial security for contractors. When the United States launched the first space shuttle mission in 1981, the space program enjoyed a brief resurgence, but this comeback ended with the explosion of the shuttle *Challenger* in 1986. NASA resumed crewed flights two years later, but the enthusiastic days of Apollo had gone forever. The increasing costs of space missions forced NASA to increase its participation in joint international missions. Despite these setbacks, contractors found new ways to remain active in space missions. In 1989, private corporations took over the launching of commercial payloads from NASA. McDonnell Douglas, Martin Marietta, and General Dynamics all sent satellites into orbit at less than half the cost of a space shuttle mission.

Commercial Aircraft

The United States' affluence and desire for travel following World War II represented an important market for aerospace manufacturers. Companies used the technology developed for the military during the war to produce faster and more comfortable passenger planes. Just as with military aircraft, the more advanced civilian designs proved more costly, and a failed project could leave a manufacturer deeply in debt. Even a successful design could require years to become profitable.

Douglas and Lockheed led the immediate postwar commercial programs. The DC family from Douglas and Lockheed's Constellation provided both intercontinental and transatlantic service and proved very popular. However, these piston-engine planes did not represent the future of the commercial airline industry. Britain's De Havilland Comet, the world's first jet airliner, entered service in 1952, proving that just as in the military sector, American companies trailed their British competitors in passenger jet technology. Unfortunately for De Havilland, a number of mysterious accidents grounded the Comets for two years, giving American manufacturers time to cut into De Havilland's technological lead.

Leaders Douglas and Lockheed did not embrace jet airliners as enthusiastically as did Boeing. The Seattle-based company realized that the company's development costs for the B-52 bomber, KC-135 tanker, and a civilian airliner would be prohibitive unless Boeing could coordinate efforts on all three aircraft. Boeing used the same basic design for both the KC-135 and what would become known as the 707, the most successful U.S. first-generation jet airliner, which entered service in 1958.

This method of combining operations helped manufacturers offset some of the risk involved in developing new aircraft. Douglas managed to lengthen its DC-8 jet by 37 feet in the mid-1960's, offering room for seventy more passengers. Boeing found that its 707 design did not allow for the same modifications, giving Douglas a significant advantage in the market. Boeing soon recaptured its position at the forefront of airliner manufacturing by developing the world's first jumbojet. Based on Boeing's failed attempt to win the Air Force's competition to build an enormous new transport, the Boeing team modified their design into the 747, which rolled off the assembly line in 1968.

These methods helped manufacturers to control costs and to insure themselves to some extent against failure. Companies also advertised their planes in travel magazines, hoping to win passenger loyalty. However, creating a new design always entailed financial risk. When Boeing began work on the new 727 in the early 1960's, the company found that it would have to sell three hundred of the planes simply to break even. The 727 became remarkably successful, but the three-hundred-plane total was the equivalent of the entire production runs of commercial airliners twenty years earlier. The enormous sums of money that aerospace companies spent on research and production of military, space, and civilian aircraft eventually came back to haunt the manufacturers. In the late 1960's and early 1970's, companies faced the twin threats of reduced military budgets and a slumping economy. Boeing had to cut its workforce by nearly two-thirds, and Lockheed, staggering under the burden of producing the massive C-S Galaxy transport and new L-1011 airliner, nearly went out of business. Lockheed remained afloat solely because the federal government guaranteed the company's credit to potential lenders. High costs also forced some companies to merge, including the 1965 merger of McDonnell and Douglas. Merger trends continued through the remainder of the twentieth century, as manufacturers found themselves unable to compete in the changing marketplace.

This time of transition and economic distress eventually passed, and the commercial sector of the industry emerged with a clear structure. Boeing led U.S. airliner producers and followed up its earlier designs with new airplanes, including the 737, 757, 767, and the next generation of airliners, the 777. McDonnell Douglas and Lockheed maintained secondary positions, while European consortium Airbus entered the U.S. market, providing stiff new competition for Boeing. The U.S. aerospace industry finished the twentieth century as the world's leader, but changing government and commercial needs forced manufacturers to cut costs in order to remain competitive.

Matthew G. McCoy

Bibliography

Bilstein, Roger. *The American Aerospace Industry: From Workshop to Global Enterprise*. New York: Twayne, 1996. A solid historical examination of corporate development in American aviation that also examines the role of general aviation manufacturers, such as Cessna and Piper.

_______. *Flight in America: From the Wrights to the Astronauts*. Rev. ed. Baltimore: Johns Hopkins University Press, 1994. A good overview of aviation and space travel that also examines technological trends in aviation.

Pisano, Dominick, and Cathleen Lewis, eds. *Air and Space History: An Annotated Bibliography*. New York: Garland, 1988. An extraordinarily thorough bibliographical guide covering a wide range of topics in flight, including economic, political, technical, and corporate subjects.

See also: Airplanes; Boeing; Lockheed Martin; McDonnell Douglas; Manufacturers; Mergers; National Advisory Committee for Aeronautics; National Aeronautics and Space Administration; Space shuttle; Spaceflight

Ailerons and flaps

Definition: Hinged sections on the trailing edges of wings.

Significance: Both ailerons and flaps can be deflected to change local wing camber and to increase or decrease local lift. Ailerons are used to control the airplane in roll, while flaps allow flight at lower speeds for landing and takeoff.

Ailerons

Early experimenters with gliders turned their vehicles by shifting their bodies so their weight was to the left or right of their wing's lifting center. This action made the glider roll or bank to help it turn. Wilbur and Orville Wright improved on this effect by twisting or warping their wood and fabric wing with ropes and pulleys so that one wingtip was at a higher angle of attack than the other and the difference in lift on the two wingtips helped it roll. This design gave their airplane much greater maneuverability than early European designs which tried to turn using only a rudder. This wing-warping control system was the essential element in the Wright patent on the first successful airplane.

Glenn H. Curtiss, another American aviation pioneer, patented a different way to control an airplane in roll, using ailerons, originally small, separate wings that were placed between the upper and lower wings of his biplane near the wingtips. The pilot could change the angle between these small wings and the flow to increase the lift on one wing and decrease that of the other. The Wrights claimed that this was a violation of their patent, and the case spent many years in the courts until the U.S. government stepped in to resolve the dispute.

Today's ailerons are built into the trailing edge of wings near the wingtips, and they work by changing the wing's camber, or curvature, instead of its angle of attack. The ailerons deflect either up or down opposite to each other to increase the lift near one wingtip while lowering lift on the other wingtip. This makes the wing roll, with one wing moving up and the other down. Usually the aileron deflecting up produces more drag than the one moving down, which helps the airplane turn. In most turns, the aileron movements are coordinated with the movement of the rudder to create a turn which is balanced so that the airplane passengers feel only a downward force and no sideward force. A coordinated turn not only feels better but also is more aerodynamically efficient.

If the pilot wants to roll the airplane without turning, the rudder must also be used to oppose the turning motion caused by aileron drag; this is called a cross-control maneuver. A similar cross-control use of rudder and ailerons can make the airplane rotate to the left or right in a sideslip motion without rolling.

Flaps

Flaps often resemble ailerons except that they are placed on the wing near the fuselage rather than near the wingtips. Flaps normally are only deflected downward since they are used to increase temporarily the wing's lift on landing and sometimes on takeoff. This maneuver allows flight at lower speeds and landing and takeoff in shorter distances.

Early aircraft did not need flaps because they flew at low speeds and could land in much shorter distances than today's planes; however, as airplanes became more streamlined and could cruise at higher speeds and altitudes, they needed higher speeds for takeoff and landing. Designers added flaps to give additional lift and drag and to reduce landing speeds. The famous DC-3 airliner was one of the first commercial planes to use flaps to combine good cruise performance with reasonable landing and takeoff distances.

There are many types of flaps, from simple plates that deflect down from the bottom of the wing to very sophisticated combinations of little wings that extend down and behind a wing. The split flap was used on the DC-3 and many World War II airplanes. Fowler flaps are more common today, but many smaller aircraft use simple hinges on the rear part of their wings to deflect a plain flap. The Fowler flap increases the wing camber while increasing the wing area. The space that opens up between the deployed Fowler flap and the wing allows an airflow that helps control the pressures over the flap and delay wing stall.

Many airliners designed in the mid-to-late twentieth century used complex flap systems that worked like the Fowler flap but had two or more flap elements that opened out below and behind the wing. These flap systems were very carefully designed temporarily to give very high lift at low speeds on sleek, modern wings that were shaped for flight near the speed of sound. They allowed airplanes that cruise at 500 to 600 miles per hour to land at low speeds and come to a stop on relatively short runways.

Today's commercial transport designs do not need these complex flap systems and tend to use simpler Fowler flaps, which are lighter and easier to build and maintain. This shift is partly because of improvements in wing and airfoil design and partly because most major airports now have longer runways.

Flaperons and Slats

Occasionally, an airplane design needs extra flap area to get lower landing speeds and the ailerons are also used as flaps. This kind of aileron is called a flaperon, and it requires a more complex hookup to the aircraft controls than a standard aileron and flap system.

Some aircraft have flaps on the front of their wings that can also be deflected downward to increase the wing camber. These leading edge flaps, or slats, help control the flow over the wing at high angles of attack and allow the wing to go to a higher angle of attack before it stalls.

James F. Marchman III

Bibliography

Barnard, R. H., and D. R. Philpott. *Aircraft Flight*. 2d ed. Essex, England: Addison Wesley Longman, 1995. An excellent, nonmathematical text on aeronautics. Well-done illustrations and physical descriptions, rather than equations, are used to explain virtually all aspects of airplane flight.

Docherty, Paul, ed. *The Visual Dictionary of Flight*. New York: Dorling Kindersley, 1992. A profusely illustrated book showing the parts and the details of construction of a wide range of airplane types, old and new. An outstanding source of information about what airplanes and their parts look like.

Wegener, Peter P. *What Makes Airplanes Fly? History, Science, and Applications of Aerodynamics*. New York: Springer-Verlag, 1991. A well-written and well-illustrated historical but slightly technical review of the development of aerodynamics and airplanes.

See also: Aerodynamics; Airplanes; Glenn H. Curtiss; Forces of flight; Wing designs

Air Canada

Definition: Major airline company of Canada.

Significance: Air Canada is the major airline of Canada, supported by both private investment and government subsidies.

Government Air Service

Canadian air history began in 1909, when John McCurdy piloted his famous "Silver Dart" on its first flight. After World War I, small so-called bush airlines introduced commercial air flight into the country, and some of these evolved into the modern Canadian lines. James A. Richardson, a Winnipeg businessman, started Western Canadian Airlines, which later became Canadian Pacific Airlines and then Canadian Airlines International. The Canadian parliament passed the Trans-Canada Airline Act on April 10, 1937, creating Trans-Canada Air, which began with a new Lockheed 10A Electra, two used Electras, and a Stearman Model 4. The new company hired the bush pilots, who had to learn instrument flying on the Electras. At first, Trans-Canada Air served as an airmail carrier flying from Vancouver to Seattle, and only began regular commercial passenger service in 1939. The line accepted applications for stewardesses. A thousand applied; twelve were hired.

The postwar period represented an era of continued growth and expansion. The carrier transported more than 180,000 passengers in 1945 and employed more than 32,000 people, compared to 21,000 passengers and less than 500 employees in 1939. In 1945, the airline bought its first Douglas DC-3, which flew until 1983.

Trans-Canada enjoyed a government monopoly on all domestic Canadian air routes from 1937 to 1959, but then the government granted other Canadian companies the right to compete. Many remote northern areas of Canada were accessible only by air, and the country required a broad range of air services that could be met by smaller and intermediate-sized lines in addition to Trans-Canada. Canadian Airlines and Canadian Pacific Airlines (CPA) emerged as major rivals. Four other important regional airlines and hundreds of smaller companies competed as well. On January 1, 1965, Trans-Canada Air changed its name to Air Canada.

Throughout the post-World War II years, the line endured numerous labor and financial problems. Furthermore, it had a difficult time trying to expand into the U.S. market and complained that American government officials favored American companies. In response, Air Canada sought partners in other countries. In 1966, the company signed a key agreement with the Soviet airline Aeroflot, becoming the first North American airline to do so, and setting up routes for both carriers from Moscow to Canada.

Air Disasters

The worst disaster of Trans-Canada Air occurred on November 29, 1963, at St. Thérèse de Blainville, north of

Technical Characteristics of Air Canada's DC-9-32 Fleet

Timetable Code: D9S
Period of Use: 1966-present
Manufacturer: McDonnell Douglas
Number of Aircraft: 17
Engines: JT8D-7B
Typical Number of Seats: 91
Cargo Capacity: 3,410 pounds
Fuel Capacity: 4,260 U.S. gallons
Overall Length: 119.3 feet
Wingspan: 93.3 feet
Top of Fin from Ground: 27.5 feet
Cruise Speed: 490 miles per hour
Range (Full Passenger Capacity): 1,265 miles
Typical Cruise Altitude: 33,000 feet

Source: Data taken from (www.aircanada.ca/about-us/our-fleet/au303k.html), June 5, 2001.

Montreal, when Flight 831, a DC-8F, went down, killing all 111 passengers and 7 crew on board. This was the third fatal crash on the line's passenger flights. The first occurred at Armstrong, Ontario, in February, 1941, when a Lockheed 14 crashed, killing twelve (nine passengers and three crew). In 1947, a Lockheed 18 went down near Vancouver, killing twelve passengers and three crew, and in 1954, at Moosejaw, Saskatchewan, a training plane crashed into a Trans-Canada North Star DC-4M, killing thirty-one passengers and four crew, as well as the pilot of the trainer and a woman on the ground. In June, 1983, a fire in the washroom of Air Canada Flight 797, a DC-9, forced the plane to land at Cincinnati Airport in Covington/Hebron, Kentucky. Eighteen passengers and five crew escaped but twenty-three passengers died in the fire.

Privatization

In 1989, the Canadian government privatized Air Canada, but problems from competitors continued. The airline replaced its Boeing 727's with Airbus A300's and Boeing 767's. In 2000, Air Canada and Canadian Airlines International, which had replaced CPA after the latter ceased operation in 1989, merged after complex negotiations. The merger, however, did not bring all the hoped-for benefits. Air Canada still suffered stiff competition from airlines with low fares and better service. By 2001, Canada 3000 Airlines and Westjet Airlines joined Air Canada as the three largest airlines in Canada. (Canada 3000 would file for bankruptcy by the end of the year, however.) While the national line suffered heavy losses, which they blamed on business layoffs in an economic slump, Westjet Airlines showed a profit. Air Canada bought out and closed another new airline, Roots Air, which also threatened competition. In July, 2001, the airline was fined by the Ontario Securities Commission for stock irregularities by giving information to market analysts in advance of public release.

In 1997, Air Canada, along with United Air Lines, Lufthansa, Thai Airways International, and Scandinavian Airlines System (SAS) had formed the Star Alliance, the foremost international air alliance group joined subsequently by other airlines. By 2001, Air Canada was the seventh largest airline in North America and the twelfth largest in the world. It carried thirty million passengers annually and employed forty-five thousand people. Through its own lines and connecting flights it reached more than ninety airports in Canada and the United States, in total serving 188 destinations directly on five continents. Its major hubs were located in Toronto, Montreal, and Vancouver, and had a fleet of 375 planes.

Frederick B. Chary

Bibliography

National Transportation Safety Board. *Aircraft Accident Report: Air Canada Flight 797*. Washington, D.C.: Author, 1986. A technical report on the fire aboard Flight 797 in 1983.

Noble, Kimberly, et al. "Air Gerry." *Maclean's* 112, no. 36 (September 6, 1999): 42-45. An article describing the attempt of Toronto financier Gerry Schwartz to take over Air Canada.

Smith, Philip. *It Seems Like Only Yesterday: Air Canada, the First Fifty Years*. Toronto: McClelland and Stewart, 1986. A history of the airline written for the general public.

See also: Air carriers; Airline industry, U.S.; Airplanes; Safety issues

Air carriers

Definition: That part of the commercial system of air transportation consisting of airlines certified to provide domestic and international service.

Significance: Air carriers are an integral part of the air transportation system. Air travel is vital to the economic and social welfare of most nations. It facilitates business activity by allowing rapid and frequent personal contact between companies, as well as allowing for the rapid shipment of goods worldwide. Air travel links distant communities to important public and private centers of operation. It also supports a number of other travel-related industries such as hotels, rental cars, and travel agencies.

Air Carriers

In 1903, Orville and Wilbur Wright flew the first aircraft for a total of twelve seconds and reached a speed of 31 miles per hour. While many people were fascinated by the accomplishment, few could have guessed the impact that powered flight would have on the world or the progress that would be made in less than a century of innovation. The use of aircraft in World War I gave governments worldwide reason to consider aviation as more than a hobby of the wealthy or a barnstorming circus event. Aircraft now offered serious military possibilities. The years immediately after World War I found airplanes taking on another role as mail carriers. However, commercial development of the aviation industry did not really take off until the end of World War II. Maintaining an accessible, affordable, and safe air transportation network is now considered vital to any nation's infrastructure.

Early Years

At the end of World War I, Europe was faced with extensive damage to the railroad system that had once linked the continent. This provided an excellent opportunity for the fledgling aviation industry. However, early efforts to establish a viable passenger airline proved unsuccessful due to the high operating costs. In order to improve the odds of success, the governments of the major European nations, France, Great Britain, and Germany, became actively involved in the promotion of air operations through direct subsidies to national airlines.

In the United States, the political climate following World War I did not support such direct action to promote commercial aviation. Instead, the aviation industry received assistance from the U.S. Postal Service. The first private efforts to demonstrate the potential value of aircraft for mail delivery were conducted in 1911 and 1912. In 1916, the U.S. Congress made the first appropriation of money to be used to support airmail delivery. A second appropriation was passed in 1918 and the first official airmail route was flown on May 15, 1918, between Washington, D.C., and New York City. On February 2, 1925, the Air Mail Act of 1925, also called the Kelly Act, was passed. The Kelly Act, entitled "An Act to Encourage Commercial Aviation and to Authorize the Postmaster General to Contract for the Mail Service," was intended to get the U.S. government out of the airmail business and support private companies wishing to enter it.

Walter Folger Brown, appointed postmaster general by President Herbert Hoover, was determined to provide greater direction to the aviation industry. To this end, he initiated passage of the Air Mail Act of 1930, also known as the McNary-Watres Act. Officially, the act gave the postmaster general the power to award airmail contracts without the competitive bidding process required by the Kelly Act and lowered the rate paid for airmail delivery by paying carriers for space used rather than weight of mail carried. The McNary-Watres Act had two major effects on the shape of the aviation industry. First, it encouraged consolidation of many small companies after Brown made it clear that airmail contracts would be granted to companies that were large, financially stable, and capable of conducting transcontinental routes. Second, the reduction in airmail rates encouraged air carriers to consider expanding operations to passenger service. Brown did get the consolidation that he wanted as smaller companies began to merge, forming the airlines now known as American Airlines, United Air Lines, and Trans World Airlines (TWA). He also got complaints from smaller carriers passed over for airmail contracts.

While other carriers vied for the domestic airmail routes, Pan American Airways was interested in the international market. With the Foreign Air Mail Act of 1928, Pan American began to establish itself as the foreign airmail carrier. The company went on to establish themselves as the premier U.S. international carrier. Although Pan American would lose its monopoly over international flights after World War II, it would continue to dominate international travel until the 1980's.

When Franklin D. Roosevelt became president in 1933, he named James A. Farley postmaster general. After a U.S. congressional investigation into the awarding of contracts under Postmaster Brown, Farley canceled all contracts on February 3, 1934, and announced that the Army Air Corps would now deliver the mail. In a little over five months of service, the Air Corps recorded sixty-six crashes and twelve deaths, leading Roosevelt to order a halt to the operation. The Air Mail Act of 1934 reauthorized the postmaster general to award contracts based on competitive bidding, created a Federal Aviation Commission to study aviation policy, and required the separation of airline companies and aircraft manufacturers.

Regulating the Industry

With the passage of the Air Mail Act of 1934, there were three federal agencies responsible for regulating the aviation industry in the United States. The Postal Service reviewed applications for airmail routes, while the Interstate Commerce Commission set and reviewed the rates to be paid. In addition, the Air Commerce Act of 1926 had given the Department of Commerce responsibility for insuring safety through the registration and certification of aircraft and the certification of pilots. After several months of study, the Federal Aviation Commission recommended the Civil Aeronautics Act of 1938. The act created the Civil Aeronautics Authority to establish policies relating to the safety and economics of air transport. The 1940 Amendment to the Civil Aeronautics Act gave the Civil Aeronautics Board (CAB) legislative and accident investigation authority, while a separate Civil Aeronautics Administration was charged with promoting and enforcing air safety.

The CAB Years

The CAB was authorized to issue Certificates of Public Convenience and Necessity to any person or business it deemed fit, willing, and able to perform public air transport. No one would be permitted to engage in public transport without such a certificate. These certificates were often very specific in their limitations on the types of services the air carrier was allowed to provide. They might designate certain allowed routes, intermediate stops, through service, and prohibited stops or types of traffic. Passenger fares and cargo rates were also strictly regulated. Carriers wishing to change fares or rates were required to file a petition arguing the just and reasonable nature of the increase. Although route and pricing assignments were the most visible of the CAB's activities, the board also exercised a broad range of other controls over U.S. air carriers, including setting standards for record keeping, maintenance scheduling, mergers, and intercarrier agreements. Labor issues were subject to the Railway Labor Act, which prescribed a detailed system of dispute resolution between parties, including intervention by the U.S. president if all other steps failed.

The Civil Aeronautics Act included a clause, Section 401e, which stated that all carriers that had provided adequate and continuous airmail service between May 14 and August 22, 1938, would be granted a permanent Certificate of Public Convenience and Necessity. The first airline to receive this certificate was Delta Air Corporation. Fifteen other carriers received such a certificate, including American Airlines, Braniff International Airways, Chicago and Southern Airlines, Continental Airlines, Northwest Airlines, Pennsylvania-Central Airlines, Transcontinental & Western Air, Eastern Air Lines, Inland Airlines, Mid-Continent Airlines, National Airlines, Northeast Airlines, United Air Lines, Western Air Express, and Wilmington-Catalina Airlines. Under the CAB, five mergers were eventually approved among this original group. Chicago and Southern and Northeast Airlines merged with Delta. Inland merged with Western. Mid-Continent merged with Braniff. Pennsylvania-Central became Capital and later merged with United. The airlines known as Transcontinental & Western changed their corporate name to Trans World Airlines.

The years between 1938 and 1958 were growth years for the airline industry. The Douglas aircraft, the DC-3, was introduced in 1936 and became one of the most popular passenger aircraft in history. From the airlines' perspective, the two-engine aircraft, which seated twenty passengers, substantially lowered their operating costs, making passenger operations more attractive. The introduction of the first jets into commercial service in 1956 was a watershed event for aviation. Almost overnight, the speed of air travel doubled, but the jet also created problems for airlines not used to maintaining these new aircraft and air traffic systems not used to the speed with which incidents might occur. A series of accidents led to public outcry for government action to improve the safety of the skies.

The Federal Aviation Act of 1958 modified the old Civil Aeronautics Act of 1938, expanded the power of the government over matters of safety, and created the Federal Aviation Agency as an independent agency answerable only to Congress and the president. With the formation of the Department of Transportation in 1966, the agency's name was changed to the Federal Aviation Administration (FAA) and placed under the secretary of transportation. The FAA consolidated all air safety research and development under one agency and assumed responsibility for the creation of safety rules. The fact that air travel increased from just over 1 million passengers in 1938 to 267 million in 1978 meant that the FAA served a vital role in developing a safe air transportation system.

While technological developments continued to improve the safety of air transportation, a growing number of critics began arguing against the economic regulation of the industry. These critics contended that the regulation of route entries and exits as well as prices created fares that were higher than the unregulated intrastate routes. This regulation limited competition on profitable routes and prevented carriers from exiting less profitable ones. In the wake of public distrust with government caused by the Vietnam War and the Watergate crisis, the political climate was ripe for deregulation of the domestic airline industry. The Airline Deregulation Act of 1978 marked the end of an era.

Regulating International Aviation

World War II diverted the attention of many governments from issues of commercial domestic aviation and interrupted international aviation. In the closing year of the war, however, the Allied Powers, led by the United Kingdom, the Soviet Union, and the United States, were ready to begin considering the role of aviation in the postwar world. A conference was held on November 1, 1944, in Chicago, Illinois, involving representatives of fifty-four allied nations to consider matters relating to international aviation. The Chicago Conference, as it is commonly known, considered four proposals for the international aviation environment. Australia and New Zealand suggested creating an airline with international ownership and management. The United States proposal favored individual ownership and unrestricted rights to all international markets with the fares and frequency of flights determined by free market forces. The British also wanted individual ownership, but they favored tight regulation of market entry, fares, and frequency. They suggested the creation of an international body to regulate matters relating to international aviation. The Canadian proposal represented a compromise between the U.S. and British positions. It called for limited competition and a multilateral organization to allocate routes and review questions of fares and frequency. Unfortunately, the representatives were not able to reach agreement on any of the proposals.

Pan American was one of the earliest airlines to shift its focus from mail delivery to world tourism. (Library of Congress)

In 1946, the United States and Great Britain met again to address the question of international aviation. The result of the meeting was an agreement commonly called Bermuda I, after the site of the meeting. This agreement became the first bilateral air service agreement, that is, the first agreement between two countries on the air rights granted to each party. Bermuda I set the pattern that all other agreements would follow. It established the principle of reciprocal rights or the exchange of air rights in air service agreements. Bilateral agreements would also designate the number and sometimes the name of the carriers from each side allowed to operate to each country and would establish limits on capacity and route frequency. As a result of Bermuda I, an international body was established to allocate routes and set fares. This organization, the International Air Transport Association (IATA), was made up of representatives from all airlines providing international service. A second organization, the International Civil Aviation Organization (ICAO), was formed to deal with technical and safety-related aviation matters. Over time, bilateral agreements became the key to international aviation. Agreements would be negotiated by governments, often with the input of their major carriers. These agreements were designed to protect the national carriers from international competition.

The decision by the United States to begin deregulation of its own markets in 1978 also affected international aviation. Invoking U.S. antitrust law, which prohibited competitors from colluding to set price and allocate supply, the U.S. government issued a notice to IATA that it was considered an illegal cartel. U.S. airlines were notified that they could no longer participate in the fare- and route-setting mechanisms of IATA. The U.S. government later announced its intention to pursue a strategy of liberalizing aviation markets by pursuing new bilateral agreements, known as Open Skies Agreements, that reduced or eliminated route restrictions and freed carriers to offer fares responsive to consumer demand.

Era of Aviation Liberalization

Efforts to free air carriers from economic regulation proceeded on both the domestic and international fronts after 1978. Domestically, the Airline Deregulation Act resulted in the phase-out of the CAB and the establishment of the FAA as the chief agency concerned with matters relating to aviation. Although carriers would still be required to file for a Certificate of Convenience and Necessity, they were now free to enter and exit markets based on consumer demands. Fares would now be determined by the market forces of competition and consumer demand.

People Express in many ways typified the fate of the many carriers that arose after 1978. It was the first carrier established after deregulation and one of many carriers to begin operation as a no-frills, low-fare, regional carrier. The demand for its service was tremendous, leading to explosive growth. In only six years, it became the tenth-largest carrier in the nation. However, to achieve this rate of growth it began acquiring new aircraft and eventually whole carriers, including Frontier Airlines, Britt Airways, and Provincetown Boston Airlines (PBA). Its rapid growth and high debt, along with renewed competition from the large, prederegulation carriers, eventually sent People Express to the brink of bankruptcy and resulted in its acquisition by another carrier. Since the beginning of deregulation, over two hundred carriers have started and failed. Most never achieved the size of People Express before high start-up costs and intense competition forced them into bankruptcy. The fate of the established carriers was often not much better after deregulation. Under the CAB, no major U.S. carrier had ever gone bankrupt, because fares were set and new routes allocated to ensure that airlines made a profit. The cost of operation was not a critical factor to success. The new, deregulated start-up carriers would challenge the older carriers with lower costs and lower fares. In 1982, Braniff International became the first major airline in the United States to file for bankruptcy. It would not be the last. Pan American and Eastern eventually ceased operations entirely. Other carriers faced periods of serious financial crisis, as fuel hikes, economic downturns, and low-cost competition threatened profits and forced cutbacks.

On the international scene, the U.S. government began pursuing a strategy of encirclement to encourage liberalization. In effect, the United States would attempt to sign Open Skies Agreements with countries surrounding major restricted markets, such as Great Britain and Japan, in an effort to divert traffic and revenues from them. It was hoped that this pressure would encourage more rapid liberalization. The final step in the liberalization of the international aviation markets would be to allow foreign carriers to fly domestic passengers, known as cabotage. For instance, in 2001, British Airways had the right through bilateral agreement to fly passengers into New York City, but the airline could not pick up passengers in New York City and fly them to another destination in the United States. One of the features of the European Union (EU) has been to allow carriers from any EU state to fly within the territory of another member state. There are proposals to establish such provisions across the Atlantic between Europe, the United States, and Canada.

While there is still debate about the effects of deregulation on national and international markets, there is evidence that fares have declined. In the light of declining fares, carriers have focused on cutting costs, creating hubs to funnel traffic from smaller markets into larger ones, and utilizing technology to manage revenues. More affordable airfares also means that more people are likely to travel by air. In 2001, it was forecast that more than one billion passengers a year would board an aircraft in the United States in 2011. (This prediction was made, however, before the terrorist attacks of September 11, 2001, resulted in a drastic reduction in U.S. air travel due to security concerns.) Internationally, the rate of growth in passenger traffic is higher in the Latin American and Asia-Pacific markets, where developing nations are turning to air transportation to link their countries and support their industries.

Dawna L. Rhoades

Bibliography

Cappelli, P., ed. *Airline Labor Relations in the Global Era: The New Frontier.* Ithaca, N.Y.: Cornell University

Press, 1995. A collection of articles written by industry experts that focuses on the labor issues involved in deregulation and international aviation.

Heppenheimer, T. A. *Turbulent Skies: The History of Commercial Aviation.* New York: John Wiley & Sons, 1995. A very readable history of the development of aviation up to the deregulation of the U.S. aviation market.

Kane, R. M. *Air Transportation.* 13th ed. Dubuque, Iowa: Kendall/Hunt, 1998. A widely used introductory text for airline management.

Sochor, E. *The Politics of International Aviation.* Iowa City: University of Iowa Press, 1991. An excellent, insider view of the development of the international aviation environment.

Toh, B. "Towards an International Open Skies Regime: Advances, Impediments, and Impacts." *Journal of Air Transportation World Wide* 3, no. 1 (1998). A very good background on the goals and tactics used by the United States government in pursuing more liberalization in the international air transportation industry.

Wells, A. T. *Air Transportation: A Managerial Perspective.* 3d ed. Belmont, Calif.: Wadsworth, 1994. One of the most widely used introductory texts in the air transportation industry.

See also: Aer Lingus; Aeroflot; Aeromexico; Air Canada; Air France; Airline industry, U.S.; Airmail delivery; Alitalia; American Airlines; British Airways; Continental Airlines; Delta Air Lines; EgyptAir; El Al; Food service; Frequent flier miles; Iberia Airlines; Japan Airlines; KLM; Korean Air; Lufthansa; Northwest Airlines; Pan Am World Airways; Pilots and copilots; PSA; Qantas; SAS; Singapore Airlines; Southwest Airlines; Swissair; Trans World Airlines; Transatlantic flight; Transcontinental flight; Transglobal flight; United Air Lines; US Airways; Virgin Atlantic; World War I; World War II

Air Combat Command

Also known as: ACC

Date: Activated on June 1, 1992

Definition: Major command responsible for providing air combat forces to the United States' unified combatant commands, including the U.S. Strategic Command, U.S. Atlantic Command, U.S. Central Command, U.S. Southern Command, U.S. European Command, North American Aerospace Defense Command, and U.S. Transportation Command.

Significance: The Air Combat Command provides the air power flexibility required to respond rapidly for defense, peacekeeping, and humanitarian missions globally.

History

The creation of the Air Combat Command (ACC), officially activated on June 1, 1992, occurred as the result of the reorganization of the U.S. Air Force in the post-Cold War period. Prior to 1992, the primary commands of the Air Force included the Strategic Air Command (SAC) and Tactical Air Command (TAC). SAC exercised control over the nation's nuclear arsenal and strategic defense planning, while TAC functioned as the command for tactical, or specific mission, coordination. During the Vietnam conflict, SAC performed many tactical missions while TAC engaged in strategic bombing. The overlapping responsibilities continued through Operation Desert Storm in 1991. After U.S.-Soviet relations improved at the end of the Cold War, Air Force commanders reevaluated the need for the two distinct commands.

Air Force Chief of Staff General Merrill A. McPeak, Vice Chief of Staff General John M. Loh, and SAC Commander in Chief General George L. Butler advocated the restructuring of the Air Force commands in a manner that would eliminate repetitive functions and provide an efficient allocation of resources. Military officials, after reviewing the proposed changes, agreed to merge most of SAC and TAC under a new command, the Air Combat Command (ACC), and to reorganize the Military Airlift Command (MAC). The goal of the ACC was to provide the Air Force with the power to implement policy on a global basis.

On June 1, 1992, ACC officially replaced TAC in a brief ceremony conducted at Langley Air Force Base, with the former commander of TAC, General Loh, becoming the new commander of ACC. SAC was then deactivated and the U.S. Strategic Command assumed responsibility for the nuclear weapons of the U.S. Air Force and U.S. Navy. The newly created ACC initially assumed control over all fighter resources within the forty-eight contiguous states, as well as bombers, reconnaissance platforms, intercontinental ballistic missiles (ICBMs), and battle-management resources.

Continuing efforts to streamline and reorganize the Air Force commands resulted in numerous changes within ACC during its first few years of operation. Shortly after the activation of ACC, all combat search-and-rescue units transferred from the Air Mobility Command (AMC) to

ACC. In February, 1993, the Air Rescue Service (ARS) was reassigned to ACC. Several months later, it was renamed the U.S. Air Force Combat Rescue School and assigned to Nellis Air Force Base, Nevada. While ACC gained some resources, it lost others, as the 58th and 325th Fighter Wings, responsible for F-16 and F-15 training, were reassigned to the Air Education and Training Command (AETC) on July 1, 1993. At the same time, ACC lost two of its numbered units when the Twentieth Air Force, responsible for ICBMs and consisting of six missile wings, one test and one training wing, and the base at which they were stationed, F. E. Warren Air Force Base, Wyoming, transferred to the Air Force Space Command. That same day, the Air Force also deactivated the Second Air Force, a unit responsible for reconnaissance missions that ACC had acquired from the merger between SAC and TAC.

Operations

ACC operations include global military and humanitarian missions as well as the peacetime defense of the United States. ACC provided troops, both active-duty and reserve units, during Operation Desert Storm and Operation Southern Watch in the Persian Gulf. ACC has also been involved in the war against drugs and continues to provide Airborne Warning and Control Systems (AWACS), reconnaissance, and fighter aircraft to prevent the transportation of illegal substances into the United States. In Eastern Europe, ACC enforced the no-fly zone against the Serbians in Operation Deny Flight. ACC has also participated in numerous humanitarian missions throughout the world. In Eastern Europe, ACC units supplied aid to Bosnian civilians in Sarajevo. In Turkey, ACC deployed active-duty personnel for Operation Provide Comfort, rendering humanitarian assistance to the Kurdish inhabitants of the northern portion of Iraq who were being persecuted by the Iraqi government. In Africa, ACC troops participated in Operation Restore Hope in Somalia, providing food for the country's starving population, as well as providing similar aid in Operation Support Hope in Kenya and Uganda. ACC also provided relief for the victims of a bloody civil war in Rwanda. Closer to the United States, ACC personnel were involved in Operation GTMO out of the United States Naval Base at Guantanamo Bay, Cuba, for the purpose of providing relief for Haitian refugees. In 1994, ACC assisted in the processing of Cuban refugees during Operation Safe Haven.

ACC continues to defend the United States from foreign enemies while expanding its role to assist people around the world. Quick response and constant preparedness allow ACC to exercise global power for the United States. In August, 1993, two B-52's stationed out of Ellsworth Air Force Base, South Dakota, set a record flying time on an around-the-world flight and in 1994, two B-52's from Barksdale Air Force Base, Louisiana, established the longest jet flight in history during their 47.2-hour flight around the world. By the year 2000, Air Combat Command's original mission had been altered to reflect streamlining within the Air Force and the need for flexible response to rapidly changing worldwide circumstances. Although no longer responsible for many of the duties formerly performed by Strategic Air Command, such as control of ICBMs, ACC maintains a prominent position in the Air Force and is vital for combat, rescue, and theater airlift missions.

Organization

Headquartered at Langley Air Force Base, Virginia, ACC operates fighter, bomber, reconnaissance, battle-management, rescue, and theater airlift aircraft, in addition to command, control, communications, and intelligence systems. ACC operates with 102,000 active-duty personnel and civilians and can mobilize an additional 64,400 members of the Air National Guard and Air Force Reserve during times of national emergency. ACC's numbered air force units consist of 775 aircraft, with the

Air Combat Command (ACC) Bases

ACC Base	*Air Force*
Barksdale AFB, Louisiana	Eighth Air Force
Beale AFB, California	Twelfth Air Force
Cannon AFB, New Mexico	Eighth Air Force
Davis-Monthan AFB, Arizona	Twelfth Air Force
Dyess AFB, Texas	Eighth Air Force
Ellsworth AFB, South Dakota	Eighth Air Force
Holloman AFB, New Mexico	Twelfth Air Force
Langley AFB, Virginia	Ninth Air Force
Minot AFB, North Dakota	Eighth Air Force
Moody AFB, Georgia	Ninth Air Force
Mountain Home AFB, Idaho	Twelfth Air Force
Nellis AFB, Nevada	Twelfth Air Force
Offutt AFB, Nebraska	Twelfth Air Force
Seymour Johnson AFB, North Carolina	Ninth Air Force
Shaw AFB, South Carolina	Ninth Air Force

number of ACC-accessible aircraft totaling 1,700.

The strength of the ACC includes four numbered air forces and two direct reporting units. The First Air Force exercises oversight of the air defense forces for North America under the North American Aerospace Defense Command. It is headquartered at Tyndall Air Force Base, Florida, with additional units stationed at Northeast Air Defense Sector in Rome, New York, and Western Air Defense Sector at McChord Air Force Base, Washington.

The Eighth Air Force, headquartered at Barksdale Air Force Base, Louisiana, controls ACC forces in the central United States, with the direct responsibility for war fighting under the U.S. Atlantic Command, as well as functioning as the Joint Task Force/Bomber for U.S. Strategic Command. The Eighth Air Force consists of units stationed at ten bases. The Second Bomb Wing of B-52's operates out of Barksdale Air Force Base. The Twenty-seventh Fighter Wing with its EF-111's and F-16's is based at Cannon Air Force Base, New Mexico. The Seventh Wing with its B-1 and C-130H aircraft is stationed at Dyess Air Force Base, Texas. Ellsworth Air Force Base, South Dakota, is home to the Twenty-eighth Bomb Wing of B-1 bombers. Lajes Field, Azores, is the home base for the Sixty-fifth Air Wing. The 314th Airlift Wing with its C-130E/H's is based at Little Rock Air Force Base, Arkansas. The Fifth Bomb Wing, composed of B-52 bombers, is stationed out of Minot Air Force Base, North Dakota. Whiteman Air Force Base, Missouri, is home for the 509th Bomb Wing of B-2 and T-38 aircraft. The Eighty-fifth Group, stationed at Keflavik Naval Air Station, Iceland, and the Third Air Support Operations Group out of Fort Hood, Texas, also operate under the Eighth Air Force.

The Ninth Air Force, which is responsible for ACC forces in the eastern United States under U.S. Central Command, is headquartered at Shaw Air Force Base, South Carolina. Units from the Ninth Air Force are stationed at five bases along the Atlantic seaboard. Forces at Langley Air Force Base, Virginia, home of the First Fighter Wing, include the C-21A and the F-15C/D as well as Forty-first Rescue Squadron and Seventy-first Rescue Squadron at Patrick Air Force Base, Florida. The 347th Wing out of Moody Air Force Base, Georgia, includes F-16's, C-130E's and A-10's. At Pope Air Force Base, North Carolina, the Twenty-third Wing includes A-10's, F-16C/D's and C-130E's. The Fourth Fighter Wing of F-15E's is stationed at Seymour Johnson Air Force Base, North Carolina, and the Twentieth Fighter Wing, with its A-10's and F-16C/D's, is located at Shaw Air Force Base, South Carolina. In addition, the Ninth Air Force includes the Thirty-third Fighter Wing, with its F-15C's at Elgin Air Force Base, Florida; the Ninety-third Air Control Wing, with its E-8 Joint STARS, and the Fifth Combat Communications Group at Robins Air Force Base, Georgia; the 823d Red Horse Civil Engineering Squadron at Hurlbert Field, Florida, and the Eighteenth Air Support Operations Group at Pope Air Force Base, North Carolina.

The last of the numbered air forces within the Air Combat Command is the Twelfth Air Force, headquartered at Davis-Monthan Air Force Base, Arizona. The Twelfth Air Force controls ACC forces in the western United States and Panama and operates under the U.S. Southern Command and the U.S. Strategic Command. The Twelfth Air Force shares Joint Task Force/Battle Management duties with the U.S. Strategic Command. Under the Twelfth Air Force are the Ninth Reconnaissance Wing of U-2R/S and T-38 aircraft, based at Beale Air Force Base, California; the 355th Wing of A-10's, and EC-130H/E's, based at Davis-Monthan Air Force Base; the Forty-ninth Fighter Wing of F-117A's, T-38's, HH-60's, and German F-4E's, based at Holloman Air Force Base, New Mexico; the Twenty-fourth Wing of C-21's, C-27's, and CT-43's stationed at Howard Air Force Base, Panama; the 366th Wing out of Mountain Home Air Force Base, Idaho, with its F-15C/D/E's, F-16's, KC-135R's and B-1's; and the Fifty-fifth Wing with its C-21, E-4, RC-135 S/U/V/W, EC-135, KC-135, TC-135 S/W, WC-135, and OC-135 aircraft stationed at Offutt Air Force Base, Nebraska. The Twelfth Air Force also has units stationed at four other bases, with the 388th Fighter Wing at Hill Air Force Base, Utah; the 820th Red Horse Civil Engineering Squadron at Nellis Air Force Base, Nevada; the 522d Air Control Wing and the Third Combat Communications Group at Tinkler Air Force Base, Oklahoma; and the First Air Support Operations Group at Fort Lewis, Washington.

Direct reporting units under the ACC include the Fifty-seventh Wing, with its A-10A, F-15 C/D/E, F-16, HH-60, and Predator uncrewed vehicles, as well as the Air Warfare Center, the U.S. Air Force Air Demonstration Squadron (Thunderbirds), the Ninety-ninth Air Base Wing, and the U.S. Air Force Weapons School located at Nellis Air Force Base, Nevada. The Fifty-third Wing at Eglin Air Force Base, Florida, is assigned to the Air Warfare Center and is responsible for several subordinate units, including the Sixth Electronic Combat Group and the Seventy-ninth Test and Evaluation Group at Eglin Air Force Base, the 505th Command and Control Evaluation Group at Hurlburt Field, Florida, and the 475th Weapons Evaluation Group at Tyndall Air Force Base, Florida. The second direct reporting unit is the Aerospace Command and Control, Intelligence, Surveillance and Reconnaissance Center. The ACC also operates the Aerospace Expeditionary Force Center and is responsible for Air Force search-and-rescue missions within the continental United States.

Cynthia Clark Northrup

Bibliography

Hanser, Lawrence M., Maren Leed, and C. Robert Roll. *The Warfighting Capacity of the Air Combat Command's Numbered Air Forces*. Santa Monica, Calif.: RAND, 2000. This study, funded by the RAND Corporation, analyzes the fighting capacity and preparedness of Air Combat Command forces. Statistical information and assessments indicate the need for strengthening the forces.

Logan, Dan. *ACC Bomber Triad: The B-52's, B-1's and the B-2's of Air Combat Command*. Atglen, Pa.: Schiffer, 1999. An excellent source that provides photographs and histories of all 208 bombers under ACC command since 1992. Information on weapons systems, unit and special mission objectives, and technical line drawings detail the strength and power of ACC resources.

McFarland, Stephen L. *A Concise History of the U.S. Air Force*. Washington, D.C.: Air Force History and Museums Program, United States Air Force, 1997. This work provides a general overview of the structure and organization of the Air Force, including the formation of ACC. General responsibilities and mission objectives are provided.

Air Force, U.S.

Definition: One of the primary components of the United States armed forces, with the key responsibility for air warfare.

Significance: The United States Air Force, officially established as a separate military service on July 26, 1947, is responsible for the effective prosecution of offensive and defensive air operations.

Foundation of the Air Corps

The U.S. Army Signal Corps organized an aeronautical division on August 1, 1907, only three years after the Wright brothers made their famous flight at Kitty Hawk, North Carolina. Until 1914, however, this division was interested more in balloons and dirigibles than in flying machines. The Army had used manned balloons for the observation of enemy lines during the American Civil War (1861-1865) and the Spanish-American War of 1898. The Wright brothers sold their first airplane to the Army in 1909 but it was used only for experimental purposes. The Wright brothers taught the first officers how to fly. The first operation unit, the First Aero Squadron, was formed in December, 1913, under the command of Captain Benjamin D. Foulois.

Congress provided additional funds in 1914, just a few weeks before World War I began in Europe, and the Army established the Aviation Section of the Signal Corps to test planes and train pilots. In April, 1917, the United States entered the war on the side of the Allied Powers of England, France, Italy, and Russia. Both the Allies and the Central Powers of Germany, Austria-Hungary, and Turkey had developed air forces that were bigger and better equipped than that of the United States.

Despite additional money from Congress, the United States never caught up with the European nations in aviation technology. The vast expansion required mass production of pilots, observers, and mechanics. A network of schools was established for advanced training in aircraft engineering and the aviation section sent officers to the Massachusetts Institute of Technology and similar universities to learn the required skills. On May 24, 1918, President Woodrow Wilson created the Army Air Service within the War Department to develop plans to catch up with the nations of Europe.

When World War I ended on November 11, 1918, the Air Service had 19,000 officers and 178,000 enlisted men. The American aircraft industry had built more than 11,500 planes (mostly training craft) during the fifteen months the United States fought in Europe, but as soon as the war stopped, the number of personnel was quickly reduced as the nation demobilized its military and ended the draft. The production of aircraft came to almost a complete halt.

While in France, American pilots flew mostly French-made planes under the command of Brigadier General William "Billy" Mitchell. Although the air war played only a minor role in the war's outcome, air power had shown its potential for providing cover and protection for ground troops. The British, recognizing the importance of its air power, created an independent Royal Air Force (RAF) in 1918.

Between World War I and World War II (1941-1945), General Mitchell led an unsuccessful and bitter struggle within the War Department and in congressional hearings to create a separate air force. The Army Reorganization Act of 1920 made the Air Service part of the Army, and in 1926, the name of the Army Air Service was changed to the Army Air Corps. In the 1930's, as the Fascist powers of Germany, Japan, and Italy spent heavily to build up their military forces, the United States kept only a small Army, Navy, and Air Corps. Congress appropriated only small amounts of money on the growth or modernization of the armed services, principally because the Great Depression of the 1930's had had a devastating impact on the American economy.

Creation of the Air Force

The horrible battles of World War I, which killed more than twenty million soldiers, led some planners to see the advantages of airplanes as weapons of war. With proper use of airplanes, the bloody massacres of trench warfare, in which armies attacked each other by running across open fields trying to dislodge the enemy from its trenches, could be avoided.

Advocates of air power believed that airplanes would change the nature of warfare by carrying the war to the enemies' factories and supply stations. Bombs would destroy cities and buildings and cause panic among the civilian population. Intense bombing would quickly weaken the morale of the people and destroy their will to fight. Air power would shorten war, would make warfare less expensive, and would save the lives of a nation's ground and naval forces. The bombing of civilians, as savage as that might be, was more merciful than gassing and machine-gunning vast numbers of soldiers.

General Henry Harley "Hap" Arnold became chief of the Air Corps in 1938. A graduate of West Point who had been trained in flying by Orville Wright, Arnold pushed for increased appropriations, but not until after the Japanese attack on Pearl Harbor on December 7, 1941, did Congress respond. Arnold and Mitchell had tried to teach these lessons to War Department observers in the 1930's, but only a real war would convince doubters that air power was a crucial factor in victory.

In 1941, the Air Corps began an expansion in response to events in Europe and the Pacific. After American entry into the war, all Army air units were merged into the Army Air Force (AAF) under General Arnold. The AAF quickly grew into a powerful organization composed of 16 air commands (12 of them overseas), 243 combat groups, 2,400,000 officers and men, and almost 80,000 aircraft.

During the war, two U.S. air commands, the Eighth and the Fifteenth, joined with the British RAF Bomber Command in the strategic bombing of Germany. Strategic bombing was defined by Mitchell as the use of air power against the enemy's heartland and industrial base. Although the night-and-day bombing of Germany destroyed its war industries, the bombing proved far less accurate than expected. Because bombardiers could not see through the cloud cover over northern Europe, precision was lost. Radar helped, but pinpoint targets often disappeared in the cluttered radar images of large cities. Another problem for

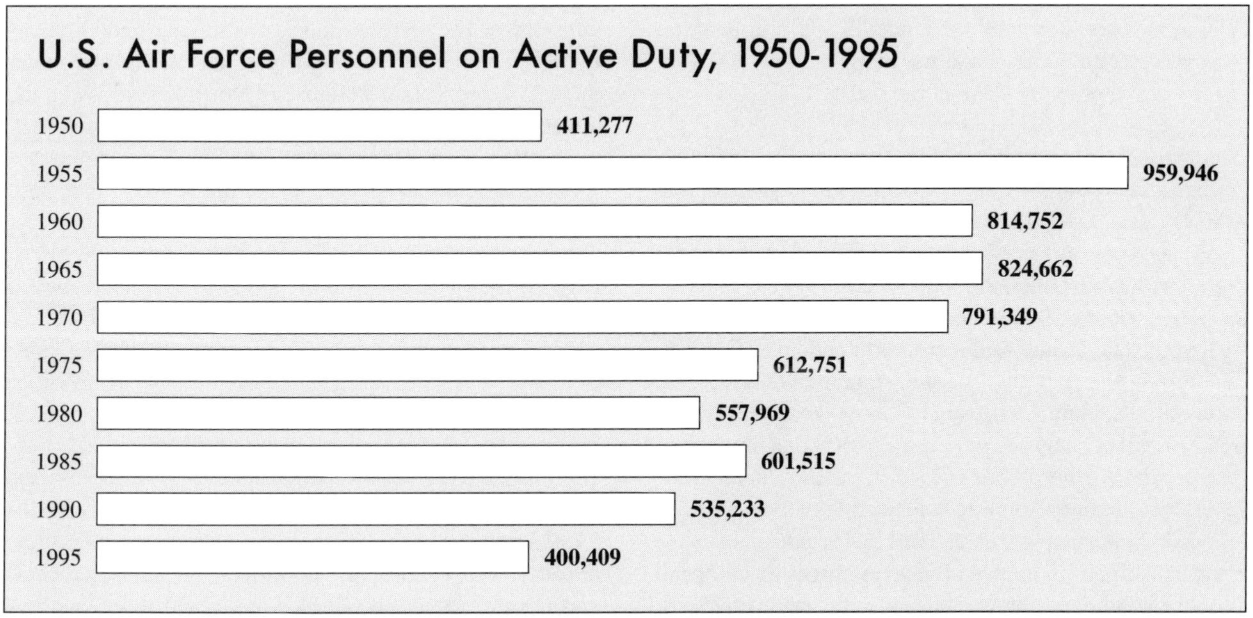

Sources: The data used in this graphic element are based on information found in *Historical Statistics of the United States: Colonial Times to 1970* (2 vols., Washington, D.C.: Government Printing Office, 1975) and *The Time Almanac 2000*, edited by Borgna Brunner (Boston: Information Please, 1999).

the Allies was their lack of air defense; during long-range strikes, unescorted Allied bombers were knocked out of the sky with appalling accuracy by German ground fire. Almost 75 percent of Allied crews never returned from bombing missions. Not until the introduction of a new escort fighter, the North American P-51 Mustang, were the skies reclaimed from German control to become safer for American and British pilots.

In the Pacific, the Tenth Air Force in the China-Burma-India theater and the Fourteenth in China supported the British and Chinese against the Japanese. The Fifth, Seventh, and Thirteenth Air Forces joined with the Army and Navy in the series of island invasions and conquests that were the stepping-stones to the attack of Japan. The Tenth Air Force carried supplies over the Himalayan Mountains, playing a key role in preventing a Japanese victory in China.

Because of the vastness of the Pacific Ocean, strategic bombing of Japan remained impossible until after the B-29 bomber was produced and bases were established close enough to Japan to allow U.S. forces to strike at Japanese industrial cities. When sustained bombing of Japan did begin in May, 1944, the Japanese military leaders recognized that they were close to defeat. From the Mariana Islands, the B-29 bombers of the Twentieth Air Force carried out a bombing campaign against Japan that ended with the atomic bombing of Hiroshima and Nagasaki in August, 1945. The Pacific war was ended on August 12, and the armed forces began a swift demobilization. Despite the reduction in force, however, the United States still had the most powerful military in the world, with about 300,000 officers and men in the AAF at the end of 1947.

Organization

In March, 1946, the basic pattern of unit organization for the AAF was established in descending order as follows: command, air force, air division, wing, group, squadron, and flight. On July 26, 1947, the National Security Act created the independent U.S. Air Force. Stuart Symington became the first secretary of the Air Force, and he appointed General Carl Spaatz the first chief of staff. One month later, Air Force pilot Captain Charles E. "Chuck" Yeager flew the Bell X-1 rocket plane beyond the speed of sound. From this time on, the Air Force was crucial in the development of supersonic flight and atomic bombs and other weaponry.

The U.S. Air Force was charged by President Harry S. Truman to organize, train, and equip air forces for air operations, including joint operations with the Army and Navy; to gain and maintain air superiority over other nations; to develop a strategic air force and conduct air reconnaissance operations; to provide airlift and air support for airborne operations; to furnish air support to land and naval forces, including occupation forces; and to provide air transport for the armed forces, except as provided by the Navy for its own use. The first opportunity the Air Force had to display its power came in 1950.

The Air Force in the Korean War

During the early hours of June 25, 1950, Communist North Korean soldiers launched a surprise attack across the thirty-eighth parallel, the dividing line between North Korea and the Republic of South Korea. Within three days, the Communists had advanced to Seoul, the South Korean capital, and the U.S. ambassador requested the immediate evacuation of all American civilians in the city. Three days after the invasion, the United Nations declared war on the North because it had violated the U.N. Charter, which outlawed wars of aggression. U.N. forces, composed primarily of American and South Korean forces, went to war the next day.

The U.S. Air Force soon employed new jet fighters, such as the F-86 Sabre, to gain control of the skies. It also helped protect U.N. ground forces with close support and successful raids to destroy North Korean reinforcements and supplies. U.S. Air Force B-29's were first used in August, 1950, for the purpose of knocking out North Korea's ability to build guns and tanks, a goal that was accomplished by the end of September. By late October, the U.N. troops had advanced far into North Korea, almost reaching the Yalu River, which forms North Korea's northern boundary with China. The first appearance of a Russian-built MiG-15 fighter on November 1 showed that China was about to enter the war to prevent the collapse of North Korea. When the Chinese attacked late in November, the U.N. forces retreated to South Korea under the protection of the U.S. Air Force.

The battle for air superiority continued to escalate until July, 1953, when an armistice was signed. In air-to-air combat, U.S. Air Force fighters had shot down 792 MiGs, losing only 78 American craft. After the war ended, the Air Force kept a large number of units in the Pacific to defend against any future Communist invasions in Asia.

The Cold War and the Arms Race

In the 1950's, the invention of the hydrogen bomb, 100,000 times more powerful than the atomic bombs used against Japan, inaugurated an arms race between the United States and the Soviet Union. By the middle of the decade, both nations had developed long-range rockets and were working on missiles that could reach halfway around the world in about thirty minutes. Under the command of General Curtis LeMay, the Air Force's Strategic Air Command (SAC) became the key instrument of American defense strategy. SAC planes were in the air twenty-four hours a day, always flying along the Russian border within thirty minutes of Moscow. Another major part of the Air Force's fleet was the long-range B-52 Stratofortress, which would remain the principal bomber in the U.S. Air Force for more than forty years.

In the 1960's, intercontinental ballistic missiles (ICBMs), such as the Atlas, Titan, and Minuteman, were added to the U.S. arsenal. With the development of satellites, the Air Force also expanded its mission into space. Together with the Navy's missile-launching submarines, bombers and land-based ICBMs made up the triad of the U.S. nuclear deterrent force. With nuclear weapons in the air, on the land, and under the sea, the United States felt assured that the Russians would never be able to destroy America's entire nuclear force. At least one arm of the triad would remain, in the event of a devastating nuclear attack, to launch a nuclear attack against the initial aggressor.

In October, 1962, a U.S. Air Force U-2 reconnaissance plane took photographs of Russian missile bases being built in Cuba. The missiles launched from these sites would be within striking distance of Washington, D.C., New York City, and other targets along the Atlantic Coast. President John F. Kennedy ordered the armed forces on alert for the possibility of nuclear war. On October 22, the president declared a strict quarantine of all offensive military equipment under shipment to Cuba. The U.S. Air Force kept the island of Cuba under constant surveillance, providing the Navy with constant information on Russian ships at sea apparently on their way to Cuba. On October 28, only minutes before American and Soviet ships prepared to fire on each other, Soviet premier Nikita S. Khrushchev agreed to remove the missiles as well as a unit of Russian bombers from the island. In the event that shooting had begun, President Kennedy had been ready to launch nuclear missiles toward the Soviet Union and to order SAC planes to attack Russian military bases. The Cuban Missile Crisis represented the closest the world had yet come to a nuclear war.

The U.S. Air Force in Vietnam

Communist expansion in Southeast Asia created new challenges in the mid-1960's. In 1964, the United States became directly involved in a war in the former French colony of Vietnam. The war began when the Communist North invaded South Vietnam. In 1965, after two deadly Communist attacks on American military advisers in South Vietnam, President Lyndon B. Johnson authorized Operation Rolling Thunder, which called for continuous bombing of the North's main cities and military bases.

U.S. Air Force and U.S. Navy airplanes supported more than 500,000 American ground troops in a difficult and unpopular war. The F-4 Phantom II provided close air support for Army and Marine units in the fields and jungles, while the F-105 Thunderchief led hundreds of bombing raids against North Vietnam. SAC B-52 bombers dropped millions of tons of bombs on remote strongholds of the North Vietnamese Army and were refueled by KC-135 Stratotankers.

In 1972, the U.S. Air Force launched Operation Linebacker, which produced the heaviest bombing of the entire war. This intense campaign forced the North to sign a peace treaty in January, 1973. U.S. land forces were withdrawn a few months later and were no longer available to help South Vietnam when the North launched another invasion in 1975. This time, the Communist forces were successful, and South Vietnam was absorbed into the North.

In the 1970's, despite budget reductions, the U.S. Air Force spent as much as it could to modernize its aircraft and missiles while also becoming more heavily involved in space exploration. Meanwhile, the Soviet Union continued to produce new weapons at a faster rate than the United States and built up its ground forces in Europe and the Far East to great levels. After the Soviet Union's disastrous war in Afghanistan (1979-1989), however, the military balance shifted back to the United States.

The U.S. defense buildup of the 1980's allowed the U.S. Air Force to expand its arsenal of aircraft and to deploy a wide range of new weapons systems, including the Lockheed F-117A stealth fighter, which was developed in response to the need for an aircraft capable of attacking targets without being detected by radar. The F-117A was first used in the 1989 U.S. invasion of Panama, known as Operation Just Cause, in which Manuel Noriega, the self-appointed president of Panama, who had been engaged in drug smuggling, was arrested and brought to the United States for trial.

New American technologies, such as radar-proof bombers and space-based weapons systems, helped bring an end to the Cold War. The leaders of the Soviet Union realized that they could no longer compete with American military advances. The tearing down of the Berlin Wall in 1989 marked the final days of the long Cold War and a few years later the Soviet Union itself began to break apart.

U.S. Air Force Actions in the 1990's

The 1990-1991 conflict in the Persian Gulf was in many ways the test case for modern air power in the precision-

The F-16 Fighting Falcon is one of many fighter jets deployed by the U.S. Air Force in peacekeeping missions around the world. (AP/Wide World Photos)

weapons era. After Iraqi armies invaded the neighboring state of Kuwait on August 2, 1990, President George Bush imposed an immediate economic embargo against Iraq and its leader, Saddam Hussein. On August 7, Bush ordered the launch of Operation Desert Shield to free Kuwait and drive out the Iraqi invaders. A coalition of Allied forces under the command of General H. Norman Schwarzkopf was mobilized to carry out the effort against Iraq. When Hussein refused to leave Kuwait peacefully, Operation Desert Storm was begun on January 16, 1991.

After ten days of continuous bombing by coalition air forces against Iraqi military targets, the ground war began. Within two days, the Iraqi army had been crushed, and a cease-fire went into effect. The Gulf War demonstrated the dominance of air power, a key factor in the Allied victory. The Iraqis had been defeated by air strikes that totally destroyed their ability to fight.

In the early 1990's, during conflicts in the former republic of Yugoslavia, U.N. peacekeeping forces undertook an air campaign in an attempt to stabilize the region. The campaign in Bosnia included U.S. Air Force planes and offered more evidence that precise attacks were possible and could shatter air defenses and land forces. In 1995, after eleven days of air attacks brought the Serbian government to the peace table in Dayton, Ohio, the Dayton Accords were signed, helping to bring the promise of peace to the region. Once again, it was proven that air power could make the major difference in victory or defeat.

Humanitarian Aid

Since 1947, the U.S. Air Force has participated in almost six hundred humanitarian relief efforts in response to floods, fires, hurricanes, and earthquakes. Nearly one hundred of these relief efforts have taken place since 1987. Immediately after the Gulf War, the U.S. Air Force used its capabilities to provide critically needed food supplies to the people of the Soviet Union. This effort was called Operation Provide Hope. A few years later, Operation Provide Promise brought humanitarian aid to Bosnia and Serbia. With the ending of the Cold War, the need for the Air Force to respond to humanitarian crises has become a major part of its mission.

Leslie V. Tischauser

Bibliography

Benson, Lawrence R. *Golden Legacy, Boundless Future: A Brief History of the United States Air Force*. Washington, D.C.: Secretary of the Air Force, Office of Air Force History, 1997. A well-written, fiftieth-anniversary history of the U.S. Air Force.

Dick, Ron, and Dan Patterson. *American Eagles: A History of the United States Air Force*. Charlottesville, Va.: Howell Press, 1997. A comprehensive history of the U.S. Air Force, featuring photos of the airplanes and materials at the Air Force Museum at the Wright-Patterson Air Force Base in Ohio.

Hallion, Richard P., and Bernard C. Nalty. *History of the United States Air Force*. Washington, D.C.: Office of Air Force History, 1999. An official but critical analysis of air power, strategy, and key personalities in U.S. Air Force history.

See also: Air Force bases; Antiaircraft fire; Dogfights; Gulf War; Korean War; Military flight; Royal Air Force; Strategic Air Command; Tactical Air Command; Vietnam War; World War I; World War II

Air Force bases

Definition: A U.S. Air Force facility that serves as a base of operations for military aircraft.

Significance: Air Force bases are an important strategic component of the U.S. military

The U.S. Air Force maintains bases in the United States and overseas. Within the continental United States, Texas holds the greatest number of bases with eight, followed by California with six and Florida with five. Overseas, Japan ranks first with three bases, followed by Germany, the United Kingdom, and South Korea, each with two.

Overseas, the bases are operated under bilateral agreements and Status of Forces agreements that reflect the United States' rights and use.

The total worldwide cost of government investment in the bases (original acquisition plus improvements as of September 30, 2000) was $50,483,479,441. The oldest active Air Force base is the Francis E. Warren Air Force Base (AFB) in Laramie County near Cheyenne, Wyoming, which was established in 1867 by the U.S. Army as Fort D. A. Russell. The name of the 5,866-acre facility was changed by presidential degree to Fort Francis E. Warren in 1930. Warren, a former senator and governor, won the Medal of Honor in the Civil War. The Army relinquished jurisdiction of the facility to the Air Force in 1947. It now serves as home to the Ninetieth Space Wing, part of the Air Force Space Command, Peterson Air Force Base, Colorado.

The second oldest installation is Offutt Air Force Base, Nebraska, which was activated as Ft. Crook in 1896. Dating from 1916, Kelly Air Force Base, Texas, was the oldest continuously active air base in the United States. Some other longstanding Air Force bases are Scott Air Force Base, Illinois, established in June, 1917; Bolling Air Force Base, Washington, D.C., established in October, 1917; Brooks Air Force Base, Texas, established in December, 1917; and Pope Air Force Base, North Carolina, established in 1919.

A recent "designated" base is Buckley Air Force Base, Denver, Colorado, in October, 2000. A recent "established" base is Schriever Air Force Base (formerly Falcon Air Force Base), Colorado Springs, Colorado, named for General Bernard A. Schriever, which was activated in September, 1985, and renamed in June, 1998.

The largest Air Force bases in terms of acreage are Eglin Air Force Base, Florida, located on Choctahatchee Bay a few miles from the Gulf of Mexico near the towns of Fort Walton Beach and Destin. Eglin is the sprawling home of the Air Warfare Center, which covers 463,452 acres. It was activated in 1940 and is named for Lieutenant Colonel Frederick Eglin, who was killed in an air crash near Anniston, Alabama, in 1937.

The largest base in terms of personnel is Lackland Air Force Base, in San Antonio, Texas, with approximately 23,500 military, Department of Defense civilians, and students.

The air bases are controlled by eight different Air Force commands. Two bases, the Air Force Academy in Colorado Springs, Colorado, and Bolling Air Force Base, in the District of Columbia, are considered Direct Reporting Units (DRUs).

Twenty-six Air Force installations were closed and six were realigned as a result of the disposal authorities contained in the Base Closure and Realignment Acts of 1988, 1991, 1993 and 1995, following action by the Defense Base Closure and Realignment Commissions (BRAC).

Air Guard and Reserve

In addition, there is the Air Force Reserve Command (AFRC) at Robins Air Force Base, in Macon, Georgia. The AFRC was established on February 17, 1997, and is staffed by 173,725 personnel. The Air National Guard, established on September 18, 1947, and composed of

Major Active-Duty Domestic Air Force Bases, 1998

State	*Air Force Base*	*Command*	*Major Units*
Alaska	Eielson AFB	PACAF	354th Fighter Wing
	Elmendorf AFB	PACAF	Headquarters, 11th Air Force; 3d Wing
Alabama	Maxwell AFB	AETC	Air University
Arizona	Davis-Monthan AFB	ACC	Headquarters, 12th Air Force
	Luke AFB	AETC	56th Fighter Wing
Arkansas	Little Rock AFB	AETC	314th Airlift Wing
California	Beale AFB	ACC	9th Reconnaissance Wing
	Edwards AFB	AFMC	Air Force Flight Test Center
	Los Angeles AFB	AFMC	
	McClellan AFB	AFMC	Sacramento Logistics Center
	Travis AFB	AMC	Headquarters, 15th Air Force
	Vandenberg AFB	AFSPC	Headquarters, 14th Air Force
Colorado	Schreiver AFB	AFSPC	Space Warfare Center
	Peterson AFB	AFSPC	Headquarters, Air Force Space Command; 21st Space Wing
	USAF Academy	DRU	
District of Columbia	Bolling AFB	DRU	11th Wing
Delaware	Dover AFB	AMC	436th Airlift Wing
Florida	Eglin AFB	AFMC	Air Armament Center
	Hurlburt Field	AFSOC	Headquarters, Air Force Special Operations Command; 16th Special Operations Wing
	MacDill AFB	AMC	6th Air Refueling Wing
	Patrick AFB	AFSPC	45th Space Wing
	Tyndall AFB	AETC	Headquarters, 1st Air Force (Air National Guard); 325th Fighter Wing
Georgia	Moody AFB	ACC	347th Wing
	Robins AFB	AFMC	Headquarters, Air Force Reserve Command; Warner-Robins Air Logistics Center
Hawaii	Hickam AFB	PACAF	Headquarters, Pacific Air Forces; 15th Air Base Wing
Idaho	Mountain Home AFB	ACC	366th Wing
Illinois	Scott AFB	AMC	Headquarters, Air Mobility Command; Tanker Airlift Control Center
Kansas	McConnell AFB	AMC	22d Air Refueling Wing
Louisiana	Barksdale AFB	ACC	8th Air Force
Maryland	Andrews AFB	AMC	89th Airlift Wing
Massachusetts	Hanscom AFB	AFMC	Electronics Systems Center
Mississippi	Columbus AFB	AETC	14th Flying Training Wing
	Keesler AFB	AETC	2d Air Force
Missouri	Whiteman AFB	ACC	509th Bomb Wing
Montana	Malmstrom AFB	ACC	341st Space Wing, 819th Red Horse Squadron
Nebraska	Offutt AFB	ACC	55th Wing

State	*Air Force Base*	*Command*	*Major Units*
New Jersey	McGuire AFB	AMC	Headquarters, 21st Air Force; 305th Air Mobility Wing
Nevada	Nellis AFB	ACC	Air Warfare Center
New Mexico	Cannon AFB		
	Holloman AFB		
	Kirtland AFB		58th Special Operations Wing
North Carolina	Pope AFB		
	Seymour Johnson AFB		
North Dakota	Grand Forks AFB		
	Minot AFB	ACC	5th Bomb Wing
Ohio	Wright-Patterson AFB	AFMC	Headquarters, Air Force Materiél Command; Aeronautical Systems Center
Oklahoma	Altus AFB	AETC	97th Air Mobility Wing
	Tinker AFB	AFMC	Oklahoma Air Logistics Center
	Vance AFB	AETC	71st Flying Training Wing
South Carolina	Charleston AFB	AMC	437th Airlift Wing
	Shaw AFB	ACC	Headquarters, 9th Air Force
South Dakota	Ellsworth AFB	ACC	28th Bomb Wing
Tennessee	Arnold AFB	AFMC	Arnold Engineering Development Center
Texas	Brooks AFB	AFMC	Human Systems Center
	Dyess AFB	ACC	7th Bomb Wing
	Goodfellow AFB	AETC	17th Training Wing
	Kelly AFB	AFMC	San Antonio Air Logistics Center
	Lackland AFB	AETC	369th Recruiting Group
	Laughlin AFB	AETC	47th Flying Training Wing
	Randolph AFB	AETC	Headquarters, Air Education and Training Command; Headquarters, 19th Air Force; Air Force Recruiting Office
	Sheppard AFB	AETC	80th Flying Training Wing
Utah	Hill AFB	AFMC	Ogden Air Logistics Center
Virginia	Langley AFB	ACC	Headquarters, Air Combat Command
Washington	Fairchild AFB	AMC	92d Air Refueling Wing
	McChord AFB	AMC	62d Airlift Wing
Wyoming	Francis E. Warren AFB	AFSPC	Headquarters, 20th Air Force; 90th Space Wing

111,633 personnel, is overseen by the National Guard Bureau in the Pentagon but is commanded by the governor in each state or territory and by the commanding general in the District of Columbia. Guard and Reserve units operate from active Air Force bases as well as from commercial airport facilities.

Because those two elements of the nation's total military force play such an active part in fulfilling day-to-day as well as short- and long-term active duty requirements for the active Air Force, their more than eighty locations also might well be considered bases. For example, the Air Guard provides 100 percent of the Air Force interceptor force, 9 percent of the B-1B bomber force, 43 percent of the tactical airlift, 27 percent of the air-rescue capability, 30 percent of the tactical fighters, 25 percent of the tactical air support, 41 percent of the KC-135 refueling capability, and 9 percent of the strategic airlift capability, plus provides six aircraft for the Air Force's special operations missions. On any given day, 95 percent of the reserve units are rated ready for combat. Of its unit-owned aircraft, in times of war or other special needs, 98 percent would be gained by the Air Combat Command or Air Mobility Command.

James C. Elliott

Bibliography

Cragg, Dan. *Guide to Military Installations*. Mechanicsburg, Pa.: Stackpole Books, 1997. A thorough review of U.S. military installations at home and abroad.

Crawford, William "Roy," L. Ann Crawford, R. J. Crawford, and J. J. Caddell. *Military Space: Opportunities Around the World*. Falls Church, Va.: Military Living, 1998. A listing of military air installations offering space-available flights to military personnel as well as facilities at each base.

Evinger, William R. *Directory of U.S. Military Bases*. 3d ed. Phoenix, Ariz.: Oryx Press, 1998. An excellent directory of all military establishments in the United States and overseas, with information about their history, size, and assigned units.

Air Force One

Date: Beginning in 1933

Definition: Any one of several different aircraft on which the president of the United States is traveling.

Significance: One of the most recognizable and functional symbols of the American presidency, the planes that serve as *Air Force One* are equipped to support a wide range of missions. They function as command and communications centers, flying White Houses, and a secure and efficient means of transportation for the U.S. president, the commander in chief of the U.S. armed forces.

The Origins of a Presidential Plane

The first presidential plane was not called *Air Force One*, because an independent air force did not exist until 1947. Had the first presidential plane, delivered to the U.S. Navy on June 6, 1933, to transport President Franklin D. Roosevelt, been similarly named in 1933, it might have been called *Navy One*. This plane, a Douglas Dolphin Amphibian accommodating five passengers, served as the presidential plane until December 4, 1939.

Although Roosevelt often flew on commercial aircraft, the U.S. Army Air Force contracted with Douglas Aircraft to build a C-54 Skymaster to serve as the presidential plane. With a C-54A fuselage and C-54B wings, this VC-54C became known as the *Sacred Cow* and was equipped with a secret elevator extending from the aircraft's belly to lift the wheelchair-bound Roosevelt into the plane.

The First *Air Force One*

The *Sacred Cow* served President Harry S. Truman for the first two years of his administration. In 1947, on board the *Sacred Cow*, Truman signed an act creating an independent U.S. Air Force. After the *Sacred Cow* was retired in 1961, it was obtained by the U.S. Air Force Museum in 1985. Following a ten-year restoration effort, it was put on display at the U.S. Air Force Museum at Wright-Patterson Air Force Base near Dayton, Ohio, along with several of its successors.

The Second *Air Force One*

Douglas Aircraft also built the next *Air Force One*, a VC-118, commissioned into service on July 4, 1947. The aircraft was also known as the *Independence*, after Truman's Missouri hometown. At a cost of more than $1 million, the VC-118 was a military version of a Douglas DC-6. It carried Truman to Wake Island to discuss with General Douglas MacArthur the escalating problems in Korea. The VC-118 was retired from presidential service in 1953 and from Air Force duty in 1965.

Subsequent *Air Force One* Planes

President Dwight D. Eisenhower's aircraft was the C-121 military version of the Lockheed Constellation, affectionately known as the "Connie" in commercial transport. This presidential plane, named the *Columbine III*, was Eisenhower's third Connie. He had flown two C-121's as the Supreme Allied Commander in Europe. The *Columbine III* left presidential service when Eisenhower left office in 1961 and was retired to the Air Force Museum in 1966.

President John F. Kennedy's first presidential plane, a Douglas C-118, was the first plane to display the distinctive presidential paint scheme. It was followed by Special Air Missions (SAM) 26000, the first plane actually to use the call sign *Air Force One* when the president was on board. SAM 26000, a Boeing VC-137C, the military version of a 707, was the first jet designed exclusively for presidential use. Purchased in 1962 at a cost of $36.6 million, with its distinctive blue-and-white color scheme, the airplane became a symbol of American power.

SAM 26000 served as the presidential plane during some of the most important events in twentieth century American history. It carried Kennedy to Berlin for his famous speech at the Berlin Wall, to Dallas on November 22, 1963, and then home again, for the last time, only a few hours later. Before the plane could take off on the day of Kennedy's assassination, Lyndon B. Johnson was sworn in as president.

Johnson reconfigured SAM 26000 to suit his preferences. Because he wanted to see what others were doing, he jettisoned wooden cabin partitions in favor of clear plastic. He also required a chair and desk that could be raised or lowered by pushing a button. SAM 26000 carried Johnson to Vietnam at the height of the war, a destination shared by President Richard M. Nixon on his first SAM 26000 trip.

Soon after Nixon took office, SAM 26000 went back to Boeing for a complete overhaul, which included the removal of the taping system that recorded all incoming and outgoing calls on the flying White House. Beginning in 1970, SAM 26000 was used on thirteen secret missions to meet with North Vietnamese officials. In 1972, Nixon flew the plane on his historic mission to normalize relations with China. In 1971, Nixon renamed SAM 26000 *The Spirit of 1976*. In December, 1972, a similar 707, with a cost of $36.2 million and the tail number 27000, took over as the presidential plane. SAM 26000 served as the backup presidential aircraft for five more presidents: Gerald R. Ford, Jimmy Carter, Ronald Reagan, George Bush, and Bill Clinton. When, in January, 1998, Clinton's 747 *Air Force One* 28000 ran off a runway in Champaign, Illinois, SAM 26000 was sent to fetch the president, once again becoming *Air Force One*. SAM 26000 served for thirty-six years and retired to the U.S. Air Force Museum in 1998 with its last flight on May 19.

Other Presidential Aircraft

Because the call sign *Air Force One* is used to identify whatever plane is carrying the president, many less well-known aircraft have served as *Air Force One*. A Beech VC-6A turboprop transported Johnson from Austin, Texas, to his family ranch. Eisenhower used an Aero Commander U-4B twin prop, the smallest *Air Force One*. Johnson used a T-39A Sabreliner in Texas. A VC-140B Jetstar carried presidents Nixon, Ford, Carter, and Reagan. All have retired to the U.S. Air Force Museum. Other aircraft available for use by the president can serve as *Air Force One*, should the president be on board. The fleet includes five 707's, several C-9's and Gulfstreams, all based at Andrews Air Force Base in Maryland. *Marine One* is the name of the Marine helicopter that lands on the White House lawn to ferry the president literally from the doorstep.

The Modern *Air Force One*

The modern primary presidential plane, a VC-25A, is not one, but two, Boeing 747's, with the tail numbers 28000 and 29000. Deployed on September 6, 1990, and March 26, 1991, at an ultimate cost of $626 million for the two planes, spare parts, and a hangar, the planes are marvels of modern technology and comfort. The VC-25A has self-contained baggage loaders and front and back stairs. It has a range of 9,600 statute miles, or 8,348 nautical miles, but is capable of being refueled in flight. With a wingspan of 195.7 feet, it has a maximum takeoff weight of 833,000 pounds. It can operate at altitudes as high as 45,100 feet and attain speeds of 701 miles per hour. Four General Electric jet engines provide the thrust for the 231.8-foot-long aircraft. The VC-25A has armor that protects it from nuclear explosions and the ensuing electromagnetic pulses. It has jamming systems to deflect antiaircraft missile attacks and can make evasive maneuvers to avoid enemy aircraft. It has at least eighty-seven telephones and sixteen video monitors and can carry two thousand meals. It can stay aloft for a week. It has seven bathrooms, a bedroom suite for the president, and full office facilities. It has a total capacity of 102, with 26 crew.

The presidential seal is emblazoned on many items, such as the door, seats, blankets, glasses, silverware, note-pads, and boxes of cigarettes, before these were banned during the Reagan administration. Guests aboard *Air Force One* are presented with a certificate upon deplaning attesting to the fact they were passengers on the presidential plane.

Current *Air Force One* Specifications

Builder: Boeing
Power Plant: 4 General Electric CF6-80C2B1 jet engines
Thrust: 56,700 pounds per engine
Length: 231 feet, 10 inches
Height: 63 feet, 5 inches
Wingspan: 195 feet, 8 inches
Speed: 630 miles per hour
Ceiling: 45,100 feet
Maximum Takeoff Weight: 833,000 pounds
Range: 6,800 nautical miles
Passenger/Crew Capacity: 102 (26 crew)
Introduction Date: December 8, 1990 (No. 28000); December 23, 1990 (No. 29000)
Date Deployed: September 6, 1990 (No. 28000); March 26, 1991 (No. 29000)
Inventory: Active force, 2; ANG, 0; Reserve, 0

Source: Data taken from (www.af.mil/news/factsheets/VC_25A_Air_Force_One.html), June 6, 2001.

Extraordinary Plane, Ordinary Problems

For all its modern marvels and protections against attack, *Air Force One* must fly in crowded airspace and deal with commonplace aviation problems. On May 27, 1997, *Air Force One* and a United Parcel Service 747 had a near-midair collision, coming within 500 vertical feet of each other when mandatory ATC separations were not maintained. In 1995, an investigation by the Inspector General of the U.S. Department of Transportation into bogus parts on commercial jetliners led to the conviction of emergency-oxygen and fire-suppression equipment suppliers for *Air Force One*. In 1997, *Air Force One* was inspected for center-wing tank problems after the explosion of TWA 800, an earlier model 747. In 1998, *Air Force One* repeatedly disappeared from the Federal Aviation Administration's air traffic control radar screens.

As *Air Force One* is not safe from aviation problems, neither is it immune from bad press. When, on May 18, 1993, Clinton idled *Air Force One* on a Los Angeles International Airport tarmac while receiving a haircut from a Hollywood hairstylist, the negative publicity about the misuse of *Air Force One* was blistering.

Air Force One in Film

In 1997, *Air Force One* won a better place in Hollywood with the release of the blockbuster action film of the same name, filmed in cooperation with the U.S. Air Force and starring actor and real-life pilot Harrison Ford as the U.S. president. Audiences were treated to a glimpse inside the world's most famous aircraft. The filmmakers improved upon reality by adding a presidential escape pod, an oversized conference room, and a virtual arsenal of weaponry that does not actually exist on the real *Air Force One*. The weapons and vehicles that the Secret Service uses to protect the president are actually carried on support aircraft, one of which arrives well in advance of *Air Force One* to complete security arrangements.

Mary Fackler Schiavo

Bibliography

Francillon, Rene J. *McDonald Douglas Aircraft Since 1920*. Rev. ed. Annapolis, Md.: Naval Institute Press, 1988. A reference series detailing virtually every Douglas aircraft, including the presidential planes.

Schiavo, Mary. *Flying Blind, Flying Safe*. New York: Avon (Hearst), 1998. Describes investigating bogus aircraft parts cases, including the *Air Force One* case.

TerHorst, Jerald F., and Ralph Albertazzie. *The Flying White House: The Story of Air Force One*. New York: Coward, McCann & Geoghegan, 1979. The story of earlier *Air Force One* aircraft.

See also: Air Force, U.S.; Airplanes; Military flight

Air France

Also known as: Compagnie Nationale Air France
Date: Founded in 1933
Definition: The major international airline of France.
Significance: Based in Paris, Air France serves thousands of people every year in international flights to the United States, Canada, and all major European points. Air France serves China as well, a particularly important service in terms of the globalization of world business.

History

Air France was founded in 1933 through the merger of five companies originally founded between 1919 and 1929, Société Centrale pour l'Exploitation de Lignes Aériennes, Compagnie Internationale de Navigation, Air Union, Air Orient, and Compagnie Générale Aéropostale. The new company negotiated with the French government to become the country's national air carrier. World War II nearly destroyed the company, but on October 11, 1945, Paris-to-London service was resumed. In 1948, the new Compagnie Nationale Air France was reincorporated, with 70 percent of the company owned by the government. On January 12, 1990, all four of France's government-owned airlines, Air France, Air Inter, Air Charter, and UTA, were merged into the Air France Group.

Transatlantic flight was initiated in 1946 with a Paris-to-New York route, and thirty years later, Air France inaugurated supersonic transatlantic flight with the Concorde. The first Concorde flight was between Paris and Rio de Janeiro, but ultimately the only profitable supersonic route was between Paris and New York, and in 1982 the company made that its only Concorde route.

In the realm of subsonic flight, however, Air France, by the end of the twentieth century, served more than 230 cities in 88 countries. Its fleet is composed of Boeing 737's, 747-200's, 747-300's, and 747-400's, 767's and 777's, and Airbus A310's, A320's, and A340's.

Corporate Divisions

Air France is a major employer within the airline industry. In the United States alone, it employs nine hundred people

Events in Air France History

1933: A group of airlines collectively named the Société Centrale pour l'Exploitation des Lignes Aériennes (S.C.E.L.A.) is renamed Air France.

1938: Air France becomes the world's third-largest airline network, with one hundred aircraft. Its expansion is subsequently interrupted by World War II.

1946: The airline inaugurates Paris-to-New York service.

1947: Air France cooperates with the French Postal Service to establish night-mail service, giving the airline the largest network in the world.

1959-1960: Air France begins implementing Caravelle and Boeing 707 jet aircraft, cutting flight times in half.

1966: Air France's last remaining long-haul, propeller-driven aircraft are removed from service.

1968: Air France adopts the medium-haul Boeing 727 and, two years later, the long-haul Boeing 747, which carries approximately five hundred passengers.

1974: Air France flies the first Airbus A300 aircraft.

1976: Air France introduces the Concorde for supersonic travel along the airline's Paris-Dakar-Rio de Janeiro route.

1978: The airline establishes a cargo division.

1981: The airline introduces business-class service.

1990: Responding to increasing competition in the airline industry, Air France merges with UTA and Air Inter to become the Air France Group, one of the world's largest air transport groups.

1995: Air France launches the Airbus A340 wide-body aircraft.

2000: Air France enters into a global alliance with Aeromexico, Delta Air Lines, and Korean Air. An Air France Concorde jet crashes shortly after takeoff from Paris's Charles de Gaulle Airport, killing all on board and four on the ground.

to serve Air France customers. The Air France Industries Division is responsible for complex maintenance activities that include full checks on all equipment, and required major overhaul operations. The Air France Maintenance Division, operating from two main bases, at the Paris-Charles de Gaulle and Orly airports, is the hands-on and online maintenance service. They handle preflight checks, random daily inspection checks, and the entire range of minor to major repairs, up to and including all overhaul work to the Concorde fleet. The Industrial Logistics Division combines all of the Air France groups' industrial and repair activities.

Air France has developed a customized fleet service, which offers a range of services tailored to individual needs. This service is particularly directed to young airlines or others unwilling or unable to develop their own customized aeronautical maintenance structure. Air France markets all its industrial and aircraft maintenance services under the Air France Industries brand, which ranks as the second worldwide carrier in terms of aircraft maintenance and employs approximately nine thousand staff members. Air France Industries also offers an extensive range of skills and available training, particularly in Boeing fleets powered by General Electric and CFM International engines, as well as in various components used in Boeing fleets and the Airbus. Because of extensive experience with the Boeing 747, Air France is considered a world leader in that aircraft's overhaul, completing more than three hundred B-747 overhaul checks a year. Combining this skill with the handling of more than 300 engines and 62,000 components, Air France is considered the major service-oriented airline. Developing a concept of customized fleet service and having developed a global quality approach, Air France generated a 29.8 percent increase in operating revenues between 1999 and 2000, amounting to approximately 3.3 billion francs, or 497 million euros.

In 2000, Air France launched a very special globalized service called SkyTeam with three partners—Aeromexico, Delta Air Lines, and Korean Air—which focuses entirely on customer service. Air France retains leadership in the French market, Europe's largest domestic air-transport market, as well as leadership of world tourism.

Safety Record

Given their global network of flights, Air France's safety record is good. Two hijackings, on June 27, 1976, in Entebbe, Uganda, and on December 24, 1994, in Algiers, Algeria, resulted in passenger deaths, and crashes in 1988 and 1992 in France, and 1998 in Columbia also had fatalities.

However, Air France's most dramatic disaster was the crash of the Concorde on Tuesday, July 25, 2000. The Concorde's first fatal accident occurred when Concorde Flight 203, bound from Paris to New York, crashed within sixty seconds of takeoff, killing all 109 persons on board. Another four people were killed in a local hotel on the ground. Both Air France and British Airways Concorde flights were temporarily grounded. Air France suffered some very negative publicity as a result of the crash.

According to the official version presented by a French accident investigation, the crash occurred when a tire hit debris on the runway and burst. This, in turn, caused chunks of tire rubber to puncture under-wing fuel tanks and led to a loss of thrust from an engine on the left wing, which veered the Concorde to the edge of the runway. Proceeding too quickly to abort, the pilot apparently took off with engines not functioning properly, thereby losing control and crashing into the Paris suburb of Gonesse. Air France mechanics are charged to have disassembled the undercarriage for service, reassembling it without the part that keeps the wheels in correct alignment. The missing spacer was found on the original part in the workshop, according to the *Paris Observer.* Thus, it was not a loss of power that caused the fatal accident. The theory of the missing spacer has been controversial, discussed, refuted, and validated. While Concorde compensation settlements were reached in Germany for families of seventy-five passengers, it was unclear whether the settlement ultimately included Continental Airlines, which investigators believe may have been the source of a metal strip on the runway that may have caused the accident.

Pamela M. Gross

Bibliography

Baker, Colin. "The Quiet Revolutionary: Despite His Success in Turning Round the Fortunes of Air France, Jean-Cyril Spinetta Prefers to Remain out of the Spotlight." *Airline Business* 17, no. 8 (June, 2000): 44-49. Profile of Air France's chairman and CEO.

Gallacher, Jacqueline. "Mission Impossible? Can Air France Banish Its Inefficient State-owned Structure to the Past and Implement a Radical Turnaround?" *Airline Business* (February, 1994): 28-31. Discusses the effects of deregulation and privatization on Air France.

Lefer, Henry. "Air France Concorde: A Valuable Symbol." *Air Transport World* 23, no. 10 (January, 1986): 46-49. An assessment of the Concorde on its tenth anniversary.

Sparaco, Pierre. "Ailing Carriers Expect Air France Support." *Aviation Week and Space Technology* 154, no. 26 (June 25, 2001): 58. Covers Air France's relationship with Air Afrique.

_______. "Air France Rescue Draws Fire." *Aviation Week and Space Technology* 141, no. 5 (August 1, 1994): 24-25. Coverage of Air France finances and management.

See also: Air carriers; British Airways; Concorde; Hijacking; Supersonic flight

Air rage

Definition: Uncontrolled verbal abuse or physical violence caused by frustration connected to problems related to commercial flying.

Significance: During the late 1900's and early 2000's, airline passengers and crews were sometimes put at risk by violent passengers who acted out their frustrations when the airlines did not deliver the service they anticipated and demanded. Ground rage, related to air rage, has also become a problem affecting airline employees who deal with the flying public on the ground.

Manifestations of Air Rage

Air rage takes a number of forms. Frequently it results in unpleasant, abusive verbal exchanges between passengers and airline personnel. Even when the manifestations of air rage are merely verbal, they create problems by distracting airline employees from the jobs they should be doing and leaving them unnerved as they go about serving other passengers. Verbal exchanges also occasionally escalate into physical assaults, so once passengers show signs of acute frustration through shouting or speaking rudely, those who deal with them must be vigilant to ensure that the situation does not escalate into greater violence. Most air rage occurs in the confined space of an aircraft that is often flying 5 or more miles above the earth at speeds of just under 10 miles a minute. These factors imbue such situations with the potential for extreme danger.

Air rage is usually verbal rather than physical. This type of anger may attract considerable attention because it frequently involves raised voices, loud shouting, acrimonious outbursts against the airline and its personnel, and the unbridled use of vulgar language. It may be accompanied by intoxication, a condition heightened when people drink in the controlled environment of an airliner. Most people get drunk more quickly at an altitude of 30,000 feet than they would at sea level.

Although air rage is usually directed at airline employees, particularly flight attendants, it may also be directed at fellow passengers. When this is the case, flight attendants become involved quite quickly. They can often control the situation by moving unruly passengers to seats away from the object of their anger, although this is not consistently a fail-safe solution.

Some Causes of Air Rage

During 2000, 615 million ticketed passengers flew on

the airlines of the United States. Airport facilities have become so overtaxed by the dramatic increase in air travel that the once-friendly skies are perceived by many travelers as being quite unfriendly. Among the major complaints voiced by those who fly are late departures and arrivals, the cancellation of scheduled flights because of bad weather or mechanical problems, overbooking, substantial differences in air fares on the same flight, and lost luggage.

Crowded on-board conditions, especially when flights are full, adds to the frustration of fliers, particularly if they have been flying uninterruptedly for many hours as long-distance travelers are sometimes forced to do. Many airlines attempt to avoid serving in-flight meals, which are expensive. Passengers who drink and do not eat become intoxicated quickly, particularly in the controlled atmosphere of an airplane and particularly if they have been without sleep for extended periods. Intoxication can cause some drinkers to become hostile and aggressive.

Flying, especially on flights lasting more than three hours, causes many people to become disoriented, a condition that drinking can intensify dramatically. Another factor in some incidents of air rage is the airlines' prohibition of in-flight smoking. People who are dependent on nicotine are forced to undergo extended periods without it from the time they enter an airport until the time, often as much as ten hours later, when they leave it. This deprivation can lead to irritability in otherwise serene people.

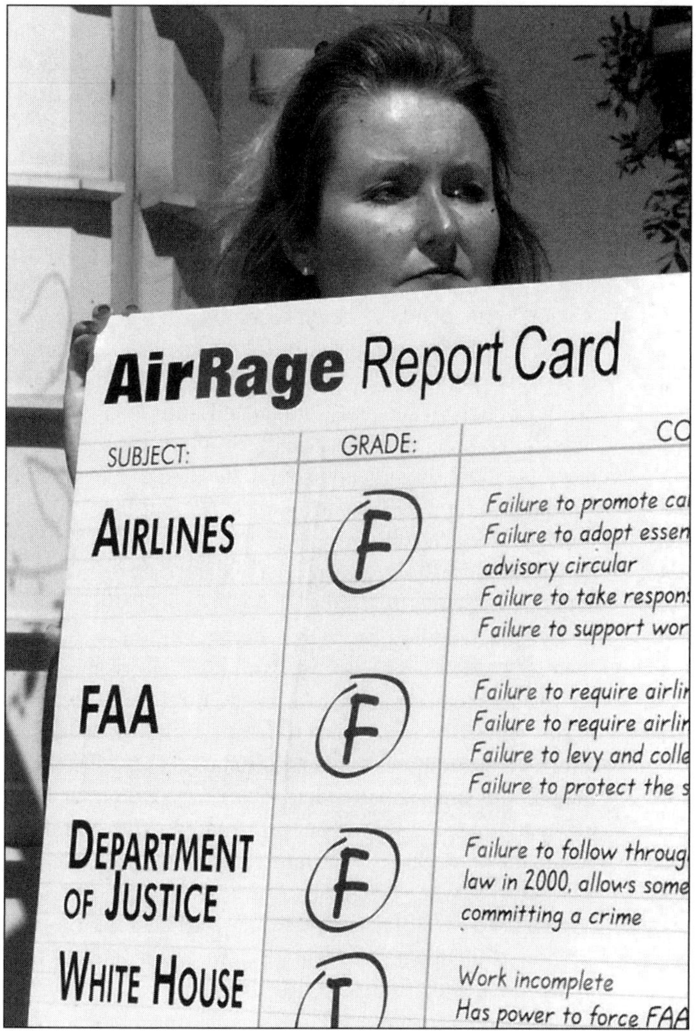

A flight attendant holds a "report card" summarizing her union's discontent with official reaction to the increase in so-called air rage incidents. (AP/Wide World Photos)

Air rage is not new. In 1969, rock musician Jim Morrison, of the band The Doors, and a traveling companion were ejected from a flight because they were drunk, smoking cigars, and making very loud obscene statements. The Federal Bureau of Investigation (FBI) investigated the behavior of the "long-haired hippies," thereby inaugurating what became a rather substantial FBI file on the singer. Air rage is, however, on the increase and increasing attention has been focused on it and on such related issues as road rage and ground rage.

Ground rage is usually directed against those who check passengers in for their flights. Often passengers become frustrated by long check-in lines. If, after waiting thirty minutes, passengers are told that flights on which they have reservations have been canceled or oversold, so they cannot be accommodated, they may be pushed beyond the limits of their endurance and may direct their anger toward the ticket agent, even though this person had nothing to do with creating the situation that understandably frustrates the passenger.

Other situations can provoke violent behavior, as was the case when the small child of one passenger wandered from the boarding area onto the boarding platform. The child's father tried to push past the ticket kiosk to retrieve his errant child, but was restrained. He struck out at the airline personnel who were trying to hold him back. Eventually, he pushed one of them with such vigor that the agent's neck was broken.

In this case, the angry father was arrested and brought to trial, an outcome that most airlines prefer to avoid. The court found against the plaintiffs, Continental Airlines and the injured agent, and acquitted the defendant of the charges against him, citing extenuating circumstances.

The Extent of Air Rage

Statistics on air rage vary considerably. U.S. senator Dianne Feinstein of California estimated that some five thousand cases occur annually, whereas airline officials estimated the number at about three thousand. In contrast, flight attendants, who must deal with it frequently, suggest that nine thousand is a more accurate number.

The Federal Aviation Administration (FAA) officially recorded 314 cases in 2000, and 592 cases in 1998 and 1999 combined. However, United Air Lines alone reported 1,075 cases during those two years. The incidents of air rage reported by United dropped from 310 in 1999 to 266 the following year, probably because of a dramatic increase in the legal penalties that air rage can incur.

The Legal Consequences of Air Rage

In October, 2000, a notable air rage incident occurred on American Airlines Flight 67 from London to Chicago. Jorgen Kragh, a fifty-three-year-old Danish businessman, became incensed when the passenger in front of him reclined her seat. He pushed the seat forward and banged on it to the point that its occupant had to get up. A flight attendant attempted to control the situation, but Kragh became irrational and threatening.

The flight captain, informed of the situation, radioed ahead to the Bangor International Airport in Maine, where he subsequently landed. A specially trained team of police officers boarded the plane and arrested Kragh. Bangor's airport, which has eight to twelve such emergency landings in a typical year, maintains a special team to deal with them. It handles the necessary paperwork expeditiously, so that the landing planes may continue without undue delays. In this case, the plane was en route to Chicago in less than an hour and a half.

Kragh, confined to jail, the next day entered a no-contest plea upon arraignment in Bangor. He was sentenced to twenty-one days in prison and ordered to pay a $5,000 fine. Legally, he might have been jailed for a considerably longer time and fined $25,000 for each incident of violence.

Jeff Russell, the marketing manager for Bangor's airport, notes that during the peak summer season, between four hundred and seven hundred international flights a day pass over Bangor, so that the percentage of flights diverted to Bangor because of air rage is minuscule. In one five-week period in the summer of 2000, three flights landed because of air rage incidents.

Ways to Deal with Air Rage

As air rage became an increasing concern in the last half of the 1990's, it grew increasingly apparent that measures had to be taken to control it. In April, 2000, the U.S. Congress passed legislation that increased the civil fine from $1,100 to $25,000 for each violation. A single incident can involve several violations, each punishable by a separate fine. To this can be added $10,000 in civil penalties and the possibility of up to twenty years in prison.

This legislation apparently has restrained some impassioned air travelers, although air rage is not under complete control. Frank Del Gandio, manager of the FAA's Recommendation and Safety Analysis Division, fears that if the matter is left unchecked, air rage will eventually result in the loss of one or more commercial aircraft. Brian Poole, head of the Safety Analysis branch of the FAA's Office of Accident Investigation, cautions, however, that in dealing with air rage, one must differentiate between what is rude and what is dangerous.

Alarmed at the danger air rage poses, Senator Feinstein in July, 2000, urged airline officials to limit the number of drinks flight attendants could serve to each passenger on a single flight. She suggested a two-drink limit, although those on flights that require a change of planes might still drink enough to become intoxicated and abusive under Feinstein's plan. Some people have urged the airlines to ban all alcoholic beverages from planes just as they banned smoking in the 1990's. Airline officials fear that if this is done, passengers will bring their own liquor onto the plane and drink more than they would have, had they been able to order drinks from the flight attendants.

Certainly alcohol has been a contributing factor in more than one case of air rage. In April, 2001, a twenty-two-year-old woman on United Air Lines' Flight 857 from San Francisco to Shanghai drank too much. When the flight attendant refused to serve her additional drinks, adhering to the practice among bartenders who observe their customers becoming drunk, the young woman punched the flight attendant in the face. The flight was diverted to Anchorage, Alaska, where the assailant and her twin sister were arrested. The U.S. District Attorney in Anchorage sought over $50,000 in damages from the two sisters and recommended jail time for each of them.

The diversion of this flight greatly inconvenienced over two hundred passengers, causing one passenger to miss her brother's funeral, and causing countless others to miss appointments, connections, and events dependent on their timely arrival. Certainly any penalties exacted from those whose air rage causes flights to be diverted can in no way compensate the scores of passengers whose plans are compromised by such a diversion.

The FAA has prepared a brochure that defines air rage and outlines the penalties for those who engage in it. The fifty-thousand-member Association of Flight Attendants urges better training of airline personnel to deal with air rage. It also has asked that the penalties for air rage be posted in airport bars and restaurants.

Statistically, the incidents of air rage are small compared with the large numbers of people who fly every year. Nevertheless, air rage is a present and real danger to all who fly and must be dealt with effectively if the skies are to be safe.

R. Baird Shuman

Bibliography

Curtis, Wayne. "Uncivil Aviation: How a Small City's Airport Became the Capital of Air Rage." *Atlantic Monthly* 287 (April, 2000). A thorough discussion of how Bangor, Maine, has become a major facility for expelling unruly airline passengers.

Newman, Maria. "Man Found Not Guilty of Attack on Airline Worker." *The New York Times*, August 25, 2000. Newman provides details about the acquittal of an airline passenger who broke the neck of a gate agent attempting to restrain him.

Tanz, Jason. "FAA Faces Air Rage." *Fortune* 143 (April, 2001). Tanz provides telling statistics about the prevalence of air rage on American air carriers.

See also: Air carriers; Airline industry, U.S.; Commercial flight; Flight attendants; Passenger regulations; Safety issues

Air shows

Definition: Events featuring the exhibition of aircraft and the demonstration of aviation skills.

Significance: Early air shows helped to promote aviation and increase public awareness about the excitement of flying. Air shows continue to display the latest in aviation techniques and development.

History

The first airplanes had more value as curiosity pieces than as means of transportation. For ten years after the Wright brothers' flight of 1903, aviation was kept alive by devotees who toured the country while performing at circuses, fairs, and anywhere else people would pay to see them. These daredevils performed aerobatic feats, walked on airplane wings, made parachute jumps, and took paying customers for joyrides. Many of these pioneer pilots died in pursuit of their aerial adventures, but they lent an air of romance and danger to the new field of aviation.

World War I-era pilots often had little or no training, flying instead by instinct and sheer courage. During the war, these daring pilots flew into combat zones with courage and determination.

After World War I, the U.S. government offered thousands of surplus airplanes, most of them Curtiss Jennys, for sale at bargain prices. Although these airplanes were stronger than those that had been built before the war, they were not always safe. Made mostly of wood and cloth, they also lacked satisfactory navigational equipment. However, many former military pilots bought these airplanes and used them for an exciting and dangerous type of flying called barnstorming.

Barnstormers toured the United States in the 1920's and put on daring air shows at county fairs and other events. Audiences were thrilled to watch. The pilots flew the airplanes in wild aerobatics and daring stunts. Performers, called wing-walkers, stepped from wingtip to wingtip in midair or leaped from the wing of one flying airplane to another. There were many accidents, some fatal.

Highly skilled World War II pilots were used to faster, more technically advanced airplanes than those of World War I. Although World War I dogfights had spurred aviators to postwar displays of courage and craziness with aerobatics, barnstorming, and cow-pasture thrill shows, post-World War II pilots had more venues in which to display their skills, including air races, air shows, carnivals of the sky, and precision flying. The air shows of the 1940's and 1950's were also showcases for new and sometimes customized aircraft. Parachuting and mock dogfights remained popular parts of air show activities.

Aerobatics

The first recorded aerial stunt was performed in 1913 by French aviator Adolphe Pégoud. Pégoud flew his Blériot monoplane upward in an ascending arc until he was flying upside-down and then dove to close the circle. He had, unwittingly, invented the loop. When American daredevil Lincoln Beachley heard of Pégoud's stunt, he jumped into his Curtiss biplane and completed three loops. In a final gesture of victory, he landed his airplane inside San Francisco, California's immense Machinery Palace.

In 1915, Ruth Law was the first woman to loop-the-loop. She made her first exhibition of this stunt in Daytona Beach, Florida. After World War I, Law and her husband and business manager, Charles Oliver, formed Ruth Law's Flying Circus. The circus toured Asia performing aerobatic exhibitions. Law was famous for racing her airplane against cars on a racetrack and for making car-to-plane transfers in which a stuntperson would leap off the car and grab a rope ladder hanging from the moving airplane. This stunt was dangerous, full of risks and excitement. The audiences at the air shows responded in amazement. In another dangerous trick, Law would fly with a copilot, climb out of the open cockpit, and stand on the wing while the copilot looped-the-loop as many as three times. This stunt terrified and thrilled audiences.

On October 1, 1910, Blanche Stuart Scott made her public debut as part of the Curtiss Exhibition Team at an air meet in Chicago, Illinois. She was given the nickname "Tomboy of the Air" for performing daring stunts, such as flying upside-down and under low bridges. In her most famous stunt, the Death Dive, Scott would dramatically climb to a height of approximately 4,000 feet, level the airplane for a moment, and then nose-dive. Audiences screamed and waited breathlessly, while the airplane plunged straight down to a level of 200 feet above the ground before snapping out of the dive and landing to thunderous applause.

Barnstorming

When highly skilled military pilots returned to the United States after World War I, they found hundreds of surplus airplanes, but few available flying jobs. Not wanting to give up flying, many of them turned to exhibitions, air shows, crop dusting, fire spotting, air racing, or any aerial activity that would attract a crowd.

Barnstorming—so named because some ex-war pilots and flying pioneers actually flew through barns, in one door and out the other—became an American craze. These skilled aviators were able to display at home the skills they had acquired flying over hostile German terrain, now winning income and applause. Barnstormers took the loops and stunts of earlier pilots and added even riskier tricks. Some of the tricks required more than one person aloft. One pilot would fly the airplane, while the other would walk on its wings or hang beneath the airplane, holding on to a rope by the teeth. Pilots captivated crowds when they crawled from the wing of one airplane to another in midflight. These daredevils of the air, in their Jenny aircraft, continuously worked to develop newer and more daring exploits.

Many women also found post-World War I opportunities in barnstorming. After years of struggling to be recognized as serious pilots, they discovered a chance to become involved in flying, even if in a more theatrical and unusually dangerous way. Many women entering aviation through the air show or barnstorming circuit were seen as renegades and breakers of tradition. Women found this entry into aviation was not only a chance to earn money but also a possible way to prove themselves as pilots. Many began as wing-walkers or parachutists, working their way up to the airplane controls. In October, 1928, Florence "Pancho" Barnes opened Pancho Barnes' Mystery Circus of the Air, a stunt barnstorming troupe.

In 1923, Emerson Lockhart, an American aviator, bought a Jenny for $175. Determined to create the best aerial thrill show, he made a sign that said simply, "Aero Thrill Show, A Stupendous Exhibition of Flying Skill." His airplane attracted the curious but it was the sign that hooked them. Lockhart eventually got permission to use a local farmer's pasture alongside the highway. He parked the airplane and, as expected, began to draw a large crowd. It was estimated that there were about two hundred people along the highway when he took off. He dazzled his audience doing loops and buzzed the parked cars many times. His grand finale was a dive from 3,000 feet, pulling out about 100 feet above the pasture, after which Lockhart did a sharp bank, a roll, and brought the airplane down in the pasture.

Lockhart finally realized his dream when he arrived in California ready to be an air-show star. After his arrival, he found out that California was loaded with air shows but overloaded with barnstormers. He adjusted his act into actual aerial stunt flying and joined a show-business-oriented flying team. He was paid the then-generous sum of twenty-five dollars per week.

Motion Pictures

The burgeoning motion-picture industry was a magnet for many former military pilots, recreating the glorious days of battle and providing new career opportunities. Even pilots who had neither seen combat nor been in the military took to the air in the numerous postwar films then being made in Hollywood. A special group of cinematic war aces was created, perhaps the best-known of whom was Art Scholl.

When Scholl performed in his Pennzoil Chipmunk, a small aircraft that he had designed himself, he became an accomplished aerobatic pilot who represented the United States in competitions around the world. He became highly sought after in the choreography of aerial stunts for film productions and, later, television commercials.

Scholl did not start out to be a stunt pilot. After graduating from high school in Brown Deer, Wisconsin, he moved to California to enroll as a student of engineering at the Northrop Institute. He later left Northrop and enrolled at Mount San Antonio College, from which he earned his degree in aeronautics.

Air Derbies

With performing pilots across the United States, there became a need for public competition and a platform for pilots to display their skills, abilities, and showmanship. Cross-country racing became the event that fired the imagination of the public and fueled the imagination of pilots.

The 1929 Women's Air Derby was a cross-country race that included such high-profile pilots as Amelia Earhart, Ruth Elder, and Bobbi Trout. The first race did more than display women's ability to compete as pilots, it allowed women to realize that they were not alone in their dreams of careers and accomplishments as pilots.

Marvel Crosson was a veteran commercial pilot from Alaska; she and her brother, Joe Crosson, had taught themselves to fly after piecing together an airplane using surplus parts from World War I airplanes. Favored to win the 1929 Women's Air Derby, Marvel experienced engine trouble over the mountains near the Gila River in Arizona. She bailed out too low for her parachute to open properly and was killed, thus becoming the first casualty of the Women's Air Derby. The winner of the flight, Louise Thaden, dedicated her Symbol of Flight trophy to Crosson and sent it to Crosson's mother. Crosson's death created a rash of editorials against women air racers and even against women fliers, but Women's Air Derby continued until 1976.

Pilots of both genders thrived in their competitive race to professional recognition and eventually competed against each other. It was not long before air-show performers, racers, and record setters were recognized for their skills regardless of their gender.

Modern air races and air shows, organized by branches of the military or by independent event sponsors, continue to be held throughout the year across the United States. These events usually incorporate past, present, and future aircraft and continue to feature exciting exhibitions of pilots' flying skills.

Lori Kaye and Maureen Kamph

Bibliography

Bruno, Harry. *Wings over America: The Inside Story of American Aviation.* New York: Robert McBride, 1942. A classic work detailing the early history of aviation and air shows in America.

Solberg, Carl. *Conquest of the Skies: A History of Commercial Aviation in America.* Boston: Little, Brown, 1979. A comprehensive history of U.S. commercial aviation featuring illustrations, a bibliography, and an index.

Yount, Lisa. *Women Aviators.* New York: Facts on File, 1995. A collection of profiles of famous women in aviation history, including those who flew in the pioneering Women's Air Derby.

See also: Aerobatics; Barnstorming; Amelia Earhart; Jenny; Military flight; Ninety-nines; Wing-walking; Women and flight

Air traffic control

Definition: Air traffic control (ATC) uses technology and trained staff to assure safe movement of aircraft in airspace and at airports.

Significance: ATC continually monitors every instrument flight rules (IFR) flight from takeoff to landing, as well as visual flight rules (VFR) flights upon pilot request and controller availability, enabling reliable, efficient transportation of people and goods by airlines.

History and Evolution

Air transportation is essential to modern life, and it requires that passengers feel safe during air travel. The features of aviation that distinguish it from other transportation forms are its high speed and vertical operation. Crashes are devastating because of their intense impact, created by the heights from which aircraft can fall and the speeds at which they may be traveling. The potential danger is alarming to consumers, especially because the frequently high death tolls from single crashes make a strong impression in public awareness, while many people are unaware of the great overall safety of air transportation compared to other forms of travel.

The potential for severe injury or death to aircraft passengers has led to air traffic control (ATC) systems that have evolved from early traffic control with signal flags in the 1920's to the sophisticated systems using advanced technology and specially trained staff of the twenty-first century. Current ATC assures the safe movement of virtually all aircraft operating in airspace and at airports. Its objectives include giving pilots all the data and control ser-

Air traffic controllers spend much of their time tracking the complicated movements of flights on radar screens. (AP/Wide World Photos)

vices needed to maximize safe, efficient aircraft operation; maximizing safe air traffic at airports; and minimizing unavoidable flight arrival and departure delays. It is ATC, a product of the National Airspace System, that makes air transportation the safest means of mass transportation in the United States.

Commercial airplanes generally travel airways, which are analogous to roads, although they are not physical structures. Airways have fixed widths and defined altitudes, which separate traffic moving in opposite directions. Vertical separation of aircraft allows some flights to pass over airports while other processes occur below. Air travel usually covers long distances, with short periods of intense pilot activity at takeoff and landing and long periods of lower pilot activity while in the air, the portion of the flight known colloquially as the "long haul." During the long-haul portion of a flight, pilots spend more time assessing aircraft status than searching out nearby planes. This is because collisions between aircraft usually occur in the vicinity of airports, while crashes due to aircraft malfunction tend to occur during long-haul flight.

Flying Rules

The main rule systems governing flights are instrument flight rules (IFR) and visual flight rules (VFR). The minimum instruments needed for VFR are an airspeed indicator, an altimeter, and a magnetic direction indicator. In VFR, pilots fly using visual ground references and a "see and be seen" rule. The minimum requirements for VFR vary, but often include cloud ceilings of 1,000 feet and visibility of three miles.

IFR are used if aircraft operate above 18,000 feet, an area known as Class A airspace. Outside this airspace, any aircraft may use VFR, although only slow-moving, low-flying aircraft or small jets on short flights routinely do so. In some conditions, such as congested airspace around medium and large airports (Class B, C, and D airspace), in poor visibility, and at night, pilots must obtain permission from ATC controllers to fly by VFR and usually are only granted that permission if they are instrument rated and there is at least one mile visibility.

At airports with control towers, all air and ground movements are subject to permission from and instruction by ATC. This is because major airport peak traffic may involve three to four landings or takeoffs per minute. With dense aircraft concentration around airports, maintaining acceptable collision risk is not possible without strict adherence to procedures set out and monitored by ATC.

All pilots wishing to utilize IFR must demonstrate their ability through detailed testing, and all aircraft must have adequate flight instruments. For each flight, a detailed flight plan must be filed with the Flight Service Station, part of the Federal Aviation Administration (FAA); flight clearance must be received from Clearance Delivery or Ground Control (or from Approach Control if the pilot files while in the air); and ATC directions must be followed throughout the flight. Such directions usually depend upon radar surveillance, including the use of airborne radar beacon transponders that allow ATC facilities to identify each aircraft in flight.

Airspace is divided into classes designated A through E. Class A airspace is all airspace above 18,000 feet; Classes B, C, and D are designated around specific airports rated by their size and amount of traffic, and Class E covers all other airspace between 14,500 and 17,999 feet. In addition to vertical airway spacing, horizontal spacing is important. This is created by planned time intervals (often ten minutes) between aircraft on the same track, with lateral spacing of ten miles.

Landing, Takeoff, and En Route Procedures

Air terminal ATC, the element of air traffic control that is most familiar to the public, involves aircraft departures and arrivals. Its procedures include issuing instrument flight rules route clearances and communicating departure runways, taxi instructions, and definition of climb and altitude routes. These operations assure passengers of safe, speedy air traffic patterns.

A departing aircraft enters the taxiway as instructed by the ground controller and the pilot awaits being fitted into the pattern of incoming and outgoing flights. ATC controllers allocate available departure runways to enable safe aircraft separation. Once the aircraft climbs to its initial altitude on an ATC-instructed heading, departure control makes sure that radio contact with the departing pilot is established before allowing a new takeoff. More instructions clear the aircraft for its final climb to the en route segment of the flight and for transferring the pilot to the next control facility.

Air traffic controllers relay descent instructions to incoming aircraft to keep them separated by five-mile intervals. As the aircraft approaches an airport, its speed is adjusted and its flight path altered to maintain an aircraft separation of over three miles within airport boundaries. ATC controllers determine aircraft landing sequences, stacking plans, and takeoff adjustments to handle aircraft flow. To simplify this flow, all commercial aircraft remain at their origin airport until it is confirmed that a landing site will be available at their destination airport at the planned arrival time. Travelers often become frustrated when a pilot cannot obtain a landing slot after leaving the gate at the origin airport, but the practice maximizes safety since flight delays are safer when spent on the ground than in the air.

The last part of descent control transfers the aircraft to the approach controller. Data from radar surveillance determine the final landing directions. Radar monitors optimize landing, and once on a runway, the pilot and the ground controller interact to prevent aircraft movement conflicts on the field. This controller also tells the pilot how to reach the craft's apron or parking position at the airport.

En route ATC includes monitoring the routes between terminals granted to individual pilots. A flight follows a predetermined path in a defined airway corridor. Effective en route ATC instructs pilots when and how to avoid nearby aircraft. During most flights, a given ATC facility periodically transfers control of each flight within its jurisdiction to the next facility on a flight plan. For this reason, ATC gives pilots radio-frequency changes that occur as they are passed on to the next controller along their flight paths.

The availability of inertial navigation units for commercial aircraft has reduced the need for this communication. In an inertial navigation unit, a computer and gyroscope are oriented to true north, while speed sensors track the aircraft's direction and the distance to its destination. Although inertial navigational units can fly virtually automatically until the aircraft reaches an airport, en route information is provided for safety and to warn of impending delays or other dangers. As a result, all IFR aircraft are monitored continuously throughout each flight. In addition, pilots must get ATC approval before making any flight path alterations. Required en route progress reports are tracked on air route surveillance radar, which monitors aircraft in each sector of the air route ATC system.

Craft-to-Ground Communication

Navigation within a designated—and desired—airspace requires pilots to identify the exact position of their aircraft and assure that they are in the airway assigned in their flight plan. This depends upon ground beacons and electronic equipment in airplanes. The most widely used ground system is the very high frequency omnidirectional range beacon (VOR). VOR stations operate on noise-free radio frequencies and provide fine accuracy. On airplanes, visual displays indicate the course needed to travel directly to a VOR station.

Most stations have distance-measuring equipment, which gives pilots the distance to VOR stations. VOR and distance-measuring equipment service all aircraft. Other methods used for navigation are Doppler navigators and inertial navigation systems, which do not require ground stations or radio beams. Another navigational tool is the Global Positioning System (GPS). It is composed of GPS satellites and provides worldwide position ability accurate to 350 feet horizontally and 500 feet vertically. GPS is available for all phases of flight.

Pilots and ATC controllers communicate by radio en route and at airports. This helps ATC to make continuous updates of the positions of all planes in its operation area and provides an unambiguous means of instant flight instruction communication. All ATC surveillance of aircraft uses radar. Each radar system possesses a primary beacon that sweeps its coverage area and transmits images of all objects encountered back to a radarscope. The primary beacons are synchronized with a secondary radar system that uses automatic aircraft transponders to identify each flight in a given jurisdiction. Such radar systems are found in many air route traffic control centers and air terminal radar control facilities, providing sophisticated ATC.

The operation of the ATC system utilizes computer-assimilated flight information such as the position, course, airspeed, and transponder number of all craft in a jurisdiction. This enables controllers to determine the safest, most direct flight plan for each aircraft and to make continual updates. ATC also incorporates traffic alert and collision avoidance systems in aircraft. ATC technology in the United States is located at a national air traffic control command center at about twenty-four regional control centers, nearly six hundred terminal control facilities, and 250 flight service stations. All facilities interact to provide a nationwide weather profile, data on all airborne flight positions, and information on airport flight acceptance. Data are analyzed by a national computer and automatically circulated to regional facilities and airports. Regional air traffic control facilities are also computerized and automated.

Recognition of aircraft under IFR is essential at all points in a flight, especially when aircraft enter an airport terminal control area. Recognition is achieved through cross talk between aircraft flight transponders and surveillance radar beacons. For this reason, the FAA requires that all aircraft flight transponders are kept on from takeoff until landing is complete. A flight transponder provides several levels of information to ATC. When turned on, it continually transmits a radar symbol showing the geographic position of the aircraft relative to ATC facilities. The level of information transmitted is determined by a pilot's responses to queries from ATC facilities. Most flight transponders also alert ATC facilities of radio failure, skyjacking, and other emergencies by pilot input to transmit specific code numbers.

Terminal Control

Terminal ATC is found in most public airports. Control facilities are divided into two parts, an ATC tower (ATCT) and approach and departure control. The tower controls approaching or departing aircraft in the five-mile radius of the airport traffic area around the airport. Approach and departure control covers a radius between five and thirty-five miles from the airport, beginning where tower control stops.

ATC controllers identify and sequence all aircraft in the airport traffic area, expedite aircraft arrivals into and departures from the airport traffic area, control all ground aircraft movement, cancel flight plans, and provide other use-

Top Ten FAA-Operated Air Traffic Control Towers, 1996

Tower	*Rank*	*Annual Operations*
Chicago O'Hare International, Illinois	1	909,186
Dallas/Ft. Worth International, Texas	2	869,831
Atlanta International, Georgia	3	772,597
Los Angeles International, California	4	764,002
Miami International, Florida	5	546,487
Phoenix Sky Harbor International, Arizona	6	544,363
Van Nuys, California	7	532,221
Detroit Metro Wayne County, Michigan	8	531,098
St. Louis International, Missouri	9	517,352
Oakland International, California	10	516,498

Source: Federal Aviation Administration, *Statistical Handbook of Aviation*, 1996.

ful information. In both landing and takeoff control, IFR aircraft may be provided with horizontal and vertical path guidance. In contrast to ATCT control, which regulates the flow of traffic within the airport itself, approach and departure control regulates the safe and efficient flow of air traffic around airports.

This regulation begins as soon as an aircraft reaches the perimeter of the area under approach and departure facility control. At that time air traffic controllers begin to fit the aircraft into the pattern of current air traffic. They define the course, altitude, and speed changes that each aircraft needs for a safe landing within the framework of existing arrivals and departures.

Terminal ATC is based mostly on data obtained from ATC radar beacon systems. Major air terminals are equipped with computerized automated radar terminal systems (ARTS), which accept radar inputs from all surveillance systems. They automatically show individual aircraft on ATC video displays, where a plane is seen as a radar screen data block in front of the controller. The block shows a symbol for each aircraft, along with its identifying call sign, speed, and altitude. The systems also warn of possible collisions or instances where aircraft approach minimum safe altitudes.

Using ARTS, all flights are kept at specific distances, horizontally and vertically, from one another. Flight plans are fed into computers and updated as flights progress. ATC controllers watch carefully to prevent collisions. As aircraft converge on an airport to descend for landing, congestion may occur. In this situation, IFR planes are instructed to circle a specific location designated by an IFR intersection. As more planes arrive, they are held increasingly farther out from the airport.

For instrument landings, pilots use an instrument landing system (ILS) similar to VOR. Cockpit instruments indicate deviations to either side of a localizer beam leading to the runway, and guidance information from a glide slope beam warns if the plane is too high or low on the approach.

The State of Flux in ATC

Air safety is created by the efficient operation and continued updating of ATC. ATC operation modes change quickly and extensively, since the system operates at near capacity, while the total number of operating aircraft is increasing. As a result, the FAA is continuing an automation and modernization program addressing aspects of en route air navigation methods, weather identification and dissemination, flight service stations, and facets of airport terminal ATC. Goals of this National Airspace System Plan include meeting continued increased airspace demands, reducing operational errors, collision risk, and weather-related accidents, and minimizing escalating ATC operation and maintenance costs. ATC's main need is to increase its efficiency, because while industry demand will grow, major airports do not have the ability to significantly expand, due to physical or environmental constraints. The prospect for construction of new large airports is also dim.

Sanford S. Singer

Bibliography

Buck, Robert N. *Weather Flying*. New York: Macmillan, 1988. Describes bad weather flying and related ATC, weather checks and information, equipment needs, VFR, and takeoff and landing in bad weather.

Cronin, John. *Your Flight Questions Answered: By a Jetliner Pilot*. Vergennes, Vt.: Plymouth Press, 1998. Discusses airports and ATC from a pilot's perspective.

Garrison, Kevin. *Flying in Congested Airspace: A Private Pilot's Guide*. Blue Ridge Summit, Pa.: Tab Books, 1989. Covers ATC history, U.S. airspace, airport radar service and terminal control areas, VFR and IFR safety techniques, and departure and arrival regulations.

Illman, Paul E. *The Pilot's Air Traffic Control Handbook*. Blue Ridge Summit, Pa.: Tab Books, 1993. Provides pilot information on ATC, its history, use of airspace, terminal ATC, approach and departure control, flight service stations, and air traffic controllers.

Massie, David. *Airline Pilot: Let the Pros Show You How to Launch Your Professional Piloting Career*. New York: ARCO, 1990. Covers pilot demographics, education, qualifications, occupational ratings, training and experience levels needed, and testing.

Mathews, James A. *How to Prepare for the Air Traffic Controller Exam*. Hauppauge, N.Y.: Barron's Educational Series, 1997. Gives insights on the air traffic controller's skills and education needs, as well as qualifying exams.

Rowberg, Richard E. *Safer Skies with TCAS*. Washington, D.C.: U.S. Government Printing Office. Report on Office of Technology Assessment thoughts on traffic alert and collision avoidance systems.

Airbus

Definition: One of the two largest global manufacturers of heavy commercial aircraft.

Significance: The Airbus consortium of European aircraft manufacturers is a successful example of regional and global economic cooperation in the production of a highly valuable, strategically significant, and high-technology product.

Corporate Information

Airbus, headquartered in Toulouse, France, is owned by two leading European aerospace companies. One is the European Aeronautic Defense and Space Company (EADS), born of the merger between Airbus consortium partner companies Aerospatiale-Matra of France, Daimler Chrysler Aerospace of Germany, and CASA of Spain; the other is BAE Systems of the United Kingdom. In June, 2000, EADS and BAE Systems announced the creation of Airbus Integrated Company, intended to consolidate Airbus Industrie resources and practical knowledge in existing locations around Europe into a single entity. As a result, all Airbus-related design, engineering, and manufacturing assets located in France, Germany, Spain, and the United Kingdom became part of a new Airbus company under the day-to-day control of a single management team. As of 2001, the company employed some forty thousand people throughout Europe. The consortium members are both industrial participants and shareholders. Their role is to carry out most aircraft design and all manufacturing under Airbus's management.

Each partner company operates under the laws of the country in which it is incorporated. The partners are responsible for their own financing of the research, development, and production phases of the aircraft programs. Airbus Industrie's production system is flexible and appears to be quite effective and efficient, as evidenced by the fact that approximately 96 percent of all aircraft work is performed in plants operated by the partner companies. Fully equipped sections of Airbus Industrie aircraft are produced in factories throughout Europe and transported to Toulouse, France, or Hamburg, Germany, for the final assembly. The production network is set up in an innovative way that uses the specialized skills of each partner and associate.

Airbus Industrie has more than 1,500 suppliers in twenty-seven countries and cooperative agreements with aerospace industries in nineteen countries. More than 35 percent of the components for the company's aircraft are supplied by over five hundred United States companies. Numerous suppliers are also located in the Asia-Pacific region, such as Singapore Technologies Aerospace, which produces wing ribs and passenger doors for the A320, A321, A319, and A318 and engine mounts and thrust reverser doors for the A340 and A330; and the Indian company Hindustan Aeronautics, which also builds A320 passenger doors.

One of the keys to Airbus's sales success has been the flight operational commonality that exists among all the company's fly-by-wire, or fully automated and computerized, aircraft. The Airbus philosophy has been to develop families of fly-by-wire controlled aircraft with similar cockpits and flight handling characteristics and common systems and hardware. As a result, pilots trained to fly any Airbus fly-by-wire aircraft feel equally at home in any of the single-aisle models in the A320 family, such as the A318, A319, A320 and A321, and the wide-body A330 and A340 models. This commonality may result in millions of dollars of savings for airlines. It reduces training costs, increases crew productivity, and provides pilots with the flexibility of flying a wide range of routes, from short-haul to ultra long-haul.

Organizational Structure

Airbus Industrie was created on May 29, 1970, and was formed as a public interest group on December 18, 1970. The company was formed under French law, in the absence of a functional legal framework accepted throughout the European Union, then known as the European Economic Community. The public interest group is a form of business organization that permits participating firms to integrate their activities in certain domains while preserving their individual identities. The French public-interest law was used as an appropriate legal framework for the company as it was beneficial to Airbus Industrie's goals in establishing itself in the market and managing its risk, at least initially. Originally, two partners, Aerospatiale and Deutsche Airbus, had equal ownership of the company. Each partner assumed equal unlimited liability relative to the project. Because the company was a public interest group, new members could be admitted with the consent of both partners. To provide oversight of the entire project, an organizational structure was formed in December, 1970. This department dealt directly with third parties to sell aircraft and provide pilot and crew training. There were two representatives from each industrial partner in the assembly of members. A supervisory council was organized to administer the assembly. This structure was intended to act as a true multinational collaboration.

Airbus is an outstanding example of successful multinational cooperation in the large commercial aircraft sector of the aerospace industry. Airbus was developed with the support and cooperation of the governments of the European Union member states with companies in the consortium (France, Germany, Spain, and the United Kingdom).

Airbus Development

Model	*Program Launch*	*First Flight*	*In Service*
A300	May, 1969	October, 1972	May, 1974
A310	July, 1978	April, 1982	April, 1983
A300-600	December, 1980	July, 1983	June, 1984
A310-300	August, 1983	July, 1985	December, 1985
A320	March, 1984	February, 1987	March, 1988
A300-600R	March, 1987	December, 1987	May, 1988
A330-300	June, 1987	November, 1992	December, 1993
A340-200/300	June, 1987	April, 1992/October, 1991	March, 1993
A321	November, 1989	March, 1993	January, 1994
A300-600ST	November, 1991	September, 1994	January, 1996
A319	May, 1993	August, 1995	April, 1996
A330-200	November, 1995	August, 1997	April, 1998
A340-500/600	December, 1997	May, 2001	2002
A318	April, 1999	2001	2002
A380	December, 2000	2004 (projected)	2006 (projected)

This cooperation greatly contributed to Airbus's success. Even though cooperation within the consortium took place among technical experts, it was the governments' willingness to create a large producer of commercial aircraft that provided the impetus for such cooperation to occur. Airbus's strategy was to develop large civil aircraft that were both distinguished and economically attractive.

The explicit and systematic arrangement of the Airbus project began with the structure of its management. True collaboration, the goal of the new European industry, required joint financing, marketing, and work-sharing agreements, and thus some sort of transportation decision-making and administrative structure. To define the mutual rights and responsibilities to which the collaborative agreement would be subjected, Deutsche Airbus (now Daimler Chrysler Aerospace Airbus, a member of EADS) had a limited liability of DM 100 million (approximately 55 million U.S. dollars), while both the Spanish and French members were liable for the entirety of their resources. Deutsche Airbus was privately owned and consisted of independent firms. In contrast, Aerospatiale, the French participant, could rely on the assistance of the state for its liability.

Aircraft Projects

Airbus's first project was the development of the A300, envisioned as a short- and medium-range aircraft. By 1971, two basic designs had been decided upon, the A300B2 and the A300B4. Both were wide-body, twin-aisle, two-engine aircraft having a capacity of 220 to 270 passengers for air travel over 1,200 nautical miles. The first fuselage was completed in September, 1971. In November, 1971, the first two wings were shipped from Britain to Toulouse. The landing gear was attached in January, 1972, engines were mounted in April, 1972, and the systems testing progressed throughout the year. On September 28, 1972, the aircraft was rolled out and one month later, on October 28, 1972, it flew its first flight.

In July, 1978, the A310, a shortened version of the A300 seating 218 passengers in a standard two-class configuration, was launched. Airbus was set to expand from a sound base and to create a complete range of airliners with a common theme. Following the A310 project, British Aerospace, which had taken over Hawker-Siddley, became a full partner on January 1, 1979.

Airbus brought the new A320, a single-aisle, 130- to 170-seat aircraft, into the family during the same year. The launch of the A320 filled out the product line. The A320 was revolutionary, incorporating the very latest technology and, as a result, providing better operating efficiency and better performance. The flight deck set the standard for all subsequent Airbus cockpits, with obvious advantages to pilots and operators. Among the innovations installed were fly-by-wire controls, which removed cumbersome mechanical controls.

The A320 itself was followed in 1989 by the A321, a lengthened version seating from 180 to 200 passengers, and in 1992 by the A319, a 120-seat version. The family was completed in 1999 with the introduction of the 107-seat A318.

In 1987, Airbus launched two larger aircraft in a single program: the A330, a 235-seat, twinjet, medium-haul aircraft, and the A340, an ultralong, four-engine, 295-seat jetliner. The two new airliners shared the same airframe, the same wing design, and the same popular twin-aisle cross-section of the A300/A310. The proven fly-by-wire controls of the A320 were extended to both the new aircraft. The A340 entered service in 1993 and the A330 joined it one year later, the first commercial transport jointly certificated by European and U.S. aviation authorities. The twin-engine A330-200 and A330-300 carry 253 passengers and 335 passengers respectively, with the A330-200 capable of a 12,000-kilometer range.

Funding and Trade

Airbus as a consortium had a unique funding mechanism compared to those of other commercial producers in the market. The supervisory council approved Airbus Industrie's routine payments for the purchase of major equipment items and overhead expenses. The partners incurred nonrecurring costs resulting from the development process as well as production funding expenses. Each partner was responsible for financing research, development, and production. The industrial firms involved in the Airbus program have historically looked to their respective governments to secure this money. Usually this has been accomplished through low-interest loans repaid through sales proceeds.

The manner in which the Airbus consortium secured its money through government subsidies and loans, especially during its first years, has been a major issue of friction between Airbus and its competitors, particularly the Boeing Company of the United States. Airbus's competitors argued that the company's project financing practices were in direct violation of fair trade rules set by the General Agreement on Tariffs and Trade (GATT), specifically as they applied to fair trade in the commercial aircraft manufacturing industry. These rules were later more formally institutionalized by the World Trade Organization (WTO).

In 1985, the GATT agreements were updated, further limiting the ability of governments to provide financial assistance by requiring governments to lend money to companies such as Airbus at the same rates that would be charged to consumers taking out bank loans, thus preventing such companies from having their loans subsidized through lower interest rates. Airbus was accused of receiving over $13 billion of subsidies from European Union governments between its inception and 1990. European Union governments were also accused of providing loans to Airbus at much lower than market rates, and in some instances at free rates. It has been estimated that the subsidy amount, if compounded at commercial rates, would amount to over $25 billion.

Significance and Position

Airbus is an example of regional and global economic cooperation to produce a valuable high-technology product. By practicing cost diversification, Airbus has engaged in prudent risk management. They have hedged against downturns in the financial cycle and in the long selling cycles that prevail in the industry. Airbus was successful because it was able to develop a unique corporate structure and culture, which were to a large extent independent from the influence of politicians and were developed based on business principles and economic planning, rather than political necessities.

On its World Wide Web page, Airbus states its corporate philosophy: "Setting the standards' means anticipating the market, offering innovation and greater value, focusing on greater passenger comfort, and creating a true family of aircraft." Airbus also claims that "Real competition always creates a better product," and Airbus's two aircraft manufacturing partners brought European competition to the forefront of the world market. At the start of 2001, when its turnover reached $17.2 billion, Airbus had received more than 4,200 orders in total and had over 2,400 aircraft in service with 176 operators.

Triantafyllos G. Flouris

Bibliography

Addisson, Colin. *Airbus*. Shepperton, Surrey, U.K.: Ian Allan, 1991. A well-organized historical and introductory account of Airbus to the early 1990's.

Collision Course in Commercial Aircraft. Cambridge, Mass.: Harvard Business School Publishing, 1991. A well-written case study on the fundamentals of the international trade aspects of commercial aircraft manufacturing.

Lynn, Matthew. *Birds of Prey: Boeing Versus Airbus*. New York: Four Walls Eight Windows Publishers, 1997. Highlights contentious issues in the business of commercial aircraft manufacturing trade as told by both contenders, Boeing and Airbus.

Subcommittee on Technology and Competitiveness. *Airbus Industrie: An Economic and Trade Perspective.*

Washington, D.C.: U.S. Government Printing Office, 1992. A U.S. government publication outlining the competition issues involved in commercial aircraft manufacturing.

Thornton, David. *The Politics of an International Industrial Collaboration*. New York: St. Martin's Press, 1995. A comprehensive account of the relationship between politics and economics in the development of Airbus Industrie and an excellent introduction to its business organization.

See also: Aeronautical engineering; Airline industry, U.S.; Airplanes; Boeing; Manufacturers

Aircraft carriers

Also known as: Carriers, flattops

Definition: Large naval warships whose purpose is to project a nation's air force by sea into virtually any other part of the world. To that purpose, carriers are distinguished by their flat landing and takeoff decks and their complement, carried on deck and in internal hangar spaces, of fighter, attack, reconnaissance, antisubmarine, electronic warfare, and other aircraft essential to the mission of the carrier.

Significance: Aircraft carriers are the most important and complex ships in a modern navy. Their air power allows them to project a nation's presence, influence, and power almost anywhere in the world. Aircraft carriers are powerful instruments of technology, combat, and diplomacy around the world. A carrier sent to a trouble spot in the world focuses attention and power on that spot.

Description

The first prototype aircraft carriers entered the British navy during World War I, but were first used significantly in battle in World War II by the British, American, and Japanese navies. Aircraft carriers have essentially two components: the ship and its systems, and the air wing, or complement of aircraft it carries. Aircraft carriers are the largest military ships in the world; a modern U.S. carrier displaces more than 97,000 tons of water, travels at a speed in excess of 30 knots, and carries a complement of more than 5,000 Navy and Marine personnel. A navy captain usually commands the ship; another captain on board commands the air wing.

Aircraft carriers are extremely expensive, large, and potentially vulnerable to attack. They operate at sea as part of a carrier battle group, which also includes guided missile cruisers, guided missile destroyers, frigates, attack submarines, and replenishment/resupply ships. These combatant ships have antisubmarine, antiair, and antiship roles. The carrier, when on active patrol, also operates a combat air patrol of its own fighter aircraft. The carrier's only other armaments are missile launchers and Phalanx radar-guided 20-millimeter cannon for antiaircraft and antimissile defense. It carries a wide array of sophisticated intelligence-gathering equipment, radar, sonar, countermeasures, and other electronic systems.

The aircraft of a modern carrier are many and diverse. A U.S. carrier in 2001 carried one-half dozen different types of aircraft, including the supersonic swept-wing F-14 Tomcat, an air superiority, strike, and fleet defense fighter aircraft that carries both missiles and cannon. The F/A-18 Hornet (C/D models) and F/A-18 Super Hornet (E/F models) are all-weather fighter and attack aircraft. These supersonic aircraft carry both missiles and cannon. The EA-

Aircraft carriers provide mobile platforms from which military aircraft can take off and land for air raids or reconnaissance. (Digital Stock)

General Characteristics of U.S. Aircraft Carriers

	Nimitz Class	*Enterprise*	*John F. Kennedy*	*Kitty Hawk* Class
Builder(s)	Newport News Shipbuilding Company	Newport News Shipbuilding Company	Newport News Shipbuilding Company	New York Ship Building Corporation; New York Naval Shipyard
Power Plant	2 nuclear reactors, 4 shafts	8 nuclear reactors, 4 shafts	8 boilers, 4 shafts	8 boilers, 4 geared steam turbines, 4 shafts
Overall Length (feet)	1,092	1,101	1,052	1,062.5
Flight Deck Width (feet)	252	252	252	252
Beam (feet)	134	133	130	130
Displacement, fully loaded (tons)	97,000 (approximate)	89,600	82,000	80,800 (approximate)
Speed (miles per hour)	34.5+	34.5+	34.5+	34.5+
Aircraft	85	85	85	85
Ship(s)	USS *Nimitz*, USS *Dwight D. Eisenhower*, USS *Carl Vinson*, USS *Theodore Roosevelt*, USS *Abraham Lincoln*, USS *George Washington*, USS *John C. Stennis*, USS *Harry S. Truman*, *Ronald Reagan* (under construction)	USS *Enterprise*	USS *John F. Kennedy*	USS *Kitty Hawk*, USS *Constellation*
Crew (Ship's Company/ Air Wing)	3,200/2,480	3,350/2,480	3,117/2,480	3,150/2,480
Armament	2-3 NATO Sea Sparrow launchers, 3-4 20-millimeter Phalanx CIWS mounts	2 Sea Sparrow missile launchers, 3 20-millimeter Phalanx CIWS mounts	Sea Sparrow missiles with box launchers, 3 20-millimeter Phalanx CIWS mounts	Sea Sparrow launchers, 3 20-millimeter Phalanx CIWS mounts
Date Class First Deployed	May 3, 1975	November 25, 1961	September 7, 1968	April 29, 1961

Source: (www.chinfo.navy.mil/navpalib/factfile/ships/ship-cv.html), June 6, 2001.

6B electronic warfare and countermeasures aircraft are subsonic, achieving speeds of more than 500 knots per hour. They carry countermeasures equipment and anti-radar missiles. The subsonic S-3 Viking antisubmarine plane carries missiles, rockets, mines, torpedoes, and depth charges. The propeller-driven E-2C Hawkeye airborne early warning and command plane is unarmed. The SH-60 Seahawk antisubmarine, search-and-rescue, and special operations helicopter carries machine guns, missiles, and torpedoes. All but three U.S. carriers have nuclear propulsion. The ships' reactor plants are capable of carrying them many times around the world without refueling.

Carriers of other countries carry different aircraft. Several nations' carriers embark the British Aerospace Harrier, capable of vertical and short takeoffs and vertical landings (V/STOL). The Royal Navy currently flies the Sea Harrier FA2, which carries bombs, missiles, and cannon, and flies at about Mach 0.9. France's carrier carries the Super Etendard attack and reconnaissance aircraft. France also operates the Rafale M multirole fighter aircraft, capable of Mach 2, which carries missiles and cannon.

Early History

The first instance of a heavier-than-air craft taking off from a warship was the flight of a Curtiss airplane from a ramp mounted on the U.S. cruiser *Birmingham* in 1910. The first landing on a warship was on the USS *Pennsylvania* in 1911. Several other experiments of this type continued in the United States and Great Britain around this time. Seaplane carriers, which simply carried seaplanes in their hulls, launched them on catapults, and then winched them aboard after they had made a water landing, were developed in the first twenty years of the 1900's as well. The first true aircraft carrier, which enabled aircraft to take off and land, was the HMS *Furious* in 1917.

World War II

During the interwar years, the United States, Japan, France, and Great Britain built large fleet aircraft carriers. Earlier carriers were built on the hulls of former cruisers, battleships, or battlecruisers; later ships were built from the keel up as aircraft carriers. In World War II, these aircraft carriers came into their own and proved their worth as offensive and defensive strike weapons, primarily with the navies of the United States, Great Britain, and Japan. Germany had begun the construction of at least two aircraft carriers before World War II. Only one came near completion; named the *Graf Zeppelin*, it was never finished, scuttled at the end of the war, and later sunk. Fleet aircraft carriers effectively spelled the end of then-conventional surface warfare, wherein capital ships attempted to destroy each other with long-range gunnery, in World War II. The British Royal Navy carrier raids on the Italian port of Taranto and the 1941 Japanese carrier attack on Pearl Harbor, based in part on the Taranto raid, showed conclusively that carrier-based aircraft could, under the right conditions, destroy capital ships in port. The battles of the Coral Sea and Midway, in May and June, 1942, definitively demonstrated that carrier battle groups were capable of searching out and sinking each other while hundreds of miles apart. At Coral Sea, the American carrier force, though losing a carrier, stopped the Japanese approach to Australia; at Midway, some weeks later, the American force destroyed all four Japanese carriers and effectively turned the tide of the Pacific war. In the 1944 battle known as the Marianas Turkey Shoot, American fighter planes destroyed most of what was left of the Japanese carrier air component.

In addition, smaller carriers, called escort carriers, or jeep carriers, did important service in the war by performing much-needed convoy escort protection in the Atlantic Ocean, as well as performing antisubmarine duty and support of amphibious landings, as well as strikes against land targets.

Modern Carriers

After World War II, the aircraft carrier moved from a largely battlefleet role to a variety of roles in both peacetime and war. Technical innovations in carriers included the embarcation of jet aircraft; nuclear weapons capability; steam catapults, which made jet operations possible; and angled flight decks, which allowed the simultaneous takeoff and recovery of multiple airplanes. In Korea, navy planes launched from carriers flew combat sorties and search-and-rescue missions; the same was true for carriers positioned in "Yankee Station" in the South China Sea during the Vietnam conflict. Carriers were used to retrieve the Mercury, Gemini, and Apollo astronauts from their space missions in the 1960's and 1970's.

After the war, the United States embarked on the idea of the "supercarrier," a ship which would be larger and more powerful than previous ships. The original supercarrier, the *United States*, was canceled before being built. However, subsequent carriers incorporated supercarrier designs, such as increased space for aircraft, fuel, and munitions.

Modern Supercarriers

Currently, the United States Navy is by far the largest operator of aircraft carriers in the world, with thirteen in service: *Ronald Reagan* (under construction), *Harry S. Truman*, *John C. Stennis*, *George Washington*, *Abraham Lincoln*, *Theodore Roosevelt*, *Carl Vinson*, *Dwight D. Eisenhower*, *Nimitz*, *Enterprise*, *John F. Kennedy*, *Constellation*, and *Kitty Hawk*. Other nations operating carriers as of 2001 included Great Britain (*Invincible*, *Illustrious*, and *Ark Royal*); France (*Charles de Gaulle*); India (*Viraat*); Thailand (*Chakri Nareubet*); Italy (*Giuseppe Garibaldi*); Spain (*Principe de Asturias*); Brazil (*São Paulo* and *Minas Gerais*); and the Russian Republic (*Kuznetsov*). Of the U.S. carriers, only the *Kennedy*, the *Constellation*, and the *Kitty Hawk* are conventionally powered; the remaining ten have nuclear propulsion. France's *Charles de Gaulle* is nuclear powered. The carriers of India, Great Britain, Spain, Italy, and Thailand are designed to operate helicopters and the Harrier V/STOL fighter/attack aircraft.

In addition to aircraft carriers, several other types of warships conduct air operations, largely with helicopters; these are used by several nations. The U.S. Navy operates several amphibious warfare ships, sometimes called helicopter carriers, whose flat decks are for the purpose of embarking and debarking troop concentrations, such as Marine Expeditionary Units (MEUs).

American aircraft carriers are ships with multiple roles in the twenty-first century. Apart from their immediately obvious tactical military role, they are often used as strategic tools of diplomacy and national will when needed. American carriers have been sent to "show the flag" and project American power in world trouble spots since the 1950's. American carriers were sent to the Persian Gulf to safeguard shipping during the Iran-Iraq War of the 1980's, and during the Desert Shield campaign of 1990 and 1991.

Future

The future of the aircraft carrier in the short and middle term seems assured. Despite comments from critics that aircraft carriers are too expensive (costing billions of dollars) and too vulnerable to either military or terrorist attack, the navy's newest carriers are being built with an expected useful service life of over thirty years (taking into account service life extension programs to update structures and systems). In addition, the commitment of the United States to thirteen active aircraft carriers (though not all are at sea at the same time) indicates America's commitment to the concept of anytime, anywhere naval carrier operations for some time to come.

Several aircraft carriers of World War II and later eras have been permanently docked and serve as museums, including the *Hornet* in Alameda, California, *Intrepid* in the New York City harbor, *Lexington* in Corpus Christi, Texas, and the carrier *Yorktown* in Charleston, South Carolina. Many active carriers have their own World Wide Web sites.

Robert Whipple, Jr.

Bibliography

Allard, Damien. "French Fleet Air Arm." (frenchnavy.free.fr/main_menu_english.htm) A good English description of the French navy's ships, history, and aircraft.

Clancy, Tom. *Carrier: A Guided Tour of an Aircraft Carrier.* New York: Penguin, 2000. A definitive descriptive treatment of a modern American carrier and its political and military role, as well as its technological and human systems.

Galuppini, Gino. *Warships of the World: An Illustrated Encyclopedia.* New York: Times Books, 1983. An illustrated guide to world warships from ancient times to the present, with several illustrations of modern and early aircraft carriers.

Royal Navy. (www.royal-navy.mod.uk) The official World Wide Web site of the British Royal Navy, including information on carriers and aircraft.

Toppan, Andrew. "Haze Gray and Underway: World Aircraft Carriers and Lists." (www.hazegray.org/navhist/carriers/) An extensive World Wide Web site that includes a detailed discussion of the rise and rationale of the modern supercarrier, an up-to-date list of all aircraft carriers currently serving in the world, as well as lists of all other aircraft carriers which have served in the world's navies.

United States Navy. "The Carriers." (www.chinfo.navy.mil/navpalib/ships/carriers/) An official U.S. government Web site with descriptions, rationale, and history of U.S. carriers.

Wukovits, John F. "Greatest Aircraft Carrier Duel." (www.thehistorynet.com/WorldWarII/articles/1999/03992_text.htm) A detailed description of a classic U.S.-Japanese carrier aircraft battle from World War II, the Marianas Turkey Shoot.

See also: Gulf War; Harrier jets; Hornet; Landing procedures; Military flight; Navy pilots, U.S.; Takeoff procedures; Tomcat; Vertical takeoff and landing; World War II

Airfoils

Definition: A two-dimensional, front-to-back section or slice of a wing.

Significance: The shape of a wing's airfoil section or sections determines the amount of lift, drag, and pitching movement the wing will produce over a range of angles of attack and also determines the wing's stall behavior.

The shape revealed if a wing were to be sliced from its leading edge to its trailing edge is called the wing's airfoil section. Although airfoils come in many different shapes, all are designed to accomplish the same goal: forcing the air to move faster over the top of the wing than it does over the bottom. The higher-speed air on the top of the airfoil produces a lower pressure than the flow over the bottom, resulting in lift. The shape of the upper and lower surfaces of the airfoil and the angle that it makes with the oncoming airflow, or angle of attack, determines the way the flow will accelerate and decelerate around the airfoil and, thus, determines its ability to provide lift.

Flow around the airfoil also causes drag, and an airfoil should be designed to get as much lift as possible while at the same time minimizing drag. The shape of the airfoil then determines the balance of lift and drag at various angles of attack. An airplane designer tries to select an airfoil shape that will give the best possible lift-to-drag ratio at some desired optimum flight condition, such as cruise or climb, depending on the type of aircraft. The amount of pitching movement, or tendency for the airfoil to rotate nose up or down, is also a function of the airfoil's shape and the way lift is produced. Pitch must be evaluated along with the forces of lift and drag.

Camber and Thickness

Early airfoil shapes were thin, essentially cloth stretched over a wood frame, a type of airfoil sometimes seen today in the wings of ultralight or hang glider-type aircraft. Usually the frame for such an airfoil was curved, or cambered. The camber line, or mean line, of an airfoil is a curved line running halfway between its upper and lower surfaces. If the airfoil is symmetrical, in other words, if its upper surface is exactly the inverse of its lower surface, then the camber line is coincident with its chord line, a straight line from the leading edge to the trailing edge of the airfoil. A symmetrical airfoil is said to have zero camber. The amount of camber possessed by an airfoil is defined by the maximum distance between the chord and camber lines expressed as a percentage of the chord. In other words, an airfoil has 6 percent camber if the maximum distance between its chord and camber lines is 0.06 times its chord length.

Experimenters in the late 1800's tried wings built with airfoils with different amounts of camber and different positions of maximum. They found that the location of maximum camber affected both the amount of lift generated at given angles of attack and the airfoil's stall behavior and that too much camber can give high drag. Later researchers learned to create temporary increases in camber by using flaps.

Later aircraft used thicker airfoils with both upper and lower surfaces covered first with fabric and then with metal. The thicker airfoils allowed a stronger wing structure as well as a place to store fuel. They also proved able to provide good aerodynamic behavior over a wider range of angle of attack as well as better stall characteristics, but excessive thickness made for increased drag.

NACA Airfoils

In the 1920's, the National Advisory Committee for Aeronautics (NACA) began an exhaustive study of airfoil aerodynamics, examining in detail the effects of variations in camber and thickness distributions on the behavior of wings. This systematic study of variations in the amount and position of maximum camber and thickness resulted in the wind-tunnel tests of hundreds of airfoil shapes. NACA also developed a numbering system, or code, to describe the shapes. In the first series of tests, each of the numbers in a four-digit code was used in a prescribed set of equations to draw the airfoil shape. For example, the NACA 2412 airfoil had a maximum camber of 2 percent of its chord with the maximum camber point at 40 percent of the chord from the airfoil leading edge, and the maximum thickness was 12 percent of the chord.

Many other series of NACA airfoils were developed and tested. The 6-series airfoils were designed to provide very low drag over a set range of angle of attack by encouraging a low-friction laminar flow over part of the surface. Other series of airfoils were developed for use on propeller blades. NACA's successor, the National Aeronautics and Space Administration (NASA), has continued to test and develop airfoils including a series of supercritical shapes that give lower drag near the speed of sound, as compared to older designs.

Modern Airfoil Design

Throughout the twentieth century, airfoil design was essentially a matter of creating a shape based on desired camber and thickness distributions, testing it in wind tunnels and then in flight. Today, airfoils can be selected from hundreds of past designs or custom-developed by specifying a desired distribution of pressure around the surface and using computers to solve for the shape that will give those pressures. Then wind-tunnel tests are done to validate the computer solution. The result is that every airplane can have a wing with a unique distribution of airfoil shapes along its span, all designed for optimum performance. The basic idea is the same as it has always been, to find the combination of camber and thickness which will give the best available mix of lift, drag, and pitching movement for the task at hand.

James F. Marchman III

Bibliography

Abbott, Ira H., and Albert E. Von Doenhoff. *Theory of Wing Sections*. New York: Dover, 1959. The classic and still most complete reference of NACA airfoil section information.

Anderson, John D., Jr. *A History of Aerodynamics*. Cambridge, England: Cambridge University Press, 1998. A thorough examination of the development of aerodynamic theory and application.

Barnard, R. H., and D. R. Philpott. *Aircraft Flight*. 2d ed. Essex, England: Addison-Wesley Longman, 1995. An excellent, nonmathematical text on aeronautics. Well-done illustrations and physical descriptions, rather than equations, are used to explain all aspects of flight.

See also: Aerodynamics; Forces of flight; Hang gliding and paragliding; National Advisory Committee for Aeronautics; National Aeronautics and Space Administration; Roll and pitch; Ultralight aircraft; Wing designs

Airline Deregulation Act

Date: October 30, 1978

Definition: A federal law passed in order to eliminate the U.S. government's control of airline regulations, routing, fares, and schedules.

Significance: In an effort to create a freely competitive airline industry, the Airline Deregulation Act revolutionized the way airlines were allowed to do business.

History

The freedom of flight must have been one of humankind's earliest dreams. It must have been wondrous to watch birds soar, swoop, and land at will, and then lift off again to fly. When humans finally began to take to the skies in flying machines, they little dreamt the skies would one day need controlling, scheduling, and regulating. The idea of free and open skies for anyone who could fly became unrealistic in the increasingly commercial world of aviation, where business and safety were primary concerns.

World War I, which began in 1914, helped to educate the public about aviation and its possibilities. Warring nations began manufacturing war planes with metal bodies, which had previously been made of wood, and developed more powerful engines to increase speed and flying distance. After the war, organized air service developed rapidly in Europe.

By 1921, government subsidies for the development of new aircraft were in place, and most of Europe's major cities were linked by air service. In 1926, the U.S. government began awarding airmail delivery contracts to private air carriers and subsidizing private air carriers to transport the mail. This development helped to spawn the airline industry. As passenger travel on these early airlines increased, the government quickly became involved with enforcing economic and air safety regulations, as well as awarding mail contracts.

Civil Aeronautics Act

The Civil Aeronautics Act of 1938 established federal control over airline regulation. This influential legislation strengthened the government's power and led to the creation of the Civil Aeronautics Board (CAB) and the Air Safety Board, a forerunner of the Federal Aviation Administration (FAA).

The CAB took on a role of responsibility and importance. Its four main points of control were to review route requests and grant authorizations; to establish a uniform system of rates and fares; to approve airline mergers, acquisitions, and new entrants; and to rule on unfair competition. The CAB also awarded Certificates of Convenience and Necessity to domestic airlines and assigned carriers to specific routes. Airlines were required to request permission from the CAB, a complicated procedure, for any route expansion or airfare increase. The CAB simplified the fare system and regulated airfares, so that all carriers flying between the same cities had the same fares. Any requests for changes were processed, reviewed, debated, and, in many cases, eventually denied.

The CAB was originally established to ensure orderly competition and growth within the airline industry. Many policymakers believed that if the U.S. airline industry were permitted to operate unregulated, then unprofessional and perhaps unethical competition might develop, leading to the survival of only a few large airlines. If airlines were allowed to choose their own routes, they would avoid servicing small, less profitable communities and cities. Until 1978, the federal government—through the CAB and the Air Traffic Conference, a nongovernmental airline group—controlled all U.S. airline transportation. There was little or no competition among commercial airlines.

Airline Deregulation Act

On October 30, 1978, President Jimmy Carter signed the Airline Deregulation Act, which scheduled the shutdown of the CAB for January 1, 1985. Between 1978 and 1985, the CAB assisted in the transition process. With the impending elimination of the CAB, the domestic airline industry began its own restructuring for the future.

Deregulation of the airlines had many goals, some of which were to allow airlines to set their own fares, thus stimulating competition and lowering fares; to allow airlines to determine their own routing, thereby expanding flight options available to passengers; to allow airlines to discontinue unprofitable routes; and to stimulate growth within the industry by allowing for the establishment of new carriers.

Deregulation led to changes for the entire travel industry as well as the traveling public. Thousands of fares change daily in response to changing competitive conditions, in stark contrast to the fare stability that prevailed prior to deregulation. Deregulation has also given the airlines the freedom to fly wherever they wish. This freedom has spurred the creation of new route structures.

Travelers have benefited from deregulation because of the greater number of lower discount fares available. These new lower fares, fare wars, and promotional airfares have enabled more people to use air transportation than ever before. Competition between the airlines has become intense. Deregulation permitted hundreds of new airlines to start. Many new airlines such as People Express threw the industry into turmoil by offering super bargain fares. As had been expected, many of the major carriers lost money trying to compete at that level, leading some airlines to merge while others just faded away. There were bankruptcies and downsizing. Low-cost, no-frills airlines were established to tap into high traffic corridors and regional market segments. Major airlines have made dramatic efforts to increase customer loyalty with such incentives as frequent-flier programs. They have developed aggressively competitive strategies involving promotions, destination, partnering, code sharing, and as always, the special low-fare promotion.

Airlines will continue to compete fiercely for passengers. New categories of budget airfares with numerous travel restrictions are constantly being developed. Although passengers benefit from reduced prices, travel and ticketing agents must work with complex ticketing and reservation situations. Because printed tariffs can become obsolete before reaching a travel agency, travel agents rely on their computer systems to access fares and information regarding their related purchase requirements.

Deregulation allowed airlines to govern aspects of their commercial dealings. Airlines could establish their own routes and airfares. They could create their own packages and fare plans to effectively compete for more passengers. They may now function as tour operators, providing packaged tours directly to the public. In addition, they may own and operate travel agencies. Thus, they have developed new methods of selling tickets outside the existing travel agency system.

Since the dismantling of the CAB on December 31, 1984, the Department of Transportation (DOT), which was created in 1966, has watched over the U.S. airline industry. The powers of the DOT are more limited than those of the CAB. In writing the Airline Deregulation Act of 1978, Congress had intended that the federal government alone would regulate the airline industry. Federal regulations on any airline industry issue take precedence over the rights of the states or individuals to legally challenge any carrier's business practices. The DOT regulates and monitors all transportation industry and safety issues in the United States. Because this large task encompasses so many industries, there are specialized administrations under the DOT whose authority is specific to one type of transportation. One branch of the DOT, the Federal Aviation Administration (FAA), plays an important role in the continued regulation of air safety. The FAA has absolute authority over flights passing through U.S. airways.

Hub-and-Spoke System

One result of deregulation was that airlines were granted the new freedom of choosing their own routing, spurring the creation of new route structures. The airlines have basically abandoned point-to-point route systems, in which cities were connected by nonstop flights, in favor of hub-and-spoke networks. A hub is an airport through which an airline schedules the majority of its flights, like a wheel with many spokes radiating from the center of the hub. An airline schedules most of its flights from various cities, or spokes, to arrive at approximately the same time at a designated hub. The airline then schedules most of its flights to depart one to three hours later to other destinations, or spokes, along its routes. This enables flights from the various spokes to be routed through the central hub, where passengers are combined to fly on to a common destination. Today, passengers can change planes at a hub airport on the way to their eventual destinations.

Hub-and-spoke routings benefit the airlines because they lose fewer of their passengers to other airlines through interline connecting services. The hub system has proven beneficial to passengers because it allows them to remain on line, or on the same carrier, all the way through to their final destination. They can change planes at the hub airport without scrambling to change airline carriers and with less possibility of lost luggage and missed connecting flights.

A hub airport is a major connecting center, as well as the typical location of the airlines' administrative offices. A gateway is a hub for international flights and serves as an arrival and destination point for international travelers.

International Air Transportation Competition Act

In line with deregulation of the domestic airline industry, Congress passed the International Air Transportation Competition Act of 1979. The act, passed in February, 1980, encompassed ten goals. The first was to strengthen the competitive position of U.S. air carriers and to increase profitability. The second was to ensure air carriers the freedom to offer fares and rates that corresponded with consumer demand. The third was to reduce restrictions on charter options. The act also allowed multiple carrier designations for U.S. airlines with permissive route authority so that carriers could respond swiftly to shifts in demand. It eliminated operational and marketing restrictions with respect to capacity and flight frequency. It sought to integrate domestic and international air transportation and to increase the number of nonstop U.S. gateway cities. It provided opportunities for foreign airlines to increase their access to U.S. points if exchanged for benefits of similar magnitude for American carriers with permanent linkage between rights granted and rights given away. It sought to eliminate discrimination and unfair competitive practices against U.S. air carriers in foreign air transportation. Finally, it pledged to promote and develop civil aeronautics and a viable, privately owned U.S. air transport industry.

The passing of the International Air Transportation Competition Act of 1979 reaffirmed the U.S. policy of competition. The established carriers jumped at the chance to expand. They wanted long-haul flights between major cities, where they could attract the most money. What they had not expected was that the introduction of more carriers in the large markets simply meant that the planes were flying without full passenger loads. Because pricing had been placed in the control of the carriers, they did what all businesses do when supply exceeds demand: They lowered their fares. During the fierce price wars of the early 1980's, it was often less expensive to fly to a destination than it was to drive.

In 1997, the U.S. airline industry launched more than 22,348 flights a day, employed more than 586,509 people, carried 1.6 million people each day, and recorded $109.5 billion in revenues. The 1997 survey of air travelers by the Gallup organization revealed that a record 80 percent of the entire adult population of the United States had flown.

Alliances of the Future

Under deregulation, the airline industry has undergone dramatic change, leading to significant consolidation, hub systems, low airfares in competitive situations, and high airfares where competition was lacking.

The future holds alliances that could involve the largest carriers in the United States. Alliances are being negotiated with an overwhelming range, from equity positions, to code sharing, to frequent-flier programs, reciprocity, and other joint marketing arrangements. International alliances have been debated since KLM (Royal Dutch Airline) and Northwest Airlines linked in 1992. United Air Lines established the Star Alliance, which included Lufthansa, Air Canada, Thai Airways, SAS, and Brazil's Yang Airlines.

Authorities in both the United States and the European Union are analyzing how to deal with the many major airline alliances. The decisions they make will shape the future of airlines around the world. It remains to be seen whether these alliances will benefit consumers through greater choice, lower fares, greater convenience, and frequent-flier miles or whether they will create monopolies, higher fares, and new noncompetitive situations. Perhaps government action may be needed to call for reregulation.

Lori Kaye and Maureen Kamph

Bibliography

Gidwitz, B. *The Politics of International Air Transport.* Lexington, Mass.: D. C. Heath, 1980. An examination of the political and legal aspects of international air travel, featuring maps, a bibliography, and an index.

Kane, R. M. *Air Transportation.* 13th ed. Dubuque, Iowa: Kendall/Hunt, 1999. A classic text on the commercial airline industry in the United States.

Solberg, Carl. *Conquest of the Skies: A History of Commercial Aviation in America.* Boston: Little, Brown,

1979. A history of the U.S. commercial aviation industry, published shortly after deregulation.

Waters, S. R. *Travel Industry World Yearbook: The Big Picture, 1994-1995*. New York: Child & Waters, 1994. A compendium of travel industry statistics.

See also: Air carriers; Airline industry, U.S.; Airmail delivery; Airports; Federal Aviation Administration; Frequent-flier miles; KLM; Lufthansa; Mergers; Northwest Airlines; SAS; Ticketing; United Air Lines

Airline industry, U.S.

Definition: The system of airline carriers and aircraft supplying national and international air travel services for passengers in the United States.

Significance: The U.S. airline industry has evolved to become the most highly developed transportation network in the world, comprising more than 5,000 aircraft valued at more than $65 billion. U.S. airlines schedule more than 12,000 domestic and international flights per day, with more than 200,000 available seats. These flights carry a major portion of the world's air commerce.

History

Scheduled air transportation actually began in Germany, where, as early as 1910, zeppelins carried 35,000 passengers per year. The first recognized attempt at establishing a scheduled air passenger service in the United States occurred in 1914, with the short-lived St. Petersburg-Tampa Airboat Line, which consisted of one Benoist Type XIV flying boat. The carrier's 18-mile route across Tampa Bay was covered in twenty minutes at a price of $5.00 for the one passenger per flight. The airline contracted with the city of St. Petersburg for two daily round trips. This endeavor lasted for four months and carried 496 passengers. There would not be another regularly scheduled airline in the United States until well after the end of World War I.

After World War I, Europe again led the way in developing scheduled airlines. With the advances in design and construction, large military aircraft could be converted for civilian uses. Geographic features such as the English Channel made scheduled flights feasible and flights between London and Paris began while the Treaty of Versailles was still being negotiated. Because most of Europe's transportation and communications infrastructure had been destroyed by the war, government assistance was available to fledgling airlines of European countries. These airlines would be used to connect the outlying outposts of the various empires.

In contrast, the U.S. transportation systems within the United States had not been damaged by the war. Indeed, the most highly developed railroad system in the world remained intact. The long distances involved in U.S. travel lent themselves more comfortably to Pullman-type sleeper cars and well-stocked dining cars than to open-cockpit biplanes. At the time, trains were as fast as most of the existing airplanes and could continue to travel throughout the night while passengers slept. Unlike the governments of European nations, the U.S. government displayed little inclination to subsidize the development of air transportation.

The next serious attempt to establish a regularly scheduled air service in the United States had more to do with politics than with aeronautics. In 1919, Congress passed the Volstead Act, which made illegal the selling and consumption of alcoholic beverages. However, the institution of Prohibition did not quench the thirst of the American public. Lying conveniently off the coast of Florida were the independent nations of Cuba and the Bahamas, which were beyond the jurisdiction of U.S. laws. In order to meet the demands of American consumers, Aeromarine Airways developed a route that came to be known as the Highball Express. Using fourteen-passenger flying boats, Aeromarine began a scheduled service from New York City to Havana, Cuba. Aeromarine added flights to Nassau in the Bahamas and expanded its service to Cleveland and Detroit. The airline continued to operate until September, 1923, ultimately carrying 17,000 passengers.

The Role of the U.S. Post Office

Although the U.S. government's official position was that air transportation should be developed without government subsidies or direct government involvement, the U.S. Post Office had begun experimenting with airmail delivery as early as 1911. In 1918, the Post Office began to establish a scheduled airmail delivery system. This activity culminated in the development of a transcontinental route structure with numerous connecting routes. With the passage of the Air Mail Act of 1925, also known as the Kelly Act, the Post Office turned these routes over to private operators through competitive bids. Many, if not most, of the existing airlines in the United States were established with the goal of obtaining a Post Office airmail contract. The airmail contract would ensure that the fledgling airlines could generate enough income to remain in existence. The goal of the contract airmail service was to subsidize the private

Immediate Impact of the Terrorist Attacks of September 11, 2001, on Major U.S. Carriers

Air Carrier	Financial Results	Announced Job Cuts (% of workforce)	Expected Federal Compensation	September, 2001, Traffic	
				Number of Passengers (% decrease from 9/00)	Load Factor (% in 9/00)
Alaska Airlines	2000: $70 million net loss 2001: $3 million net loss through 9/30/01	None	$100 million	805,000 (22.0%)	60.7% (62.7%)
American Airlines	2000: $813 million profit 2001: $964 million net loss through 9/30/01	~20,000 (14%)	~$1 billion	61.3 million (6.9%)	59.6% (69.8%)
Continental Airlines	2000: $342 million net income 2001: $54 million net income through 9/30/01	12,000 (~20%)	$458 million	2.9 million (32.0%)	61.4% (72.4%)
Delta Air Lines	2000: $928 million net income 2001: $482 million net loss through 9/30/01	Up to 13,000 (15%), 11,000 from retirement, voluntary severance	$690 million	5.87 million (35.0%)	56.2% (68.5%)
Northwest Airlines	2000: $256 million net income 2001: $207 million net loss through 9/30/01	10,000 (19%)	$500 million	3.1 million (33.9%)	63.8% (76.5%)
Southwest Airlines	2000: $603 million net income 2001: $47.6 million net income through 9/30/01	None	$280 million	3.6 million (28.7%)	53.4% (65.7%)
United Air Lines	2000: $50 million profit 2001: $1.84 billion net loss through 9/30/01	~20,000 (20%)	$800 million	4.2 million (34.6%)	61.1% (69.9%)
US Airways	2000: $269 million net loss 2001: $960 million net loss through third quarter	11,000 (24%), 2,000 voluntary	$331 million	2.92 million (39.0%)	56.1% (66.9%)

Source: Data taken from "Ailing Airlines: An Industry in Crisis," MSNBC.com, November 1, 2001 airlines in order eventually to develop a viable passenger service.

By 1927, the Post Office had contracted all the airmail routes to private companies. The original transcontinental route was eventually awarded to what is today United Air Lines. Mergers forced by the postmaster general resulted in the formation of American Airlines. American received the newly established southern transcontinental route. Transcontinental Air Transport, the company that had pioneered the idea of combining aircraft and passenger trains for transcontinental travel, was forced to merge with Western Air Express. The resulting airline, Transcontinental and Western Air, eventually became Trans World Airlines (TWA). Other airlines were awarded routes in support of the transcontinental routes. Eastern Air Lines received a north-south route along the East Coast; Northwest Airlines received a route northward to Minneapolis. Many other smaller airlines also received contracts, although within a few years most of these companies had been absorbed by the larger operators.

The U.S. Post Office also awarded mail contracts for international airmail. Pan American Airways, under the leadership of Juan Trippe, won the vast majority of these contracts. Although there were accusations of favoritism and unfair practices associated with Pan American's unparalleled success, these were never proven. Pan American became the chosen instrument of the U.S. State Department and remained virtually the only international airline in the United States until the advent of World War II. Pan American absorbed any potentially competitive company. By the time the United States entered World War II, Pan American had developed into the largest, most successful airline in the world. Not only did the company fly throughout Central and South America and across the Atlantic Ocean to Europe, but it also covered the Pacific Ocean, with service to exotic destinations such as Hong Kong.

Although the efforts of the U.S. Post Office resulted in a strong network of air carriers, not everyone was pleased with its practices. The methods employed by Postmaster Walter Folger Brown had alienated many of the small operators who had not been awarded airmail contracts. With the election of President Franklin D. Roosevelt in 1932, many complaints surfaced. Roosevelt ordered an investigation that led to the ill-fated cancellation of all domestic airmail contracts in 1934. The president ordered the army to carry the mail, but this arrangement proved to be disastrous, as a number of poorly trained and equipped pilots were killed in accidents. The airmail contracts were returned to the original operators shortly thereafter, but the passage of the Air Mail Act of 1934 placed severe restrictions on the airlines. The provisions of this law resulted in large financial losses, and many airlines were on the brink of bankruptcy when Congress passed the Civil Aeronautics Act of 1938, which saved the airline industry.

Although all the domestic airmail contracts had been canceled, none of the international contracts that had been awarded to Pan American were affected. Pan American, with the cooperation of the U.S. Navy, continued to expand its operations. Rates in the Pacific were actually increased above the maximum allowable rate in the name of national security. Pan American continued to operate successfully and profitably as the country prepared for World War II.

World War II

World War II had a tremendous impact on the U.S. airline industry. The United States assumed the role of supplying the worldwide war effort with supplies, material, and human resources. The aircraft manufacturing industry began to mass-produce large, long-range aircraft, operated by airline-trained crews. These crews flew to every corner of the world, establishing routes, airports, and facilities that would be invaluable at war's end. This effort, which came to be known as airlift, was one of the major military accomplishments of World War II. The experience gained by the airlines and crews, coupled with advancements in aircraft design and technology, catapulted the United States into a dominant position in international air transport. The war destroyed or dispersed the airlines of most of the countries of the world. European aircraft manufacturing was forced to concentrate on building fighters and bombers rather than large transport-type craft. Pilots were engaged in combat flying rather than long-range international flights. U.S. airlines assumed the role of international transportation.

The domestic U.S. transportation system also experienced growth during World War II. Routes were expanded to meet the needs of the government, and the number of passengers exceeded the number of airplane seats available. Although the majority of airlines still utilized the venerable Douglas DC-3 for domestic service, larger and faster airplanes began to appear. Passenger-generated income finally exceeded that generated by airmail delivery, and the structure of a mature industry began to develop. By the end of the war, air travel had become an accepted fact and was no longer considered an unusual extravagance. Industry and government began to rely more heavily on the airplane and its scheduled services.

The war had a significant impact on the structure of international air service. Prior to the war, Pan American had been the so-called chosen instrument of the United States. The demands of war had led to a number of additional airlines being granted international routes. Both American and TWA flew transatlantic routes during the war, and they were allowed to continue with these routes after the war, much to the dismay of Pan American. Political difficulties with the Roosevelt and Truman administrations further weakened Pan American's dominance as airlines such as Eastern, Braniff International, and Northwest were granted international routes, often into the heart of Pan American's territory. While Pan American was losing its international monopoly, the airline was simultaneously barred from establishing a domestic network in order to compete with the new international airlines, a prohibition that was eventually to have disastrous consequences for the once-proud Pan American.

Postwar Industry

As a result of World War II, the U.S. airline industry had evolved into a large worldwide operation. Airlines began to become differentiated into passenger airlines, all-cargo airlines, commuter airlines and air taxis. Airlines were divided into categories. Trunk airlines, such as American, Eastern, TWA, and United, were those that operated between nations and crossed oceans. Later this category was changed to include large domestic airlines, and trunk airlines became reclassified as regional airlines, which did not operate international routes and typically served four or five states. Commuter airlines were scheduled airlines that were restricted to aircraft weighing less than 12,500 pounds. This classification system lasted until the passage of the Airline Deregulation Act in 1978.

Industry Growth

Throughout the 1940's and 1950's, the Civil Aeronautics Board (CAB) continued to award routes to various airlines in an attempt to expand the transportation network and maintain an environment of controlled competition. The airlines, in turn, competed with each other primarily in terms of service and speed. This competition resulted in performance improvements in the various types of commercial aircraft. Airlines such as American favored the Douglas DC-6/7 series, whereas TWA favored the Lockheed Constellation series. Continual performance increases led to passengers enjoying pressurized comfort while crossing the country in eight hours or less. In addition, pressure from a number of nonscheduled operators charging significantly lower fares forced a number of the domestic trunk airlines to begin offering lower fares. In 1948, Capital Airlines introduced coach-class service and, by 1949, was joined by American and TWA. In May, 1950, United followed suit. The airlines, with the approval of the CAB, agreed to a mutual $99.00 coast-to-coast coach fare in 1952.

In an attempt to guarantee service to smaller communities, the CAB established the classification known as feeder airlines in 1944. Feeder airlines, such as Pioneer, Allegheny, North Central, Piedmont Airlines, and Mohawk Airlines, were heavily subsidized to supply air service to small communities. In 1955, this airline classification was renamed local service carriers. Most of these airlines relied on Douglas DC-3 aircraft that had been retired by the larger trunk carriers. Eventually, local service carriers replaced the DC-3 with more modern aircraft and introduced turboprop aircraft, such as the F-27 Friendship. By the 1960's, the CAB was coming under increasing pressure to do away with the subsidies and allow the free market to determine success. However, with shorter routes serving small cities, it would be more difficult for these smaller airlines to make a profit.

Recognizing this fact, in 1959, the CAB instigated the so-called "use-it-or-lose-it" policy. Cities were informed that if they did not have a minimum of 1,800 passengers a year, their air service would be terminated.

At the same time, the local service carriers began to pressure the CAB for the right to expand their route structures. They purchased larger aircraft, including jets such as the Douglas DC-9. Airlines such as Frontier grew to challenge the existing trunk airlines. As the expansion of the local service carriers continued, a number of mergers began to occur, with the stronger airlines devouring the weaker ones. The original thirteen local service carriers had been reduced to nine by 1968. At this time, the name of the local service carrier classification was changed to that of regional carrier. Mergers between the regionals continued, with Allegheny, the largest, merging with Mohawk, the third largest, to form what was eventually renamed USAir in 1979. North Central and Southern Airlines also merged to form Republic Airlines. These so-called regionals were now as large as the trunk airlines. As the regionals abandoned their short-haul markets, the third-tier airlines, calling themselves commuters, expanded to fill the void.

The CAB recognized this discrepancy and in 1980 established a new airline classification system. All airlines in the United States were now to be classified based on their revenues as majors, nationals, or regionals. Majors were required to generate $1 billion in annual revenues, and nationals were required to generate $75 million. Regionals were divided into two subclassifications: large, which

Large Air Carrier Classifications, 1995-1996

MAJORS

Alaska	Continental	Northwest	United
America West	Delta	Southwest	United Parcel Service
American	Federal Express	TWA	USAir

NATIONALS

Air Transport International	Business Express	Hawaiian	Rich International
Air Wisconsin Airlines	Carnival	Horizon Air	Simmons
Aloha	Contintental Express	Kiwi	Southern Air
American International	Continental Micronesia	Markair	Sun Country
American Trans Air	DHL Airways	Mesa	Tower
Arrow	Emery	Midwest Express	Trans States
Atlantic Southeast	Evergreen	Polar Air Cargo	USAir Shuttle
Atlas	Executive	Reno	ValuJet
			World

LARGE REGIONALS

Air South	Express One	Midway	Spirit
Amerijet	Fine	North American	Sun Jet
AV Atlantic	Frontier	Northern Air	UFS
Buffalo	Grand American	Pan Am	Viscount
Challenge	Kitty Hawk	Reeve	Western Pacific
Champion	Miami Air	Ryan International	Zantop

MEDIUM REGIONALS

Air 21	Florida West	Pacific International	Trans Continental
Airtran	Grand Airways	Prestige	Transmeridian
Capitol Cargo	Laker Airways	Renown	Tristar
Casino Express	Milton	Sierra Pacific	USA Jet
Custom Air	Nations Air	Sun Pacific	Vanguard
Eastwind	Omni	Sun World	
Falcon	Pace	Tatonduk	

Source: Federal Aviation Adminstration, *Statistical Handbook of Aviation*, 1996.

earned $10 to 75 million in annual revenues, and medium, which earned less than $10 million. Under this new system, former local service carriers Republic and USAir received major airline status.

The Jet Age

Perhaps the one event that had the greatest impact on air transportation was the development of the jet engine. From its inauguration in the De Havilland Comet, it was clear that the jet engine would alter the entire industry. U.S. airlines entered the jet age in October, 1958, when Pan American scheduled the first Atlantic jetliner crossing by an American airline. With the establishment of this jet service, Pan American dominated the transatlantic service for

American airline. With the establishment of this jet service, Pan American dominated the transatlantic service for a number of years. Other airlines quickly followed Pan American's lead and introduced jet service. National Airlines was, in 1958, the first airline to introduce domestic jet service. This event marked the eclipse of the large, multiengine, piston-powered airliners that had been developed since World War II. Aircraft manufacturers continued to manufacture jet-powered aircraft in different sizes to meet the needs of the evolving market. Today, many turboprop commuter aircraft have been replaced with small, fifty- to seventy-seat jet aircraft. A number of commuter airlines plan a transition to all-jet fleets just as their larger counterparts have done.

Deregulation

Claimed by some analysts to be the most important event in air transportation history, the Airline Deregulation Act was passed in 1978. Through this act, the U.S. government reduced its role in the regulation of most of the business and financial aspects of airline operations. The CAB no longer approved a new airline's entry into the industry. Neither would the CAB approve routes or set fare structures. The airlines entered into an era of virtually unrestrained competition. The immediate effect of deregulation was the appearance of a number of new airlines. Many of these new entrants to the industry were low-cost, no-frills operations. The largest of these airlines was People Express, which rapidly expanded to compete with a number of the major airlines. These new entrants typically had significantly lower operating costs than did existing airlines, due to lower employee seniority, nonunion labor, and older aircraft. Although many of these new airlines were initially very successful, the major airlines reacted to the threat by lowering fares, purchasing the competing operators, and increasing frequency of flights. The major airlines also formed what came to be known as hub-and-spoke route structures, which led to the domination of various geographical markets. Code sharing with, or the outright purchase of, commuter airlines supplied a continuous flow of passengers to the hub airports. Computer reservations systems and complex revenue-management systems allowed the largest airlines to further dominate the market. Of the large number of new airlines that entered interstate or international service immediately after deregulation, only two remain.

Deregulation also had a devastating effect on some of the oldest and largest airlines, a number of which were unable to adjust to the changing demands of a deregulated environment. Giants such as Pan American, Braniff, and Eastern ceased to exist. Others, such as Republic, were forced to merge and ceased to exist.

By 1990, 60 percent of the airline business was concentrated in the hands of the big three: American, United, and Delta. Airlines such as TWA, America West, and Continental were forced into Chapter 11 bankruptcy and reorganization. The early 1990's, in particular, were a disastrous time for the airline industry. All airlines except Southwest lost large amounts of money; some were unable to recover and return to profitability, leading to additional consolidation within the industry.

In response to the crisis of the 1990's, Northwest and US Airways looked for relief to European partnerships. While the partnership between US Airways and British Airways was less than a complete success, a similar partnership between Northwest and KLM led to a return to profitability for Northwest. Other partnerships, such as United's Star Alliance have been successful and have strengthened the competitive positions of those airlines involved. Even with such successes, the fact remains that the U.S. airline industry is subject to the fluctuations of the economy. It performs strongly in a strong economy and suffers in a weak economy. It is also very probable that the strong will continue to devour the weak and that the number of major airlines will decrease in the future. For example, in 2001, American Airlines made a proposal to buy TWA; the merger was accepted by the federal government.

September 11

The sudden, tragic events of September 11, 2001—four commercial jets hijacked by Islamic fundamentalists and then crashed, three flown into the World Trade Center's Twin Towers and the Pentagon, resulting in more than three thousand deaths—had an enormous impact on the U.S. airline industry. The government ordered an immediate national ground stop that morning until the situation could be assessed. Commercial flights resumed a few days later, but passengers and crews alike were understandably reluctant to board airplanes, given the threat of further terrorist acts.

The loss of revenue over the following weeks forced major air carriers and airports to lay off tens of thousands of employees. The longer-term picture was also grim. Tourism and business travel were curtailed. Moreover, the already fragile U.S. economy was placed firmly into recession. A federal bailout plan for the industry was devised. President George W. Bush signed into law an emergency aid package providing five billion dollars in direct federal aid and ten billion dollars in loan guarantees. The measure also offered federal help with rising insurance costs in the dustry sought ways to make flying safer and to reassure the public, but many predicted that the consequences of September 11 would be evident for years to come.

Ronald J. Ferrara

Bibliography

Davies, R. E. G. *Airlines of the United States Since 1914.* Washington, D.C.: Smithsonian Institution Press, 1998. An extremely well-written, informative, and comprehensive work on the history of the airlines of the United States.

Kane, Robert M. *Air Transportation.* 13th ed. Dubuque, Iowa: Kendall/Hunt, 1998. A well-written work that includes data on the past and present status of the airline industry.

See also: Air carriers; Airmail delivery; Airports; American Airlines; Continental Airlines; Delta Air Lines; Federal Aviation Administration; Flight attendants; Food service; Frequent flier miles; Mergers; Northwest Airlines; Pan Am World Airways; Pilots and copilots; PSA; Southwest Airlines; Ticketing; Trans World Airlines; Transatlantic flight; Transcontinental flight; Transglobal flight; United Air Lines; US Airways

Airmail delivery

Definition: The distribution of intercity and international first-class mail by aircraft rather than surface transport.

Significance: The U.S. Post Office tradition of guaranteeing the fastest possible mail delivery continued with the initiation of airmail service in 1918, when scheduled air service remained an untried and unproven concept. The airmail efforts of the U.S. Post Office also resulted in the development of the modern air transportation system, as most modern major airlines began operation as contracted airmail carriers.

Early Experiments

The earliest U.S. airmail experiment took place in 1858, when balloonist John Wise was contracted by Thomas Wood, the postmaster of Lafayette, Indiana, to carry a packet of mail to New York by air. This trip was advertised as the world's first official airmail delivery. However, the winds did not cooperate, and Wise was blown south rather than east. After a short flight, he landed a mere 30 miles from his point of departure.

Airmail delivery was one of the first commercial uses for airplanes in the early 1900's. By the end of the century, private express mail services such as Federal Express were challenging the U.S. Post Office for speedy mail delivery. (AP/Wide World Photos)

Early U.S. Airmail Routes

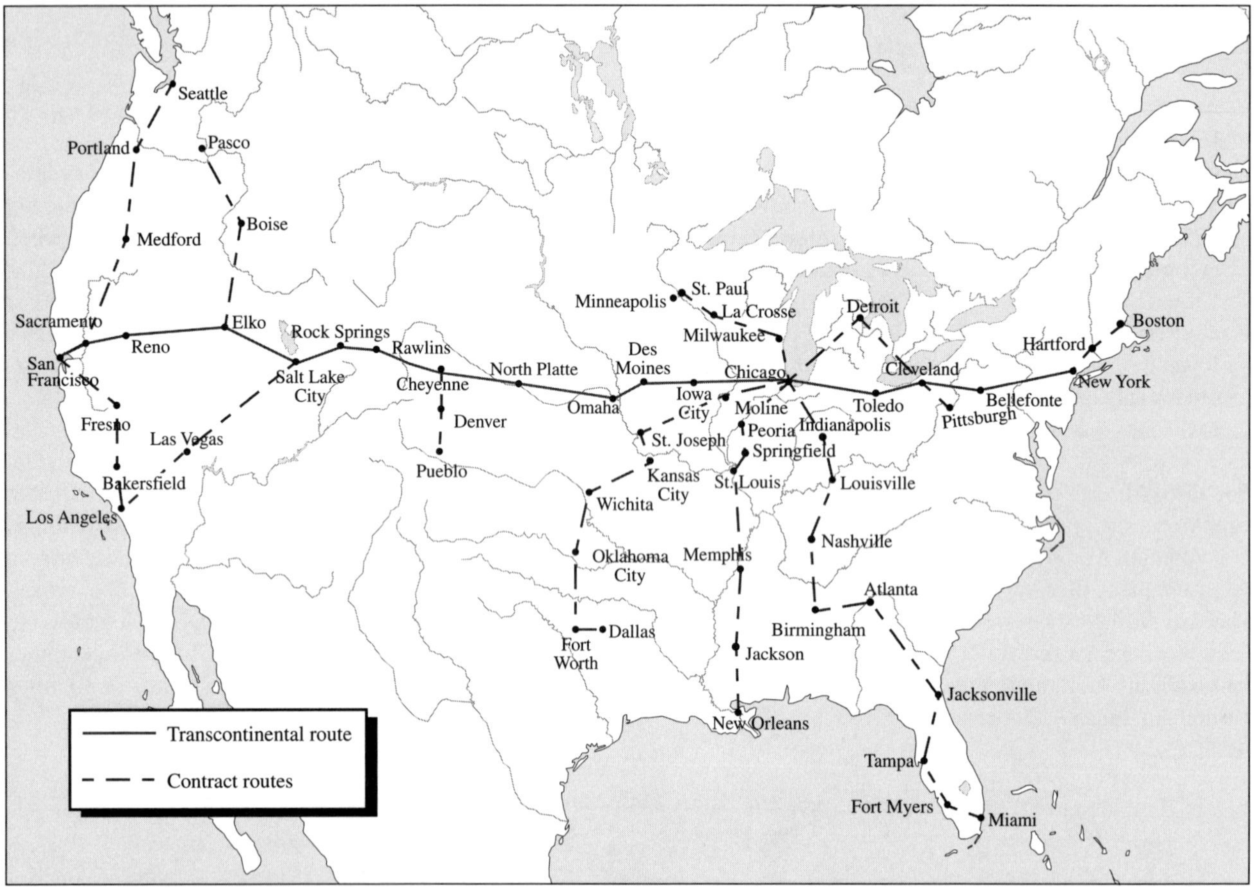

Although this first attempt was less than successful, the U.S. Post Office continued to support airmail experiments. As early as 1910, aviation pioneer Glenn H. Curtiss had unofficially carried mail in an aircraft. The first official mail to be carried in a heavier-than-air craft was carried by pilot Earle Ovington at the International Aviation Meet at Garden City, New York, in September, 1911. During the weeklong air meet, Ovington carried more than 40,000 pieces of mail by air to nearby Mineola, New York. Postmaster Frank Hitchcock was so impressed by the feat that he authorized the carriage of airmail from New York to Los Angeles. Given the state of aviation at this time, however, it would be many years before this plan became a reality.

The Air Mail Service

As early as 1912, the U.S. Post Office had requested funding to establish an experimental airmail service. The request, for a congressional appropriation of $50,000 for the establishment of the experimental program, was denied. Nevertheless, the postmaster continued to support the demonstration flights that were taking place at various air meets and renewed the request for funding each year.

In 1916, an appropriation was finally approved. The U.S. Post Office requested bids to operate experimental routes in Alaska and Massachusetts. Because the state of aircraft design at the time was such that there were no aircraft capable of successfully meeting the terms of the contract, there were no bidders.

The U.S. Post Office began negotiations with various aircraft manufacturers to design aircraft capable of meeting the demands of a scheduled airmail service. However, in 1917, the United States entered World War I, and the efforts of the Post Office were curtailed. World War I would have one effect that would later benefit the efforts of the Post Office. In response to the demands of the war, the design and performance of aircraft underwent significant improvement. By war's end, aircraft were available that could meet the demands of the postal service.

On June 30, 1918, the U.S. Post Office finally received an appropriation of $100,000 to initiate an experimental airmail service. The problem now was that the Post Office had no pilots or aircraft. U.S. Army captain Benjamin Lipsner was appointed to run the experiment, and Army pilots were assigned to fly the initial flights. The Post Office ordered six modified Curtiss aircraft to support the effort. On May 15, 1918, the first airmail flight was conducted, connecting Washington, D.C., Philadelphia, and New York. The effort was a success, even though one of the pilots from Washington, D.C., became lost and landed in Maryland, causing the mail to be put on a train for delivery. When, on August 12, 1918, the Post Office officially took over from the Army and began flying the mail with its own pilots and aircraft, the U.S. Air Mail Service was officially launched.

The original mail route between Washington, D.C., and New York was steadily expanded. The intent was to establish a transcontinental route from New York to San Francisco with feeder routes running north and south that connected with the transcontinental route. The original transcontinental route was completed on September 8, 1920. Using a combination of aircraft during the daylight hours and trains at night, airmail arrived twenty-two hours faster than the fastest train then in service. On February 22, 1921, an attempt was made to demonstrate day and night flying along the route. This day-night transcontinental demonstration was successful, and continuous flight was authorized. In addition, Congress appropriated $1,250,000 for the expansion of the service. In conjunction with this effort, portions of the airway were lighted with beacons, emergency airfields were established, and radio stations reported weather. As a result of these developments, the time for transcontinental mail delivery was reduced to twenty-six hours and fourteen minutes for eastbound travel and to twenty-nine hours and thirty-eight minutes westbound. This effort proved to be very popular, and the Post Office ordered fifty-one specially designed Douglas mail planes to replace the World War I surplus airplanes that had been operating since 1918. The arrival of these new aircraft further reduced the time required for transcontinental mail delivery. On July 1, 1924, airmail postage was set at eight cents per ounce, and regular night mail service began.

During the period from 1918 to 1927, when the U.S. Post Office operated the airmail service, the route structure increased from the original 218 miles to more than 2,700 miles. Air Mail Service planes experienced more than two hundred crashes, and more than eighty pilots were killed or injured. Of the original forty pilots hired by the Post Office, thirty-one were killed. Air Mail Service pilots flew more than 13,000,000 miles and carried 301,000,000 letters. They completed more than 93 percent of their scheduled flights. The total government expenditure for the entire period was $17,411,534. Income from the service totaled approximately $3,000,000.

The Air Mail Act of 1925

Even though the U.S. Post Office had demonstrated the feasibility of transcontinental airmail service, the ultimate goal was to turn the system over to private companies. Congress passed the Air Mail Act of 1925, known as the Kelly Act, which authorized the Post Office to solicit competitive bids for various airmail routes, culminating with the award of the transcontinental route to private companies. The first contracted airmail flights began in 1926. The last flight of the U.S. Post Office Air Mail Service took place on August 31, 1927.

The passage of the Kelly Act initiated the era of contract airmail. As the postal routes were turned over to private companies, the award of the contracts was used as a tool to encourage, and in some cases to force, operators to begin carrying passengers, as well as mail. Many modern carriers that continue to operate began as contract airmail carriers. In 1925, the initial five contracts were awarded. The awarding of routes continued, with the transcontinental route being awarded in two parts in 1927. The volume of mail increased substantially in 1928, when the airmail postage rate was lowered to five cents per ounce.

In 1929, Walter Folger Brown was appointed postmaster general under the administration of President Herbert Hoover. Brown had a vision for the air transportation system that would make it the most efficient system in the world. To support his goal, he arranged the passage of the Air Mail Act of 1930, also known as the McNary-Watres Act. This act gave the postmaster virtually total control of the contract airmail bidding process. Brown proceeded to implement a system favoring larger, more well-financed operators at the expense of the smaller operators, forcing a number of airlines to merge in the name of efficiency. He invited only the large operators to attend conferences in Washington, D.C., where contracts were awarded. These conferences became known as the spoils conferences and resulted in a scandal that led to the cancellation of all airmail contracts in 1934. Brown's actions resulted in the large airlines, such as United Air Lines, Trans World Airlines (TWA), American Airlines, Eastern Air Lines, and Northwest Airlines, receiving lucrative contracts at the expense of smaller operators. By July, 1933, twenty-three airmail routes had been established, covering 27,735 miles.

With the election of President Franklin D. Roosevelt in 1932, there were charges of collusion and graft, and the contracts came under investigation. Although no evidence was uncovered indicating illegal activity, Roosevelt canceled all existing airmail contracts and turned the airmail over to the U.S. Army, which began flying the mail on February 19, 1934. The Army was, however, ill-equipped and ill-trained to fly the mail. They were only able to service twelve of the existing routes. Immediately tragedy struck. By the end of the first week, five pilots had been killed, and a total of twelve army pilots would perish. The press blamed the Roosevelt administration for the deaths, and there was a public outcry. By June 1, the airmail contracts had been returned to the civilian operators. However, the controversy was far from over.

The Air Mail Act of 1934

In response to the allegations of corruption, Congress passed the Air Mail Act of 1934, known as the Black-McKellar Act. This act revised the airmail contract awarding process, redefined the eligibility requirements for companies bidding on contracts, and forced airlines holding airmail contracts to be independent of any other companies involved in aviation. It also separated the airlines from the cartels, or parent companies, that furnished the airlines their financial backing and, in so doing, imposed a major burden on the companies that were awarded airmail contracts. The airlines lost large amounts of money because of these restrictions, and they would continue to do so until the Civil Aeronautics Act of 1938 was passed and the United States prepared to enter World War II.

One of the original goals of the contract airmail system had been to subsidize the fledgling airlines while they developed a profitable passenger transport system. While the rates paid to the airlines for the carriage of mail varied over the years, the cost to the U.S. Post Office had been steadily decreasing. The average cost of the contracts in 1929 was $1.10 per mile, which had decreased to $.54 per mile by the time the contracts were canceled in 1934. From 1938 to 1953, the airmail rates were set by the Civil Aeronautics Board (CAB). After 1953, the CAB paid the subsidies directly to the airlines in the form of what was called a fair charge for the services rendered, rather than a fixed amount based on weight.

Beginning in 1953, the U.S. Post Office began shipping first-class mail by air on a space-available basis on regularly scheduled airline flights. This, in effect, meant that some letters were being shipped by airmail while being charged the cheaper ground rate. By the mid-1970's, the Post Office had begun exploring the feasibility of removing the additional airmail charges. After this was done, the airmail officially ended, with all first-class mail being delivered by the fastest means available. Through the utilization of regional as well as major airlines, the bulk of the intercity first-class mail now travels by air in the cargo compartments of most airline flights. The revenue provided by the mail has been reduced to a relatively minor percentage of airlines' total revenues.

Ronald J. Ferrara

Bibliography

Christy, Joe. *American Aviation: An Illustrated History.* 2d ed. Blue Ridge Summit, Pa.: Tab Aero, 1994. A classic aviation history sourcebook containing many photographs and illustrations.

Glines, Carrol V. *The Saga of the Airmail.* Princeton, N.J.: D. Van Nostrand, 1968. An informative work that traces the evolution of the airmail, with much interesting data, such as early airmail pilot reports.

Holmes, Donald B. *Air Mail: An Illustrated History, 1793-1981.* New York: Clarkson N. Potter, 1981. An interesting overview of the development of the airmail both within the United States and internationally.

Airplanes

Definition: A means of air transportation that is propelled by an internal combustion, turboprop, or jet engine.

Significance: The invention and use of airplanes defined the twentieth century, during which the world witnessed two world wars and the development of aircraft from propeller-driven airplanes to supersonic jets. Each year, airplanes are used around the world for transportation, commerce, and recreation.

Nature and Use

Airplanes fly with the help of the laws of physics and engineering. They come in all shapes and sizes and serve different purposes. Some aircraft are used for training; others are used for transporting goods and freight. Military aircraft are used in waging warfare. Passenger airliners are used for the daily transportation of travelers.

Although airplanes have different designs and functions, all airplanes share common traits. The fuselage, or body of the aircraft, carries people, cargo, and baggage. Attached to the fuselage are the wings, which provide the lift to carry the aircraft and its payload. To balance the airplane in flight, the tail, or empennage, is very important. The landing gear allows the airplane to operate on the earth's surface. The flight controls are used to maneuver the aircraft in flight. Flaps provide additional lift and drag for takeoffs and landings.

Fuselage

The primary job of the fuselage is to provide space for the flight crew and passengers. The attachment of the wings and other load-bearing structures is also an important function of the fuselage.

Depending on the size and function of the aircraft, the fuselage provides a safe haven for those inside the craft. For large airplanes that fly at high altitudes, these compartments are pressurized and air-conditioned. In smaller general aviation airplanes, the cockpits can be drafty, noisy, and either cold or hot, depending on the time of the year.

In airliners, seats are arranged to allow the greatest number of paying passengers to ride inside the fuselage. In older airliners that have been converted to cargo carriers, the fuselage is a cavernous hold without seats in the cabin.

Wings

Wings are as varied as other parts of the airplane. They come in different shapes and sizes, depending on the aircraft's speed and weight requirements. A slower airplane may have a rectangular wing or a tapered wing. A rectangular wing is one in which the chord line, or cross section, of the wing, remains constant from the root of the wing near the body of the aircraft to the wingtip. A tapered wing is one that becomes narrower toward the tip. High-speed aircraft, such as jet transports, airliners, or fighter aircraft, have swept-wing designs. The purpose of the swept wing is to allow the airplane to fly at higher airspeeds.

The size of an airplane's wing in relation to the airplane's size is important. The larger the airplane, the bigger the wing must be to support it in flight. Many factors determine how the wings work in lifting an airplane.

The first factor is that of the wingspan. This is the distance from one wingtip to the other. Small general aviation airplanes typically have wingspans from 35 to 40 feet. Larger airplanes, such as the Boeing 747, have wingspans that easily exceed 100 feet.

The second factor is that of chord. The wing chord is the distance as measured from the leading edge of the wing, or front, to the trailing edge. In a rectangular wing, the chord is constant and, as such, is a constant-chord wing. On tapered, elliptical, or other odd-shaped wings, the chord is not constant. On these wings, the average chord, or mean aerodynamic chord (MAC), is required in equations dealing with the wing.

One important equation in aircraft wing design involves the load the wing will bear while in flight. Wing loading directly relates to the size, or the wing area, of the airplane wing. The first mathematical step in determining wing loading is to determine the wing area by multiplying the wingspan by the chord, or MAC.

After the wing area has been determined, the wing loading can be determined, using the weight of the airplane. The gross weight, or GW, is the operational weight of the airplane. To determine the wing loading of a particular aircraft, the weight of the airplane is divided by wing area. For the lightest of civilian airplanes, wing loading may reach values as low as 6 pounds per square foot, whereas a tactical jet bomber will have a wing loading of more than 375 pounds per square foot.

As the wing flies through the air, it does so at a particular angle. This angle, measured by the relationship between the relative wind and the chord line of the wing, is directly related to the speed of the aircraft. An airplane flying at high airspeeds will have a small angle of attack, whereas one flying slowly will have a large angle of attack.

The lift equation aptly expresses the relationship between the speed, angle of attack, and weight of the aircraft. An airplane's lift must equal its weight in order for the airplane to remain in flight. Pilots are unable to change either the density of the air or the area of the wing. However, they are in control of the other two variables, the airplane's speed and angle of attack.

Because lift must always equal weight in level flight, if the airplane slows down, the angle of attack must increase. Accordingly, an increase in speed will require a decrease in the angle of attack.

Empennage

The empennage is the tail structure of the aircraft, which includes the vertical stabilizer and rudder, along with the horizontal stabilizers and elevator. These essential components provide stability for the airplane in flight.

The vertical stabilizer stands straight up, like a fin, from the aft portion of the airplane's fuselage. It is important to the stability of the aircraft in that it helps the airplane track a straight path. The larger the vertical fin is in area, the more stable the aircraft is around the vertical axis.

Attached to the trailing edge of the vertical stabilizer is the rudder. By way of the pedals at the pilot's feet, the rudder controls movement about the vertical axis of the airplane.

Acting in concert with the vertical stabilizer are the horizontal stabilizers. Located on each side of the fuselage and near the vertical stabilizer, they provide longitudinal stability to the airplane about the craft's lateral axis. The combination of the horizontal stabilizers and elevators re-

Acting in concert with the vertical stabilizer are the horizontal stabilizers. Located on each side of the fuselage and near the vertical stabilizer, they provide longitudinal stability to the airplane about the craft's lateral axis. The combination of the horizontal stabilizers and elevators resembles the main wing in shape. However, the function of the horizontal stabilizers and elevators is totally different from that of the wing. Whereas the wing lifts up in force, the horizontal stabilizer provides a downward force that provides longitudinal stability to the airplane.

Landing Gear

In order to move around on the earth's surface, all aircraft have landing gear. The most common arrangement of the landing gear is the tricycle landing gear, in which the aircraft has two main wheels that extend from either the wing or the fuselage and a third wheel that extends from the nose of the aircraft. The brakes are located on the main wheels, or mains, whereas the steering is the function of the nose gear. Depending on the size and model of the aircraft, nose-gear steering maneuvers the airplane on the ground. The nose wheel can be freewheeling, with the maneuvering done by differential braking. Aircraft steering can be actuated by rods, cables, or hydraulics systems.

Another arrangement of the landing gear is the conventional landing gear, typically seen on older aircraft. In the conventional arrangement, there are two main wheels in the front of the fuselage, with a smaller tailwheel located on the aft end. Conventional gear was the norm in the early period of aviation, but it fell out of fashion in the 1950's and 1960's, because the tricycle landing gear is inherently more stable.

Another type of arrangement, found on the B-52, is the bicycle landing gear, which has two sets of main landing gear centered on the fuselage, one behind the other along the centerline of the fuselage. Because there are no supporting landing gear outside the body of the craft, devices known as outriggers keep the wingtips from striking the ground.

Flight Controls

The flight control system controls the aircraft in flight and comprises the devices that command movement of the aircraft around all three axes: longitudinal, lateral, and vertical.

The elevator controls the airplane's longitudinal movement about its lateral axis. In other words, it causes the airplane's nose to go up or down. In this manner, combined with the power output of the engine, the elevator adjusts

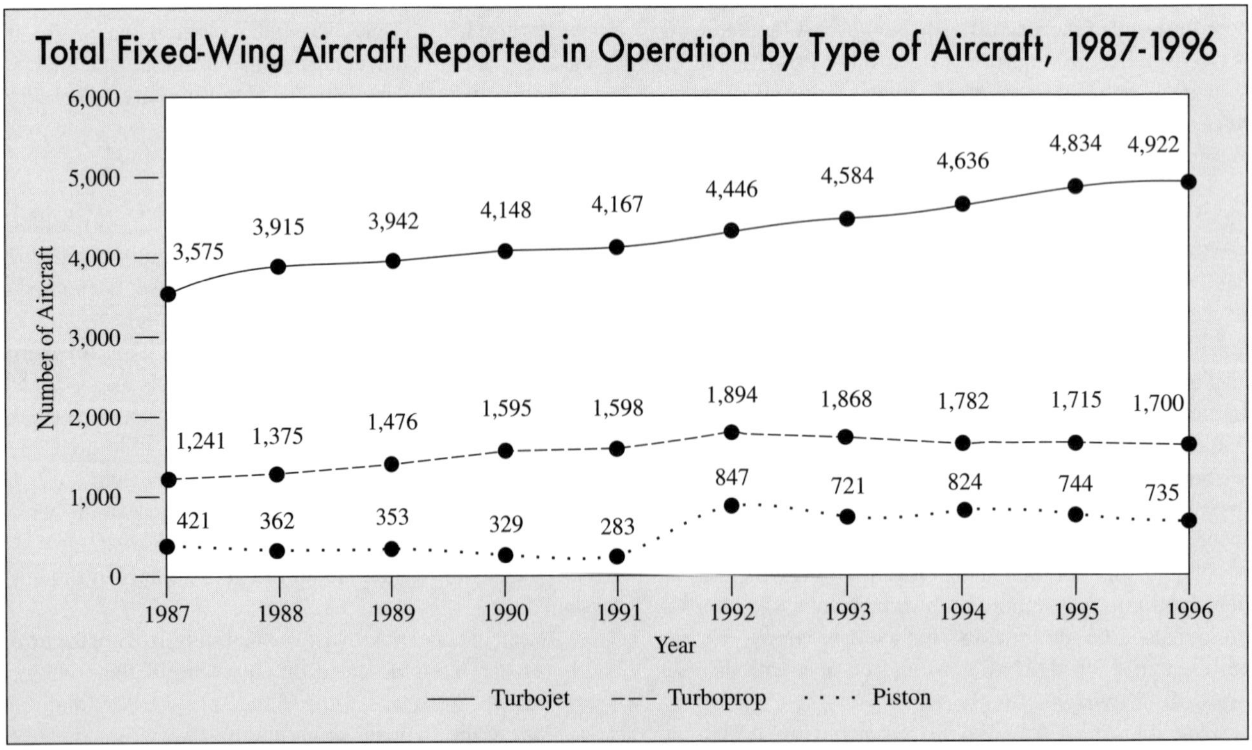

Source: Federal Aviation Administration, *Statistical Handbook of Aviation*, 1996.

opposite aileron spoils the lift on the opposite wing. This starts a rolling movement about the longitudinal axis.

Finally, the rudder controls the airplane about the vertical axis. Actuated by the pedals at the pilot's feet, the trailing edge of the vertical stabilizer moves the airplane's nose either left or right, depending on which pedal is depressed.

Flaps

Airplanes have flaps for both takeoffs and landings. Located on the inboard portion of the wing at the rear, flaps change the shape of the wing in a way that creates both lift and drag. The first half of travel, after takeoff, creates more lift than drag, whereas the last half of travel, before landing, creates more drag without a noticeable increase in lift. With the flaps partially extended for takeoff, the wing will generate more lift at lower airspeeds, allowing for shorter and safer takeoffs. On the other end of the spectrum, an aircraft approaching a landing with full flaps extended is generating more drag. This will allow the pilot to fly a steeper approach, land more slowly, and stop in a shorter distance.

There are four types of flaps: plain flaps, split flaps, slotted flaps, and Fowler flaps. Each has its own characteristics, with the first three found typically on general aviation airplanes. The fourth type, the Fowler flap, is typically found on larger air transports. The Fowler flap system is heavier and more complex than the other three types of flap, necessitating a larger aircraft.

The Power Plant

The internal combustion engine powers many of today's light airplanes. The most popular arrangement of the engine is in the horizontally opposed configuration. The engine is air-cooled and typically arranged in a flat four- or six-cylinder configuration, allowing the best cooling for all of the cylinders.

Unlike the aircraft engines built before World War II, the modern aircraft engine is highly engineered and very reliable. Although modern engines may still fail, the likelihood of complete power loss is minimal.

Most aircraft engines are four-stroke engines, which means each cylinder has an intake stroke, a compression stroke, a power stroke, and, finally, an exhaust stroke. The amount of power the engine puts out depends on the engine's size. Essentially, an engine's power increases with its size.

Aircraft engines come in all sizes, from one of the smallest, the 65-horsepower Continental A-65, to the 350-horsepower TSIO-540. As horsepower requirements reach higher than 350, many aircraft manufacturers opt to equip their high-end models with turboprop engines.

The advantages of a turboprop engine over an internal combustion engine are increased power output, smoother operation, and the ability to operate at higher altitudes. At higher altitudes, a pilot can take advantage of winds that are more favorable and realize better specific fuel consumption. Typical cruise speeds for airplanes equipped with turboprops are in the 230-knot to 350-knot range. For faster cruise speeds, a jet engine is required.

Jet engines are very simple devices. The thrust of a jet engine is determined in pounds of force rather than in horsepower, as are reciprocating engines and turboprop power plants.

The heart of the jet engine is the compressor and turbine. Linked together by a common shaft, the turbine and compressor spin at rates as high as 20,000 revolutions per minute. As the fuel and air mixture burns in the combustion chamber, the exhaust gases escape through the turbine, spinning it at high speeds. The turbine, by way of the common shaft, spins the compressor, ingesting more air into the engine. The potential power available from a jet engine is phenomenal.

More phenomenal than jet engines are future engine possibilities. Presently under development, the Stirling engine may be the most significant innovation for aviation in the near future. The Stirling engine, an external combustion engine originally invented in 1896, is on the verge of becoming the power plant of choice not only for airplanes but also for cars, boats, and many other applications.

Types of Airplanes

There are as many airplanes as there are reasons for their existence. Small, privately owned aircraft such as Cessna, Beechcraft, and Piper aircraft are used for transportation and recreation. Most privately owned airplanes are single-engine, one- or two-seaters that have a range of about 400 miles and a speed of 100 miles per hour. Higher-end privately owned airplanes are turboprops and jets that are rather expensive to acquire and maintain. Powered by two engines, these more complex airplanes require more training and certification for their operation than do smaller craft. The turboprops are capable of 230- to 275-miles-per-hour cruise speeds, whereas some privately owned jet aircraft can reach a speed of 600 miles per hour.

The cost of the privately owned aircraft varies. In 2001, small two-seaters in flying condition could be purchased by bargain hunters for less than $12,000. Such airplanes are rudimentary but capable of flight and cost effective for flight training.

At the same time, the cost of a typical four-seater family airplane began at $25,000 for an older, used craft and could reach as high as $150,000 for a new model. Still more sophisticated models could cost as much as $500,000. The smaller light twin-engine craft cost $50,000 to $75,000, on the low end of the market for used craft. On the high side, a newer aircraft cost as much as $750,000.

In 2001, the cost of turboprop aircraft and small corporate jets started at well over the $1 million mark. Depending on the make and model of the corporate jet, the cost can reach as high as $40 or $50 million. Typically used in commercial endeavors, these aircraft are a strain for one or two individual owners to manage financially.

The airliner, the type of airplane with which most passengers are familiar, flies at high speeds and altitudes. Smaller commuter airliners carry an average of fifty passengers and a crew of five or six, including the two pilots. As the airline industry moves into the twenty-first century, there is a desire to move away from the turboprop aircraft of the 1980's and 1990's, as passengers prefer the smoother, higher, and seemingly safer ride of jet aircraft.

The final category of aircraft is military aircraft. The armed forces use different types of airplanes for different jobs. The task of protecting the nation from intruders falls to fighter planes, jets that can fly at almost twice the speed of sound. Fighters carry one or two crew members, and their mission is to stop any unannounced intruder into national airspace. The military branches also operate airline-type aircraft to move personnel and cargo throughout the world.

Joseph F. Clark III

Bibliography

Bergman, Jules. *Anyone Can Fly.* Garden City, N.Y.: Doubleday, 1977. An outstanding and easy-to-understand explanation of aviation written for the beginner.

Langewiesche, Wolfgang. *Stick and Rudder: An Explanation of the Art of Flying.* New York: McGraw-Hill, 1972. Hailed as the most important book on aviation, this classic text explains basic principles of flight in a simple manner.

Stinton, Darrol. *The Design of the Aeroplane: Which Describes Common-sense Mechanics of Design as They Affect the Flying Qualities of Aeroplanes Needing Only One Pilot.* New York: Van Nostrand Reinhold, 1983. Outstanding text relating aircraft design to flying qualities; written in a technical format, with a great deal of mathematical explanation.

Van Sickle, Neil D. *Van Sickle's Modern Airmanship.* New York: McGraw-Hill, 1999. A technical work about flying and the aviation industry that extensively covers all aspects of the business.

Airport security

Definition: The use of technology and well-trained staff to protect aircraft, aircrews, and passengers from terrorist attacks, to prevent cargo theft, and to solve illegal ticket problems.

Significance: Airport security seeks to protect aircraft, carrier personnel, and passengers from acts of terrorism, prevent cargo theft, and solve ticket-associated problems such as theft and black (or gray) market sale.

Airport security includes the protection of aircraft and the people on board from terrorist attacks, the prevention of cargo theft, and ticket-associated problems such as theft and black (or gray) market sale. The most visible aspect of this security operation is the protection of aircraft from hijackings and bombings perpetrated by the world's international terrorists. In the United States, all airport security activities are overseen by the Federal Aviation Administration (FAA).

In the late 1960's, a series of aircraft hijackings alerted the U.S. government to the need for the active implementation of airport security methods and an enumeration of security requirements for U.S. air carriers. The FAA initiated an anti-hijacking program in late 1970. At that time, President Richard M. Nixon ordered air carriers to use surveillance equipment in all U.S. airports. He also instructed the Departments of Defense and of Transportation to collaborate with air carriers in identifying the utility of metal detectors and X-ray equipment—already used by U.S. armed forces—in prevention of hijackings.

Basics of Passenger Screening

By early 1972, with the utility of tested detection technology deemed probable, the FAA ruled that all air carriers must use an FAA-approved passenger-screen system to check all aircraft passengers through behavior profiles, metal and other detection methods, physical searches, and identification checks. In late 1972, in the face of continued hijackings, the FAA required the screening of all passengers and their carry-on baggage entering commercial aircraft. Hijacking and sabotage of U.S. aircraft also led to treaties between nations and regulations that established the current U.S commercial aviation security system. Air carriers, airports, and the FAA each play specific roles in assuring this security. Especially important to continued and expanded passenger screening development was the 1988 bombing of Pan American Flight 103 over Lockerbie, Scotland. This led to the creation of a Presidential Commission on Airline Security and Terrorism in 1989 and the enactment of the Aviation Security Improvement Act of 1990.

Air carriers, airports, and the FAA participate in the passenger-screen system. The air carriers provide secure travel conditions; maintain security programs; screen passengers, cargo, and carry-on and other baggage; and protect their aircraft. In turn, the airports provide safe aircraft-operating environments, develop and maintain sound security programs, and provide airport law-enforcement personnel. The FAA, for its part, is expected to furnish both the administrative and procedural guidance to identify and analyze threats, establish requirements for the activities of air carriers and airports, coordinate all crisis situations, enforce regulations, and supply any needed technical assistance.

Ensuring the safe commercial air travel, not only of domestic carriers within the United States but also of U.S. and foreign carriers leaving and entering the country, is a huge task. Currently, about two million passengers and all of their luggage are screened daily for metal weapons and other dangerous materials. The anti-hijacking and screening program currently used requires each air carrier to have a passenger-screen system capable of preventing entry of weapons, explosives, and other threat objects into the passenger compartments of its aircraft. Since this effort began in 1972, X-ray and metal-detection systems as well as the selection, training, and testing of security personnel have improved. It could be argued that the mere presence of screening systems increases aircraft security by acting as a deterrent to crime and terrorism.

The FAA continually investigates new means to enhance screening technology that detects metal weapons and other dangers, such as plastic explosives. The ideal passenger screen would quickly detect these threats with high sensitivity and very few false alarms. It would also expose, prior to their boarding aircraft, any passengers carrying weapons or explosive devices that might be used to frighten flight crews into changing the aircraft's destination.

However, the acceptance or rejection of a given screen item by airport operators, air carriers, and passengers and flight crews is just as important as is screen performance. Judgments must be made as to whether a screen harms, injures, or otherwise endangers people; violates Fourth Amendment rights; allows air carriers to screen passengers quickly enough to maintain flight schedules; or makes people believe that their privacy is being invaded or reveals too much personal information. A considerable effort is being exercised to address these problems while introducing items that optimize passenger-screen systems.

Screening Procedures

Carry-on baggage and other luggage is analyzed by X-ray

X Ray Origins and Uses

X rays are forms of penetrating electromagnetic radiation with shorter wavelengths and higher energy than visible light. X rays were discovered in 1895 by the German physicist Wilhelm Conrad Röntgen, who was studying cathode rays in a gas-discharge tube. Röntgen noticed that, although the tube was inside a black box, a nearby barium-platinocyanide screen emitted fluorescent light whenever the tube was in operation. He found that the observed fluorescence was due to an invisible radiation, more penetrating than ultraviolet light.

The first X-ray tube was the Crookes tube, a partially evacuated glass bulb holding two electrodes. When electric current is passed through the tube, the gas inside is ionized; the positive ions thus produced strike the cathode, causing ejection of fast-moving electrons. The electrons bombard the glass walls of the Crookes tube and produce low-energy X rays. Improvements culminated with the development of Coolidge tubes, commonly used in a variety of X-ray detectors.

These detectors began with photographic emulsions, which are affected by X rays in the same way they are affected by visible light. A substance's absorption of X rays depends upon its density and atomic weight. The lower a material's atomic weight, the more transparent it is to X rays of given wavelengths. Thus, when a human body is X-rayed, the bones, at a higher atomic weight than flesh, absorb X rays more effectively and cast darker shadows on photographic plates. X rays also cause fluorescence in materials such as barium platinocyanide and zinc sulfide. When a screen coated with such material is substituted for photographic film in a technique known as fluoroscopy, the structures of opaque objects, such as flesh, bones, and tumors, or guns and explosives, may be directly observed.

Such detectors, which vary in their ability to discriminate between screened objects, can be interfaced with computers to produce the X-ray scanners used by hospitals, customs inspectors, and airport security. The danger involved in using X rays to screen living beings arises from radiation's potential to cause cancer. To minimize such health risks, the X-ray scanners used in airports utilize low-energy (soft) X rays and exposures of just a few seconds.

analysis, a process called active screening. This procedure is routinely performed on baggage that is not carried in an aircraft's passenger compartment. In contrast, routine preboarding screens occur in two concurrent phases. First, all passengers place their carry-on baggage onto a conveyor belt for inspection by X-ray equipment. Then, they walk through a portal that detects metal objects. If the portal alarm sounds, the passengers involved are searched more completely to ensure that they are not carrying any threatening object. Alarm-clearing searches use handheld metal detectors and physical pat-downs. However, these procedures do not detect all possible nonmetal weapons, explosives, and other threat objects.

Few airports have routine passenger screens that can operate at the highest possible level of technology for the detection of threatening objects, where imaging shows both the body and objects carried beneath clothing. Even the most advanced systems do not produce images of photograph quality. Screen operators must view and interpret the images, and when they perceive threats, they can, together with airport police, body-search passengers. Sound operator judgment and decision-making ability are crucial to screening success, as inaccuracy can cause the passenger-screen system to fail either by missing threatening objects or by excessive numbers of false alarms.

Even assuming the best screening possible, the imaging of passengers is a complex process. Legitimately, passengers may fear health risks from X rays and also may be unwilling to have others view their bodies. An alternative process, less-often used and more expensive, is called passive screening, which analyzes natural body radiation. Both kinds of body screening produce interpretable images of threat objects, including metal and nonmetal weapons and explosives. It is important that at any screen site, people who object to a procedure, such as one displaying an image of the naked body, should be given the option of another method, such as hand-wand screening. Such options may be costly, but will alleviate legitimate concerns based on nationality, religion, and personal beliefs.

In addition to physical screening, other screening occurs behind the scenes. The FAA integrates data from intelligence agencies, air carriers, and airport surveillance crews to identify immediate threats to given aircraft. Whenever hijackers or terrorist targets are identified, a higher alert level is invoked and screening imposes additional procedures to assure timely terrorist detection and detention. Included are much more stringent baggage inspection, passenger questioning, and identification checks. The routine screen alone is used only where the probability of bombing or hijacking is deemed to be minimal.

A security system must be both effective and suitable. Effectiveness is measured by the ability to detect threat objects and depends upon system technology capability. Suitability is measured by system operation without undesired characteristics, such as excessive radiation. Performance is evaluated continually to provide feedback that will allow both the air carriers and equipment manufacturers to improve the systems. The minimum acceptable performance level is compliance with the standards set by the FAA. Evaluations test the system's ability to detect and react to a terrorist threat or action, often by using federal teams carrying mock threat items into the field.

Passenger Screen Operators

The importance of passenger-screen system personnel and the quality of their performance cannot be overemphasized. These individuals are responsible for accurately determining potential passenger-related danger to aircraft, aircrews, and passengers. The FAA has formulated criteria for their selection, training, and motivation. Many critics frequently assume that poor passenger-screen system performance is due to low wages and that higher wages alone would improve performance levels.

Although it may be argued that wage increases are always useful, other factors are more responsible for imperfect personnel performance. These relate to finding more valid modes of selecting operators, better training methods, monitoring and analyzing effectiveness of passenger-screen systems and personnel, and providing feedback to the individuals involved. Training is increasingly important as screening systems evolve, because personnel performance involves very complex tasks. Accelerated human factors programs are suggested as one overall means to aid personnel choice, training, motivation, and provide data on ergonomics of screening equipment.

Lapses in Security

On September 11, 2001, the importance of airport security became tragically apparent. On that morning, four commercial planes—two 767's out of Boston's Logan Airport and 757's out of Washington's Dulles International Airport and Newark, New Jersey—were hijacked by teams of terrorists armed with box cutters. The hijackers gained access to the cockpits, either killed or incapacitated the flight crews, switched off the transponders, and took over the controls. American Airlines Flight 11 and United Air Lines Flight 175 were each flown into the Twin Towers of the World Trade Center in New York City, causing both buildings to collapse a short time later. American Flight 77 was crashed into the Pentagon in Washington, D.C. The cess to the cockpits, either killed or incapacitated the flight crews, switched off the transponders, and took over the controls. American Airlines Flight 11 and United Air Lines Flight 175 were each flown into the Twin Towers of the World Trade Center in New York City, causing both buildings to collapse a short time later. American Flight 77 was crashed into the Pentagon in Washington, D.C. The fourth hijacked plane, United Flight 93, went down in a field near Pittsburgh, Pennsylvania, when passengers in touch with loved ones by phone realized the nature of their situation and stormed the cockpit. In total, more than three thousand people, from the planes and on the ground, died that day.

Investigations in the aftermath of the tragedy revealed many lapses in the security system at U.S. airports. As a result, an aviation security bill was signed into law on November 19, 2001. Within a year, all screening was to be done by federal employees, U.S. citizens having undergone criminal background checks. After three years, airports meeting federal standards could request that private contractors handle screening, based on the findings of pilot programs at five airports of varying size. Airports were given until the end of 2002 to install explosives detection X-ray systems for checked bags. Until then, all checked bags were to be inspected by X rays, passenger matching, or hand checking. The Computer Assisted Passenger Prescreening System (CAPPS) would be used to screen all passengers, and a database would allow cross-checking with government watchlists. Flight deck doors were to be strengthened and kept locked during flights. Pilots and crew would attend training courses on dealing with hijackers, and the Transportation Department could authorize cockpit weapons. Flight schools were to conduct background checks on foreign nationals seeking instruction in operating large aircraft. The bill created a Transportation Department agency called the Transportation Security Administration to supervise security issues.

There was an immediate expansion of the Federal Air Marshal (FAM) program. FAMs respond to criminal incidents and other in-flight emergencies on board U.S. aircraft. They are authorized to carry firearms and to make arrests in order to preserve the safety of aircraft, crew, and passengers. On September 28, 2001, an FAA amendment raised the maximum age requirement from thirty-seven to forty years of age.

Following the attacks, new regulations were put in force. Travelers were advised to allow extra time to check in, to bring government-issued identification, and to board with only one carry-on bag and one personal bag, such as a purse or briefcase. Curbside checking of luggage was banned, and several airports curtailed traffic to and from the terminals.

Air Cargo Theft

Aircraft have carried freight since airmail began in 1918, but it was not until the late 1950's that other air freight exceeded airmail as a revenue source for air carriers. Large air freighters can carry heavy loads, such as automobiles and cattle. However, the fast transportation of packages weighing under fifty pounds is also an important air carrier revenue source. The speed of aircraft makes them ideal for transporting perishable items such as flowers, fruits, and vegetables and enables manufacturers to get merchandise to destinations quickly, thus raising their profits.

A number of unique security problems are associated with carrying air cargo. Cargo often contains more expensive items than those shipped by other freight-carrying methods; hence, the potential for loss is greater. It is also more difficult to identify where losses occur. In other methods of shipment, items are simply picked up, moved, and delivered to loading docks. Air cargo movement is much more complex: cargo is first moved from freight terminals to flight terminals, then loaded onto freight aircraft before shipping, with opportunities for theft all along the way. When freight is placed on a passenger airplane, risk is increased because it must go to a passenger terminal and is exposed to additional handlers. At many airports, carts travel to and from flights along unlit routes, creating still more opportunities for theft. Moreover, 90 percent of air cargo is shipped at night, the time period when most crime occurs. Pilferage, fraudulent pickups, and theft by drivers also occur, as in other freight-carrying operations.

Air cargo theft problems first surfaced in the 1960's in New York, when racketeers infiltrated the pickup and delivery segment of the air freight industry. Eventually, the U.S. Department of Justice indicted the racketeers and corrupt truckers. These occurrences and some spectacular thefts of valuable items led carriers to establish an Airport Security Council in 1968. However, by 1970 it became clear that terminal facilities and carrier cargo handling were inadequate, losses were not reported systematically, and there was no good way of telling where losses occurred. By 1988, airlines and shippers at New York's John F. Kennedy Airport and in Los Angeles together reported $11 million of freight stolen per year and it became clear that air cargo theft was growing.

The FAA, Department of Transportation, the Treasury Department, the air carriers, and airports have taken combined action, along with legislation, to protect air cargo more effectively. Airport security personnel are trained more efforts must be made before it can be completely eradicated.

Ticket Theft and Fraud

Another aspect of airport security concerns airline tickets. Air carriers need to make a profit on the sale of tickets to survive and to subscribe to other aspects of airport security. The prices for equivalent legitimate airline tickets differ by carrier, depending on travel class, payment date, travel day, duration of stay at the destination, and other factors. Often, people sitting next to each other on a plane have paid very different ticket prices for the same flight. Everyone wants the lowest fare possible, so many passengers buy tickets from gray (black) market vendors. The most risky sources of such tickets are personal newspaper ads, because some offer stolen tickets. Furthermore, some travel agencies sell both legitimate and stolen tickets.

Computers contribute greatly to the validation of tickets. For example, the airlines enter into their reservations systems the numbers of all known stolen tickets. Ticket agents can enter the ticket number into a database to check its legitimacy and receive an immediate response. Tickets are also checked by computer after flight departure. If a stolen ticket is discovered, a phone call to the destination allows airline personnel to meet the flight, interview the passenger, and determine where and how the ticket was obtained. Because stolen tickets are so easy to sell, travel agencies are often burglarized and robbed. This causes danger to agency personnel and an average loss of $1,000 to the air carrier for each stolen ticket.

Another source of stolen or fraudulent tickets arises from frequent flier miles, programs begun in the early 1980's to reward airline brand loyalty with free flights in return for miles traveled on a carrier's aircraft. The programs are quite lucrative for air carriers, but they also attract criminals who see frequent flier ticket certificates (coupons) as a way to make dishonest money. It took some time before air carriers began to realize that frequent flier fraud cost them billions of dollars in decreased revenues. To solve the problem, the programs have been modified so that unused mileage expires after a certain date. Seats used for the program have also been reduced in number, so that passengers need to reserve months ahead of time to use the mileage. Although these measures diminish illegal coupon brokerage, it still occurs. Brokers have established prices they will pay for specific frequent flier distance awards. In some cases, employees of firms that handle frequent flier accounting for air carriers have set up fictitious accounts and sold the resultant certificates to brokers. There have also been several federal prosecutions of travel agents for frequent flier fraud. Efforts to prevent this type of illegal operation are an ongoing part of airport security.

Sanford S. Singer

Bibliography

Baldeschweiler, John D. *Determination of Explosives for Commercial Aviation Security.* Washington, D.C.: National Academy Press, 1996. Describes systems considerations, testing protocols, staff, and performance criteria to detect explosives in a timely way, maximizing commercial aviation safety.

Moore, Kenneth C. *Airport, Aircraft, and Airline Security.* 2d ed. Boston: Butterworth-Heinemann, 1991. Deals with topics in airport security including aircraft hijacking, cargo, and ticketing security.

Swenson, George, Jr. *Airline Passenger Security Screening: New Technologies and Implementation Issues.* Washington, D.C.: National Academy Press, 1996. Covers screening technology, screen operation and costs, operator selection, training and motivation, and legal issues.

Tsacoumis, Theodolfus P., ed. *Access Security Screening: Challenges and Solutions.* Philadelphia: American Society for Testing and Materials, 1992. Describes many aspects of weapons, explosives, and X-ray detection systems useful to aviation security.

Wilkinson, Paul, and Brian M. Jenkins. *Aviation Terrorism and Security.* Portland, Oreg.: Frank Cass, 1999. Describes issues such as attacks on civil aviation, trends and lessons, politics of aviation terrorism, and international aviation organizations.

Airports

Also known as: Aerodromes, airfields, landing strips

Definition: An area of land that provides for the taking off, landing, and surface maneuvering of aircraft.

Significance: Although airports mark the beginning and ending points of aircraft flights, they are more than mere runways or grass areas for takeoffs and landings. Airports are facilities that provide for the maintenance and servicing of aircraft, serve as exchange points for passengers and cargo, and host the various navigational aids used by pilots to guide an aircraft in flight.

Nature and Use

An airport is defined by the type of aircraft it serves and by where it is located. Airports range in size from large commercial air carrier airports, such as Chicago's O'Hare International Airport, with enplanements, or paid boardings, of more than 30 million passengers per year, to small, privately owned grass landing strips in rural areas with landings of only a few small aircraft each year. In the United States, there are about 15,000 airport landing facilities, only 5,000 of which are open to the public. Even fewer, about 3,000, are served by commercial air 3carrier service. The other airports are small, general aviation airports in private or public ownership.

An airport serves as the transition and exchange point for passengers and cargo between air and ground transportation. Therefore, an airport's operations include the buildings and facilities that support the transition and exchange of services. Aircraft and passenger facilities often associated with the landing facilities are maintenance, passenger terminal, cargo, fueling, parking, and hangar-storage facilities.

An airport is typically a facility that handles propeller- and jet-driven fixed-wing aircraft. In some countries, the definition for an airport can include landing areas other than on land. Specific areas on rivers and waterways are known as seaports or sealanes. A facility specifically used by helicopters is known as a heliport or helipad. Vertical takeoff and landing (VTOL) aircraft can operate out of special short takeoff and landing (STOL) facilities, heliports, or regular airports. If designed to do so, airports also have the ability to handle helicopters, airships, hot-air balloons, and ultralights.

Types of Airports

Although airports may be classified in a number of different ways, the broadest categories are general aviation and commercial service airports. General aviation airports are those that do not receive regularly scheduled passenger service but rather have a primary purpose of serving the aviation interests and needs of small or outlying communities. General aviation includes such activities as corporate and business transportation, recreational flying, aircraft instruction and rental, aerial application, aerial observation, skydiving activities, and other special uses.

Commercial service airports are those that receive scheduled passenger service. These airports can be further classified into large-hub, medium-hub, small-hub, or nonhub airports. The different classifications reflect the number of enplaned passengers boarding aircraft annually at the airport. A large-hub airport will normally have more than

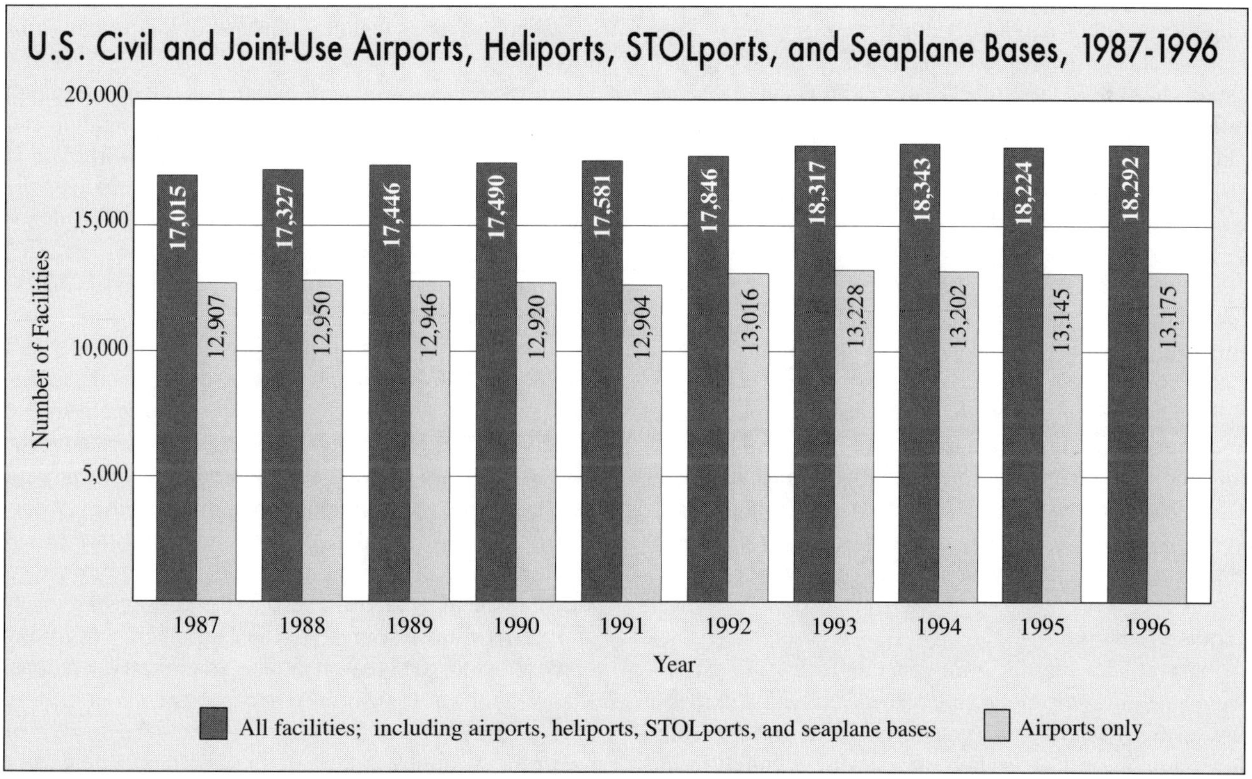

Source: Federal Aviation Administration, *Statistical Handbook of Aviation*, 1996.

five million enplanements, a medium-hub airport more than one million, a small-hub airport more than one-quarter million, and a nonhub airport fewer than one-quarter million.

The term "hub" has more than one meaning in air transportation. For instance, an air traffic hub refers not to an airport, but to the geographic and demographic characteristics of a community. A large air-traffic-hub airport would be associated with a large city from which many people have access to the air transportation system. A medium air-traffic-hub airport would similarly be associated with a medium-sized city, and so forth.

The term hub is also used to describe an airline route structure. An airline hub operation is one where a large number of an airline's flights converge from distant airports to exchange passengers and then return to those same airports. Because the operation resembles the spokes and the hub of a wheel, it has come to be known as a hub-and-spoke operation. The airports and cities at the end of the spoke served by the main hub airport are commonly known as origination or termination airports, because the majority of those airport's passenger enplanements originate from the local community.

Planning and Design

To ensure that airports are constructed with adequate safety parameters, nations develop their own guidelines in conformance with the International Civil Aviation Organization (ICAO). In the United States, ICAO guidelines are reflected in specific airport design criteria established by the Federal Aviation Administration (FAA). The design criteria are based upon the operational speed and overall size of the aircraft intended to use the airport.

An airport is normally developed according to an airport master plan. A master plan is an overall concept of the long-term development of an airport and serves as a guide for developing the physical facilities of an airport. It takes into account the environmental effects of airport construction and operations, ground-access needs, and economic and financial feasibility. It outlines schedules for the prioritization and phasing of airport improvements. The plan must also conform to the design standards of the federal government. A major product of airport master planning efforts are the airport layout plans and drawings.

Landing Facilities

An airport's landing facilities generally consist of a runway or landing strip along with related taxiways and parking areas. A runway is a graded or paved area suitable for the taking off or landing of aircraft. Although most runways in developed nations serving small to large commercial aircraft are paved, there are still many airports that are either grass or dirt strips. These types of landing strips usually serve small piston- or turbine-engine aircraft in rural or undeveloped areas of a country or in developing nations.

Runways

In the early days of aviation, dirt and grass runways were the norm. They tended to be wide open field areas that allowed pilots to take off and land in whichever direction the wind was blowing. This is because aircraft weighed relatively little and needed only a short distance to take off.

As aircraft and pavement technology developed and the weight of aircraft increased, the need for longer and stronger runway surfaces emerged. The previously open fields were soon developed into graded areas oriented in the direction of the prevailing winds. These graded areas were then paved. If strong winds occasionally blew from a direction different to that of the paved runway, crosswind runways might also be graded and paved.

Aircraft are designed to land into the wind. When winds blow from a different direction than the orientation of the primary runway, some aircraft are unable to handle the side forces of the wind when landing or taking off. A secondary crosswind runway built to accommodate the occasional crosswind is then used instead of the primary runway.

An airport's runway configuration is often based upon one of four basic layouts: single, V-shaped, parallel, and intersecting. The many airports built during World War II were arranged in a triangular pattern to accommodate the various wind directions. They were also built to accommodate aircraft that only needed 5,000 feet of land to take off. The introduction of jet aircraft required runways in excess of 7,000 feet.

Along with jet aircraft and increasing passenger and cargo usage came the demand for larger terminal and service buildings, which had to be built within the old airport fence line. These demands shaped many modern airports, where one or two runways were extended and others were closed down to accommodate building facilities. Additional runways of up to 12,000 feet in length and oriented parallel to the others might be added to accommodate larger and heavier aircraft, such as the Boeing 747.

Parallel runways, preferred in large, newly built airports, allow for the greatest number of aircraft operations. Many airports are constrained in capacity by their runway configuration. The runways cannot handle the demand of the aircraft waiting to use them at any particular time without some delay.

The configuration of runways is primarily determined by the prevailing wind direction and the type and volume of expected aircraft activity. Other considerations also affect the layout, including the availability of airspace around an airport to accommodate safe aircraft approaches, departures, and landing patterns; environmental conditions; weather conditions; topography; and the availability of land.

Any paved runway or landing strip is really two runways in one, because each can be used for landing or takeoff in two different directions. Runways are numbered according to the compass heading of the direction in which they are oriented. The numbers are rounded up or down to the nearest whole number and the last-digit zero is dropped. Taxiways are identified by a letter of the alphabet and spoken using the aeronautical alphabet, such as Taxiway Alpha or Taxiway Bravo.

Landing and takeoff surfaces at an airport fall under two general categories: flexible or rigid. Flexible pavements, such as asphalt, dirt, or grass, tend to compress under an aircraft load, whereas rigid pavement, such as cement or concrete, resists compression. The type of pavement used at an airport is determined by the weight of the aircraft expected to use it, and by other factors, such as the expected useful life, anticipated wear characteristics, cost of construction, and exposure to climatic effects.

Because aircraft are affected by the wind and are subject to human inaccuracies in flight and on the ground, safety areas are established in and around an airport to accommodate the safe passage of aircraft. Safety and protection areas help to ensure that the potential for aircraft collision is minimized. They also help prevent serious damage to an aircraft should it go off a runway or taxiway, by allowing it to come to a safe stop without hitting an obstruction.

Within the vicinity of an airport, construction of buildings or towers may be prohibited or restricted in height. Tree heights are also controlled. This is intended to help ensure the structure does not interfere with aircraft in flight and that adequate margins of safety exist.

Taxiways. Taxiways connect the ends of the runways to the main parking and building areas. They are defined pathways used by aircraft to travel on the surface of the airport from one point to another point. Stub taxiways connect the middle portion of a runway with a primary taxiway or taxi route. Stub taxiways can be set at right angles to the runway or set at other angles, in which case they are known as high-speed turnoff taxiways.

Taxilanes are routes that lead from the main taxiway into a parking or terminal area. The design of taxiways and taxilanes must take into account the wingspan and weight of the expected aircraft, so that collisions do not occur with other aircraft or buildings.

Aprons, Ramps, and Parking Stands. Aircraft parking areas are often called ramps, aprons, parking stands, or tie-downs. The various terms are often interchangeable and depend upon local usage. Technically, a ramp is a transition area from a taxiway to an apron, stand, tie-down area, or hangar. Aprons and parking stands are designed for the parking of aircraft, provide access to airport terminal facilities, and allow for the performance of aircraft services. They also accommodate ground vehicle activity.

A tie-down area designates a parking area for primarily general aviation aircraft. Ropes or chains embedded in the ground or pavement are used to secure the aircraft.

All parking areas are structurally designed to accommodate the parking and maneuvering of aircraft. They accommodate the different kinds of ground services, such as fueling, baggage handling, and deicing, provided to the aircraft. Also considered in the design are such things as provisions for electrical, pneumatic, hydraulic, water and lavatory services, access by emergency vehicles, security restrictions, and protection from jet and propeller blast.

Airfield Marking and Signs

On the airport runways, taxiways, and other areas where aircraft operate, special pavement markings help a pilot navigate properly. Runway markings are painted white, whereas taxiway and parking area markings are painted yellow.

The runway markings differ for the different types of navigational equipment used for landing. The intent behind the different markings is to help increase the visual cues a pilot receives as the weather conditions worsen. Markings on a runway identify whether the runway can be used solely by visual reference or with precision or nonprecision instruments in the aircraft.

The pavement surfaces generally have centerlines painted on them to help guide the pilot. Edge markings may also exist. To prevent pilots from mistakenly entering onto a runway from a taxiway, hold-short or stop lines are painted perpendicularly to the taxiway's centerline. A runway's centerline is dashed, whereas a taxiway's centerline is solid.

To further help guide a pilot to a destination on the airport or to distinguish a critical safety marking, signs to the side of the pavement surface accompany the pavement markings. Most signs are internally lighted for enhanced visibility.

In the twentieth century, some airports were icons of modernist architecture. The "Theme Building" at Los Angeles International Airport is an example of this type of space age design. (AP/Wide World Photos)

Airfield Lighting

Runways, taxiways, and parking areas are also identified at night by the color of lights used to outline them. The runway and taxiway lights outline the perimeter of the pavement. A runway will have primarily white lights, although green, amber (yellow), and red lights are found near the ends of a runway to clearly identify for the pilot the runway ends, or thresholds. Taxiway and parking areas are bordered by blue lights. Lights embedded into the pavement, called in-pavement lights, provide an added visual cue to assist a pilot.

Navigational Aids

An airport employs a number of different navigational aids to assist pilots in making successful landings, whether the weather is clear or poor. Some of the navigational aids found on an airport are very high frequency omnirange (VOR), localizer (LOC) and glide slope (GS) transmitters, visual approach slope indicators (VASI), Global Positioning Systems (GPS), wind cones, rotating beacon systems, and approach light systems (ALS).

When instrument meteorological conditions (IMC) require pilots to navigate solely through the use of instruments in the airplane instead of outside visual references, the navigational aids align pilots with the runway and help guide the planes down.

An approach light system assists the pilot in making the transition from the instruments in the cockpit to the runway. Newer technology has advanced such that properly equipped airports and aircraft can land in zero-zero conditions, in which there is no visibility either forward or up or down, and the pilot cannot see the lights or runway markings.

The rotating beacon for a land airport is intended to help a pilot easily identify the general location of an airport. Its alternatingly flashing green and white light helps pilots to locate the airport from among all the other lights found in or near a city. The rotating beacon can also be used to communicate with a pilot who has lost radio contact.

Other Terminal Facilities

Various buildings exist on an airport to serve and accommodate aircraft. The number and type are dependent upon the activity level of the airport, and the kind of aircraft that use it. At a large commercial air carrier airport, the passenger terminal building is the primary structure. The terminal building is designed to serve the needs of the passengers, airlines, and businesses that use it or operate within it. The primary purpose of an airport terminal is to transfer passengers and baggage between surface and air transportation with a minimum of time, confusion, and inconvenience.

The location and design of a terminal is determined by a number of factors, such as the configuration of the runway layout, access to a ground transportation network, future expansion capabilities, design requirements established by the FAA, the surrounding terrain, and environmental impact.

he terminal itself must be able to accommodate the various demands placed upon it by business, international, and leisure travelers and by nonpassengers, such as meeters and greeters, employees, and delivery personnel. All of these people have different needs. International travelers, for instance, are subject to customs, immigration, and security requirements to which local business or leisure travelers are not.

Within a large airport terminal building, there may be all types of businesses, such as car rentals, game rooms, restaurants, and retail stores. Some large airport terminals look like small shopping malls, with stores selling clothing, books, gifts, flowers, and other specialty items.

There are many other buildings situated on an airport. A service center for aircraft is called a fixed-base operator (FBO). An FBO can provide any number of aircraft services, such as fueling, maintenance, avionics, pilot lounges, weather planning, flight training, aircraft and pilot-supply sales, aircraft rental, and sightseeing, charter or air taxi flights.

Also found on airports are hangars and other types of aircraft storage facilities, aviation-related businesses, cargo terminal buildings, fuel storage tanks, airfield maintenance buildings, airport administrative offices, and manufacturing facilities. Many large airports actually resemble small cities, with their own electric power generating plants, wastewater treatment facilities, roadway system, parking facilities, and police and fire rescue buildings.

Stephen M. Quilty

Bibliography

Gesell, Laurence E. *The Administration of Public Airports*. 4th ed. Chandler, Ariz.: Coast Aire, 1999. A comprehensive textbook that provides substantive detail on the many aspects of airport management, economics, planning, operation, and liability.

Richardson, J. D., J. F. Rodwell, and P. Baty. *Essentials of Aviation Management*. 5th ed. Dubuque, Iowa: Kendall/Hunt, 1995. An excellent text for reference and overview of the operations of general aviation airports and fixed-based service operations.

Wells, Alexander T. *Airport Planning & Management*. 4th ed. Blue Ridge Summit, Pa.: Tab Books, 2000. A good general-purpose text that covers the various aspects involved in developing and operating an airport.

Alitalia

Also known as: Alitalia Linee Aeree Italiane S.p.A.

Definition: Italy's national airline, one of the leading European airlines since the 1940's.

Significance: Alitalia is the fifth-largest airline in Europe, with service across Europe, North America, Africa, Australia, the Middle East, and Asia.

History

Alitalia was started on May 5, 1947, transporting passengers and cargo from Turin, Italy, to Rome. In 1957, Alitalia merged with Linee Aeree Italiane (LAI), creating Italy's national airline. In 1960, Alitalia had the honor of being the official carrier for the Rome Olympic Games. The airline expanded its service in 1970 with the addition of an Italy-to-North America route, eventually expanding to include airports in New York, Miami, Boston, Chicago, Los Angeles, San Francisco, and Toronto.

With the opening of the Malpensa International Airport in Milan in 1998, Alitalia increased its service to include passenger flights between Milan and Rome's Leonardo da Vinci International Airport. As one of Italy's largest airlines, Alitalia operates in 133 cities in 63 countries across Europe, Africa, Australia, the Middle East, and Asia. It offers three classes, Magnifica, PrimaBusiness, and Economy. Alitalia participates in frequent flier programs with Continental and US Airways. The Italian government owns 53 percent of the airline.

Financial Trouble

In 1997, the European Union authorized three installments of aid by Italian authorities when Alitalia developed financial difficulty. In May, 1999, Alitalia made plans to join the Northwest/KLM Transatlantic Joint Venture. The three signed a commercial cooperation and integration agreement and an alliance coordination agreement.

At the same time, Italy was in the process of building a new airport in Milan: Malpensa International Airport. The new facility was intended for use as an important hub for Northern Italy. The European Commission recommended that Alitalia move from its current site at Linate International Airport to Malpensa. According to a study done by Solomon Smith Barney, the move was critical to Alitalia's operation in order for it to increase its share of the Northern Italian traffic to at least 50 percent.

Several European carriers protested this move, claiming that it would give Alitalia an unfair competitive advantage over the other airlines, as well as creating air traffic control problems. The Italian government temporarily delayed Alitalia's move. Because of these delays, KLM called off the merger in May, 2000. In August, 2000, Alitalia filed suit for compensation for breach of contract.

To further complicate matters, Alitalia was subjected to two strikes. The first was in October, 2000, by flight attendants and ground workers as part of an ongoing pay dispute. The airline was forced to cancel over two hundred flights between Milan and Rome. In March, 2001, the flight attendants and air traffic controllers threatened to strike.

In an attempt to recoup its losses and renew its reputation after the merger disaster with KLM, Alitalia investigated alliances with Sky France and Swissair. Code-share agreements were formed with other airlines, such as Qantas, Japan Airlines, and Malaysian Airlines.

Safety Record

Since 1960, Alitalia has experienced nine crashes. The worst crash was on May 5, 1972, when an aircraft crashed upon approach in Sicily, Italy. All 115 persons aboard were killed. The most recent crash was on December 17, 1991, in Warsaw, Poland. There were no fatalities in this incident. Since its beginning in 1947, there have been 425 fatalities on Alitalia flights.

Other Interests

Alitalia is the parent company of Alitalia Group, which is made up of twelve companies involved in air travel and related operations. The airline has a long-standing tradition of sponsorship and support of the arts. Displays of contemporary art can be seen in Alitalia airport lounges in Rome, Milan, and New York. Alitalia also supports the promotion of Italian traditions, and contributed to the restoration of the Upper Basilica of St. Francis in Assissi in 1998 and 1999.

Ulisse 2000 and *Arrivederci* are publications produced by Alitalia for its passengers. *Ulisse 2000* is published monthly for passengers on international and intercontinental flights and offers articles on fashion, celebrities, science, nature, people and places. *Arrivederci* is also published monthly for passengers on domestic flights and focuses each month on a particular region or town.

Alitalia's business school provides training for its managers. There is also a program for individuals who are reluctant to fly. In an effort to provide continuous improvement and customer satisfaction, Alitalia signed a training service agreement with CAE, Inc. Effective from 2001 to 2011, Alitalia will be able to utilize training devices installed at Flumicino Airport in Rome. The ultimate goal of this training center is to increase penetration into the commercial flight training market. The airline also signed a deal with Sextant In-Flight Systems for video-on-demand systems in five of its 747-400's. In addition, Alitalia signed an agreement with Mercury Air Cargo to provide cargo handling through Los Angeles International Airport.

Events in Alitalia History

1947: Alitalia makes its first flight, from Turin to Rome.
1948: Alitalia makes its first intercontinental flight, flying a thirty-six hour, Milan-to-Rome-to-Dakar-to Natal-to-Rio de Janeiro-to-São Paulo-Buenos Aires route.
1950: The airline begins employing stewardesses, who are dressed in designer uniforms.
1951: The airline begins offering hot meals on its flights.
1957: Alitalia merges with Linee Aeree Italiane (LAI) to form a national Italian airline.
1960: The airline begins jet aircraft flights, with the Caravelle and the Douglas DC-8/43.
1967: Alitalia introduces Arco, a new electronic booking system.
1969: The airline retires its last remaining turboprop aircraft, becoming the first European airline with an all-jet fleet.
1970: Alitalia employs its first B-747 jumbo jet on flights to North America.
1972: An Alitalia DC-8 crashes into a mountain near Palermo, Sicily, killing 115.
1984: The airline employs its first MD-80's for use on medium-length flights.
1998: Alitalia enters into an alliance with KLM Royal Dutch Airlines to expand its service; the alliance is terminated in 2000.

In April, 2000, Alitalia signed a deal with McDonald's in which the airlines sold advertising space on one of its aircraft. Since then, Alitalia has signed similar deals with chocolate maker Perugina, luxury goods brand Bulgari, and auto maker Renault. Alitalia's logo has even been adjusted to fit into the color scheme of the ad. The ads were priced at approximately $460,000 per plane for one year.

Other interests of Alitalia Group include airline travel services, automated ticketing services, air fire-fighting services and repair, maintenance, and overhaul operations.

Maryanne Barsotti

Bibliography

Flint, Perry. "Roman Holiday: For Alitalia, the Last Two Years Have Been Anything but a Vacation." *Air Transport World* 28, no. 9 (September, 1991): 22-26. An industry news article on Alitalia's management in the early 1990's.

Hill, Leonard. "Roman Remake: Resurgent Alitalia Hopes for Continued Profits on Low-Cost TEAM Airline Subsidiary and an Integrated Hub System with Strategic Ally KLM." *Air Transport World* 35, no. 12 (December, 1998): 37-38, 41, 70. A news article discussing Alitalia's plans—later quashed—to form an alliance with KLM.

Lyth, Peter J., and Hans-Liudger Deinel, eds. *Flying the Flag: European Commercial Air Transport Since 1945.* New York: St. Martin's, 1998. A comparative analysis of seven European national airlines, including Alitalia. The book focuses on how flag-carrier airlines have survived trends toward globalization and strategic alliances.

Altitude

Definition: A measured or calibrated height above the ground or above sea level.

Significance: Pilots use indicated altitude to maintain height separation from other aircraft and from ground obstructions. However, the actual, or true, altitude is usually not the same as the indicated altitude.

Indicated Altitude

The standard aircraft altimeter is an aneroid (without liquid) barometer that measures the ambient or static air pressure outside the airplane. It is calibrated through the use of a Standard Atmosphere model so that it presents this pressure to the pilot as an altitude. Because the air pressure on the ground varies a great deal with the movement of air masses across the country, an offset can be introduced by the pilot to make the indicated altitude equal to the actual altitude of an airport before takeoff and while approaching to land. The offset, if any, is indicated by the reading in a window, known as the Kollsman window, on the face of the altimeter.

The Standard Atmosphere

The Standard Atmosphere model is based on an arbitrarily chosen, midlatitude, average value for the pressure, temperature, humidity, and density of the air at sea level. It assumes a sea-level pressure of 29.92 inches (76.00 centimeters) of mercury, a sea-level temperature of 59 degrees Fahrenheit (15 degrees Celsius), 0 percent humidity, and an air density calculated from the ideal gas law. It further assumes that the temperature decreases linearly with an increase in altitude at the rate of 3.566 degrees Fahrenheit for every 1,000 feet for the first 36,000 feet, the troposphere, and then is constant, the stratosphere. These assumptions, along with the gravitational and thermodynamic laws, yield the Standard Atmosphere, uniquely defining the "standard" air pressure, density, and temperature at every altitude.

It should be noted that pressures expressed as a height of mercury are not using true pressure units but are reflecting a common way to measure pressures. An accurate mercury thermometer can be made by bending a 6-foot-long glass tube into the shape of an upright "U," filling it half-full of mercury, and attaching a vacuum pump to one end. Atmospheric pressure at the other end then pushes the mercury down, and the difference in mercury heights is a direct measure of the atmospheric pressure. Pressure expressed in inches of mercury can be converted to pressure expressed in units of pounds per square inch (psi) by multiplying by 70.73.

Pressure Altitude

Because the altimeter is calibrated in feet of altitude, or in kilometers in Europe and elsewhere, but is really measuring only atmospheric pressure, the actual air pressure can be obtained by consulting tables of the Standard Atmosphere. This is important because the performance of engines and airplanes depends directly on the air density, and the only way to determine air density is to calculate it from the measured air pressure and temperature, using the ideal gas law. Pilots adjust their altimeters so that 29.92 appears in the Kollsman window—that is, with no offset from the Standard Atmosphere—and the indicated altitude is then called the pressure altitude.

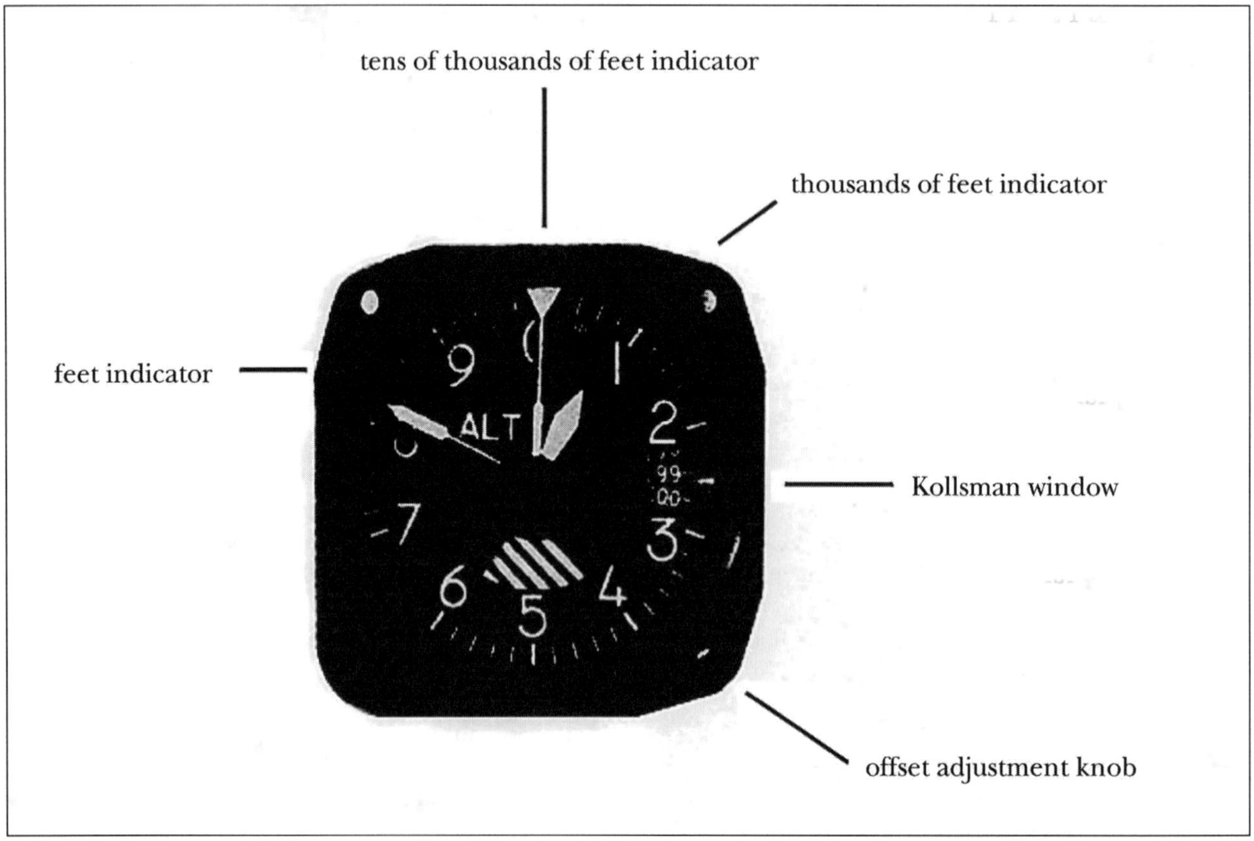

An altimeter measures an aircraft's altitude in feet, thousands of feet, and tens of thousands of feet.

Whenever a pilot is flying above an indicated altitude of 18,000 feet, the altimeter must be set to 29.92 (no offset) to simplify vertical separation of aircraft. All aircraft flying in busy airspace are also required to use transponders that report both position and pressure altitude to air traffic control.

Density Altitude

When the pressure altitude is combined with the outside air temperature through the ideal gas law, the density of the air can be calculated. It is most convenient to express this air density in terms of the altitude in the Standard Atmosphere, which is defined to have this density. This calculation is called the density altitude. The performance of airplanes, in terms of the available engine thrust or power, takeoff distance, climb rate, cruise speeds, and landing distances, depends directly on density altitude, or air density, and is specified as such in aircraft flight manuals. It is very helpful for a pilot to realize intuitively that low pressures (especially due to high elevations) and high temperatures result in very low air density (high-density altitudes) and that in high-density altitudes, aircraft performance will be greatly reduced from sea-level values.

Density altitude is easily calculated from the pressure altitude and the temperature using either an E6-B circular slide rule or an electronic calculator. The current density altitude is also broadcast at many high-altitude airports.

True Altitude

When approaching to land, it is important to enter the appropriate offset, or altimeter setting, into the altimeter, so that it will both give guidance regarding obstacle clearance on the approach and read field elevation after landing. However, usually the variation of pressure with altitude above the airport will not follow the Standard Atmosphere model, and the indicated altitudes of obstructions will still not exactly equal their true altitudes.

Air pressure varies more rapidly with altitude if the air is colder than that assumed by the Standard Atmosphere. Therefore, if a pilot flies into air that is colder than standard, or if the altimeter has not been adjusted en route while flying toward a region of lower pressures, the airplane's true altitude is lower than the indicated altitude, and safety may be compromised, especially in mountainous terrain. Unstable weather conditions and high winds around mountain ranges can also produce locally lower air pressures that result in erroneously high indicated altitudes.

Other Altitude-measuring Instruments

The radar altimeter measures an aircraft's height above the ground by measuring the time it takes a radio wave to return, using the known speed of light. Transport and other complex aircraft find the radar altimeter to be a valuable aid in avoiding obstructions during the landing approach.

Altitude can also be derived by geometry from three or more satellites in the Global Positioning System (GPS) used for navigation. This may become the preferred altimeter for high-altitude, oceanic flight.

W. N. Hubin

Bibliography

Barnard, R. H., and D. R. Philpott. *Aircraft Flight.* 2d ed. Essex, England: Addison-Wesley Longman, 1995. An excellent, nonmathematical text on aeronautics. Well-done illustrations and physical descriptions, rather than equations, are used to explain all aspects of flight.

Gleim, Irvin N. *Federal Air Regulations and Aeronautical Information Manual.* Gainesville, Fla.: Gleim Publications, 1999. A republication of official Federal Aviation Administration information for pilots, with discussion of altimeter errors and setting procedures.

U.S. National Oceanic and Atmospheric Administration. *U.S. Standard Atmosphere, 1976.* Washington, D.C.: U.S. National Oceanic and Atmospheric Administration, 1976. Covers the basics for computation of atmospheric properties, the elemental constituents of the atmosphere, and tables of atmospheric properties to 1,000 kilometers.

American Airlines

Definition: A large U.S. air carrier of passengers and cargo based in Dallas-Fort Worth, Texas.

Significance: American Airlines is considered to be the largest airline in the world based on total revenue generated in a given year.

Early Years

It is not surprising that an airline with a history going back to the airmail days of Charles A. Lindbergh would become the largest airline in the world. The ride has not always been a smooth one. Like the industry itself, American Airlines has had its ups and downs. Along the way, American Airlines and the people who ran it have helped to change the face of aviation, introducing many of the ideas and innovations that have shaped the way the airline industry does business.

On April 15, 1926, Charles A. Lindbergh, chief pilot for the Robertson Aircraft Corporation of Missouri, flew an airplane containing a bag of mail from Chicago, Illinois, to St. Louis, Missouri. Later that day, he and two other pilots flew three aircraft full of mail from St. Louis back to Chicago. Robertson Aircraft Corporation became the second company to receive a contract to fly U.S. mail and was one of nearly eighty companies that were eventually consolidated to form American Airlines. Lindbergh went on to make aviation history in May, 1927, with the first solo flight across the Atlantic.

In 1929, the Aviation Corporation was formed to begin acquiring newly developing aviation companies in the United States, including Robertson Aircraft Corporation. The airline subsidiary of the Aviation Corporation was called American Airways. The name was changed in 1934 to American Airlines, and Cyrus Rowlett Smith, more commonly known as C. R. Smith, was named president of the company. Smith was largely responsible for molding the dozens of small, poorly financed airlines purchased by the Aviation Corporation into a dominant U.S. carrier. He also became one of the leaders of the aviation industry itself and helped to develop executives who would later manage many other U.S. airlines.

Smith continued as the chief executive of American Airlines until 1968, when he was named secretary of commerce by President Lyndon B. Johnson. Under Smith's leadership, in 1932 American Airlines became the first airline to introduce a coupon program targeted at business travelers. The coupon books were sold to businesses and could be redeemed for discounted ticket purchase. In 1936, American became the first airline to fly the new Douglas DC-3. The introduction of "Admirals Clubs" VIP lounges in airports was another American first, aimed at attracting the frequent-traveling business passenger. By 1940, American Airlines had become the leading domestic air carrier in terms of revenue passenger miles.

Like many of the other large U.S. carriers, American Airlines helped the war effort during World War II by using its fleet to transport men and supplies for the Air Transport Command. Nearly half of American's fleet was eventually placed into action, along with their crews. C. R. Smith also left the airline temporarily to serve during the war. However, the company did not neglect its domestic operation. In 1942, the company established Sky Chefs, an airline catering subsidiary that remains one of the largest airline caterers in the world. It also introduced the first scheduled cargo services in 1944.

After the war, American established another first in 1949 by becoming the only airline in the United States with an entire fleet of pressurized passenger aircraft. These aircraft allowed American Airlines pilots to fly at higher altitudes, improving gas mileage and avoiding inclement weather. In 1952, American introduced the Magnetronic Reservisor to keep track of seats available on its flights. Before introducing the Reservisor, the airline had kept track of available seats through a large display board posted in each reservation office. As passenger volume expanded, this system was simply unable to keep up with demand. The Reservisor used a matrix of relays in which shorting plugs could be manually inserted to indicate a sellout. Agents worked at electrical keypads into which they keyed their requests for seating. If a light began blinking, the request was accepted. This system, designed by Teleregister Corporation, marked the first time any airline had attempted to use an electronic device to handle reservations. The Reservisor was replaced in 1960 by a new electronic system called the Semi-Automated Business Research Environment (SABRE). The SABRE system was developed in conjunction with IBM and ran off two IBM mainframe computers. The system, which cost almost $40 million, was capable of processing 84,000 telephone calls per day. By 1964, the SABRE network extended from coast to coast. Meanwhile, American Airlines introduced the first nonstop, transcontinental service in 1957 and became the first U.S. airline to fly the Lockheed Electra, a U.S.-designed turboprop aircraft, and the Convair 990, a fanjet-powered airplane.

End of an Era

With the retirement of C. R. Smith, George Spater assumed the leadership of American Airlines. Spater, a lawyer and the former general counsel of American Airlines, did not prove to be equal to the challenge of running the airline. The years between 1968 and 1973, when Spater was replaced, are considered one of the low points in American's history. The airline's profits began declining in 1968 and it posted a loss of $48 million in 1973, one of its worst years ever. Part of the problem was that American began receiving the first of its newly ordered Boeing 747's, the largest passenger plane built, at a time when air traffic volume began a sharp decline. For the airline industry as a whole, there were too many seats and not enough passengers. Although the Boeing 747 was ideal for long-haul international flights, it did not suit the shorter domestic routes of American. A second major problem was the oil crisis of the early 1970's, which sent jet fuel prices soaring. To add to American's troubles, several of its executives were charged with taking kickbacks from American suppliers. A final blow came when Spater himself resigned in 1973 after admitting that he had authorized an illegal contribution to the 1972 Committee to Re-Elect the President, Richard M. Nixon.

In February, 1974, Albert V. Casey, a former executive of the Times Mirror Company, was elected president and chief executive officer after receiving the endorsement of C. R. Smith, who had temporarily agreed to run the airline after Spater's resignation. Although Casey had no airline experience, he proved to be an excellent judge of managerial talent and a man willing to make hard choices. Casey went on to sell all of American's Boeing 747's and the Americana chain of hotels owned by the company. These actions, combined with the flight schedule reduction precipitated by the Arab oil embargo of 1973 and the fare increases authorized by the Civil Aeronautics Board, allowed American to post a $20.4 million profit in Casey's first year. However, to keep American profitable, Casey needed a long-term strategic vision for the company. One of the people who helped him shape that vision was Robert L. Crandall. Bob Crandall had left Trans World Airlines (TWA) after being passed over for the job of chief financial officer. Crandall had initially worked in American's finance department but was transferred to marketing, where he moved quickly to revive the SABRE system, which had suffered from underinvestment during the Spater years. American began marketing the SABRE system to travel agents in 1975. By 1976, the system had been installed in 130 locations, including most of the top travel agencies. In 1977, American became the first airline to offer a restricted discount fare, the Super Saver. Other airlines soon adopted this new type of fare, which offered substantial savings to passengers willing to book early and stay over for a specified period of time before returning.

Following the deregulation of the airline industry in 1978, American Airlines began a major expansion of its route structure throughout the United States and the Caribbean. It began to retire its older, less fuel-efficient aircraft. The company headquarters were moved to Dallas-Fort Worth, which became one of American's hubs in 1981. Meanwhile, in 1980 Robert Crandall was elected president and chief operating officer of the company. Crandall then became the chairman and chief executive officer in 1984 on the retirement of Al Casey.

Events in American Airlines History

1930: The Aviation Corporation, a holding company acquiring small aviation companies, is incorporated into American Airways.
1934: American Airways is renamed American Airlines.
1936: American Airlines becomes the first in the nation to fly the Douglas DC-3 for passenger service.
1940: American becomes the first in the nation in revenue passenger miles flown.
1942: Sky Chefs, an American Airlines subsidiary, begins airline catering operations.
1944: American introduces regularly scheduled freight service, the first in the United States.
1949: American becomes the first airline with a fleet consisting entirely of postwar pressurized aircraft.
1959: American becomes the first airline to offer transcontinental jet service, with the Boeing 747.
1960's: With IBM, American develops the Semi-Automated Business Research Environment (SABRE), a real-time data processing system that allows agents to track flight reservations.
1974: American introduces one-stop automated check-in service.
1977: American introduces its popular Super Saver fare on routes from New York and California.
1979: The airline moves its headquarters from New York City to Dallas/Fort Worth, Texas.
1981: American introduces its revolutionary AAAdvantage frequent-flyer award program and establishes its first hub at Dallas/Fort Worth, strengthening its hub-and-spoke network throughout the 1980's.
1984: American introduces its regional American Eagle network and retires its freight fleet.
1994: American institutes nonsmoking transatlantic service.
1999: American joins the oneworld Alliance, a global network of airlines, to greatly expand customer service.
2001: American acquires TWA.
2001: On September 11, in one of the worst acts of terrorism in world history, two American Airlines jetliners are hijacked; one is flown into the World Trade Center in New York City and one into the Pentagon in Washington, D.C. Thousands are killed. On November 12, in an unrelated incident, yet another American Airlines jet crashes, within moments of takeoff from New York's JFK Airport.

The Crandall Years

Under Crandall, a new structure was created for the airline. A holding company, AMR Corporation, was created in 1982 and became the parent of American Airlines. A second subsidiary, AMR Services, was formed in 1983 to provide aviation services to other airlines. The AMR Consulting Group was created in 1992. This group was joined in 1993 by the SABRE Technology Group, which become its own separate company in 2000.

Crandall and American were busy on other fronts as well, creating the American Eagle network and a host of new fares, including the Ultimate Super Saver and the Senior SAAVers Club. The American Eagle network of regional carriers linked small-to-medium-sized cities with the American network, while the new fares appealed to an ever-wider customer base. The American Airlines network of routes was also increasing. By 1994, American Airlines had more flights to London than any other U.S. carrier. Unfortunately, this fact worked against the airline when it announced its plan to form an alliance with British Airways. Crandall, who had once opposed alliances, had decided to bow to the inevitable. If other carriers were going to form alliances, then American would as well. In fact, American would do it better. In 1994, American Airlines reported having eight alliances in *Airline Business* magazine's first yearly alliance survey. Five years later, this number had grown to twenty-eight alliances. American sought antitrust immunity in the United States, which would have allowed American and British Airways to cooperate more closely, particularly in the area of harmonizing price. The fact that the two carriers would control the majority of the transatlantic traffic and the landing slots at London Heathrow caused both the U.S. government and the European Commission to reject the immunity. The carriers have continued in a more limited alliance, called oneworld Alliance, which also includes Japan Airlines and Cathay Pacific.

When Crandall retired in 1998, he left his successor, Donald J. Carty, a much stronger company that he himself had inherited. Carty had joined the company in 1978 to help resolve financial problems with the Americana hotels. After taking over, Carty expanded the American network through two acquisitions. The first occurred in 1998, when Reno Air, a regional carrier based in Reno, Nevada, was added to the American family. Reno Air brought to American an excellent network in the western states, particularly California. It also brought a brief strike from American Airlines pilots concerned with the pay differences between Reno Air and American pilots, as well as the merging of pilot seniority lists. In 2001, American purchased TWA's assets, which included TWA's 180 jets, its St. Louis hub, and its twenty thousand employees.

By 2001, the airline that began as a collection of small, underfunded air carriers in the 1930's was serving over fifty countries with more than 4,100 flights per day. The SABRE system it created had gone on to become a very successful company in its own right, breaking new ground when it introduced Travelocity.com, the first World Wide Web site to offer travel reservations over the Internet. The airline was tested, however, by the terrorist attacks of September 11, 2001, in which two of the hijacked planes were from American's fleet. Like all airlines, American was hurt financially but helped by the emergency aid package signed by President George W. Bush that month.

Dawna L. Rhoades

Bibliography

Bedwell, D., and J. Wegg. *Silverbird: The American Airlines Story.* Boston: Plymouth Press, 2000. The most up-to-date book on the history of American Airlines.

Braznell, W. *An Airman's Odyssey: Walt Braznell and the Pilots He Led into the Jet Age.* St. Louis: University of Missouri Press, 2001. An excellent account of the early years at American Airlines and the changes wrought by the introduction of jets.

Jones, G. *The Big Six Airlines.* Osceola, Wis.: Motorbooks International, 2001. A pictorial history of the largest U.S. carriers.

Reed, D. *The American Eagle.* New York: St. Martin's Press, 1993. A fascinating account of the Bob Crandall years at American Airlines.

Animal flight

Definition: Sustained and powered airborne travel by birds, insects, or mammals through the use of wings.

Significance: Animal flight, particularly that of birds, is important to humans, who first learned and dreamt of flight by studying flying animals. The study of animal flight remains a source of information for understanding and design of flying vehicles.

History

Animals have been flying for millions of years. The first flying animals were insects, which appeared approximately 350 million years ago. From that time, flight evolved separately among three other kinds of animals. There are four types of animals capable of flight: insects, birds, bats, and pterosaurs, the last of which are extinct. Each of these groups developed the ability to fly independently, and, in many cases, different species in each group separately evolved the ability to fly. Additionally, some mammals and birds developed the ability to glide but not to fly.

Unlike aircraft, which gain lift with wings that are either fixed or rotating, animals almost universally accomplish flight by flapping their wings. The flapping motion provides not only lift but also thrust and is referred to as ornithoptic propulsion. Both animals and manufactured aircraft using this method of achieving flight are commonly called ornithopters.

Basis of Animal Flight

The same aerodynamic laws that apply to man-made aircraft also apply to animals, and animal flight is divided into three categories, based on how it is attained. Gliding animals do not fly but trade potential energy (height) for kinetic energy (speed) to remain aloft. Gliding is only useful for small distances. Flying animals use their wings to generate both lift and thrust to remain in the air. Soaring animals, a cross between gliders and fliers, usually use wing movement only for takeoffs and landings, generally relying on subtle changes in wing geometry, thermals, and prevailing winds to gain altitude. Many large birds soar rather than fly. For an animal to remain in level and steady flight, the lift that it generates with its wings must be equal to its weight, whereas the thrust it creates must be equal to its aerodynamic drag. All flying animals generate both lift and thrust by the same method: flapping their wings.

In flapping, or ornithoptic, flight, the wing must produce lift and thrust at the same time. However, lift and thrust does not have to be produced constantly. During a single wing beat, lift and thrust vary. As long as the average lift and drag over the period of the wing motion are equal to the drag and weight, respectively, this will keep the animal in level and steady flight over time.

Aerodynamics

The primary difference in the aerodynamics of aircraft and animal flight is the slower speed and smaller size of flying animals, compared to that of manufactured aircraft. This difference is characterized by a parameter called the Reynolds number, which measures the effect of aerodynamic inertial forces compared to aerodynamic viscous, or frictional, forces. The lower the Reynolds number is, the more important the effects of fluid viscosity, or friction, become. The Reynolds value of most aircraft, whether general aviation craft, commercial airliners, or fighters, is in the millions. For birds and insects, however, the

Reynolds value is usually 100,000 or less and is sometimes even less than 1,000 for very small insects. For flying obects with a Reynolds value greater than 100,000, thick, curved, or cambered, airfoils work best, whereas those with Reynolds values of less than 100,000 tend to work best with thinner, flatter airfoils.

This difference is demonstrated by examining the value of the lift-to-drag ratio as a function of Reynolds number for a number of given airfoils. Most fat airfoils have a higher lift-to-drag ratio at high Reynolds numbers, whereas thin airfoils have a higher lift-to-drag ratio at low Reynolds numbers. This fact was originally discovered during World War I by the Germans, who determined that fat wings worked better on their faster biplane fighters. Likewise, as an animal's speed and size increases, the shape of its wings changes to reflect the increase in Reynolds number. Thus, large or fast birds, such as pigeons and falcons, have wing cross sections that look surprisingly similar to those of modern aircraft wings.

Another important aspect of speed or Reynolds number is how the roughness of a wing affects flight efficiency. The faster an object flies, the smoother the wing needs to be for maximum lift and minimum drag. At low Reynolds numbers, however, the lift dramatically drops for smooth wings, whereas it does not for rough wings. Thus, smooth manufactured wings do not operate as efficiently as rough wings, whether the animal wings are roughened by feathers, scales, or fur. Rough animal wings are most efficient for low speeds. The motion of the feathers and fur allows animals to sense when their wings are about to stall.

Whereas avian biomechanics are complex, insect biomechanics are relatively simple and easy to analyze. This simplicity lends itself well to duplication using modern mechanical technology. In the simplest of insect wings, wing motion is controlled by contraction of interior muscles. The motion in this case is indirect; other insect systems have a direct relationship between muscle movement and wing motion. Without examining complex muscle mechanics, however, one can quickly determine the limit to a flying animal's size by examining the weight in relation to the size.

Scaling determines whether ornithoptic propulsion is efficient for a given weight and length scale. The length scale is a measure of an animal's size, in either length or wingspan. A flier's weight is proportional to the length scale cubed, whereas the wing area is proportional to length scale squared. This is known as the cube-square law. Thus, one can deduce that the wing loading (weight divided by wing area) is proportional to the length scale. As size increases, the wing loading must also increase. Eventually, the wing loading will be too great for the bones and muscles of an animal to withstand, and any animal above this size will be unable to fly.

The required power or energy input for a given weight can be determined from commonly known aerodynamic relations and can be shown to increase as the 7/2 power of the length scale. Thus, if the wingspan of an animal doubles, the power required to fly must increase by more than ten times. Based on muscle-mass arguments stating that the amount of energy available is related to the amount of muscle mass, the power available to flying animals can be shown to quadruple as the wingspan doubles. Thus, as the size of a flying animal increases, required power will soon overtake available power, not only limiting the animal's maximum possible weight but also decreasing the animal's ability to take off, climb, and hover. Hence, larger flying animals tend to use soaring as the primary flight mode instead of powered flapping.

The ratio of unsteady lift, derived from flapping, compared to steady lift, derived from forward motion, shows that flapping frequency can be directly related to the size and weight of a flying animal. Using the flapping frequency as an approximate measure of this ratio and comparing it with the flier's length scale, it is shown that the frequency is inversely proportional to this length scale, which can also be related to the Reynolds number. Thus, as the Reynolds number increases, or as the speed or size of a flier increases, the frequency at which the wings flap decreases. Eventually, the flapping frequency will decrease to the point where the wings will be stationary, indicating that there is a limit to the efficiency of flapping as a flight mechanism, as size increases. On the contrary, as weight decreases, there is a limit below which flapping is a very efficient flight mode. This principle has direct applications to the development of manufactured microaerial vehicles: Instead of shrinking down conventional aircraft designs to a smaller scale, it may be more practical to design miniature aircraft that use flapping wings instead of fixed wings for lift generation and engine-propeller combinations for thrust generation. The efficiency of flapping flight for small birds may also be one reason why wings evolved over propellers for thrust generation. As the size of a flying animal decreases, the generation of unsteady lift becomes more important to its flight. This is especially true for insects that derive much of their lift from unsteady effects alone.

Bird Flight

Birds are by far the best-known animal fliers. There are more than 9,000 species of birds, of which only a handful, such as the penguin, kiwi, ostrich, and emu, are flightless.

Birds are characterized as warm-blooded, egg-laying vertebrates with feathered wings and strong hollow bones, many of which are fused together to increase strength and decrease weight. They have powerful muscles that allow for flight and require large amounts of food for energy. Birds evolved from dinosaurs approximately 150 million years ago.

Most birds appear to have evolved flight from ground-up gliding, used both to catch prey and to evade predators. Wings may also have developed as an aid to increase leaping distances and as a display to attract mates. Two scenarios for the evolution of flight include the ground-up scenario, in which running and leaping animals evolved wings, and the tree-down scenario, in which tree-dwelling creatures evolved wings to move from tree to tree. In either case, the ability to survive and gain access to unoccupied niches appear to be the greatest reasons for the development of bird flight.

Although it appears that modern birds evolved from dinosaurs, birds are not related to the now-extinct pterosaurs, or flying archosaurian reptiles. Pterosaurs were lizards and appeared to be proficient fliers with wing structures, similar to those of bats, that had an outstretched membrane over a thin upper limb. They had large heads that may have assisted their flight stability. Their wingspans ranged from a few inches to almost 40 feet. The pteranodon had a wingspan of up to 25 feet but weighed only 25 pounds. Due to their large sizes, most pterosaurs were probably soaring animals that relied on thermals to fly at high altitudes.

The oldest known bird is the archaeopteryx, named for the Greek "ancient wing," which lived around 150 million years ago. It had a wingspan of approximately 18 inches and weighed about 1 pound. With its feathers and beak, it had similarities to modern birds, and with its teeth and clawed wings, it had similarities to dinosaurs.

There is a wide variety of flying birds, including the small hovering hummingbird, the swift falcon, and the lumbering condor. Each adopted a mode of flight suited to its evolutionary niche. There are several differences between flying and flightless birds that illustrate requirements for successful bird flight. Flightless birds tend to have shorter, symmetrical wings, whereas flying birds have long, cambered wings that produce substantially more lift. To keep their weight low, flying birds tend to have fewer feathers than their grounded counterparts. Flying birds also have longer tails, or keels, that aid in flight stability.

Birds occupy almost every low-speed flight niche known. They are adept fliers, using their wing muscles and feathers to control the distribution of lift over the wings. This allows them to easily adjust to changes in ambient flight conditions, such as gusts or downdrafts. Their whole bodies are designed for flight. They have strong, hollow bones that minimize weight and withstand impacts. They have unique single-path pulmonary systems that constantly feed fresh oxygen to the lungs to maximize energy. They use their heads, tails, and feet to help control flight.

Variations across the bird species detail how well designed birds are for their particular niches. The hummingbird, for example, is well known for its ability to hover in one place in flight, beating its wings at an amazing 60 or more beats per second while feeding on the nectar of flowers. Although other birds, such as kestrels, terns, and gulls, can also hover, only hummingbirds can fly sideways and backward in hovering flight.

Insects

Insects are both the oldest and generally smallest of flying animals. The first winged insects appeared some 350 million years ago and were the first creatures to fly on Earth. There are one million species of insects, many of which fly. They range in size from barely visible to almost 1 foot in wingspan.

Insects are invertebrate arthropods with a hard exoskeleton and a three-part body consisting of head, thorax, and abdomen, three pairs of jointed legs, and two antennae. The legs and wings are attached to the thorax. Most winged insects have two sets of wings, fore and aft. Most flap their wings in synch, whereas a few, such as the dragonfly, flap their fore and aft wings asynchronously. In the former case, synchronous wing movement appears to be limited to approximately 200 beats per second, because the wing motion is related directly to the nerve inputs to the muscles. For asynchronously flapping winged insects, beat frequencies of more than 1,000 beats per second have been recorded, because the myogenic flight muscles used in asynchronous wing motion can contract more than once per nerve impulse.

Most insects cannot fly using the laws of conventional aerodynamics. Under these assumptions, lift is determined by the steady flow of air over a wing just as in aircraft flight. The wing areas of most insects are too small to obtain the required lift at their measured flight speeds, however. Much of their lift is instead derived from unsteady lift as described above. For insects, the clap-and-fling effect is used to generate the required lift. In this method, the wings are beaten together (clap) and rapidly pulled apart (fling). The air rushing in to fill the void develops a fast-moving vortex over the top of the wing that generates a large amount of unsteady lift. Insects must beat their wings rapidly and repeatedly to generate lift. Bees flap their wings more than 100 times per second. The common housefly beats its wings more than 20,000 times per minute, or about 300 times per second. A midge of the genus *Forcipomyia* has a measured wing-beat frequency of more than 1,000 beats per second.

The wings of most insects are less flexible than those of birds or bats. Most insects change direction and speed primarily by altering the motion and frequency of their wing beats. Pitch, yaw, and roll control involve changes in wing-beat amplitude on one wing with respect to the other, lateral wing twisting, or leg and abdominal movement. Some insects can twist their wings like those of a bird to control motion, such that a large area is projected on the downstroke and a small area is projected on the upstroke. These traits give great maneuverability to most insect species.

The number of flying insect species is enormous. Typical insect flight speeds range from 15 miles per hour for bees to 1 mile per hour for mosquitoes, and even less for smaller insects. The fastest flying insect may be the tabanid, with a flight speed estimated at 90 miles per hour; it has been observed to execute Immelman maneuvers while in flight. The Australian dragonfly can reach 36 miles per hour over short distances, outrunning most horses.

Some dragonflies have wingspans of up to 11 inches, and some butterflies have wingspans of up to 10 inches. Dragonflies have two sets of high-aspect ratio wings, and butterflies have two pairs of large low-aspect ratio wings covered with colorful, iridescent scales in overlapping rows. Lepidoptera, as butterflies and moths are known, are the only insects that have scaly wings. Flight speeds vary among butterfly species. The poisonous varieties fly more slowly than nonpoisonous varieties, because they do not have to fly as quickly to evade predators. The fastest butterflies can fly at about 30 miles per hour or faster, whereas slow-flying butterflies fly about 5 miles per hour.

Mammals

Only one mammal is truly capable of powered flight: the bat, of the order Chiroptera, a word that means "hand-wing." All other so-called flying mammalian species do not actually fly but rather glide. Other mammals that fly by gliding include the flying squirrel and the flying lemur, neither of which actually fly and the latter of which is not actually a lemur. Bats, however, like birds, do attain true powered flight.

Bats are vertebrates with fur that bear and nurse live young. Nocturnal animals, they are found in all regions of the world except for the North and South Poles. There are more than 900 different species of bats, ranging in wingspan from the 6-inch bumblebee bat to the 6-foot flying fox. Some bats migrate, whereas others hibernate.

Because the fossil record is limited, the origin and evolution of bats remain unknown. It is believed bats appeared around fifty million years ago. Bats are related to the colugo, or flying lemur, but their common link is a mystery. They probably evolved from arboreal ancestors related to primates that used gliding and climbing as separate means of locomotion. The fact that the earliest bats had tails supports this assertion.

Bats are divided into two suborders based upon their method of navigation. Those of the suborder Microchiroptera use a type of sonar called echolocation to navigate and search for prey. They send off high-pitched sounds beyond the range of most human hearing. These sounds echo off surroundings and other animals, and bats use the echoes to determine the size and distance of the object. Microchiropteras include the vampire bat, the only mammal to feed exclusively on blood. Bats of the suborder Megachiroptera, such as the fruit bat, use their sense of smell to find food. Both types of bats have poor eyesight.

Bats are agile fliers. The bat wing is a membrane stretched across elongated fingers of the hand which support the distal, thrust-producing portion of the wing. Bats can change the effective airfoil cross section of the wing by moving their fingers. The fingers are extremely flexible, much like those of humans, and allow a bat to create almost any desired airfoil shape. The uropatagium, a membrane stretched between the hind limbs, helps stabilize the bat during flight and is often used to capture prey. Because gliding animals incorporate their hind limbs into their wings, this membrane is believed to have evolved from gliding.

Gliding mammals include the flying squirrel and flying lemur. Flying squirrels have a fold of skin extending from the wrist of the front leg to the ankle of the hind leg that forms a winglike gliding surface when the limbs are extended. The tail serves as a control device during glides to steer and stabilize flight. Colugos, or flying lemurs, arboreal climbers and gliders with lateral skin membranes and large, webbed, clawed feet, are found in certain regions of the Pacific Rim. They resemble large flying squirrels. Like bats, they have a short tail, which is used for stability and is connected to the hind limbs by skin folds.

Jamey D. Jacob

Bibliography

Alexander, R. McNeill, and Geoffrey Goldspink. *Mechanics and Energetics of Animal Locomotion*. New York:

John Wiley & Sons, 1977. Experimental and theoretical analysis of animal locomotion including walking, swimming, and flying for birds and insects.

Goldsworthy, G., and C. Wheeler. *Insect Flight*. Boca Raton, Fla.: CRC Press, 1989. A detailed scientific analysis of insect flight.

Pringle, J. W. S. *Insect Flight*. Burlington, N.C.: Carolina Biological Supply Company, 1990. Brief introduction to insect flight.

Tennekes, Hank. *The Simple Science of Flight: From Insects to Jumbo Jets*. Cambridge, Mass.: MIT Press, 1997. An excellent introduction for the layperson to the mechanics of flight, comparing insects, birds, and manufactured vehicles, including energy requirements and flight limitations.

Antiaircraft fire

Definition: Surface-to-air weapons fire providing direct protection from aerial attack.

Significance: With the advent of aircraft as a military weapon delivering offensive ordnance, air defense weaponry has become a vital part of many nations' overall military strategies.

History

Ground-based antiaircraft gunnery began in 1871 during the Franco-Prussian War (1870-1871), when the Prussian army used a Krupp-produced weapon to shoot at message balloons being sent out from the besieged garrison of Paris. The first weapons expressly designed as antiaircraft weapons were manufactured in Germany in 1908 as quick-firing, car-mounted field pieces and were used to shoot down observation balloons.

During World War I (1914-1918), as positive results of aerial combat indicated the utility of aircraft as offensive weapons, specialized antiaircraft artillery became a priority. Such guns required mounts that allowed a high angle of fire, an all-around traverse, a high rate of fire, and a high muzzle velocity for a straight trajectory, and an improved accuracy.

By the start of World War II (1939-1945), two general types of antiaircraft gun were being produced: heavy, single-shot guns for attacking high-altitude aircraft and light, fast-firing machine guns and small-caliber cannons for low-level defense. The heavy guns could cycle as many as twenty-five rounds per minute to altitudes of 5,500 meters. The lighter weapons could fire nearly one thousand machine-gun rounds per minute. Small cannons, such as the 40-millimeter double-gun-mount Bofors, could fire 120 rounds per minute per gun.

Due to the poor general performance of early aircraft, antiaircraft gunners managed to bring down about one aircraft for every one thousand rounds of ammunition, excluding machine-gun ammunition, fired. As advances in aviation rapidly progressed, antiaircraft defense lagged behind in development. In 1939, German military planners determined that it would take fifty rounds of antiaircraft artillery to bring down one enemy airplane. On this basis, they established their air defense doctrine. Aircraft design and performance developed rapidly during World War II, outpacing antiaircraft weaponry. The German military found that it expended more than twelve thousand shells for each aircraft destroyed.

Between 1950 and 1980, surface-to-air antiaircraft missiles predominated as air defense weapons. The technology of the guided missile was thought to make other antiaircraft defenses obsolete. The reality, however, was that approximately fifty missiles were fired to bring down one enemy plane, at a cost much greater than that of the required twelve thousand artillery shells of World War II. Despite these results, heavy, single-shot artillery guns have been largely replaced by vehicle-mounted, in-place, or shoulder-held guided surface-to-air missiles. The advantage of a guided missile is its ability to completely destroy its target. Shrapnel from artillery rounds may only disable modern aircraft equipped with double- and triple-redundancy systems: A single missile hit will destroy an aircraft.

Since 1945, very high-flying bombers and guided missiles have made large antiaircraft artillery obsolete, but there remains the need to engage low-flying attack aircraft and helicopters. Although data suggest that missiles are barely as effective as antiaircraft artillery, the mere threat of surface-to-air missiles often forces aircraft to fly low enough to be shot at by smaller-caliber automatic weapons. At lower altitudes, where antiaircraft gunners can clearly see their targets, most aircraft losses to air defense happen.

During the Vietnam War (1961-1975), 80 percent of aircraft losses came from low-altitude machine-gun fire requiring about ten thousand small-caliber cannon and machine-gun rounds to down each airplane. The philosophy of rapid close-range fire as the most effective antiaircraft defense has resulted in the design and use of multi-barreled unmanned radar-controlled systems such as the 20-millimeter Vulcan-Phalanx Close-In Weapon System (CIWS). This weapon fires 6,600 rounds of uranium-

*Antiaircraft fire, often called "ack-ack," lights up the sky over London during World War II. (*Digital Stock*)*

tipped ammunition up to 5,000 meters, providing a so-called wall of steel to down approaching aircraft or missiles.

Tactics

Antiaircraft fire is known to combat pilots by slang names such as "Archie," "Ack-Ack," "Flak," and "Triple A." The goal and lethal mission of air defense antiaircraft fire is attrition and deterrence. These goals are accomplished by forcing enemy aircraft either to abort their missions or to take heavy losses.

A four-step procedure establishes the basic tactics for providing successful antiaircraft fire. First, the enemy aircraft must be detected as early as possible. Second, the aircraft must be acquired to determine its direction and possible destination. Third, the aircraft must be tracked so weapons can be quickly targeted. Fourth and finally, the target must be destroyed.

Air defense artillery is employed to protect individual rear-area installations and vital military bases. Air defense weapons are deployed according to their range and mobility. Long-range, less mobile weapons are set far back from the fighting fronts to protect rear installations and to give high-altitude protection to front-line units. Mobile, short-range weapons are deployed nearer the fighting, where they can respond quickly to battlefield dynamics. The key to any successful air defense is the layering of defense at multiple depths and altitudes. In the past, air defenses had short ranges and low altitude. Modern surface-to-air systems can protect areas more than 100 kilometers from their bases and more than 10,000 meters high. A complete system can cover an entire operational theater of more than 1,600,000 square kilometers. The performance of antiaircraft fire is dependent on the destructive power of its ammunition, gained from high-impact velocities, explosive content, shrapnel, and blast and incendiary effect.

Historically, air defense has provided an impediment, and often a deterrent, to air attacks, but it has not in the long run been able to stop aerial attacks. Although aircraft have been touted as the preeminent weapon of modern warfare, they have failed to be overpowering, not because of air defense, but rather because of limitations in aircraft performance, weapons, and piloting. Modern air warfare tactics suggest that the best defense from enemy air attack is to destroy or suppress the enemy air force through air superiority.

Randall L. Milstein

Bibliography

Hogg, Ian V. *Anti-Aircraft: A History of Air Defense.* London: Garland, 1988. A good general reference with excellent illustration for researching changes in air defense weaponry and tactics throughout the twentieth century.

Kreis, J. F. *Air Warfare and Airbase Air Defense.* Office of Air Force History: Washington, D.C.: United States Air Force, 1988. A difficult to obtain yet thorough treatment of the history of American air defense doctrine.

Werrell, K. P. *Archie, Flak, AAA, and SAM: A Short Operational History of Ground-Based Air Defense.* Washington, D.C.: Government Printing Office, 1988. A concise review of antiaircraft defense from World War I to the late twentieth century.

Apache helicopter

Also known as: AH-64, AH-64A, Longbow Apache (AH-64D)

Definition: U.S. Army antiarmor attack helicopter.

Significance: The Apache helicopter serves as the principal attack helicopter for the U.S. Army.

Equipment and Armament

The Apache is the U.S. Army's principal attack and antiarmor helicopter. It carries a crew of two, seated in tandem. The pilot sits in the rear seat, behind the copilot-gunner. Two General Electric T700-701C turboshaft engines that produce a total of nearly 4,000 horsepower provide power to the four-bladed, 48-foot diameter main rotor. Mission gross weight for the AH-64A is approximately 14,600 pounds, while for the AH-64D it is approximately 16,100 pounds. Top speed for the Apache is 167 knots.

Standard armament for the Apache consists of a 30-millimeter chain gun, located on the chin of the fuselage, and various combinations of other weapons which are attached to four weapons stations on the stub wings. These other weapons include the Hellfire and Hellfire II air-to-ground missile (four per weapons station) and the 70-millimeter Hydra rocket system (nineteen per weapons station). External fuel tanks (230 gallons each) to extend the range of the Apache may also be attached to the weapons stations.

The AH-64A Apache is also equipped with a target acquisition designation system (TADS) and pilot night vision system (PNVS), which are located on the nose of the aircraft. The TADS/PNVS are used in conjunction with the integrated helmet and display sight system (IHADSS) to allow the Apache to navigate and conduct precision attacks by day, by night, and in adverse weather conditions.

The AH-64D Longbow Apache is a remanufactured and upgraded version of the AH-64A. A remanufactured craft is completely stripped and then rebuilt with almost entirely new components. To many observers, the D model of the Apache appears to be identical to the A model. However, there are two observable differences. First, the forward avionics bays on the AH-64A are extended and expanded on the AH-64D to accommodate additional electronic components. Second, the Longbow Apache is equipped with a mast-mounted assembly, located above the main rotor, which accommodates the Longbow fire control radar. While the external differences between the AH-64A and AH-64D are minimal, the internal upgrades have made the Longbow Apache a much more capable weapon system. The Longbow fire control radar has the ability to detect, classify, and prioritize targets both on the ground and in the air. It also supports fire-and-forget Hellfire missiles, that is, missiles the gunner need not track until they hit the target. The all-new suite of digital instrumentation in the cockpit enhances the situational awareness of the pilot. Navigation is improved with a Global Positioning System (GPS).

Development

Although AH-64's are currently built by the Boeing Company in Mesa, Arizona, the Apache program has changed hands twice since it began in the early 1970's. In 1973, Hughes Helicopter Company won a contract to design an advanced attack helicopter for the U.S. Army. The Hughes Model 77 (YAH-64) made its first flight on September 30, 1975. In competition with the YAH-64 was the Model 409 (YAH-63) proposed by Bell Helicopter Company. In 1976, the U.S. Army awarded a full-scale development contract to Hughes to build the AH-64. The contract was completed in 1981 and, in 1982, the Army awarded Hughes a production contract for the aircraft now known as the AH-64A Apache.

In January, 1984, just before the February delivery of the first production Apache helicopter to the U.S. Army, McDonnell Douglas Corporation bought Hughes Helicopter Company. McDonnell Douglas then moved the Apache assembly facility from Culver City, California, to Mesa, Arizona. A total of 821 Apaches were delivered to the U.S. Army before production was transitioned to the AH-64D in 1997. International sales of Apache helicopters have included the nations of Israel, Egypt, Greece, the United Arab Emirates, Bahrain, Kuwait, and South Korea.

In the late 1980's, McDonnell Douglas began developing an advanced model of the Apache, which became known as the AH-64D Longbow Apache. The Longbow Apache made its first flight on April 15, 1992, and in 1996, McDonnell Douglas received a multiyear contract from the U.S. Army to remanufacture 232 AH-64A helicopters. The Boeing Company subsequently acquired McDonnell Douglas in August, 1997. In addition to the aircraft purchased by the U.S. Army, AH-64D Longbow Apache helicopters have been sold to the United Kingdom and the Netherlands.

Combat Performance

The AH-64A has participated in two major conflicts, as well as performing peacekeeping duties in Bosnia and Kosovo. In 1989, Apache helicopters played a key role in Operation Just Cause in Panama. Since much of the activity was at night, the advanced sensor and sighting capabilities of the Apache proved to be very effective against the antigovernment forces.

Operation Desert Storm, the liberation of Kuwait in 1991, provided an opportunity for the Apache to truly show off its capabilities. During the first hours of the war, Apaches destroyed key Iraqi radar sites, which allowed Coalition aircraft to penetrate deep into Iraqi territory without detection. In the course of the land battle, Apaches were credited with destroying more than five hundred Iraqi tanks and hundreds of armored personnel carriers, trucks, and other vehicles. Only one Apache was lost to enemy fire.

Donald L. Kunz

Bibliography

Hirshberg, Michael J. *The American Helicopter: An Overview of Helicopter Developments in America, 1908-1999.* Arlington, Va.: ANSER, 2000. Historical account of helicopter developments in the twentieth century, with pictures and descriptions of many different designs.

Jackson, Paul, ed. *Jane's All the World's Aircraft, 2000-2001.* 91st ed. Alexandria, Va.: Jane's Information Group, 2000. The definitive source for aircraft photographs and specifications.

Sweetman, Bill. *Attack Helicopters: The AH-64 Apaches.* Mankato, Minn.: Capstone High-Interest Books, 2001. Contains descriptions and photos of the Apache helicopter.

See also: Boeing; Gulf War; Helicopters; McDonnell Douglas; Military flight; Rotorcraft

Apollo Program

Date: From May 28, 1964, to December 19, 1972

Definition: American project to land humans on the Moon.

Significance: The Apollo Program, designed to ensure America's international leadership in space exploration, resulted in the first landing of humans on the Moon. Apollo astronauts performed experiments and returned rock samples to Earth that helped determine the age and origin of the Moon.

The Program's Beginnings

In 1960, planners at the National Aeronautics and Space Administration (NASA) selected a crewed lunar landing as the follow-up to the Mercury effort to place a man in Earth orbit. In December, 1960, just before leaving office, President Dwight D. Eisenhower advised NASA officials that he would not approve the lunar landing project. However, after the Soviets sent Yuri A. Gagarin into Earth orbit in April, 1961, the new U.S. president, John F. Kennedy, announced on May 25, 1961, the plan "before this decade is out, of landing a man on the Moon and returning him safely to the earth." The project to accomplish Kennedy's objective was named Apollo.

The Apollo Spacecraft

To launch Apollo, NASA designed and built a huge, three-stage rocket, the Saturn V, which stood 363 feet tall and developed 7.5 million pounds of thrust at liftoff. In order to minimize the weight of the spacecraft, the Apollo engineers planned a lunar orbit rendezvous technique, requiring the Apollo spacecraft to have a modular design, consisting of three separate units, the Command Module, the Service Module, and the Lunar Module.

The Command Module, built by North American Rockwell, served as the control center for the spacecraft and provided 210 cubic feet of living and working space for the astronauts. It was designed to carry three astronauts from the earth to an orbit around the Moon and back. It was shaped like a cone, with a height of 10 feet 7 inches, a maximum diameter of 12 feet and 10 inches, and an approximate weight of 13,000 pounds. The Command Module was pressurized, so the astronauts could live and work without wearing spacesuits. The wide end of the cone was a blunt heatshield, covered with layers of special ablative material designed to burn away during reentry, dissipating the extreme heat caused by atmospheric friction.

The cylindrical Service Module, built by North American Rockwell, had a diameter of 12 feet and 10 inches and a length of 22 feet and 7 inches. It carried the electrical power systems, most of the electronics, and the life support gases. It also carried the computer system for guidance and navigation, the communications transmitters and receivers, and the oxygen and hydrogen used by the life-support and energy-generation systems. The Service Module's rocket engine produced 22,000 pounds of thrust. This rocket engine was used to slow the spacecraft to enter lunar orbit and then to speed it up for the return to Earth. Fully fueled, the Service Module weighed about 53,000 pounds.

The Lunar Module, built by the Grumman Aircraft Engineering Corporation, was designed to detach from the Command and Service Modules while they orbited the Moon and to carry two astronauts down to the lunar surface. The Lunar Module was a two-stage rocket, with each stage carrying its own fuel supply. The lower stage carried a 9,700-pound-thrust rocket engine to slow down the Lunar Module for a gentle touchdown on the lunar surface. Four landing legs, each with a landing pad to distribute the weight of the spacecraft over a larger area of the lunar soil, were attached to the Lunar Module descent stage. One of the landing legs was equipped with a ladder to allow the astronauts to climb down to the lunar surface. The upper stage of the Lunar Module consisted of a pressurized compartment providing life support for the two-man crew and an ascent engine to return the crew compartment to lunar orbit. The lower stage of the Lunar Module served as a launching pad for the upper stage. With the landing legs extended, the Lunar Module was 22 feet and 11 inches tall and weighed about 32,000 pounds.

The Apollo Flights

On January 27, 1967, during a preflight test, a fire swept rapidly through the Apollo Command Module. The three astronauts participating in the test, Roger Chaffee, Virgil "Gus" Grissom, and Edward White, were killed in the fire. After the fire, NASA officials designated the test as Apollo 1, honoring the crew. An extensive investigation of the fire showed numerous design flaws in the Apollo Command Module, and crewed launchings were postponed for more than a year while an extensive redesign was conducted.

Apollo 7, the first manned test of the Command and Service Modules, was launched from Cape Kennedy, Florida, on October 11, 1968, on a Saturn IB rocket. Apollo 7 was the only crewed Apollo mission launched on a Saturn IB rocket, which was powerful enough to carry the Command Module and the Service Module into Earth orbit, but could not lift the full Apollo assembly, including the Lunar Module. The spacecraft crew consisted of commander Walter M. Schirra, Jr., Donn F. Eisele, and Walter Cunningham, who held the title of Lunar Module pilot despite the lack of a Lunar Module on the Apollo 7 mission. The crew orbited the earth 163 times and spent almost eleven days in space, demonstrating the reliability of the Command and Service Modules for a time comparable to that of a round trip to the Moon.

Apollo 8, launched on December 21, 1968, was the first crewed mission using the Saturn V rocket, and the first mission to take humans to the Moon and back. The three-person crew consisted of Frank Borman, the commander; James A. Lovell, Jr., the Command Module pilot; and William A. Anders, the Lunar Module pilot. Apollo 8 tested the flight path and operations for the trip to the Moon and back and demonstrated that the Apollo Command Module could successfully reenter the earth's atmosphere at the high speed of a return from the Moon.

The Apollo 9 mission, launched on March 3, 1969, was the first crewed flight employing all three components of the Apollo spacecraft. The crew, consisting of astronauts James A. McDivitt, the commander; David R. Scott, the Command Module pilot; and Russell L. Schweickart, the Lunar Module pilot, made 152 orbits of the earth. They demonstrated the crew transfer procedures and the rendezvous and docking procedures between the Command Module and the Lunar Module.

The Apollo Program had to devise spacecraft that not only would take humans out of Earth's atmosphere, but also land them on the Moon. The Lunar Module transported astronauts from their primary craft to the Moon's surface. (NASA CORE/ Lorain Valley JVS)

The final test of the Apollo spacecraft came with Apollo 10, launched on May 18, 1969. Apollo 10 was a complete Apollo lunar landing mission without an actual landing on the Moon. On the fifth day of the mission, astronauts Thomas Stafford, the commander, and Eugene Cernan, the Lunar Module pilot, descended inside the Lunar Module to within 14 kilometers of the lunar surface, while John W. Young remained in lunar orbit in the Command Module.

The Saturn V rocket carrying Apollo 11 lifted off from the NASA John F. Kennedy Space Center in Florida on July 16, 1969. The Command Module, named *Columbia*, carried astronauts Neil Armstrong, the commander; Edwin "Buzz" Aldrin, the Lunar Module pilot; and Michael Collins, the Command Module pilot, to the Moon and back. Astronauts Armstrong and Aldrin landed on the Moon in the Lunar Module, named *Eagle*, on July 20, 1969. Michael Collins remained alone in the *Columbia*, orbiting the Moon. *Columbia* served as a communications link between the astronauts on the Moon and mission control in Houston, Texas. After 28 hours on the Moon, the upper stage of the Lunar Module carried Armstrong and Aldrin back into orbit around the Moon, where they rendezvoused and docked with the *Columbia*. *Columbia*, the only part of the spacecraft to return to Earth, landed in the Pacific Ocean on July 24, 1969.

Apollo 12, launched on November 14, 1969, carried astronauts Charles Conrad, Jr., the commander; Richard F. Gordon, the Command Module pilot; and Alan L. Bean, the Lunar Module pilot. Astronauts Conrad and Bean landed on the Moon in the Sea of Storms, less than 600 feet from the site where the Surveyor 3 spacecraft had landed on April 20, 1967. The astronauts recovered pieces from the Surveyor 3 to allow scientists to assess the effects of the craft's two-year exposure to the lunar environment. They also collected 75 pounds of rocks and soil for return to Earth and deployed the Apollo Lunar Surface Experiment Package to perform scientific experiments on the Moon.

Apollo 13, carrying astronauts James A. Lovell, Jr., the commander; John L. Swigert, Jr., the Command Module pilot; and Fred W. Haise, Jr., the Lunar Module pilot, lifted off on April 11, 1970. About 56 hours into the flight, an explosion in one of the oxygen tanks in the Service Module crippled the spacecraft. The crew was forced to orbit the Moon and return to the Earth without landing. The astronauts spent much of the flight in the Lunar Module, using its oxygen and electrical supplies, because of the damage to the Service Module. The astronauts landed safely on Earth on April 17, 1970.

Apollo 14, carrying astronauts Alan B. Shepard, the commander; Stuart A. Roosa, the Command Module pilot; and Edgar D. Mitchell, the Lunar Module pilot, was launched on January 31, 1971. Astronauts Shepard and Mitchell landed on February 5, 1971, within 160 feet of the target point, in the Fra Mauro region of the Moon, the intended landing site of the Apollo 13 mission. During a 4-hour, 20-minute period of extravehicular activity, Shepard and Mitchell climbed up the side of Cone Crater, providing the first experience of climbing and working in hilly terrain in the bulky spacesuits. The astronauts collected 94 pounds of lunar soil and rocks. The upper stage of the Lunar Module lifted off from the lunar surface on February 6, 1971, after 33.5 hours on the Moon. After the crew transferred to the Command Module, the Lunar Module ascent stage was guided to impact on the lunar surface, producing a seismic signal that was recorded by instruments deployed on the lunar surface by Apollo 12 and Apollo 14. The Command Module landed in the Pacific Ocean on February 9, 1971.

By 1970, public interest in lunar exploration had waned and federal budget cuts forced NASA to sacrifice current projects in order to support future ones. Apollo 15, 16, and 17 would be equipped to travel farther, stay longer, and perform more experiments than had previous missions.

Apollo 15, carrying astronauts David R. Scott, the commander; Alfred J. Worden, the Command Module pilot; and James B. Irwin, the Lunar Module pilot, was launched on July 26, 1971. Apollo 15 was the first in a series of advanced missions, carrying the Lunar Rover (LRV), which astronauts Scott and Irwin used to explore the Hadley Rille region of the Moon. The LRV allowed astronauts to travel tens of kilometers from the Lunar Module, in contrast to the hundreds of meters traveled in previous missions. The astronauts collected 173 pounds of samples from the low lunar plains, the Apennine Mountains, and the Hadley Rille, a long, narrow, winding valley. Although Apollo 15's atmospheric entry was normal, one of the three parachutes that slowed the Command Module's descent collapsed before landing. Nonetheless, the Command Module landed safely on August 7, 1971.

Apollo 16, carrying astronauts John W. Young, the commander; Thomas K. Mattingly II, the Command Module pilot; and Charles M. Duke, Jr., the Lunar Module pilot, was launched on April 16, 1972. This mission landed in the Descartes region, where astronauts Young and Duke collected 209 pounds of soil and rocks and used an ultraviolet camera and spectrograph to perform the first astronomical measurements from the surface of the Moon. The Apollo 16 crew returned to Earth on April 27, 1972.

Apollo 17, carrying astronauts Eugene A. Cernan, the commander; Ronald E. Evans, the Command Module pilot; and Harrison H. Schmitt, the Lunar Module pilot and, as a trained geologist, only scientist to visit the Moon, was launched on December 7, 1972. On this mission, astronauts Cernan and Schmitt conducted the longest LRV traverse on a single extravehicular activity, a trip of about 100 kilometers. They collected the largest amount of lunar soil and rock ever returned to Earth. Apollo 17's return to Earth on December 19, 1972, marked the end of U.S. efforts to send humans to the Moon.

Results of the Apollo Program

The major objective of the Apollo Program was accomplished with the landing of twelve American astronauts on the Moon and their safe return to Earth. These landings demonstrated the capability of American engineering, restoring American prestige by finally beating the Soviet Union in the space race. Scientists studying lunar rock samples were finally able to determine the age and origin of the Moon, finding that the Moon formed about 4,560,000,000 years ago, probably from the debris ejected when an asteroid struck Earth.

George J. Flynn

Bibliography

Brooks, Courtney G., James M. Grimwood, and Lloyd S. Swenson. *Chariots for Apollo: A History of Manned Lunar Spacecraft*. Washington, D.C.: National Aeronautics and Space Administration, 1979. NASA's official history of the Apollo Program, focusing on the design, construction, and flight of the Apollo spacecraft.

Chaikin, Andrew L. *A Man on the Moon: The Voyages of the Apollo Astronauts*. New York: Viking Press, 1994. An extensive historical account of the Apollo Program, beginning with the Apollo 1 fire and continuing through the successful Moon landing.

Logsdon, John W. *The Decision to Go to the Moon*. Cambridge, Mass.: MIT Press, 1970. An extensive history of the Apollo Program, focusing on the decisions faced by political, industrial, and NASA officials that shaped the Apollo spacecraft and the lunar landing program. Includes a comprehensive bibliography.

Neil Armstrong

Date: Born on August 5, 1930, in Wapakoneta, Ohio

Definition: As commander of the Apollo 11 lunar landing mission in 1969, the first human to walk on the moon.

Significance: In addition to his outstanding and pioneering contributions to the National Aeronautics and Space Administration's (NASA) crewed spaceflight program, Armstrong has served with distinction as a professor of aerospace engineering, chairman and director of several corporations, and member of presidential commissions.

Early Life and Education

Born to Stephen and Viola Louise Armstrong in Wapakoneta, Ohio, in 1930, Neil Armstrong was an avid enthusiast of flying from an early age. He received his student pilot's license at age sixteen, before receiving a driver's license. In 1947, he entered the aeronautical engineering program at Purdue University with a scholarship from the U.S. Navy. Two years later, he was called to active duty and earned his pilot's wings at the Naval Air Station in Pensacola, Florida. As the youngest pilot in his squadron, he flew seventy-eight combat missions from the flight deck of the USS *Essex* in Korea in 1950. He won three Air Medals for his combat duty. At the end of the war, Armstrong returned to Purdue and received his baccalaureate degree in 1955.

Professional Activities at NASA

After graduating from Purdue, Armstrong joined NASA's Lewis Flight Propulsion Laboratory in Cleveland, Ohio. Later, he transferred to NASA's High-Speed Flight Station at Edwards Air Force Base, California. There, as an aeronautical research pilot, he flew X-15 airplanes to altitudes over 200,000 feet, at speeds up to 4,000 miles per hour. As a test pilot, Armstrong also flew the X-1 rocket airplane, the F-100, F-101, F-102, F-104, F-5D, B-47, and other aircraft. Armstrong's experience with the X-15 led to his selection as a pilot of the X-20 Dyna-Soar, an experimental craft that could leave the atmosphere, orbit the earth, reenter the atmosphere, and land like a conventional aircraft. However, the X-20 project was canceled in 1962, and Armstrong then decided to become an astronaut.

In September, 1962, Armstrong was one of the first two civilians selected for astronaut training. In his first flight assignment, he served as a backup command pilot for the Gemini GT-5 mission. On March 16, 1966, Armstrong served as the command pilot for Gemini 8, and, along with pilot David R. Scott, successfully docked two vehicles in space for the first time. The flight was terminated ahead of its three-day schedule due to a malfunctioning thruster.

Demonstrating exceptional piloting skill, the crew overcame the problem and brought the craft to a safe landing. Subsequently, Armstrong served as backup command pilot and backup commander for the Gemini 11 and the Apollo 8 missions, respectively.

Armstrong's most significant role as an astronaut occurred during his command of the manned lunar landing mission of Apollo 11 from July 16 to July 21, 1969. The crew for this historic flight consisted of spacecraft commander Armstrong, Lunar Module pilot Edwin "Buzz" Aldrin, and Command Module pilot Michael Collins. On July 20, 1969, the human race accomplished what many consider the single greatest technological achievement of all time. For the first time in human history, a man set foot on a celestial body beyond the earth. After landing on the lunar surface at about 4:18 P.M. eastern daylight time, Armstrong radioed back to mission control the now-famous words, "Houston, Tranquillity Base here. The *Eagle* has landed." Six hours later, Armstrong stepped off the Lunar Module onto the surface of the Moon. Taking his first steps on the Moon, he uttered the immortal words, "That's one small step for man, one giant leap for mankind." Shortly thereafter, he was joined by Aldrin, and the two astronauts spent twenty-one hours on the lunar surface, collecting 46 pounds of lunar rocks. Their liftoff from the surface of the Moon was partially captured on a television camera they left behind, and they successfully docked with Michael Collins, who had continued to orbit the Moon alone in the Command Module *Columbia*.

Post-NASA Activities

Following his historic walk on the Moon, Armstrong received a master of science in aeronautical engineering from the University of Southern California. In the fall of 1971, he accepted a position as professor of aerospace engineering at the University of Cincinnati, an interdisciplinary post he held until 1980. Thereafter, he served as the chairman of the board of Cardwell International Corporation in Lebanon, Ohio, until 1982, when he became the chairman of the board of Computing Technologies for Aviation (CTA) Incorporated of Charlottesville, Virginia.

In 1984, along with the test pilot Charles E. "Chuck" Yeager, Armstrong joined the National Commission on Space (NCOS), a presidential panel created to develop goals for the space program in the twenty-first century. However, the explosion of the space shuttle *Challenger* on January 28, 1986, placed the commission's report on hold. Following the *Challenger* disaster, Armstrong was named vice chairman of the Presidential Commission on the Space Shuttle *Challenger* Accident. Over the years, Armstrong, an intensely private and unassuming man, has avoided as much as possible making public appearances. On the occasion of the thirtieth anniversary of the first lunar landing on July 20, 1999, he gave a lighthearted speech before the National Press Club in Washington, D.C., on behalf of the National Academy of Engineering. He described spaceflight as one of the greatest engineering achievements and observed that while "science is about what is, engineering is about what can be."

Monish R. Chatterjee

Bibliography

Brown, Don. *One Giant Leap: The Story of Neil Armstrong*. Boston: Houghton Mifflin, 1998. A biographical picture book for children that traces Armstrong's life from his childhood to his walk on the Moon.

Kramer, Barbara. *Neil Armstrong: The First Man on the Moon*. Springfield, N.J.: Enslow, 1997. A biography of Armstrong that opens with his walk on the Moon and proceeds with a chronological presentation that details his family, schooling, training, married life, and activities after the astronaut program.

Thompson, Milton O., and Neil Armstrong. *At the Edge of Space: The X-15 Flight Program*. Washington, D.C.: Smithsonian Institution Press, 1992. A complete history of the X-15 program, including the tests, pilots, and other contributors who paved the way for space travel.

See also: Apollo Program; Korean War; National Aeronautics and Space Administration; Space shuttle; Test pilots; X planes; Chuck Yeager

Astronauts and cosmonauts

Definition: The astronauts and cosmonauts were a select group of men and women trained by the United States and the Soviet Union to travel in outer space.

Significance: The astronaut and cosmonaut programs provided the world with the talented pilots who formed the foundation of human exploration of outer space. These men and women extended knowledge of the universe and brought the human race to the surface of the Moon. Politically, astronauts and cosmonauts were also used as warrior-heroes in the Cold War.

Background

The space race was the product of the worldwide struggle between the communist and industrial-democratic nations known as the Cold War. This conflict began at the end of World War II when the antifascist alliance, led by Great Britain, the United States, and the Soviet Union, began to unravel.

Science occupied an important position in the competing philosophies of democratic capitalism and socialist communism. In the Enlightenment model that underlies American and Western European democracy, science unlocked the natural laws of nature, and through the use of human reason these universal truths could be used for the betterment of the human community. Enlightenment intellectuals believed in human equality before God, but also believed in individual inequalities in intelligence, self-discipline, and drive. This would lead to a natural aristocracy based upon merit and achievement, who would create an environment in which natural law would be used for the welfare of the entire community.

Socialist philosophers believed that the workers and the peasants had the natural right to control the means of production. In this paradigm, the proper application of science and technology would create a system that would establish an egalitarian society governed by the "Dictatorship of the Proletariat." These two different worldviews had a profound impact on the development of the astronaut and cosmonaut programs. The successes and failures of the two programs would be used as examples of the strengths or weaknesses of the two competing ideologies. This was especially true in the competition for the hearts and minds of the Third World nations that were the main targets of this struggle.

The Original Soviet Voyagers

The success of Sputnik, the first artificial Earth satellite, generated momentum within the Soviet government to initiate a program designed to send a human into space. Initially, twenty candidates were chosen to be the first cosmonauts. A special facility outside of Moscow, designated "Star Town," was constructed to carry out the necessary training. By 1960, the original twenty had been reduced to twelve, and it was from this select group that the first voyagers were chosen.

The Soviets developed a strict set of standards that each potential cosmonaut had to meet. Since the original space capsules were small, these first pilots could be no taller than 5 feet, 11 inches. The stress and strain of the rigorous training and missions necessitated that the cosmonauts be both emotionally and physically fit. The unknown aspects of spaceflight required pilots who were able to act decisively under great stress and who possessed the confidence in their own abilities to handle any problems that they might encounter. The Russians looked to their military establishment for their space program because the cosmonauts would have to be able to take great risks, yet at the same time be completely obedient to their supervisors. The Soviets wanted individuals who were intelligent in the areas of science and technology, but who were not deeply philosophical. The leadership of the Soviet Union did not welcome questions about the logic and ethics of a program that would expend billions of dollars in a race that would turn the cosmos into another Cold War battlefield.

The cosmonauts also had to reflect the ideology of socialist heroes in the struggle between the philosophies of communism and democratic capitalism. It was important that they be physically attractive, so as to optimize their impact as traditional heroes. More importantly, they had to possess the politically correct characteristics of the new Soviet citizen. The vast majority of the candidates were the children of factory workers or peasants. Their ethnicity was "pure" Russian, which reflected the centuries-old belief that the Russians were the natural leaders of the Slavic peoples. The communal nature of socialist ideology was always reflected in the cosmonauts' statements concerning the importance of their voyages to the welfare of all Earth's people.

Russia's first cosmonaut embodied all these important characteristics. Yuri Gagarin was a man of the common people, and his proletarian background was made to order for his great communist achievement. He spent the first years of his young adult life working in a Russian factory, and through hard work and determination he not only became a member of the Russian military but also attained the high status of a test pilot in the Soviet air force. The Russian government and propaganda machine focused upon his peasant heritage as an example of how the communist system, supported by a strong scientific and technological community, could transform a man of the working class into the first voyager into space. His successful flight in Vostok 1 on April 12, 1961, provided Soviet General Secretary Nikita Khrushchev with a powerful propaganda weapon in his struggle against the West. This socialist egalitarianism crossed gender lines on June 16, 1963, when Valentina Tereshkova completed forty-eight orbits in Vostok 6 to become the first woman in space. The world communist community declared that it was now perfectly clear that the socialist model was the philosophy that created true equality between men and women.

Tereshkova is a significant example of how politics and ideology played an important role in the Soviet space program. She was primarily chosen because of her proletarian heritage; in fact, Khrushchev was so intent on making an ideological statement that she was selected over a number of better-qualified women. Tereshkova's flight was extremely taxing and she suffered from the effects of space sickness, which so physically drained her that on one occasion she fell into a deep sleep and the Russian ground control had serious concerns about her health.

Khrushchev's propaganda about gender equality played far better outside the Soviet Union than it did domestically. There were numerous articles in the U.S. press about the impact of the flight on the movement for universal women's rights. Ironically, the leaders of the Russian space program used Tereshkova's weakened physical reaction as proof that women did not belong in space.

Eventually, Tereshkova's life took on the aspects of a soap opera. Her wedding to fellow cosmonaut Andrian Nekoloyev was broadcast throughout the Soviet Union. Their first child, a baby girl, was subjected to a series of biological examinations to see if Tereshkova's exposure to cosmic rays had affected her daughter's health. The pressures of a life in the spotlight eventually took its toll on her marriage and she divorced in June, 1964.

The original Russian cosmonaut program also had some important problems that reflected the dark side of the personalities of these unique individuals. The fearless aggressiveness so characteristic of these exceptional people at times erupted into antisocial behavior that on several occasions ended in death. The problem of alcoholism that has damaged large segments of Soviet society also took its toll on the first cosmonauts. Grigori Nelyubov was a superior candidate who was widely respected for his great skill and coolness under stress. One evening, when he was returning from a weekend leave, he had an altercation with local authorities that ended in a physical confrontation. As a result, he was dismissed from the cosmonaut program. Nelyubov became deeply depressed, developed a severe drinking problem, and eventually took his own life. The Soviet government was successful in its initial attempts to cover up such stories. Nelyubov became a nonperson, with all traces of his connection to the cosmonaut program erased from the official records. The Soviet propaganda machine could not allow the world to know that these serious problems existed in the socialist paradise.

The Original U.S. Astronauts

The United States decided to pursue a crewed spaceflight program in 1958 as a result of the impact of Sputnik on the U.S. political scene. The Eisenhower administration viewed the Soviet space program as another political threat to the strategic balance of the Cold War. The containment policy adopted by the United States during the Truman administration was based upon two important concepts: first, the Soviet empire had to be contained within the original East-West borders as defined in Winston Churchill's "Iron Curtain" speech, and second, that inherent philosophical weaknesses in the communist model would result in the downfall of the Soviet empire. George F. Kennan, who developed the policy, believed that one of its most important aspects would be how it enabled the West to win the struggle for the allegiance of developing nations. By 1958, it was evident to everyone in the United States that Sputnik had created the impression that the science of the new socialist order could produce profound technological achievements. If containment was to be successful, the United States would have to surpass the Soviets in the race for space.

When the Kennedy administration took office in January, 1961, the astronaut program was placed on hold because of the more pressing problems concerning the Cuban Revolution. The military disaster of the Bay of Pigs left the new president looking for a way to recover the confidence of the American people. This fact, coupled with the continued success of the Soviet space program, moved Kennedy to reinstate the astronaut program. The new program was placed before the American people in Kennedy's famous speech challenging the country to land a man on the Moon by the end of the decade.

Immediately, the nation was awash in articles describing the men who would be the first voyagers into outer space. These space travelers would epitomize the outstanding characteristics of democratic Enlightenment thought. The astronauts were perfect examples of a space-age "aristocracy of merit." They were required to hold at least a bachelor of science degree in engineering or related technology. Like the cosmonauts, every candidate had to be in excellent physical condition and possess the emotional strength to handle the hazards of spaceflight. The astronauts were required to have at least 1,500 hours of flight time and also to cut a handsome figure for media purposes. Unlike the Soviet cosmonauts, however, the United States astronaut program did not include women.

The United States was looking for national heroes who would reflect the strength of its political system. To the American public, the original seven astronauts were portrayed as the ultimate Cold Warriors. They were soldiers who risked their lives every day to make sure the United States would not fall behind the Soviet Union in the race to control outer space. In reality, these men were more like ancient Greek heroes who performed great acts of bravery but who also lived lives of physical and material excess. The one man who never strayed was John Glenn. He not only remained focused on his assignment, but on a few notable occasions, he also castigated fellow astronauts for their lack of restraint. In the late 1950's and early 1960's, the U.S. press did not pursue this type of sensational story, and the exploits of the original astronauts went unreported.

The United States began its crewed presence in outer space on May 5, 1961, when Alan Shepard took the Mercury capsule *Liberty Bell 7* for a fifteen-minute flight. The mission was a success in every possible way and was a major turning point in the U.S. space program. Technologically, it confirmed that a pilot could control a space capsule in both a state of weightlessness and at significant g forces. When Shepard returned, it also became evident that astronauts could travel in space and not experience any negative physical or emotional problems. The flight of *Liberty Bell 7* also had an important political impact. Domestically, it helped return a sense of optimism about the ability of the United States to compete in space with the Soviets. Internationally, the flight helped to emphasize the openness of the U.S. system. Unlike the Soviet launches, every minute of the flight of *Liberty Bell 7* was broadcast to the entire world. Many of the news stories in the international press praised the United States for allowing this event to be covered uncensored by the government.

On February 20, 1962, U.S. astronauts obtained full status with the Soviets when John Glenn became the first American to orbit the earth in his ship *Friendship 7*. This was the flight that showed the important difference between the two programs. During the course of his trip into space, Glenn was confronted with two potentially deadly problems, and on both occasions the U.S. decision to allow the pilot some control of the craft allowed him to avoid disaster. Early in the flight, the automatic stabilization system malfunctioned and the capsule began to drift off course, but because Glenn was able manually to maneuver the craft, he eventually brought it under control. Most important, when the capsule's heat shield malfunctioned upon reentry into Earth's atmosphere, Glenn once again was able to take steps to help his situation. These two incidents exemplified the American ideal of an "aristocracy of merit" that possessed the "right stuff" to overcome possible disaster. This differed substantially from the original Soviet system, in which the cosmonaut essentially "went along for the ride."

The Race to the Moon

The quest of the United States to land a man on the Moon required an expansion of the astronaut program in both the number of personnel and the technological skill required for this next phase of the space race. The Gemini Program was given the mission to perfect the concept of a multicrewed mission and to develop the necessary skills to successfully dock with another craft. A successful Moon mission would require three astronauts, two of whom would actually land on the Moon's surface. These two men would have to separate a smaller craft from the mother ship, land,

The Apollo-Soyuz Test Mission in July, 1975, was the first joint space project between the United States and the Soviet Union. The meeting of astronauts and cosmonauts in outer space marked the beginning of an era of cooperation rather than competition between the two countries. (NASA CORE/Lorain Valley JVS)

then take off from the Moon to dock once again with the original spacecraft. The successful completion of these tasks set the stage for the Apollo Program.

The Russian cosmonauts were trying to expand their presence in space by flying missions in the Voskhod Program, which was the Russian attempt to compete with the multicrewed missions of the United States. Unfortunately, the Voskhod was just a retooled version of the one-seat Vostok capsule, and it foreshadowed the decline of the Soviet space program. Despite these problems, cosmonauts continued to make important accomplishments in the field of space science. The most significant event of the Voskhod Program occurred on March 19, 1965, when Alexei Leonov became the first man to walk in space.

The Apollo astronauts perfected the three-man mission that put an astronaut on the Moon. The spacecraft itself was the most advanced ship to date and required technological capabilities well beyond those of either Mercury or Gemini. The worst accident in the history of the Moon program occurred on January 27, 1967, when the Apollo capsule containing Gus Grissom, Ed White, and Roger Chaffee filled with pure oxygen and exploded, killing all three men. U.S. astronauts finally reached the Moon on July 24, 1969, when Neil Armstrong spoke those famous words, "One small step for man, one giant leap for mankind."

The Moon landing also played an important role in the Cold War. By 1969, the United States found itself deep in the quagmire of Vietnam, and U.S. armed forces were suffering about five hundred casualties a week. At a time when the American people were beginning to question their country's ability to successfully carry on the war, the landing on the Moon was looked upon by the Nixon Administration as confirmation of the power of the United States. Thus it is evident that, from the very beginning of the space race, the astronauts and cosmonauts were considered soldiers in the Cold War.

Shuttles and Space Stations

The problems of the Vietnam War created a sense of despair and a loss of confidence among the American electorate, while pressing problems of race relations, poverty, and rising inflation helped extinguish the enthusiasm for further space exploration. In the Soviet Union, the communist system could no longer produce the resources needed to successfully place a cosmonaut on the Moon. Both nations decided to reorient their focus to develop programs that would continue to expand humankind's knowledge of the cosmos, while at the same time working within their drastically reduced budgets.

The United States developed a program that would create a series of reusable shuttle craft that would be launched into Earth orbit to perform duties ranging from scientific experiments to the repairs of sophisticated space telescopes. The next generation of astronauts shifted from explorers to experimental scientists. This new orientation opened the way for the first women astronauts, whose scientific and technological skills were needed for successful shuttle missions. The astrophysicist Sally K. Ride, who in 1983 became the first American woman to travel into space, exemplified the educational background of these new astronauts.

The Soviets decided to concentrate on the development of large space stations that would be used as laboratories for the development of the next generation of space scientists. Once again, the new generation of cosmonauts were oriented toward academic research and created an impressive schedule of experiments, ranging from space-based communication systems to solar research. Unfortunately, the collapse of the Soviet Union drastically reduced Soviet expenditures and the space station program was devastated by this lack of funding.

Beginning in the 1990's, the United States carried out a series of varyingly successful robotic missions to Mars that have rejuvenated interest in space exploration. A new international program consisting of members from various European, Asian, African, and Western Hemisphere states could form the foundation of humanity's next step into the universe.

Richard D. Fitzgerald

Bibliography

Burrows, William E. *This New Ocean*. New York: Modern Library, 1999. A comprehensive one-volume history of spaceflight providing a detailed, chronological account of the age of space exploration.

Harford, James. *Korolev: How One Man Masterminded the Soviet Drive to Beat America to the Moon*. New York: John Wiley & Sons, 1997. A unique and interesting look inside the Soviet space establishment as seen through the life of Russia's most important space scientist.

Heppenheimer, T. A. *Countdown: A History of Space Flight*. New York: John Wiley & Sons, 1997. An excellent one-volume history of spaceflight describing the economic, social, and political impact of the Space Age.

McDougal, Walter A. *The Heavens and Earth: A Political History of the Space Age*. Baltimore: Johns Hopkins University Press, 1985. An outstanding political history of the space race detailing the important linkage between the events of the Cold War and the U.S. and Soviet space programs.

See also: Apollo Program; Neil Armstrong; Crewed spaceflight; Yuri Gagarin; Gemini Program; John Glenn; Mercury project; Russian space program; National Advisory Committee for Aeronautics; National Aeronautics and Space Administration; Alan Shepard; Space shuttle; Spaceflight; Valentina Tereshkova; Uncrewed spaceflight

Jacqueline Auriol

Date: Born on November 5, 1917, in Challans, France; died on February 12, 2000, in Paris, France

Definition: Pioneer female test pilot and world speed record holder.

Significance: Beginning her aviation career as a stunt pilot, Auriol went on to fly more than one hundred different types of planes as one of France's most successful military test pilots of either gender. She held many world speed records throughout the 1950's and was the second woman to break the sound barrier.

Jacqueline Auriol did not start flying until she was thirty years old, when she did so only out of curiosity. The daughter of a wealthy timber importer and shipbuilder, she studied drawing and painting at the École du Louvre and psychotherapy at the Sorbonne. In 1938, she married Paul Auriol, the son of future French president Vincent Auriol, and together they were active as part of the French Resistance during World War II. The couple had two sons, Jean-Paul and Jean-Claude.

Encouraged by her husband, Auriol first qualified as a tourist pilot in 1948 and later studied aerobatics with Raymond Guillaume, considered by many to be one of France's greatest stunt pilots. As her interest grew, she realized she would need a military license if she wanted access to planes used by the Groupe de Liaisons Aériennes Ministérielles (GLAM), an elite group of military pilots.

Auriol's life changed dramatically in 1949, when a seaplane on which she was a passenger crashed into the River Seine, and she was severely injured. She underwent twenty-two operations to rebuild her face and did not permit her two children to see her for nearly two years because of her disfigurement. While in the United States for the final two operations, she earned her helicopter pilot's license in only four weeks.

Auriol did not allow her injuries to prevent her from becoming licensed as a military pilot in 1950. She was accepted as a test pilot at the French Flight Test Center in Bretigny, France. In 1951, she reached a speed of 507 miles per hour in one of the first Vampire jets, breaking American aviator Jacqueline Cochran's speed record. For this, the first speed record attained by a French pilot since World War II, she received the French Légion d'Honneur and the American Harmon Trophy.

Overall, Auriol held the women's world speed record five times between 1951 and 1964. In 1953, she became the second woman to break the sound barrier and was one of the first pilots of either gender to pilot the Concorde. Auriol later worked with the French Ministère de la Coopération, locating water and mapping crop species by using remote sensing techniques. For her agricultural work, Auriol received the Ceres Medal of the United Nations Food and Agriculture Organization.

P. S. Ramsey

Bibliography

Auriol, Jacqueline. *I Live To Fly.* Translated by Pamela Swinglehurst. New York: Dutton, 1970. Jacqueline Auriol's autobiography, describing her childhood, marriage, wartime activities, and her many aviation experiences.

Cadogan, Mary. *Women with Wings: Female Flyers in Fact and Fiction.* Chicago: Academy Chicago, 1992. Profiles a wide variety of women in aviation, from eighteenth century balloonists to twentieth century astronauts.

Welch, Rosanne. *Encyclopedia of Women in Aviation and Space.* Santa Barbara, Calif.: ABC-Clio, 1998. A reference work containing a broad overview of the role played by women in the fields of aviation and space.

See also: Aerobatics; Concorde; Military flight; Sound barrier; Test pilots; Women and flight

Autopilot

Also known as: Automatic flight control systems or integrated flight control systems

Definition: A device used to control an aircraft in flight automatically.

Significance: Autopilots are equipped on large commercial, military, and many small aircraft. By reducing pilot workload, autopilots greatly increase flight safety.

Nature and Use

Many aircraft are equipped with autopilots that will fly an aircraft automatically while the pilot accomplishes other tasks. These systems vary greatly in sophistication, from simple wing levelers to completely integrated flight control systems.

The simplest autopilot is a single-axis system. Most single-axis autopilots are designed to control the motion of the aircraft around the aircraft's longitudinal axis, passing from the front of the aircraft to the rear. When movement around the longitudinal axis becomes unstable, then the aircraft will roll, or tip, from side to side. In its simplest form, the single-axis autopilot may be referred to by pilots as a wing leveler. Upon activation, a wing leveler will stabilize the aircraft by leveling the wings. By adding features such as turn, heading, and navigational control, pilots can use a single-axis system throughout most of the flight.

Another common type of single-axis system is known as the yaw damper. This autopilot maintains control of the aircraft around the vertical axis, running through the aircraft from top to bottom. When movement around the vertical axis becomes unstable, the aircraft is considered to be slipping or skidding sideways. This motion is known as yaw. Yaw dampers are designed to prevent slipping and skidding.

A form of autopilot commonly used on medium-sized aircraft is the dual-axis system. A dual-axis autopilot will maintain control of the aircraft around both the lateral and the longitudinal axes. The lateral axis of an aircraft is an imaginary line passing from wingtip to wingtip. Movement around the lateral axis causes the front of the airplane to move up or down.

For example, a dual-axis autopilot will be able to keep both the wings and the nose of the aircraft level. Pilots may use the dual-axis system to hold a particular direction, follow commands from a navigation system, maintain an altitude, and climb or descend at a specified rate.

The three-axis autopilot is a combination of a dual-axis system and a yaw damper. Airliners and large business aircraft are normally equipped with a three-axis autopilot. Three-axis systems are connected with navigation and flight-management systems. In addition, they may include features such as throttle control and ground steering.

Integration

Many autopilots can connect to, or be integrated with, a navigation system. In a single-axis autopilot, this may merely be a connection to the directional gyro. In a complex three-axis system, all of the navigation devices may be connected to the autopilot. In this case, the autopilot could be considered an integrated flight control system.

Most integrated flight control systems include a special attitude indicator known as a flight director indicator. In addition to the symbolic airplane and horizon reference line found in most attitude indicators, a flight director indicator includes a special set of needles called flight director, or command, bars. The flight director bars will move up, down, right, and left to indicate where the autopilot intends to fly. Often, these bars are operated by a special computer running in parallel with the autopilot computer. In case of an autopilot failure, the flight director computer will still be able to manipulate the flight director bars. Pilots can manually fly a precise flight path by keeping the bars centered. By allowing the flight director computer to make the complex calculations involved in flying a precise flight path, pilots are still able to reduce their workload.

How the System Works

In order to control the aircraft, an autopilot must be able to sense attitude. To do this, autopilots rely on gyroscopic instruments, or accelerometer-based sensors. Often, the attitude gyro is used to transmit information regarding pitch and roll attitude to the autopilot computer. A turn and bank indicator or a turn and slip indicator can be used to supply yaw information. The autopilot computer will compare the actual flight attitude of the aircraft with the desired flight attitude and, if necessary, move the appropriate control surface.

The device that operates the control surfaces of the aircraft is called a servo. A servo converts electrical energy into mechanical energy. Servos may be electric, hydraulic, or pneumatic. Electric and hydraulic servos are quite common. Electric servos are widely used on aircraft with mechanical or fly-by-wire controls, and hydraulic servos are widely used on aircraft with hydraulic controls.

Electric servos contain a small, electric motor. In this type of system, the computer sends a voltage to the servo, causing the motor to rotate. The motor is connected to the aircraft controls, and as the motor turns, the controls are moved.

Hydraulic servos contain a small, electrically controlled, hydraulic actuator. In this type of system, the computer sends a voltage to the actuator. Valves within the actuator channel hydraulic fluid in and out of small cylinders containing pistons. The pistons are connected to the control surface, and, as they move, the surface moves.

Pneumatic servos contain electrically operated valves. These valves channel air into bellows that are connected to the aircraft controls. The inflation and deflation of the bellows causes the controls to move.

Thomas Inman

Bibliography

Brown, Carl A. *A History of Aviation*. 2d ed. Daytona Beach, Florida: Embry-Riddle Aeronautical University, 1980. A well-illustrated book that covers the history of flight from ancient times to the space age.

Eismin, Thomas K. *Aircraft Electricity and Electronics*. 5th ed. Westerville, Ohio: Glencoe, 1994. A beginner's text starting with the fundamentals of electricity and ending with electric instruments and autoflight systems.

Helfrick, Albert. *Principles of Avionics*. Leesburg, Va.: Avionics Communications, 2000. A very complete avionics text that includes history.

Jeppesen Sanderson. *Instrument Rating Manual*. 7th ed. Englewood, Colo.: Jeppesen Sanderson, 1993. A textbook designed to assist pilots to prepare to add an instrument rating to their pilot license.

See also: Airplanes; Avionics; Flight control systems; Instrumentation; Pilots and copilots; Roll and pitch

Avionics

Also known as: Aviation electronics

Definition: A combination of the words "aviation" and "electronics."

Significance: Many aircraft cannot fly without avionics. Avionic equipment includes a variety of systems designed to assist pilots, aviation maintenance technicians, and passengers.

History

From the time avionics were invented in 1903 until approximately 1930, pilots rarely used them, navigating instead by known landmarks on the ground. In the 1930's, however, engineers began installing communications and navigation equipment in airplanes. The first system designed for airplane navigation was the direction finder (DF), also known as a homing beacon. In the late 1930's, the government began installing the first range stations, which allowed pilots to follow a specific course. Before World War II (1939-1945), electronic equipment was large, heavy, and often required an extra person to operate; therefore, only large aircraft used avionics.

During World War II, both Allied and Axis forces developed radio detection and ranging, or radar. In addition, the Allies developed the identification, friend or foe (IFF) system. The IFF system became the air traffic control (ATC) transponder. Throughout the 1940's, engineers made many improvements in the size and reliability of avionics. During the late 1940's and early 1950's, the very high frequency omnidirectional range beacon was developed, which was a great improvement to the original range stations.

In the 1960's, radios became lighter and smaller, mostly due to the application of the transistor to avionic equipment. The first avionics to use transistors were hybrids, or radios containing both vacuum tubes and transistors. In the 1970's, manufacturers introduced the first reliable solid-state avionics, using semiconductor devices rather than electron tubes. Simultaneously, avionics using digital systems were introduced. These developments allowed for even smaller, lighter, and easier to use systems. Consequently, small personal aircraft of the 1970's were able to have more complex avionics than could the large airliners of the 1950's.

The introduction of the microprocessor and database technology in the 1980's created a revolution in the avionics industry. For the first time, pilots could use long-range navigation systems, such as loran-C and Omega, for aircraft navigation. This new technology also allowed for increasingly smaller, lighter, and even easier to use avionics.

The 1990's brought the introduction of satellite navigation, known as the Global Positioning System (GPS). By the end of the decade, the U.S. government decommissioned the Omega navigation system, which GPS had made obsolete.

In the early twenty-first century, improvements in microprocessors allowed many more improvements in avionics systems. Three-dimensional moving map displays and low-cost electronic flight instrumentation are a few of the improvements to come about in the first decade of the third millennium.

Navigation

Avionics assist the pilot to navigate the aircraft in several ways. Many different navigation systems help pilots find their way across the globe and locate runways.

The automatic direction finder (ADF) indicates the direction of special radio navigation stations and AM broadcast stations. This system receives radio signals in the low- and medium-frequency bands. An indicator in the instrument panel simply points toward the source of the radio signals.

The very high frequency omnidirectional range beacon system provides the pilot with directional information relative to a course. This system receives radio signals in the very high frequency range from a station on the ground. The system is made up of a radio receiver connected to a device that converts the radio signal to visual information. The pilot chooses a bearing to fly, and a special indicator in the panel shows whether the airplane is to the left or right of a course, also known as a radial, that passes through the navigation station.

Loran-C provides pilots with long-range area navigation. The name "loran-C" is an abbreviation of "long-range navigation," with the "C" representing the fact that the current system is the third generation of loran. Originally, loran-C worked as a maritime navigation system; however, with microprocessor and database technology, it became available to pilots. Loran-C does not require the pilot to use a navigation station as a reference point, as do the very high frequency omnidirectional range beacon and the automatic direction finder. Instead, the pilot simply chooses an origin and destination within the loran-C coverage area, and the loran-C guides the pilot directly from the origin to the destination. The system consists of a low-frequency receiver, computer, database, and an indicator. The receiver listens for pulses from a set of transmitting stations, and the computer measures the time delay between pulses to determine position.

The Global Positioning System (GPS) provides pilots with a worldwide area navigation system. Although GPS is similar in design to the loran-C, it is much more accurate. Twenty-four GPS satellites orbit the earth and provide pilots with three-dimensional navigation signals. Often, the GPS system will work with a moving map display to show exactly where the airplane is. The system consists of an ultrahigh frequency receiver, computer, database, and indicator. The receiver listens for pulses from the satellites, and the computer measures the time delay between pulses to determine position. With wide- and local-area augmentation systems, GPS can be used as the sole means of navigation.

The Instrument Landing System (ILS) gives pilots guidance toward runways and consists of three major components. The first, the aircraft's localizer transmitter, is integrated with the VHF omnirange. When the pilot selects a special ILS channel, the VHF omnirange system switches to localizer mode. Now, instead of having several courses to choose from, the pilot has only one, which will lead to the end of the runway. The course directing indicator (CDI) will indicate whether the course is to the pilot's left or right.

The second ILS component, the glide slope, provides pilots with vertical guidance to the end of the runway. The glide slope consists of a UHF receiver and circuitry that converts navigation signal information to visual information. When the pilot selects an ILS channel with the VHF omnirange system, the glide slope automatically becomes active and provides information on the CDI to indicate whether the pilot is above or below the proper glide path.

The final ILS component, the marker beacon, then turns on a light in the cockpit as the aircraft passes over certain checkpoints during the approach to the airport. A special receiver in the airplane is tuned to 75 megahertz and will listen for special signals from marker transmitters placed along the localizer course.

Distance measuring equipment (DME) uses radar principles to measure the distance between the aircraft and special navigation stations on the ground. The DME displays distance, speed, and time to or from the navigation station. The aircraft system consists of a transmitter and a receiver. The UHF transmitter sends pairs of pulses to a ground station, which the ground station then sends back to the aircraft. The DME will measure the time elapsed from when the pulses were sent to when they return and will calculate distance, speed, and time.

Communication

There are many communications systems on board aircraft. In small airplanes and helicopters, the system will consist of a VHF transceiver for the pilot to communicate with air traffic controllers. Similar to a citizen's band radio, this more powerful system can have up to 2,280 channels. Many aircraft also have an intercom with which to communicate with other crewmembers and passengers.

In addition to the VHF transceiver and intercom, some aircraft may have high-frequency transceivers or satellite transceivers to allow long distance communication on transcontinental flights. Although similar in purpose, the design of these two systems is quite different. The high-frequency (HF) transceiver transmits and receives frequencies between 3 and 30 megahertz. Radio frequencies within this range have the ability to stay in the earth's atmosphere and travel around the world. The satellite system uses ultrahigh frequencies and an antenna that swivels to remain pointed at a communications satellite in orbit above the earth. The signal travels from the airplane to the satellite and is then relayed to any place on Earth.

Another communications system is the aircraft communications and reporting system (ACARS), a private, low-speed, digital communications system used by the airlines to communicate between the aircraft and the operations center. Aircraft may also include passenger address systems that allow the pilots to speak to passengers and a radio telephone system that allows passengers to call friends, relatives, and business associates.

Surveillance

Air traffic controllers use two systems to track the movements of aircraft: the primary surveillance radar and the secondary surveillance radar. The primary surveillance radar uses a powerful transmitter and a large rotating antenna to send strong bursts of microwave energy into the air. The microwave energy reflects off the aircraft, returns to the large antenna, and shows up as a dot on the air traffic controller's radar display. However, not all aircraft reflect microwaves well, and such aircraft may not show up on the radar display.

For this reason, all private and commercial aircraft are required to have special equipment on board that acts as part of the secondary surveillance radar system. Secondary surveillance radar sends a pulse code to a special radio in the aircraft called an ATC transponder. The ATC transponder replies with its own pulse code, which may contain a variety of information, such as altitude, speed range, and assigned codes, that will show on the air traffic controller's radar screen.

Aircraft can also perform surveillance on each other. Airliners and large business aircraft use a system called transponder-based collision avoidance system (TCAS). A TCAS-equipped aircraft sends a pulse code to which other aircraft with ATC transponders reply. A special instrument in the first aircraft displays the location of the second, indicates collision threats, and recommends a flight direction to avoid collision.

Many aircraft are equipped with weather-surveillance systems. These come in two varieties, active and passive. The active system uses radar. Mounted in the nose of the aircraft, the antenna points forward and sweeps back and forth. The radar transmits energy in front of the aircraft, and water droplets reflect this energy back to the radar antenna. Rain will display on a screen in the instrument panel of the aircraft.

The passive weather-surveillance system uses a special loop antenna to detect the electrical activity associated with thunderstorms and air turbulence. The activity is shown on an indicator in the instrument panel of the aircraft. Both systems help pilots avoid dangerous weather and, in some cases, can be combined into a single, comprehensive weather-avoidance system.

Autopilots

Many aircraft are equipped with autopilots, which fly an aircraft automatically while the pilot accomplishes other tasks. The simplest autopilot is a single-axis system. The single-axis autopilot controls the airplane on only one of the axes of flight. For example, a wing leveler will keep the wings level, but the pilot will be responsible for keeping the nose level, and keeping the tail in line. The dual-axis autopilot controls two axes of flight, keeping both the wings and the nose of the aircraft level, for example.

The three-axis autopilot maintains control of the aircraft in all axes or directions. Often, two- or three-axis systems are interconnected with navigation and flight-management systems, and may include features such as throttle control and ground steering. In these cases, the autopilot is considered an integrated flight-control system.

Passenger Entertainment and Convenience

There are many systems designed for passenger entertainment and convenience. Many aircraft have special telephones that passengers may use to make telephone calls. In addition, multichannel sound systems deliver several styles of music from which passengers may choose. In larger airliners, video systems allow passengers to watch movies or play video games. In some aircraft, passengers can keep track of the flight's progress by viewing a moving map display. In addition, business jets may have a local area network, printers, and modems to allow passengers to work while in flight.

Thomas Inman

Bibliography

Brown, Carl A. *A History of Aviation*. 2d ed. Daytona Beach, Fla.: Embry-Riddle Aeronautical University, 1980. A well-illustrated book that covers the history of flight from ancient times to space flight.

Eismin, Thomas K. *Aircraft Electricity and Electronics*. 5th ed. Westerville, Ohio: Glencoe, 1994. A beginner's text that starts with the fundamentals of electricity and ends with electric instruments and autoflight systems.

Helfrick, Albert. *Principles of Avionics*. Leesburg, Va.: Avionics Communications, 2000. A very complete avionics text that includes history.

B

Baggage handling and regulations

Definition: Airline procedures and processes for carrying passengers' baggage from their points of departure to their final destinations.

Significance: Airline baggage handling procedures and regulations help to ensure that passengers' baggage will accompany them on their flights and be at their destinations when they arrive. Airline procedures also ensure that all items accepted as baggage meet safety and security guidelines and regulations to protect passengers and airline personnel.

The Process

The baggage handling process begins when passengers present themselves to check in for their flights. Much like the passengers themselves, who receive seat assignments and boarding passes, baggage is also checked in. At baggage check-in, either computerized or handwritten baggage or destination tags are attached to each bag and a claim check is given the passenger. The baggage tag specifies the passenger's airline, flight, connecting cities (if any), and final destination. Computerized tags may also display the passenger name, date, time, and reservation information.

At some airports, passengers can have their baggage checked and tagged in either one of two places, at skycap locations or at airline ticket counters. Skycaps are individuals stationed at curbside locations in front of airport terminals. They offer the convenience of immediate baggage checking, enabling passengers to proceed directly to their departure gate for boarding passes. Curbside checking was banned at many airports for security reasons in the aftermath of the terrorist attacks of September 11, 2001. Airline ticket counters offer all baggage, passenger, boarding, and ticketing services.

Both skycaps and airline ticket counter agents must follow certain precautions and procedures when checking baggage. For safety and security, they must ask all passengers whether anyone unknown to them has asked them to carry any items on their flight and whether any of the items with which they are traveling have been out of their immediate control since the time they were packed. To ensure that all of a passenger's baggage arrives at its destination, skycaps and airline ticket counter agents use a procedure called "Ask, Tag, and Tell." "Ask" reminds the skycap or agent to ask the passenger's final destination and the number of bags being checked. "Tag" reminds the skycap or agent to produce the correct number of tags to the correct destination and to affix the tags to the bags. "And" is a reminder to ask whether each bag has a separate passenger identification tag. "Tell" reminds the skycap or agent to tell the passenger how many bags have been checked to their final destination.

After the bags are checked and tagged, they are usually put on a baggage belt conveyor system. Bag belt systems take bags from skycaps or ticket counters and transport them to the baggage makeup area. Airline employees in a baggage makeup area sort baggage by flight numbers and destinations and place them into carts or other conveyor systems to transport the baggage to the aircraft. At the aircraft baggage is further sorted for loading and unloading purposes. Bags going only to the aircraft's destination are loaded in one section, usually called local baggage. Bags that are going to a connecting flight are loaded into another section, usually called connect bags. Messages are sent to the destination city after the plane takes off, telling where the different bags have been loaded.

Upon landing, the process is reversed. Local baggage is unloaded into specific carts or conveyor systems that transport it to the baggage makeup area. There it is placed onto other conveyor systems that transport it to the baggage claim area. In baggage claim areas, passengers pick up their bags. Connect bags are unloaded into other specific carts or conveyor systems that transport them either to the baggage makeup area to be brought back out to the connecting airplanes or directly to the connecting airplanes.

Missing or Lost Baggage

In the event that a passenger's bags do not arrive with the passenger or a bag arrives without a passenger, airline baggage service offices in baggage claim areas handle missing, lost, or found baggage reports. These reports document how many bags are missing or found, the tag and flight numbers from the claim checks, descriptions of the

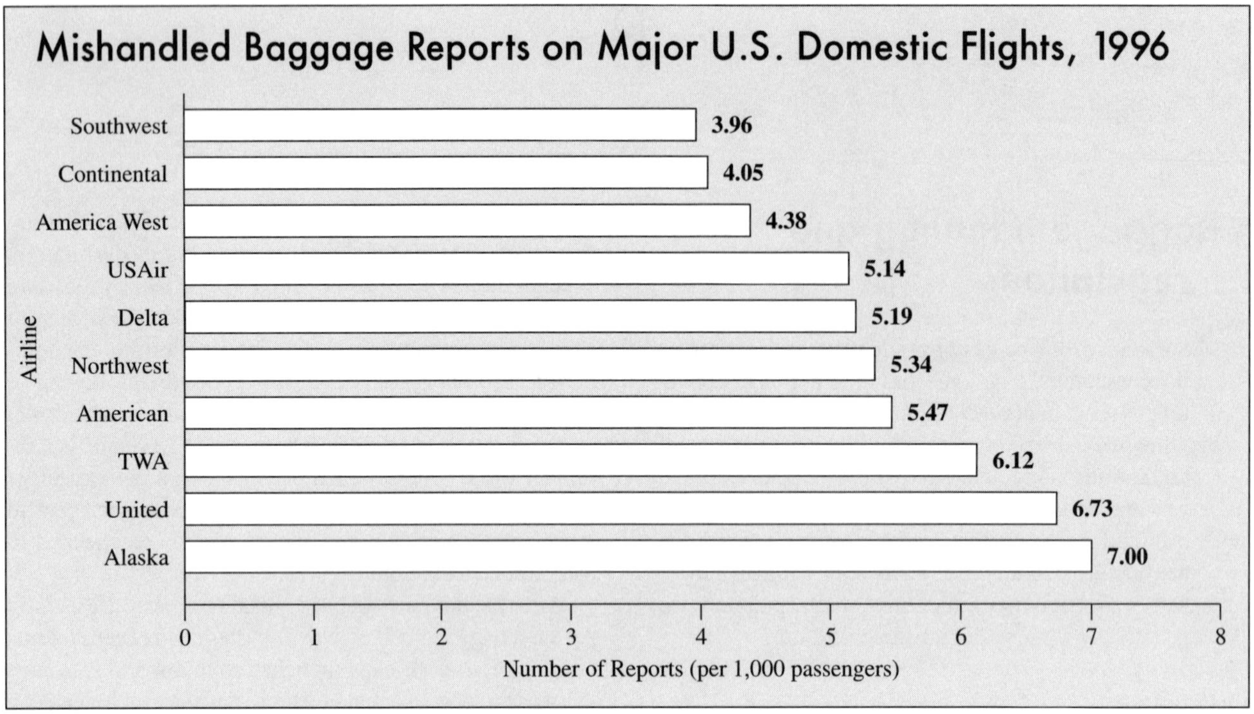

Source: Data taken from U.S. Department of Transportation.

bags and their contents, and passenger contact information. This information is entered into bag tracing systems that are intelligent databases. These systems constantly search themselves to identify and match missing or found bags to the passengers who checked them. The industry average of missing or lost bags is approximately 3 percent of every one thousand bags transported. Of that 3 percent, most are located and reunited with passengers within forty-eight hours.

Baggage Acceptance Guidelines

For the safety and security of an airline and its passengers, airlines have established baggage acceptance guidelines. These guidelines concern themselves with baggage contents, how the contents are packed, and liability for their damage or loss.

Baggage acceptance guidelines address what are known as acceptable, conditionally acceptable, and unacceptable articles. Acceptable articles are considered to be the personal property necessary or appropriate for the purposes of the passenger's travel. Typical acceptable articles are clothes, shoes, personal, or business items. Airlines accept a liability of $2,500 per bag for damage or loss. Conditionally acceptable articles are those items considered irreplaceable, fragile, perishable, or improperly packed. Conditional acceptance also addresses the condition or quality of the receptacle containing a passenger's contents. Any suitcase or box must be of reasonable durability, must stay closed or sealed, and must be able to withstand normal handling. Conditional acceptance limits the liability of an airline for damage or loss. Unacceptable articles are those considered hazardous to passengers or aircraft. At no time are they ever accepted for transport.

Other acceptance guidelines address airline and aircraft security. These include requirements that passengers cannot check a bag onto a flight for which they do not have a ticket. Bags may not be checked to a different destination than that of the passenger. For most international flights, bags are not loaded until it is known that the passenger checking them has boarded the aircraft. There are also time requirements that specify how early or how late passengers may check their bags.

Other acceptance guidelines address how many, how heavy, or how large baggage may be. Although each airline has its own specific guidelines, generally speaking, most airlines allow three bags per passenger, including carry-on items passengers keep with them in the aircraft. Most airlines do not allow any bag that weighs more than 70 pounds or that exceeds 60 to 65 inches in outside linear measurement. When any of these guidelines are exceeded, extra baggage charges, which may be significant, are incurred. There are exemptions to allow for special items, such as wheelchairs, or other devices a passenger may require.

Jim Oppermann

Bibliography

America West Airlines. *Basic Ramp Service.* Phoenix, Ariz.: America West Airlines. Chapter 3 of this America West Training Manual outlines how to read, interpret, and handle different baggage tags and the order in which baggage should be loaded and unloaded.

Nichols, Wendy, and Stefano Sala. "Minimizing Connecting Times a Must for Airline Competitiveness." In *Handbook of Airline Operations*, edited by Gail F. Butler and Martin F. Keller. New York: McGraw-Hill, 2000. An article presenting a model to reduce connecting times and outlining steps for ground handling companies to adequately schedule personnel and equipment.

See also: Air carriers; Airline industry, U.S.; Ticketing

Balloons

Definition: Fabric containers holding a lighter-than-air gas so that the containers and any payload are buoyed up and float in the sky.

Significance: One of humanity's oldest dreams has been to float in the sky. Balloon flights first transformed that dream into adventure, and they continue to do so. For more than a century after the first balloon flight, balloons were the cutting edge of science and aviation technology, and they remain the best craft for scientific missions operating in altitudes of roughly 10 to 30 miles, which are too high for airplanes and too low for orbiting vehicles.

Nature and Use

The term "balloon" may refer to the gas bag, or envelope, or to the balloon and any additional objects attached to it, which are usually hung below. Objects are attached to smaller balloons by a single line, but larger balloons require netting to spread the load over the entire gas bag. A large cargo below is called a gondola or basket, often a large wicker basket.

Buoyancy is the key to balloon flight. The ancient mathematician Archimedes stated that a body immersed in a fluid is buoyed up by a force equal to the weight of the displaced fluid. For balloons, a lighter-than-air (LTA) gas provides buoyancy to lift the balloon containing it as well as any payload. LTA gases include gases with densities lower than that of air and warmed air that has expanded and is thus lighter than the surrounding air. The two low-density gases used for balloons are hydrogen and helium, which require the balloon to be sealed so that they do not mix with the heavier air.

Air is usually heated for buoyancy by burning propane or kerosene. Warmed air rises through an open base, and air that has cooled drains out of that same orifice. Warmth is constantly drained away at the surface of the balloon so hot-air balloons require frequent firings of their burners. Consequently, they tend to have shorter range than balloons with low-density gas.

More importantly, hot air has less lifting capacity than hydrogen or helium. Typically, hydrogen has a net lift of 60 pounds per 1,000 cubic feet, but hot air provides only 17 to 20 pounds of lift. Thus, hot-air balloons must be three times larger to lift the same payload. However, hot-air balloons are less expensive to operate, because they do not have to accommodate the complexities of hydrogen and helium.

Although helium lifts 14 percent less than hydrogen (53 rather than 60 pounds) per 1,000 cubic feet, it has the major safety advantage of being nonflammable, whereas hydrogen can ignite explosively.

As a balloon increases its altitude, the density and pressure of the surrounding air decreases, meaning there is less lift available per unit volume, so the balloon must be larger to carry a given payload to higher altitudes. A partially compensating factor is that the buoyant gas also grows less dense as the pressure decrease allows it to expand, but the trend is toward miniscule lift per unit volume, as most of the atmosphere is left below. Balloon builders can compensate with lighter payloads, such as remotely controlled instruments, but at some point, the weight of balloon fabric alone matches the lift from the gas volume, and even the largest balloons can go no higher.

Balloonists have two other ways to vary the buoyancy of their craft. They can descend by decreasing buoyancy or land by valving out some of the lifting gas. They can increase buoyancy by dropping ballast, which is water, sand, or other material carried along for that purpose. In extreme conditions, balloonists have dropped all articles in the gondola and even the gondola itself.

History

For centuries, the Chinese made toy hot-air balloons of a design that could and might have been scaled up to carry passengers. There are accounts from twelfth century B.C.E. China of people in balloons, but the records are too old and incomplete to be confirmed. Likewise, drawings on pottery associated with the Nazca Lines, constructed more than two thousand years ago in southern Peru, suggest that these massive earthen line drawings were made with overhead direction from hot-air balloons.

Confirmable accounts begin in the eighteenth century. In 1782 and 1783, Joseph-Michel and Jacques-Étienne Montgolfier, two French brothers, flew hot-air balloons larger than toys, with animals as their first passengers.

Ironically, in their first balloons, the Montgolfiers had wanted to use hydrogen, which British chemist Henry Cavendish had discovered in 1776, noting in his experiment reports that this "inflammable air" was lighter than ordinary air. The French Academy in Paris was working toward a rubberized or varnished fabric to contain the troublesome gas that seeped through ordinary fabrics and escaped. When Joseph-Michel Montgolfier experienced the same problem, he noted that scraps of paper in a fireplace rose up the chimney. Paper, which the Montgolfier family manufactured, could contain smoke, so the Montgolfiers' made hot-air balloons and successfully flew three animal passengers: a rooster, a sheep, and a duck. King Louis XVI's permission was required for people to fly because it was not known whether leaving the ground might be harmful to people.

Jean-François Pilâtre de Rozier, a young doctor who wanted to take the risk of human flight, recruited the Marquis François d'Arlandes to serve as copilot and, more importantly, to secure the king's permission. On November 21, 1783, having obtained permission, the two men flew over Paris for twenty-five minutes while desperately stoking their lifting fire and sponging out fires in their rigging caused by sparks. Below them, nearly the entire populace of Paris watched.

Only a few days later, on December 1, 1783, Jacques-Alexander-César Charles, of the French Academy, flew a hydrogen balloon. The preparation required the production of large quantities of hydrogen gas and the careful varnishing of cloth to render it relatively airtight. The flight illustrated the advantages of hydrogen balloons over hot-air balloons. Because hydrogen is more buoyant than hot air, the hydrogen balloon could be a third the size of a comparable hot-air balloon. Charles flew for two and one-half hours, dropped off his passenger at sunset, and then rose high enough to be the first person to see the sun set twice in one day.

Shortly thereafter, balloonists began attempting not only to fly but also to reach destinations. Jean-Pierre Blanchard, another Frenchman, and John Jeffries, an American, decided to be the first aeronauts to fly across the English Channel to France, which they did on January 7, 1785. However, they were somewhat humbled upon arrival, because they had jettisoned most of their clothes, along with the gondola and the articles within it, in order to avoid falling into the Channel.

A rivalry ensued with the French wanting to have a flight from France back to England. Pilâtre de Rozier, who had piloted the first hot-air balloon, had another balloon made that advanced balloon technology. This hybrid de Rozier balloon had a hydrogen balloon that rode over a hot-air balloon. The pilot could vary the balloon's buoyancy, and thus its altitude, by adjusting the fire under the hot-air balloon instead of valving out hydrogen gas or dropping ballast, neither of which could be replenished. On June 15, 1785, Pilâtre de Rozier and his copilot floated in this balloon toward England.

Unfortunately, Pilâtre de Rozier fell prey to two problems of early balloonists. He had no reliable weather reports, and when the wind changed, he floated back toward France. Worse, his hydrogen lifting gas was very flammable and the varnished cloth only slightly less so. As the audience who had joyously witnessed the launch watched, the balloon caught fire, lost buoyancy, and plummeted the two aeronauts to their deaths. After Pilâtre de Rozier's death, hot-air balloons fell out of favor until experiencing a renaissance in the 1960's.

Similar problems plagued ambitious balloon flights for the next century. Attempts at crossing the Atlantic Ocean lost credibility as balloonists waited vainly for suitable weather. Balloon flights crossed the Alps in Europe, and John Wise crossed a third of North America, but the final destination of long-distance flights was always a surprise, and disaster was always just a spark away.

Despite its shortcomings, the balloon went to war in 1793 when revolutionary France was attacked by a number of neighbors. At the Battle of Fleurus in 1794, the fledgling French balloon corps fielded a single reconnaissance balloon tethered on a line several hundred feet above the ground. Observers in the balloon, who could see several miles past the line of battle, provided tactical reports via notes dropped from the gondola. The most important observation was that the attacking Austrian army had pitched an empty tent city in an effort to overawe the French commander into retreating. Because of that vital bit of intelligence, the French did not retreat but rather fought on with their exotic new technology looming over and unnerving their opponents, until they eventually won the battle.

Aerial reconnaissance was reinvented in the American Civil War (1861-1865), in which several groups operated observation balloons. The most successful was inventor Thaddeus Sobieski Coulincourt Lowe, who organized an aeronautic corps of balloon observers for the Union. Although balloon technology had not advanced tremendously, communications technology had. Lowe's observers transmitted their reports either by signal flags or by telegraph wire running down to the ground.

The potential of Lowe's reports is shown by accounts of one 1862 engagement, in which Lowe directed Union cannon fire at an area invisible to Union guns because it was behind a hill. When the Confederate horsemen rode away from the shell impacts, Lowe had the guns redirect their fire. After the war, Confederate accounts revealed that one of the horsemen was showered by so much dirt from a near-miss that his colleagues feared he had been hit. That horseman was Jefferson Davis, the Confederate president. Nearby, also in danger, was Robert E. Lee, Davis's commander of all Confederate armies. Despite Lowe's successes, a change in the Union Army high command caused Lowe to fall from favor and come under stringent control by an unsympathetic regular officer. Lowe ultimately resigned from his post, and his entire corps withered away.

Military ballooning was next reinvented by the French during the Franco-Prussian War of 1870-1871. The Prussians smashed the regular French army and surrounded the French capital city of Paris. The plucky Parisians responded by raising a militia to hold off attacks and launching balloons to carry observers and send messages out of the city, rallying the countryside. Fifty-four of sixty-two balloons got through, carrying one hundred people and two and one-half million pieces of mail. Although France eventually accepted harsh peace terms, the utility of war balloons was established.

By the time World War I began in 1914, observation balloons were in use by both sides. By the end of the war, they were being replaced by heavier-than-air airplanes.

Surprisingly, small balloons did evolve into an important military and civilian use during World War II. The development of small radio transmitters combined with remotely operating weather instruments made possible

Balloons, such as this one photographed by Matthew Brady, were used during the American Civil War and other nineteenth century wars for military observation. (Corbis)

balloon-borne radiosondes that reported temperature, pressure, and relative humidity. Angle data from antennas tracking the radiosondes yielded the more important factors of wind speed and direction at different heights. The use of weather balloons has continued into the twenty-first century to help predict weather, to plot sky conditions for aircraft, and to fire artillery more accurately.

Finally, in the 1990's, tethered balloons returned to service as aerostats, providing platforms at altitudes as high as 10,000 to 15,000 feet for radar stations and communications repeater stations.

Scientific Applications

Some have said that balloons are a pacifist technology. They are big, slow, and cannot be piloted accurately, particularly when the wind changes. Yet, balloons do a number of things well. They move gently and can carry large payloads that would not fit in an airplane fuselage. Most importantly, they can reach high enough altitudes to perform many of the research tasks generally performed by spacecraft. However, balloons can accomplish this research more cheaply and quickly than can spacecraft, and without the vibration and acceleration forces of a rocket launch into space.

In the twentieth century, a number of supporting technologies radically improved, allowing the balloon to become much more practical for research applications. Most importantly, helium became widely available as a nonflammable lifting gas. Synthetic fibers, such as nylon, polyethylene, and Kevlar, supplied fire-resistant materials with the lightness of silk and strengths approaching steel. Vulcanized rubber allowed light, cheap, disposable balloons. Virtually all measuring instruments shrank in size. Finally, worldwide weather databases and communications links made it more possible to guide balloons on long voyages.

The quest for altitude began as both an adventure and a science. The first balloonists had no idea whether the atmosphere continued indefinitely or became lethal a short distance above the ground. They soon discovered that pressure and temperature decreased with increasing altitude. Those who attempted altitude records discovered temperatures tens of degrees below the freezing point of water. However, the greatest risk was hypoxia, or oxygen deprivation, which causes weakness, shakiness, mental confusion, and eventual death. To deal with these problems, balloonists developed oxygen-supply systems and learned to use equipment that would not freeze in the bitter cold.

However, even the breathing of pure oxygen was found to be insufficient above altitudes of 49,000 feet. Swiss balloonist Auguste Piccard surmounted this problem with a pressurized cabin that essentially represented the first space capsule. On May 27, 1931, Piccard and an assistant launched from southern Germany and reached 51,793 feet, making them the first fliers ever to reach the stratosphere. More importantly, they discovered that cosmic rays increased with altitude, proving that these rays came from somewhere in space and not from radioactivity within the earth.

From 1933 to 1935, the governments of the United States and the Soviet Union foreshadowed the space race that would begin nearly a quarter-century later. Balloon flights carried personnel and instruments to steadily greater heights and developed many technologies that were later used in the space race. For example, on May 4, 1961, the American Stratolab High V balloon reached a world-record-breaking altitude of 113,600 feet, with an open gondola so that the two pilots could test space suits in near-space conditions for the Mercury orbital-flight program.

In retrospect, the best high-altitude science data began to be collected in the 1960's, after improved robotic instrumentation allowed shedding the weight of the balloonists and their life-support gear. Over the closing decades of the twentieth century, astronomic balloon-borne instruments conducted sky surveys in a number of frequency bands that cannot penetrate the lower atmosphere and provided valuable weather data from the lower stratosphere.

By the late twentieth century, the National Aeronautics and Space Administration (NASA) began using superpressure balloons for relatively small payloads of several tens of pounds. Balloons called zero-pressure balloons expand when warmed by the sun and contract at night when cooled. When warmed at high altitude, they must vent excess helium to prevent bursting. This gas loss limits mission duration to only several days. Superpressure balloons, in contrast, keep the same maximum shape when the balloon is warmed. Because no gas is lost, such balloons can operate for weeks or months, and some of these balloons have circled the globe one or more times. By the early twenty-first century, NASA had begun flying large superpressure balloons in a program called the Ultra Long Duration Balloon (ULDB). These large balloons could carry several tons of instrument payload for weeks at a time. Less well documented are flights of small superpressure balloons by U.S. intelligence agencies since the 1950's.

Recreational Applications

Although ballooning is no longer the world's primary means of aviation, a balloon ride remains a beautiful and awe-inspiring experience. Balloonists enjoy panoramic views that float by below them and sounds that float up from the ground.

However, a recreational balloon ride was a rare experience until the so-called renaissance of hot-air ballooning, which was started by American balloonist Edward Yost. While Yost was developing high-altitude balloons for the U.S. government in the 1950's, it occurred to him that polyethylene-coated nylon would be a lighter, less flammable material than that used for the Montgolfiers' balloons. He used an acetylene welding torch as a less labor-intensive source of hot air than that used by the Montgolfiers. After some development, such as replacing the welding torch with a propane burner, Yost made the first "modern" hot-air balloon launch from Bruning, Nebraska, on October 10, 1960.

Beginning in the 1960's, the new hot-air balloons radically reduced the cost and complexity of supplying buoyant gas. Thus were born ballooning clubs, competitions, and tour services. Hot-air balloons have been flown, primarily for advertising, in whimsical shapes, including those of spark plugs, light bulbs, human faces, and even a mansion.

A combination of Yost's hot-air technology, lightweight insulating material lining the gas bag, and helium made the de Rozier balloon practical for more ambitious, long-distance flights. Varying the amount of heat in the inner balloon provides altitude control for hunting favorable winds. That capability, along with worldwide weather reports, made it possible to make balloon flights across the Atlantic and Pacific Oceans. In March, 1999, another Piccard, Auguste's grandson, Bertrand, and Brian Jones spent twenty days flying 30,000 miles to make a complete circumnavigation of the globe.

Roger V. Carlson

Bibliography

Piccard, Bertrand, and Brian Jones. *Around the World in Twenty Days*. New York: John Wiley & Sons, 1999. The two authors describe the adventures and mechanics of their successful around-the-world balloon flight in March, 1999, highlighting the challenges of ballooning and the technological advances that permitted their success.

Ryan, Craig. *The Pre-Astronauts: Manned Ballooning on the Threshold of Space*. Annapolis, Md.: Naval Institute Press, 1995. A description of the lives spent and the lives lost working at progressively higher altitudes to develop equipment that was later used in space flight.

Smith, I. Steve, Jr., and James A. Cutts. "Floating in Space." *Scientific American* 281, no. 5 (November, 1999): 132-139. A description of the scientific uses of superpressure balloons at high altitudes.

Wirth, Dick, and Jerry Young. *Ballooning: The Complete Guide to Riding the Winds*. New York: Random House, 1980. A summary of the history and methods of ballooning, with many detailed diagrams and illustrations.

See also: Richard Branson; Buoyant aircraft; Lighter-than-air craft; Montgolfier brothers; Auguste Piccard

Barnstorming

Also known as: Gypsy flying, air circus

Date: Beginning in the 1920's

Definition: Originally a theater term, "barnstorming" refers to pilots and aerial performers who traveled between small, rural U.S. towns putting on air shows and selling plane rides.

Significance: Barnstormers introduced the concept of air travel to rural Americans, providing aerial displays and airplane rides. Many barnstormers were World War I veterans flying war surplus planes.

Beginnings

In the early part of the twentieth century, most people had only heard of airplanes. World War I created the first demand for planes and pilots, and although more than nine thousand men trained to fly, fewer than eight hundred of them actually saw combat. When the war ended in 1918, surplus planes, mostly Curtiss JN-4D biplanes, also known as Jennys, were both available and affordable.

During the postwar years, the civil aviation industry was in its infancy, and a pilot's license was little more than an honorary certificate. Anyone who wanted to fly an airplane could do so. Veteran aviators purchased these surplus planes and became the first barnstormers, itinerant pilots who provided many rural Americans with their first experience of flight. Barnstormers thrived in the open farmlands of the American Midwest, where a typical barnstorming season lasted from May to October. Weekends were ideal times to sell rides, and some barnstormers planned their routes to coincide with county fairs or local holiday celebrations.

Operations

Barnstorming pilots circled a town a few times to advertise their presence before landing in a nearby field and waiting for the customers to find them. On average, five minutes of flying time cost fifty cents in gas and earned the pilot five dollars. A small percentage of that sum went to the farmer who owned the field, and the rest paid for the pilot's gasoline, maintenance, and expenses. The average barnstormer made between thirty and one hundred dollars per week.

As the novelty of air flight began to wear off, barnstorming pilots found they needed to do more than give rides, and stunt-flying became part of the performance. One of the most common airborne stunts was the inside loop, in which the pilot flies upward then arches back over, while centrifugal force holds the pilot in the seat. More unusual was the outside loop, in which centrifugal force worked against the pilot. Other stunts included rolls, stalls, and reverses.

As barnstormers continued to expand their acts by adding additional performers, the air circus was born. Exhibition jumpers demonstrated parachutes, which were first deployed by a cable attached to the plane's wing. The invention of the ripcord allowed jumpers to control the length of their free fall, making the plunge appear even more dangerous.

Wing-walkers, many of whom were women, stepped out of the cockpit and into specially constructed harnesses that allowed them to defy gravity while the pilot performed a series of aerial maneuvers. Famed aviator Charles A. Lindbergh got his start traveling with a barnstormer, working for free and paying his own expenses while performing as both a parachutist and a wing-walker.

Developments

Stunts became more dangerous as spectators grew used to the sight of airplanes. The Locklear Flying Circus advertised stuntmen who would change planes in midair, and the fee per performance grew to three thousand dollars. Daredevils jumped from plane to plane, from planes to trains, climbed ladders from automobiles, motorboats, motorcycles, and horses. Repairs, part changes, and refueling operations were all performed in midair. Although parachuting had been an early part of barnstorming performances, air shows began advertising that their pilots and stuntmen performed without parachutes.

Publicity was everything. The growing motion picture industry fueled the fire, as cameras could be attached to spinning airplanes. One pilot, Leslie Miller, arranged a demonstration in which he would "loop" a Florida bridge, agreeing beforehand to fail on his initial attempt, so that on the next day, the crowd would swell in the expectation of seeing him crash.

Regulations

The Federal Air Commerce Act of 1926 marked the beginning of the end for the barnstormers. Under these new regulations, pilots had to be both licensed and medically approved for flying, air schools had to be certified, and planes had to be inspected for airworthiness and then registered and marked. Stunt flying, especially near populated areas, was severely restricted.

Although enforcement was lax, most pilots balked at the restrictions. Those who continued to fly either confined their flying to less populous areas or moved to Mexico, where the regulations held no sway. Many small-town airports were established by retired barnstormers who had settled down to give flying lessons. The air circuses gradually faded away as well. Denied their death-defying stunts, most barnstormers had gone out of business by 1930.

P. S. Ramsey

Bibliography

Caidin, Martin. *Barnstorming: The Great Years of Stunt Flying*. New York: Van Rees Press, 1965. An anecdote-driven account of barnstormers, stunt pilots, and the early days of aviation.

Collar, Charles S. *Barnstorming to Air Safety*. Miami, Fla.: Lysmata, 1998. A history of the early days of American aviation, with an emphasis on the evolution of safety regulations.

Tessendorf, K. C. *Barnstormers and Daredevils*. New York: Atheneum, 1988. A history of barnstorming and stunt flying throughout the 1920's, filled with photographs and anecdotes about early aviators and their adventures.

Van Stynweek, Elizabeth. *Air Shows: From Barnstormers to Blue Angels*. New York: Franklin Watts, 1999. A youth-oriented history focusing on the early days of air shows and aerial stunts.

See also: Aerobatics; Air shows; Crop dusting; Jennys; Charles A. Lindbergh; Safety issues; Wing-walking; Women and flight; World War I

Bats

Definition: Flying mammals that can be found in nearly every habitat, except extremely hot deserts and extremely cold polar regions.

Significance: Worldwide, there are almost 1,000 species of bats, the only type of mammal that can fly.

Bats are beneficial in nature because they eat insects, pollinate flowers, and help scatter the seeds of many plants.

Background

The fact that bats can fly makes them unique among mammals. Flying squirrels and flying lemurs, despite their names, do not really fly. Because they do not have wings, they merely glide through the air as they jump from trees. Bats, however, have wings that are the result of millions of years of evolutionary development.

Bats belong to the order Chiroptera, a word that means hand-wing. There are two suborders: Megachiroptera, which includes 42 genera and 173 species of flying foxes and Old World fruit bats, and Microchiroptera, made up of 144 genera and 813 species. The members of the first suborder are found only in Europe and Asia. Flying foxes have foxlike faces with very large eyes, but, unlike other bats, they do not use echolocation, a kind of natural radar for finding insects and other objects.

Echolocation is a process used by bats and a few other animals to identify objects in their environment and measure their distance from them. This ability involves listening to echoes of sounds produced sent by the hunter to bounce off his prey. A person who shouts in a tunnel and listens for the echo is using a simple form of this process. Because animals of the suborder Microchiroptera have this natural radar, they are much more successful and widely distributed than animals of the suborder Megachiroptera. Bats are among the most numerous mammals in the world's rainforests. In some forests, bats pollinate or disperse the seeds of more than 30 percent of all trees.

Evolution

The oldest bat fossils date to about 60 million years ago, and show that bats are native to North America. Scientists believe that the ancestors of bats were tiny, mouse-like mammals that ate flying insects and lived high up in trees. Over time, these animals developed membranes between their forelegs and their bodies that gave them the ability to glide from branch to branch, much like modern flying squirrels. This membrane gradually transformed into a moveable wing that gave bats a major advantage over their rivals: the ability to fly above the trees to catch their prey.

Students of bat evolution speculate that early species had developed the ability to echolocate, which gave them the ability to catch flying insects at night, but the exact origin of this power is unknown. Bats are sometimes compared with birds, because both can fly. However there are about eight times more species of birds than bats. This is because bats are a younger group than birds, which means they developed their flying ability long after birds took to the air. Another difference between birds and bats is that bats cannot use their legs, which are attached to their wings, to swim, dive, run, or dig. Birds' legs are separate from their wings and can thus be used for food gathering and other activities.

Flight

True flight offers many advantages to creatures, such as bats, birds, and insects, that have the ability. Flight gives these creatures access to many new sources of food and is a very effective way to evade enemies and predators. Flight also makes long-distance migrations possible and allows animals the ability to get past obstacles, such as rivers, oceans, and mountains, which flightless species cannot cross without difficulty or assistance. Flying also requires less energy than other modes of movement, such as walking or swimming. Bats and birds use less than one-fourth of the energy used by land-based animals for locomotion.

Most animals have not developed the complex structures necessary for flight. The basic requirements for flight include a method of keeping the body above the ground (lift), a means of moving the body through the air (thrust), and a design that minimizes air resistance. The bat's wing provides both lift and thrust. It contains the same basic arm and hand bones found in all mammals, except that the bat's five finger bones are very long and slender.

The membranes used for flight are extremely thin sections of skin, which are stretched between the arms, fingers, body, legs, and feet. Although they appear delicate, the membranes are actually stronger than rubber gloves and can be torn only with great force. The muscles that move the wing are located on the chest, back, and shoulder. This arrangement allows bats to fly with less energy than if the muscles were on the wing. Because bats' legs are used more for flight than for moving around on land, their pelvises and legs are very small, giving their bodies a streamlined, slender body shape that cuts down on air resistance.

Bats achieve lift and propulsion by the downstroke of their wings. The lift is caused by air moving faster over the top of the wing than under it. To increase their speed, both bats and birds increase the speed of air moving past their wings by changing the angle of the downstroke and changing the curvature of the wing. Most bats begin flight by taking off from a roosting site. Bats spend more than one-half of their lives roosting in places, such as caves and trees, in which they are protected from both weather and predators. They hang upside-down resting in their roosts, usually during the day.

When hunting, bats eat a wide variety of foods, including pollen, fruits, leaves, mosquitoes and other insects, scorpions, fish, frogs, birds, and sometimes even other bats. Most bats are intelligent and can be trained to fly or walk through mazes. They can also be taught to respond to commands. The activities of bats are regulated by light. Because they are active mainly at night, they avoid competition with birds for food. They can also escape from being captured by owls, hawks, and falcons.

Leslie V. Tischauser

Bibliography

Altringham, John, Tom McOwat, and Lucy Hammond. *Bats: Biology and Behavior.* Reprint. London: Oxford University Press, 1998. An indispensable reference covering the natural history of bats with up-to-date information and fine-line illustrations.

Hill, J. E., and J. D. Smith. *Bats: A Natural History.* Austin: University of Texas Press, 1984. A description of all bat species in the world and their habitats.

Tuttle, Merlin D. *America's Neighborhood Bats.* Rev. ed. Austin: University of Texas Press, 1997. A popular natural history of bats, including facts on their behavior and biology, with identifying photos, keys, and range maps for common species.

See also: Animal flight; Birds; Evolution of animal flight; Forces of flight; Insects

Battle of Britain

Date: From July 10, 1940, to October 31, 1940

Definition: A series of aerial bombings made by the Germans over British cities during World War II.

Significance: The Battle of Britain, designed to completely demoralize the British by destroying the nation's industrial and military infrastructure, was the first major battle to be fought almost entirely in the air.

Background

By the end of June, 1940, the German army had conquered almost every country that had opposed it. Only Great Britain, protected by the English Channel, remained in the fight, even though it had lost much of its army on the Continent in fruitless support of its allies. Thus, when German chancellor Adolf Hitler offered peace to Britain, much of the world thought his offer would be accepted. When Britain refused, Hitler issued orders for an invasion, a vital preliminary to which would be the elimination of the British Royal Air Force (RAF).

Protagonists

To carry out the destruction of the RAF, the German Luftwaffe had 1,050 fighter aircraft and 1,600 bombers, based on airfields from Norway to the Atlantic coast of France. The actual number of these craft that were serviceable and available for operations varied from day to day. Against these, the RAF could field 550 single-seat fighters immediately available and serviceable, in about fifty squadrons stationed on airfields from the north of Scotland to the west of England. The figures of available aircraft for both sides varied as the battle progressed, but the proportions remained much the same.

Aircraft

Although the German bombers were the Luftwaffe's main agents of destruction, the fighters were the most important, because they could sweep away the RAF fighters to allow the bombers clear passage. Similarly, the fighters in service with the RAF were the only weapons that could stop the German bombers from ranging over the country.

The Luftwaffe's main fighter, the Messerschmitt Bf-109E, could reach a speed of 355 miles per hour and an altitude of 36,000 feet, but it had an operating range of little more than 400 miles. This limitation meant that the Bf-109E could spend a very small amount of time over the target area if it was to have sufficient fuel to return to base. The Luftwaffe also used the Messerschmitt Bf-110, a large twin-engine aircraft that was designed to fly long distances but could also take on defending fighters like its single-engine cousin, the Bf-109. However, the Bf-110 was quickly found to be more of a liability than an asset when confronted by the RAF's more nimble Spitfires and Hurricanes.

In defense of Great Britain, the RAF employed two types of single-engine fighter, the Spitfire and the Hurricane, of which the latter made up about three-fifths of the total. The Hurricane had a top speed of 330 miles per hour and could reach altitudes of 34,000 feet. It had a range of 500 miles and could absorb a great deal of battle damage while serving as a good, steady gun platform. The Spitfire was a slightly younger aircraft and benefited in its construction from slightly newer technology. It could reach a speed of 360 miles per hour and an altitude of 32,000 feet and had a range of 400 miles. Possessing great maneuverability, the Spitfire was armed much like the Hurricane, with eight 0.303-inch machine guns. Although these were

Germany's first mass air raid on London on September 7, 1940, marked the beginning of the Battle of Britain. (Digital Stock)

perhaps outclassed by the armament of the Bf-109 and Bf-110, which carried 20-millimeter cannon, they were sufficient to shoot down Luftwaffe bombers. The range of the British fighters was not as critical as that of the Bf-109, because the Spitfires and Hurricanes had a critical advantage in the RAF's advanced and efficient fighter control system.

Fighter Control

The best fighter aircraft flown by the finest pilots would have been to no avail if they had not known where their enemy was. The RAF, however, relied upon a combination of radar to warn of enemy formations approaching the coast, an observing organization to track the enemy over land, a system of control rooms, each responsible for a certain area, with radio communications between them, and the airborne pilots themselves to make the most efficient use of its resources. In this way, the outnumbered Spitfires and Hurricanes were able to intercept the incoming Luftwaffe directly, without wasting their efforts in flying patrols simply looking for the enemy.

Convoys

The English Channel, one of the busiest waterways in the world, in 1940 served innumerable convoys carrying materials and supplies along the British coast. In early July, the Luftwaffe began attacking these convoys to force the RAF into battle to protect them. Over the next four weeks, the Spitfires and Hurricanes were in combat almost daily with the Luftwaffe. After a month, losses were almost two to one in favor of the RAF. However, the Luftwaffe's large superiority in numbers meant that it could hold out longer than the RAF, which would eventually lose. Shipping in the Channel was reduced, but never completely halted, in order to remove the potential target from the Luftwaffe's sight.

Attack of the Eagles

To some extent, the Luftwaffe's convoy battles had been merely a means of distracting the RAF while the Luftwaffe prepared and positioned its resources for the main battle. On August 13, called Eagle day by the Germans, the main attacks started, beginning four weeks of concentrated bombing designed to destroy the RAF and generally weaken the country's ability to resist an invasion. Attacks on airfields, ports, and dockyards by large formations of German aircraft set the scene for the next weeks and betokened hard, intense fighting for both sides. At the end of the day, the Luftwaffe had lost forty-six aircraft; the RAF had lost only thirteen fighters, but many of the fighter squadrons' airfields and communications were damaged. The Luftwaffe hit more and more airfields and also damaged several radar stations but, apparently not realizing the importance of the radar system, failed to follow up on these particular attacks.

The airfield damage, however, was soon felt by the RAF squadrons, and a reduction in their fighting strength and efficiency became apparent. It was clear that if the Germans continued to attack in this fashion, they might achieve victory, an outcome which had not previously been considered by the British. The RAF continued to shoot down German aircraft at a greater rate than it lost its own aircraft and achieved notable success on some occasions. German bombers based in Norway attacked northern England without a fighter escort, on the assumption that all RAF fighters would have been drawn south to the Channel coast. At a cost of fifteen bombers, the Germans discovered they were wrong.

After another month, the Luftwaffe had lost some 670 aircraft, and the RAF had lost about 400 fighters. Damage to British airfields increased, whereas production of Spitfires and Hurricanes began to fall behind their losses. Pilots also were being injured and killed faster than the training system could replace them. It could be only a matter of time before the RAF became exhausted.

Air Raids on London

On September 7, the Luftwaffe changed tactics, turning their bombers away from British airfields and factories and heading for London. Nearly 1,000 German aircraft crossed the Channel and headed for the capital, to be met by some 250 British fighters that struggled to break through the fighter escort and attack the bombers. Many German bombers did get through, however, and heavily bombed London's East End, starting many fires in the docklands area. The RAF fighters had some success, shooting down thirty-six of the German raiders, but they also lost twenty-six Spitfires and Hurricanes of their own. Similar raids continued for another week, and the RAF used the time to repair and strengthen itself while intercepting the Luftwaffe at every opportunity. Then, on September 15, the Luftwaffe attacked with the largest number of aircraft ever, more than 1,000 aircraft headed for London once more, only to be intercepted and their formations broken up before they reached the city. In the fighting, the RAF again lost twenty-six fighters, but the Germans lost sixty aircraft. This date was the high point of the battle for the RAF and has since been known as Battle of Britain Day. Two days later, Hitler postponed the invasion of Britain indefinitely.

Later Stages

The battle continued through the remainder of September and most of October, as the Luftwaffe increasingly turned its efforts toward night bombings. From time to time, it mounted large raids during the day but mainly flew small, high-altitude raids, often with bomb-carrying Bf-109's rather than bomber aircraft. These stood a much greater chance of hitting their targets and flying away again without being shot down, but their effect was minimal. Finally, at the end of October, the battle fizzled out as autumn rain set in, but the people of Britain still had months of night bombing to endure. The German effort to defeat the RAF, however, had failed.

As the air battle progressed, the Germans prepared finally to invade Britain. As part of this effort, they had gathered from the canals of Europe a vast number of barges in which to transport their troops across the Channel. These barges, assembled in the Channel ports of France, were quickly spotted by RAF bombers, who regularly attacked them, causing considerable damage both to the barges themselves and to the port facilities that would be needed to mount the invasion. The bombers were also active against the Luftwaffe, attacking airfields from which the German aircraft flew, often at considerable loss to themselves.

Losses

From the beginning of July to the end of October, the two air forces had fought a massive battle, which neither had anticipated and which only the RAF had been designed to fight. Although both sides suffered severely, the Luftwaffe's losses were sufficient to make it realize it could not achieve its objectives. The RAF was able to absorb its losses and inflict upon the Germans their first defeat of the war.

The RAF lost 1,023 aircraft, including aircraft destroyed in air raids, and 537 men. The Luftwaffe's losses were much higher: 1,887 aircraft and 2,662 men. The differing ratios of aircraft to men is accounted for by the fact that the RAF losses were almost exclusively single-seat fighters, whereas the Luftwaffe losses included many bombers carrying crews of four or five. Also, an RAF pilot who bailed out unhurt was over his own country and might be back in the air the next day, whereas any Luftwaffe airman who bailed out was inevitably taken prisoner.

Hugh Wheeler

Bibliography

Bungay, Stephen. *The Most Dangerous Enemy.* London: Aurum Press, 2000. A modern history that examines new information together with a fresh interpretation of old sources.

Mason, Francis K. *Battle Over Britain.* London: McWhirter Twins, 1969. Probably the best overall account of the battle to be compressed into one book with a good background to developments in scientific aids used.

Overy, Richard. *The Battle of Britain: The Myth and the Reality.* New York: W. W. Norton, 2001. A modern debunking of some of the popularly held notions of the Battle of Britain and celebrating the very real accomplishments of the RAF.

Ramsey, Winston G., ed. *The Battle of Britain: Then and Now.* London: Battle of Britain Prints International, 1989. A very detailed, day by day diary of the battle showing losses for both sides with, in many cases, photographs of the men concerned.

See also: Bombers; Fighter pilots; Luftwaffe; Messerschmitt aircraft; Royal Air Force; Spitfire; World War II

Beechcraft

Definition: A Wichita, Kansas-based producer of small aircraft, including both propeller- and jet-driven models.

Significance: Beech Aircraft, often shortened to "Beechcraft," developed a reputation for high-performance, luxurious airplanes during the 1930's. The company continued this legacy, eventually expanding into business jets. Beech also became a mainstay in Wichita's economy.

Beginnings

Walter Beech, founder of Beechcraft, trained to be a pilot during World War I. Beech did not let the fact that he was not an engineer deter him from pursuing aviation as a career after the war. After a stint as a barnstormer, he joined the E. M. Laird Airplane Company of Wichita in 1921 as a pilot and salesman. Beech and Laird engineer Lloyd Stearman left the company in 1925 to form a new company, TravelAir, taking on another partner, Clyde Cessna. TravelAir produced several well-respected aircraft, most notably the Model 5000, but Beech's partners both decided to leave the company. In 1929, TravelAir became part of the Curtiss-Wright Company, and Beech moved to new executive offices in St. Louis. The United States' enthusiasm for aviation ended with the onset of the Great Depression, and TravelAir went out of business in 1932.

Birth of Beechcraft

After TravelAir folded, Beech and his wife Olive Ann moved back to Wichita determined to reenter the aircraft manufacturing market. Beech wasted little time, establishing the Beech Aircraft Company the same year TravelAir ceased to exist. Beech's chief engineer Ted Wells developed a masterful new design, the Model 17 Staggerwing. The plane carried five people at the remarkable speed of nearly 200 miles per hour. The Model 17, advertised as the "Beechcraft," and the successor, the Model 18, sold well given the economic conditions of the 1930's, and Beech enjoyed its first $1 million sales year in 1938.

War Years

Beech's successful Models 17 and 18 ensured the company a prominent place in the United States' defense expansion leading up to World War II. The company sold Model 18's to the Philippines and China, and General Henry Harley "Hap" Arnold ordered 150 modified Model 18's for the U.S. Army Air Corps in 1941. During the war, Beech produced some 7,400 twin-engine aircraft for the military, and only 22 for the civilian market. The company was among the 3 percent of manufacturers to earn the prestigious Army-Navy "B" production award five consecutive times. Beech also produced wings for the A-26 Invader during the war.

Postwar Aircraft

Beech, like its Wichita competitor, Cessna, looked to take advantage of an expected boom in private plane ownership following World War II. Some experts believed that airplanes would become nearly as ubiquitous as automobiles for family transportation. Both Wichita companies held excellent positions for competing in this new market. Following traditions established before the war, each focused its efforts in a different direction. Cessna emphasized a low-priced, efficient model, while Beech looked to attract a more affluent customer by offering greater luxury. In 1947, Beech introduced the Model 35, better known as the Bonanza. The Bonanza utilized a distinctive

Beechcraft Specifications

	Performance			*Engines*	*Weights*		*External Dimensions*			*Internal (Cabin) Dimensions*		
Aircraft	*Maximum Cruise Speed (miles per hour)*	*Certified Ceiling (feet)*	*Maximum Range (nautical miles)*		*Basic Empty Weight (pounds)*	*Useful Load (pounds)*	*Wingspan*	*Maximum Length*	*Maximum Tail Height*	*Length*	*Width*	*Height*
Beechjet 400A	539	45,000	1,742	Pratt & Whitney Canada JT15D-5	10,250	5,650	43 feet, 6 inches	48 feet, 5 inches	13 feet, 11 inches	15 feet, 6 inches	4 feet, 11 inches	4 feet, 9 inches
King Air 350	362	35,000	1,806	Pratt & Whitney Canada PT6A-60A	9,440	5,460	57 feet, 11 inches	46 feet, 8 inches	14 feet, 4 inches	19 feet, 6 inches	4 feet, 6 inches	4 feet, 9 inches
King Air B200	336	35,000	1,807	Pratt & Whitney Canada PT6A-42	8,420	3,970	54 feet, 6 inches	43 feet, 10 inches	14 feet, 10 inches	16 feet, 8 inches	4 feet, 6 inches	4 feet, 9 inches
King Air C90B	284	30,000	1,267	Pratt & Whitney Canada PT6A-21	6,810	3,150	50 feet, 3 inches	35 feet, 6 inches	14 feet, 3 inches	12 feet, 7 inches	4 feet, 6 inches	4 feet, 9 inches
Baron 58	232	20,688	1,569	Teledyne Continental Motors IO-550-C	3,890	1,634	37 feet, 10 inches	29 feet, 10 inches	9 feet, 9 inches	12 feet, 7 inches	3 feet, 6 inches	4 feet, 2 inches
Bonanza B36TC	230	25,000	1,169	Teledyne Continental Motors TSIO-520-UB	2,740	1,126	37 feet, 10 inches	27 feet, 6 inches	8 feet, 7 inches	12 feet, 7 inches	3 feet, 6 inches	4 feet, 2 inches
Bonanza A36	203	18,500	930	Teledyne Continental Motors IO-550-B	2,530	1,133	33 feet, 6 inches	27 feet, 6 inches	8 feet, 7 inches	12 feet, 7 inches	3 feet, 6 inches	4 feet, 2 inches
Beech 1900D	326	25,000	1,505	Pratt & Whitney Canada PT6A-67D	10,485	6,375	57 feet, 11 inches	57 feet, 10 inches	14 feet, 11 inches	25 feet, 3 inches	4 feet, 6 inches	5 feet, 11 inches

Source: Data taken from (www.raytheon.com/rac), June 6, 2001.

V-shaped tail configuration, and carried its passengers in quiet comfort at an impressive 175 miles per hour. To demonstrate the plane's reliability, a Bonanza flew nonstop from Honolulu to Teterboro, New Jersey, a distance of 5,273 miles, with no maintenance problems and a cost of only $75 in fuel. This performance made the Bonanza famous, and despite being relatively expensive, the plane was an enormous success for Beech, which sold ten thousand Bonanzas by 1970.

The company followed up with variations of the Bonanza, as well as several different models, before developing the twin-engine turboprop King Air 90 in 1964. This plane fit into the niche between the truly private planes and the luxurious corporate jets that began appearing in the early 1960's. The King Air created such enthusiasm that the company had a $28 million backlog of orders when the first plane came off the assembly line. Purchasers of the King Air included such notables as Volkswagen, Walt Disney Productions, and Art Linkletter. By 1984, half of the twin-engine turboprop planes delivered were King Air models. During the 1960's and 1970's, Beech looked for ways to enter the growing business jet market. The company's Wichita competitors, Cessna and Learjet, dominated the field, but Beech had trouble developing its own model. Ultimately, Beech abandoned its own design and purchased the established but struggling Mitsubishi Diamond 2 business jet program. Beech moved production of the Diamond 2 from Texas to Wichita and redesignated the plane the Beechjet 400. The 400 series did not match Beech's competitors in terms of performance, but an Air Force order for 211 400's helped attract attention to the model. Beech also worked to convince owners of the King Air to purchase the 400 models, rather than competitors' offerings. Despite performance shortcomings, the 400 series became a formidable presence in the business jet market by the early 1990's, thanks to Beech's aggressive marketing efforts.

Corporate Changes

Walter Beech guided Beech Aircraft until his death in 1950. Fortunately for the company, Beech's wife, Olive Ann, proved to be an outstanding leader. Mrs. Beech guided the company for eighteen years before handing it over to her nephew Frank Hedrick. In 1979, with Hedrick ready to retire, the company merged with Raytheon, a manufacturer of missiles, electronics, and appliances. In 1982, Raytheon removed Mrs. Beech and Hedrick as managers of the company, prompting Beech to resign from the board of directors and marking the end of an era.

Matthew G. McCoy

Bibliography

McDaniel, William Herbert. *The History of Beech*. Wichita, Kans.: McCormick-Armstrong, 1982. This is a long account of the first fifty years of Beechcraft's existence. At more than 500 pages, this book covers nearly every aspect of the company's history.

Philips, Edward H. *Beechcraft: Staggerwing to Starship*. Eagan, Minn.: Flying Books, 1987. This is a short pictorial history of Beech. It is less than 100 pages long, but does provide useful information and good pictures of Beechcraft models.

Rowe, Frank Joseph, and Craig Miner. *Borne on the South Wind: A Century of Aviation in Kansas*. Wichita, Kans.: Wichita Eagle and Beacon, 1994. This book covers the development of aviation in the state of Kansas. It does not go into great depth, but it does explain the role of Beech and its impact on aviation and the economies of both Wichita and Kansas. It is also well illustrated.

See also: Aerospace industry, U.S.; Airline industry, U.S.; Airplanes; Corporate and private jets; Manufacturers

Bell Aircraft

Date: Founded in 1935

Definition: The United States' most important developer and manufacturer of helicopters, aircraft, and rocket engines, with twenty aviation development firsts to the company's credit from World War II to the present

Significance: Bell Aircraft developed some of the most important aircraft of the twentieth century, including fighter planes, booster rockets for spacecraft, and helicopters.

Lawrence Dale Bell's interest in flight was first sparked by his older brother, Grover, when the two were teenagers in Santa Monica, California, in 1910. After Grover's death in a plane crash in 1913, Larry Bell renounced his interest in aircraft, but was persuaded to join the fledgling Martin Company, quickly rising to vice president and general manager. He left Martin in 1928 to work for Consolidated Aircraft in Buffalo, New York, and when that company relocated to California in 1935, Bell decided to form his own company. Bell Aircraft had a slow start, but has been a leader in rotorcraft, or helicopter, design, and development since 1941, when Bell opened a research facility in Gardenville, New York, headed by Arthur Young and his assistant Bartram Kelly.

Events in Bell Aircraft History

1941: Larry Bell, entrepreneur and founder of Bell Aircraft Corporation, encourages inventor Arthur Young in helicopter development.
1946: Young's Model 47 helicopter becomes the first commercially licensed helicopter in the world and Bell delivers its first unit to the U.S. Army.
1950-1953: The military use of helicopters for medical evacuation increases during the Korean War, in which 80 percent of helicopters used are of Bell design.
1951: Bell Aircraft creates a separate helicopter division in Fort Worth, Texas, to accommodate the overwhelming demand for production.
1957: After the death of Larry Bell, the helicopter division is reorganized as Bell Helicopter Corporation.
1960: Textron purchases several Bell Aircraft companies, including the Bell Helicopter Corporation.
1961-1975: The military use of helicopters is cemented during the Vietnam War.
1976: Bell Helicopter becomes Textron's largest division.
1982: Bell Helicopter is incorporated as a subsidiary of Textron, now officially known as Bell Helicopter Textron.

Bell Aircraft developed the Airacuda, the first World War II twin-engine, multiplace escort fighter, with 37-millimeter cannon and flexible gun turrets. Bell also developed the XP-77, the first all-wood modern fighter aircraft, and the P-59, the United States' first jet-propelled fighter aircraft. On October 14, 1947, Bell's X-1 piloted by Chuck Yeager broke the sound barrier at 662 miles per hour, an accomplishment followed by development of the X-1A, which in 1953 set a world speed record of 1,650 miles per hour. In 1957, Bell developed the Agena rocket engine, known as the "workhorse of the space age," with a 99.7 percent reliability record. The Agena was used on the Thor, Atlas, and Titan booster rockets in the Discover, Ranger, Mariner, and Gemini space programs.

Bell Helicopters

Bell Aircraft started the U.S. commercial helicopter industry when the Bell Model 47 was granted the first commercial license issued by the Civil Aeronautics Administration on March 8, 1946, and awarded the first Helicopter Type Certificate on May 8, 1946, shortly after Bell delivered its first production-line helicopter to the military. The Bell Model 47 became the foundation of the helicopter industry in the United States and is used for police work and in the medical, mining, and farming industries. The Model 47 was used for medical evacuation during the Korean War and in the United States. Five thousand Model 47's in twenty different configurations were built before Bell stopped production in 1973.

Bell Helicopter has remained the leader in medical evacuation helicopters, with the Model 206 widely used by police, fire departments, and medical ambulance services. Having built more then thirty-four thousand helicopters since 1946, Bell is the world's most prolific manufacturer of rotorcraft.

In 1951, the Bell Aircraft Corporation created a separate helicopter company which was headquartered in Fort Worth, Texas. This corporation was bought by the global conglomerate Textron in 1960 and became its subsidiary. Bell Helicopter Textron has eight thousand employees scattered among ten plants, including the state-of-the-art Bell Helicopter Textron Canada facility at Mirabile, Quebec, with 1,800 employees.

Twenty-first Century Rotorcraft

The company's newest helicopters are the Model 427 and 407 LongRanger. In 2001, Bell Helicopter Textron's current military production was the AH-1W Super Cobra, for the U.S. Marine Corps, and the OH-Kiowa Warrior. The company completed a 137-aircraft order for the TH-67 Creek trainers for the U.S. Army. Other contracts included one hundred CH-146 Griffons, which are highly modified 412-EP's, for the Canadian Forces utility tactical helicopter program.

With Boeing Vertol, Bell also produces the V-22 Osprey tilt-rotor aircraft for the U.S. Marine Corps and Special Operations Command. The V-22, another Bell first, can take off, hover, and land like a helicopter and can fly forward with the speed and range of a high-speed turboprop fixed-wing aircraft.

Kenneth M. Krongos

Bibliography

Matthews, Birch. *Cobra! The Bell Aircraft Corporation, 1934-1946*. Atglen, Pa.: Schiffer, 1996. A meticulously researched account of the planes produced by Bell during the Great Depression and World War II.

Norton, Donald J. *Larry: A Biography of Lawrence D. Bell*. Chicago: Nelson-Hall, 1981. A biography of the founder of Bell Aircraft.

Rotundo, Louis C. *Into the Unknown: The X-1 Story*. Washington, D.C.: Smithsonian Institution Press, 1994. An in-depth, behind-the scenes look at the development of Bell's X-1 supersonic airplane.

See also: Airplanes; Helicopters; Manufacturers; Missiles; Osprey helicopter; Rockets; Sound barrier; Vertical takeoff and landing; X planes; Chuck Yeager

Bermuda Triangle

Also known as: Devil's Triangle, Limbo of the Lost, Hoodoo Sea, Port of Missing Ships, Twilight Zone

Definition: A triangular section of the Atlantic Ocean roughly defined by a line connecting the tip of Florida, the Bermuda Islands, and Puerto Rico, that is an area notorious for unexplained disappearances of boats, ships, and aircraft.

Significance: Although reports of unexplained disappearances in the Bermuda Triangle have been made since the mid-nineteenth century, the lack of substantial evidence ensures that the area's significance lies mainly in the popular imagination.

Origins of the Bermuda Triangle

A September 16, 1950, Associated Press dispatch by reporter E. V. W. Jones contains the first recorded mention of mysterious disappearances between Bermuda and the Florida coast. The dispatch ran in various newspapers within the next few days. Two years later, in October, 1952, *Fate* magazine published an article by George X. Sand on the same subject that defined the targeted area as a triangle bounded by Bermuda, Puerto Rico, and Florida.

Morris K. Jessup's *The Case for the UFO: Unidentified Flying Objects* (1955), Donald E. Keyhoe's *The Flying Saucer Conspiracy* (1955), and Frank Edwards's *Stranger Than Science* (1959) furthered speculation about the area's disappearances, blaming them on aliens from outer space. The first published use of the name "Bermuda Triangle" appears in Vincent H. Gaddis's article "The Deadly Bermuda Triangle" in the February, 1964, edition of *Argosy*. Gaddis's article, along with his 1965 book, *Invisible Horizons: True Mysteries of the Sea*, brought widespread popular attention to the region for the first time. Since then, the Bermuda Triangle has gained global renown, but the U.S. Board of Geographic Names neither recognizes the name officially nor maintains an official file on the area.

Unexplained Disappearances

The unexplained disappearances of vessels or crews that have given the Bermuda Triangle its mysterious reputation are said to have happened without warning during fair weather and have left no traces of either wreckage or bodies. When each incident is examined, however, mundane causes are often obvious, with facts frequently embellished or omitted for dramatic effect. Because the Bermuda Triangle's shipping lanes are busy, and because aircraft crisscross its skies in large numbers every day, it is not surprising that, over the years, many disasters have occurred in its waters.

Probably the most famous and dramatic disappearance is that of Flight 19. At 2:10 P.M. on December 5, 1945, five Avenger torpedo bombers took off from the Fort Lauderdale, Florida, Naval Air Station on a routine two-hour training mission in good weather. At 3:45 P.M., the flight leader and flying instructor reported that neither of his compasses was working. Voice communication stopped at 4:25 P.M., and the last radio signal was received at approximately 7:00 P.M. The most likely contributing factors to the disaster, aside from the malfunctioning compasses, were the flight leader's unfamiliarity with the area, the lack of clocks in the planes to keep track of time, few clear radio signals, and four inexperienced pilots who were unwilling to openly contradict their superior. In addition, as the weather worsened throughout the afternoon, extreme turbulence and unsafe flying conditions were reported.

The flight leader had been asked to switch to the emergency channel, but he refused, because he did not want to risk losing contact with the other four Avengers. Thus, his initial assumption that they were flying over the Florida Keys instead of the Bahamas could not be corrected by direction-finding stations. Ironically, when the flight leader first reported himself lost, he was probably right on course above the Bahamas. Although the search continued for weeks, it turned up no sign of the bombers or the fourteen men aboard. It is likely that the planes went down in the ocean by 8:00 P.M., somewhere east of the U.S. coast and north of the Bahamas, after flying around lost for four hours.

Adding to the magnitude of the tragedy was the fact that another plane was lost that night. However, its fate is more certain. A Martin Mariner seaplane with a crew of thirteen was one of several planes sent out on the search mission. An explosion was observed from a ship shortly after the plane's takeoff. Martin Mariners were nicknamed "flying gas tanks" because they tended to leak fumes. It is likely that a spark from some source ignited the plane's fuel and caused the explosion.

Many other aircraft have disappeared over the Bermuda Triangle. In 1948, a British Tudor IV airliner, the *Star Tiger*, was en route to Bermuda from the Azores when it vanished without a trace some time after the pilot radioed Bermuda to ask for a bearing. Because the plane was never recovered, no cause for the disaster could be determined, but unpredictable winds could have driven the aircraft off course after contact was lost at 3:15 A.M., giving the sea time to sweep crash debris from the scene. A Douglas DC-3 also disappeared in 1948 while flying from San Juan, Puerto Rico, to Miami, Florida. After the pilot reported being only 50 miles from the airfield in Miami, contact was lost, and the plane was never found, possibly sinking in the 5,000-foot depths of the Straits of Florida. In this case, the plane had been having trouble with its landing gear, batteries, and transmitter when it landed in San Juan and continued to have transmission problems as it left for Miami.

Although seagoing vessels have been lost in the Bermuda Triangle since the time of Christopher Columbus, most have vanished during severe weather or after a history of mechanical or personnel problems. Moreover, many disappearances associated with the area have actually occurred elsewhere. For example, the famous case of the *Mary Celeste*, encountered drifting without its crew in 1872, is often cited in stories about the Bermuda Triangle, but the ship was actually found near the Azores. The USS *Cyclops*, which sank in 1918 on a voyage from Barbados to Norfolk, Virginia, most probably lies, wrecked either by a storm or by its heavy load of manganese, on the ocean floor near Norfolk.

Possible Causes of Disappearances

Many fantastic theories exist to account for the disappearances, ranging from alien abductions to black holes to mysterious magnetic anomalies. Ivan T. Anderson claims that "vile vortices" caused by magnetic aberrations create "time slips" that convey the disappeared to other locations on Earth, including an advanced civilization allegedly lurking under the sea. Charles Berlitz has identified this civilization as the mythical lost Atlantis mentioned in the writings of the Greek philosopher Plato. Vincent Gaddis suspects that small black holes may pull ships and planes into other times or universes. However, there is no proof to substantiate any of these theories.

One recent theory has suggested that some unexplained disappearances might have been caused by large bubbles of methane hydrate, found during exploratory oil drilling in the Bermuda Triangle in 1995. Methane hydrate is a gas created when ice and methane are mixed together under conditions of high pressure far below the seabed. When the temperature rises or the pressure is released by a seaquake or underwater slide, the bubbles of gas rapidly expand at a rate of about 1 liter of hydrate to 45 gallons of methane. If the methane bubbled up under a ship, it would create a huge hole that would cause the ship to drop suddenly and sink. Aircraft might also be affected. Because methane is lighter than air, it would continue to rise in the atmosphere, causing potential problems for anything flying through it. Engine failure, explosions, or other problems could occur. However, this theory, too, remains unproven.

Although many boats, ships, and aircraft have disappeared in the Bermuda Triangle over the years, Lloyd's of London, an insurer of some of the missing vessels, has stated that there is no evidence that more disappearances occur within the Triangle than in any other similar expanse of ocean. The U.S. Navy and Coast Guard maintain that environmental causes, mechanical failures, and human errors are to blame.

However, the Bermuda Triangle does have some unique environmental characteristics that are likely to have contributed to both the area's disasters at sea and its eerie reputation. When Christopher Columbus sailed toward the vicinity of the Triangle in 1492, he became the first to record many of its most notable phenomena, among them the Sargasso Sea, an area almost the size of the continental United States, centrally located in the North Atlantic Ocean. The Sargasso Sea's name is derived from the Portuguese word for the seaweed that clogs its waters. The seaweed is inhabited by unusual species of animals adapted to life on the weeds. This strange region is isolated by strong currents that cause it to slowly rotate clockwise and leave it with more salt and less wind, clouds, and rain than the rest of the North Atlantic. It is a repository for debris, wreckage, and spilled oil drifting in from all over the world, including derelict ships that have made it known as a ships' graveyard. Because sailing ships could become stranded there for months, it is understandable that sailors learned to fear the region.

Columbus also noted curious compass variations in the Bermuda Triangle. Alleged magnetic variations have been blamed for some of the disappearances in the Triangle, but there is some compass variation almost everywhere, ranging from 0 up to 20 degrees, depending on longitude. During Columbus's time, it was assumed that the compass pointed to the North Star, but Columbus realized that the compass must be attracted to something else, later thought to be the North Pole, but eventually found to be the north magnetic pole.

Navigators are trained to compensate for variation between the magnetic pole and the true pole as a matter of routine. Close to Florida, however, it is not necessary to compensate for magnetic variation because Florida happens to be in line with both the magnetic pole and the North Pole. With a magnetic variation of zero, getting lost is actually less likely than it would otherwise be. Mysteriously spinning compasses have also been implicated in disappearances, but compass needles frequently swing or spin with the motion of a boat or plane. Compass headings are calculated by averaging the high and low readings of the swinging needle.

Thunderstorms, tornadoes, waterspouts, and hurricanes can develop very suddenly in the Bermuda Triangle and are often more violent there than anywhere else on the globe. Contrary to the claims that the disappearances have occurred during fair weather, the reality is that most occurred during severe conditions. Columbus documented some of the storms common to the Bermuda Triangle, including a hurricane in 1502 in which ten ships were lost. Strong turbulence in and around storm clouds can cause aircraft to disintegrate, and freak waves up to 115 feet high can capsize and break apart even the largest of ships.

Unusually strong ocean currents like the Gulf Stream flow swiftly through the Bermuda Triangle. These currents frequently thwart successful search-and-rescue missions and add to the mystery of the Triangle by quickly dispersing wreckage. Many of the disappearances have occurred at night or near dusk, giving the currents hours to sweep away evidence of disaster. Unpredictable currents are also caused by the region's topography, varying from some of the world's deepest marine trenches to extremely shallow shoals that surround the islands, creating tricky navigational hazards.

Mechanical failure and human error can have even more disastrous results when compounded by severe weather conditions. Every year, countless inexperienced vacationers pilot boats and small aircraft between the islands off Florida's coast. Simple navigational errors can cause these craft to become hopelessly lost at sea. Small boats can be capsized easily in even moderately bad weather, hence the frequency of small craft warnings. At night, small boats can be run over by large ships and sink without being noticed. Even large ships can capsize in high seas or because they are overloaded or top heavy. Boats can suffer hull damage in collisions with other ships, reefs, and other obstructions. Corrosion and metal fatigue can cause ships to break apart. Similar problems or structural failures in aircraft, such as a jammed rudder or loss of an engine or wing, are even more deadly. In addition, faulty wiring, leaking fuel, and combustible cargo can cause fires and explosions. Catastrophic equipment failures or damaged communications can make calling for help impossible. Even hijacking, sabotage, and insurance fraud are suspected of causing some unexplained disappearances.

Sue Tarjan

Bibliography

Dennett, Michael. "Bermuda Triangle, 1981 Model." *The Skeptical Inquirer* 6, no. 1 (Fall, 1981). Debunks claims made by Charles Berlitz regarding twelve incidents linked to the Bermuda Triangle.

Innes, Brian. *Unsolved Mysteries: The Bermuda Triangle.* Austin, Tex.: Raintree Steck-Vaughn, 1999. One of a series for young readers that encourages critical thinking about unexplained phenomena.

Kusche, Lawrence David. *The Bermuda Triangle Mystery—Solved.* 2d ed. Buffalo, N.Y.: Prometheus Books, 1986. A fascinating investigation into the creation of the legend of the Bermuda Triangle. The author's background as both reference librarian and pilot lends credence to his efforts to untangle years of garbled accounts of disasters at sea.

Oxlade, Chris. *Can Science Solve? The Mystery of the Bermuda Triangle.* Chicago: Heinemann Library, 2000. An excellent and well-illustrated account of the Bermuda Triangle phenomenon, one of a series for young children focusing on the role of science in explaining the mysteries of nature.

See also: Accident investigation; Hijacking; Instrumentation; UFOs; Weather conditions

Biplanes

Definition: An airplane with two levels of wings.

Significance: Most early aircraft utilized the biplane configuration, and biplanes remain popular for sport flying and aerobatics.

Reasons for the Biplane Configuration

From the early, pioneering flights of Orville and Wilbur Wright in 1903 through the 1920's and 1930's, biplanes represented the most practical aircraft configuration for both structural and maneuverability reasons. By the 1940's, they remained a practical choice only for training aircraft. Since then, biplanes have retained a certain popularity as sport and aerobatic and air show aircraft.

Until sufficiently light and powerful aircraft engines were developed, a large wing area was required to keep an aircraft aloft, and the biplane structure provided the most strength with the least weight. It was initially thought that thin wing sections were necessary for efficient generation of lift. In a biplane configuration, interplane struts and wire bracing provide a bridge-like strength and rigidity to the wing. Biplanes can use lesser wingspans, and both wings can use ailerons, resulting in the added advantage of maneuverability. Thus, for the first few decades of flight, biplanes were the configuration of choice for training aircraft, sport aircraft, military fighters, military bomber aircraft, and transport aircraft.

Famous Biplanes

America's best-known aircraft during World War I was the Curtiss JN "Jenny" trainer. It used a four-bay wing with eight interplane struts and many bracing wires, but it could fly two people with only a 90-horsepower OX-5 engine. After the war, Jennys were surplused and became the barnstormer's choice of airplane. The most famous World War I fighters were biplanes. In England, the De Havilland Tiger Moth was the trainer of choice between the world wars. In the 1930's, the Curtiss P-6E Hawk fighter biplane delighted the eye. In World War II, the best-known U.S. trainers were the Piper Cub monoplane and the Boeing-built Stearman PT-17 biplane. The Stearman had a reputation for indestructibility in the air and remains a popular sport biplane. When the Stearman was declared to be surplus after the war, it became the favorite of crop dusters, who took advantage of its great strength and load-carrying ability. It also survives as a popular sporting aircraft.

The biplane flowered in the interwar period. Travelair, which began producing biplanes in 1925, bettered the Jenny in control, comfort, speed, and safety. The Travelair D-4D is arguably the best-looking open-cockpit biplane ever built. During the 1920's and 1930's, the Waco Aircraft Company of Troy, Ohio, was by far the largest airplane manufacturer in the United States, building thousands of open-cockpit and cabin biplanes. The first Waco biplanes were built in 1922, but the Waco 9 appeared at the same time as the first Travelair and was highly regarded. The Waco Taperwing, using a tapered wing planform on both wings, remains popular.

Disadvantages of the Biplane Configuration

One disadvantage of the biplane is related to the extra drag of its wires and supporting struts and the interference drag between its two wings, which result in reduced cruising and top speeds for a given engine power. Another disadvantage is a poor lift-to-drag ratio that results in poor glide angles. By the 1920's, the most efficient aircraft were monoplane designs, such as Charles A. Lindbergh's *Spirit of St. Louis*. A monoplane is more simple and less costly to build. When aircraft designers learned how to make strong, internally braced aircraft entirely of aluminum, and when powerful and relatively light engines became available, the monoplane replaced the biplane as the configuration of choice for all high-speed aircraft.

The primary lifting surface of a wing is its upper surface, so the lower wing suffers the most from this; the gap between the wings is therefore usually made at least as large as the wing chord. If the wings are set at different angles (decalage), the relative loading of the wings and stall characteristics can be adjusted. Often the upper wing is mounted ahead of the lower wing, an arrangement known as positive stagger. This is particularly true for open-cockpit biplanes in which the front cockpit is under the wing and the rear cockpit, for stability reasons, is not placed too far back on the fuselage. However, the famous Beechcraft Staggerwing has a closed cabin and uses negative stagger. A biplane that has a smaller lower wing is known as a sesquiplane.

Most biplanes use the lighter tailwheel configuration for their landing gear, but the higher center of gravity and the poor view for the pilot upon landing mean that the directional instability of the tailwheel configuration requires significantly more pilot alertness and skill. Usually, the lower wing has a dihedral angle to provide lateral stability and to keep the tips farther from the ground, whereas the upper wing is straight, to simplify its construction.

Sport, Aerobatic, and Air-Show Biplanes

The biplane configuration has long been preferred for aerobatics because of its inherently good roll rate and because the extra drag of brace wires and struts prevents a rapid buildup of speed in the diving aspect of maneuvers. The 1920's and 1930's Great Lakes Trainer biplane was considered the best aerobatic aircraft of all U.S.-manufactured aircraft until the arrival of the Pitts Special. When the Great Lakes Trainer was first flown, it was found that its center of gravity was too far aft. This problem was corrected most simply by giving the upper wing rearward sweep. This correction had the side benefit of making the airplane a better snap-roll performer.

Biplanes remain favored aircraft for many air-show pilots, because of their extra visibility to spectators and because of the additional possibilities for wing-walkers. Only in the last decade of the twentieth century did monoplanes begin to dominate aerobatic competition at the highest levels. The appeal of the open-cockpit biplane, a sort of motorcycle of the air, will live on indefinitely, as pilots feel the sheer joy of flying between two wings in warm summer air.

W. N. Hubin

Bibliography

Bowers, Peter M. *Boeing Aircraft Since 1916*. 2d ed. London: Putnam, 1966. Reprint. Annapolis, Md.: Naval Institute Press, 1989. Covers Boeing-built biplane trainers, transports, seaplanes, and fighters, including the famous F-4B and P-12 models.

_______. *Curtiss Aircraft, 1907-1947*. London: Putnam, 1979. The definitive history of Curtiss aircraft, including Glenn Curtiss's pioneering early biplanes, the World War I "Jenny" trainer, interwar civil aircraft, military biplanes, seaplane racers, and the famous Hawk fighter series.

Bowman, Martin, and Jim Avis. *Stearman: A Pictorial History*. Osceola, Wis.: Motorbooks, 1997. A gorgeously illustrated history of the famous Stearman biplanes.

Boyne, Walter J. *De Havilland DH-4: From Flaming Coffin to Living Legend*. Washington, D.C.: Smithsonian Institution Press, 1984. The DH-4 was a British design that was adopted, with the U.S.-designed Liberty engine, as the standard U.S. fighter aircraft of World War I and was used decades thereafter for military training and mail carrying.

Jarrett, Philip. *Biplane to Monoplane: Aircraft Development, 1919-1939*. London: Putnam Aeronautical Books, 1997. An excellent, somewhat technical description of why and how the biplane was superseded by the monoplane for most applications.

Jerram, Michael F. *Tiger Moth*. Newbury Park, Calif.: Haynes, 1984. A well-illustrated description of the Tiger Moth's origins, development, use as a military trainer, and current use as a sport aircraft.

Kobernuss, Fred O. *Waco: Symbol of Courage and Excellence*. Terre Haute, Ind.: Sunshine House, 1992. The definitive account of the origins of the Waco Aircraft Company and its owners, designers, pilots, and early biplanes.

See also: Aerobatics; Air shows; Airplanes; Barnstorming; Jenny; Charles A. Lindbergh; *Spirit of St. Louis*; Wing designs; Wing-walking; Wright brothers; Wright *Flyer*

Biplanes are so called because they have two parallel levels of wings; these offered early aviators the largest wing area and strength for the least weight. (Library of Congress)

Birds

Definition: Warm-blooded organisms capable of flight.
Significance: Birds have provided humans with much information about heavier-than-air vehicle design.

About 8,800 species of birds make up most living organisms capable of flight. Believed to be evolved from reptiles, their weights vary from a few ounces, for tiny, flying hummingbirds, to to 300 pounds, for flightless ostriches. Most birds, however, fly well, and humans learned a lot about heavier-than-air vehicle design from observing them. Birds differ from heavier-than-air aircraft primarily in that their wings are movable, or flappable. Most aircraft have fixed wings, which do not move.

Birds' bodies are specially engineered for flight. Their skeletons are light, often weighing less than their feathers. Feathers combine the qualities of lightness, strength, and flexibility; a feather, bent double, quickly regains its shape upon release. Made of keratin, feathers also keep birds warm, dry, and protected from injury.

Bird lungs and hearts are designed for the high metabolic rates needed to produce the huge amounts of energy required by all flying machines, biologic or manufactured. Birds' respiratory systems allow for a much larger oxygen uptake than that of earthbound animals. Birds also have relatively large hearts, capable of passing all the oxygen needed for energy metabolism through the circulatory system to the other tissues.

Aerodynamics and Birds

To understand flight requirements, a background in aerodynamics, a branch of fluid dynamics that studies movement of bodies, such as birds or aircraft, through gases such as air, is essential. For example, the fifteenth century Italian artist and engineer Leonardo da Vinci studied bird flight and proposed to enable human beings to fly with flappable wings. His ideas failed because da Vinci knew nothing about aerodynamics, a science which did not exist then.

Any heavier-than-air flying vehicle must conquer gravity before it can climb into the air in controlled flight. Three main forces, exclusive of weight, are involved. The first is thrust, which birds produce by flapping their wings. Flapping merely enables a bird to move forward as long as its design allows enough thrust to exceed the drag caused by the viscosity of the air through which the bird moves. Drag diminishes the speed of moving objects due to air resistance. In vehicle design, thrust-to-drag ratios can be increased by streamlining to minimize drag.

The third aerodynamic force, lift, is the key to flight. Lift, enabling an object's rise into the air, operates upward perpendicular to the direction of forward motion, and is supplied in both birds and aircraft by wings and tails (airfoils). Bird wings are designed so the angle at which they meet air passing them causes it to flow much more rapidly past the upper airfoil surface than past its lower surface. This design lowers air pressure above the airfoil compared to that under it and engenders the lift that raises a bird into flight. In birds, this unsymmetrical airflow is produced by muscle movement that changes both the positions of wing feathers and the angle at which wings meet the air, known as the angle of attack.

The importance of the angle of attack is demonstrated in aircraft by a pilot's use or misuse of the angle during flight. An aircraft's angle of attack is changed by altering its position in space. Angles of attack of up to 15 degrees increase lift and enable faster climb rates while also slowing airspeed. If the angle is too steep, decreased lift occurs, making the aircraft drop toward the ground, or stall. When pilot misjudgment causes a stall, an aircraft will crash unless the angle of attack is adjusted to a safe value. Birds, unlike aircraft, constantly make quick wing adjustments, moving their wing muscles to prevent stalls.

Wing Design and Flight

Birds create lift with downstrokes of their wings, attached by flight muscles to a large breastbone. Birds contract flight muscles to cause this downstroke, during which long primary and secondary flight feathers spread out to provide the maximum possible surface area to push against air below. The downstroke is followed by an upstroke in which the feathers fold to minimize air resistance while positioning the wings for the next downstroke.

Bird wings have a short upper arm bone that moves up and down during flapping. They also have rigid elbow and wrist joints that move horizontally to spread or fold the wing. Furthermore, the wrist and hand bones are a carpometacarpus, derived from palm bones, a one-boned thumb, a two-boned second finger, and a one-boned third finger. Flight feathers are attached to wing bones. The primary flight feathers, most essential to flight, attach to the carpometacarpus, second finger, and third finger. Up to forty somewhat less important secondary flight feathers attach to the ulna, one of the forearm bones.

When a bird opens its wings, the bones straighten, the primary feathers spread as the elbow joint extends, and the wrist stretches. Wingspread is limited by a tendon running from shoulder to wrist. The bases of the flight feathers interconnect via a ligament running from the elbow to the second fingertip. Spreading a wing stretches the ligament, moving flight feathers into positions perpendicular to the bones to which they are attached. While the wing is spread, muscle action can either spread primary feathers further or fold them back. The greater importance of primary feathers is clear, because removing even their tips prevents flying, while more than one-half of each secondary feather must be removed to do this.

There are four basic types of bird flight. In skimming flight, birds such as albatrosses use winds to stay aloft. In soaring flight, birds such as eagles, hawks, and vultures can remain aloft for long periods of time, seeking prey below. In active flight, birds such as swallows fly all day, flapping their wings continuously. Finally, game birds such as quail conceal themselves and, when endangered, burst into the sky. They pick up speed quickly and fly short distances before landing and hiding again.

There is a wing shape most efficient for each flight type. Skimming birds have wings that are long, slender, and ribbon-shaped, with parallel edges and many secondary feathers. Skimming wings are the most highly developed, helping such birds ride the winds. Soaring birds have wings that are large, broad, almost square, and rich in primary feathers. Swallows and other birds engaging in active flight have long, tapering, pointy wings with broad bases and slender tips. Finally, game birds have short wings that beat rapidly, enabling them to get to speed quickly. However, these wings are not useful in long flights.

No bird has wings designed entirely for one type of flying. However, in gliding, birds use gravity as thrust to overcome drag and move forward, as their wings produce lift to hold them up. Drag slows down a gliding bird and causes it to sink earthward. To maximize glide time, or soar, a glider sets its wings at the angle of attack giving a good lift-to-drag ratio. Low forward speed helps, and, to alter speed, such a bird spreads its wings to increase their area and reduce glide speed or closes them to cause the opposite effect. Long-winged birds glide by adopting positions with small glide angles, avoiding stall by twisting the wings to reduce the angle of attack. This angle can also be varied along the length of each wing. For example, gliding birds may have their secondary feathers at a high angle of attack and their primary feathers flat.

Active flight requires thrust force and expenditure of energy sufficient to overcome drag and keep the bird on course. This is achieved by flapping the wings for lift and propulsion. The wing parts function differently at each stage of a wing beat. For example, many fast-flying birds start downstrokes with wings fully extended and well above the horizontal. As they flap down vertically, forward movement through the air generates lift along the entire wing. At a downstroke end the wings fold and primary feathers close. No propulsion is generated in the upstroke, at the end of which the primary feathers produce enough lift to raise and extend the wing, preparing for the next downstroke.

Body Design and Flight

A second group of characteristics enabling bird flight is the design of the bird's body. Body weight is important to flight: The heavier an object is, the larger its wings need to be to enable liftoff and maintain flight. In birds this problem is met by their relatively small, light bodies. For example, hawks and eagles have cat- or even dog-sized bodies, but they weigh only 25 to 35 percent as much as the earthbound mammals. Birds' light weight is due to several factors. First, under their feathers, birds have relatively small bodies. Second, although their feathers are bulky, they are also very light. In addition, birds have fewer bones compared with other animals, and their bones are thinner, or even hollow. This special anatomy, combined with wings that engender appropriate amounts of lift, allows birds to fly. Depending on their wing size and shape, birds can fly, soar, or skim.

Energy Needs

To meet the energy needs of flight, birds must eat a relatively large amount of food each day. For their muscles to work well, birds need efficient blood circulation to quickly supply fuel and oxygen and to remove wastes. In both birds and mammals, the blood circulatory system has a four-chambered heart that directs blood to the lungs, where the blood picks up oxygen and then travels on to the muscles and other organs, where the oxygen is used. Carbon dioxide is picked up at the same time and carried, via blood, to the lungs for disposal. The difference between bird and mammal circulatory systems is the relatively larger size and greater power of a bird heart, which is two to three times heavier, compared to body weight, than that of a mammal. Bird heartbeat rates are also much faster than those of mammals, usually from 200 to 1,000 beats per minute, compared to 80 in humans. The combination of a large heart and a faster pulse rate results in a blood-pumping capacity for birds that is relatively much greater than that of mammals.

A bird's respiratory system is very different from the bellows-type lungs of mammals. Bird lungs are relatively small, but they connect to inflatable air sacs located throughout the body, even in bones and breast muscles.

These sacs are thought to cause very efficient exchange of oxygen and carbon dioxide with the bloodstream via one-way airflow through the lungs. When a bird inhales, air enters the lungs, posterior air sacs, and anterior air sacs. Exhalation causes air from posterior sacs to enter the lungs, and air from anterior air sacs is exhaled. Thus the air constantly passes through the lungs, ensuring a more efficient absorption of oxygen and removal of carbon dioxide compared to that of mammal lungs, in which only a fraction of the air is flushed out at each breath. Bird lungs are not worked via diaphragm. The air sacs are pumped by rib movement.

Thus, with its wings; its small, light body; its superbly useful feathers; and its high-capacity heart and lungs, a bird is superbly designed to be airborne.

Sanford S. Singer

Bibliography

Allen, John E. *Aerodynamics: The Science of Air in Motion*. New York: McGraw-Hill, 1982. Discusses aerodynamic principles, including some history. Its text and diagrams clarify many issues important to understanding lift, drag, and other issues essential to understanding heavier-than-air flight.

Brooks, Bruce. *On the Wing: The Life of Birds from Feathers to Flight*. New York: Charles Scribner's Sons, 1989. Discusses aspects of bird life, including feathers, eating without teeth, and flight.

Chatterjee, Sankar. *The Rise of Birds: 225 Million Years of Evolution*. Baltimore: Johns Hopkins University Press, 1997. Describes the evolution of birds, the fossil remains of their ancestors, and avian flight.

Freethy, Ron. *How Birds Work: A Guide to Bird Biology*. Poole, Dorset: Blandford Press, 1982. Thoroughly addresses the biology of birds, including their flight.

Harrison, Colin, and Howard Loxton. *The Bird: Master of Flight*. London, England: Blandford Press, 1993. Covers avian flight completely.

See also: Animal flight; Bats; Evolution of animal flight; Forces of flight; Insects

Black Sheep Squadron

Date: From April, 1943, to January, 1944

Definition: U.S. Marine Squadron 214, one of the most renowned U.S. fighting units of World War II, which fought against the Japanese in the Pacific theater.

Significance: The Black Sheep Squadron, composed of young and inexperienced replacement fliers, received publicity not only because of its success in warfare but also because of the war record of their leader, fighter ace Gregory "Pappy" Boyington.

Pappy Boyington

At thirty years of age, Gregory Boyington was referred to by his younger fliers as "Pappy," "Gramps," and "Skipper," because he appeared old for his age. Through his off-duty indulgence in alcohol, gambling, and fighting, he had earned a dubious reputation as a troublemaker, much to the disapproval of his superior officers.

By 1940, before the United States entered World War II, Boyington, a Marine pilot, resigned from the Marines and signed up as a paid mercenary flier for the American Volunteer Group, known as the Flying Tigers, in China. Led by General Claire Lee Chennault, the Flying Tigers successfully used their P-40 fighter planes against the Japanese. During this tour of duty from November, 1941, to July, 1942, Boyington was officially credited with shooting down six Japanese aircraft.

After the United States entered World War II, Boyington returned to the mainland and was reinstated into the Marines. He was transferred to the Pacific theater, where his first job, administrative in nature, was primarily to find replacements for American flier casualties. In his autobiography, *Baa Baa Black Sheep* (1958), Boyington claimed to have provided strategic input to the secret air mission that intercepted, shot down, and killed the commander in chief of Japan's navy, Admiral Isoroku Yamamoto, over Bougainville in the Solomon Islands on April 18, 1943. Yearning to return to combat duty, he convinced his superiors that he could make a greater contribution to the war effort by training and leading a newly formed squadron.

The Squadron's Formation

The popular history of the Black Sheep Squadron primarily centers on the period from the squadron's initial formation at the Russell Islands, off New Guinea, to the date several months later when Boyington was shot down by a Japanese plane. During that time, the unit as a whole would be credited with downing ninety-four Japanese airplanes, strafing and disabling a large number of enemy aircraft parked on the ground, and successfully protecting many U.S. bomber aircraft missions.

The Black Sheep Squadron's third mission on September 16, 1943, was to escort U.S. bombers and torpedo planes attacking the Japanese airfield at Ballale Island, near Guadalcanal. On this mission, twenty Corsair aircraft of the Black Sheep Squadron engaged about forty Japanese fighter planes. In the ensuing air battle, Boyington shot down five enemy planes but had to make an emergency landing at Munda, in the Solomon Islands, because he was low on fuel. Soon after returning safely to home base, the squadron had a meeting to evaluate the mission. At this meeting, the young men wanted informally to name the squadron "Boyington's Bastards," but Boyington insisted on the more polite "Black Sheep Squadron."

Boyington's Capture

On January 3, 1944, on a mission from Bougainville to Rabaul in the Solomon Islands, Boyington shot down three enemy aircraft to bring his war total to twenty-eight, a new U.S. combat record. However, on this same mission, he was himself shot down, by a Japanese fighter aircraft. After parachuting from his aircraft, he survived for several hours in the cold waters of St. George Channel before being captured by a Japanese submarine. For the remaining eighteen months of the war, he remained a prisoner, suffering frequent beatings and interrogations, starvation, and unsanitary conditions. Because the Japanese would neither inform the neutral Swiss government of Boyington's capture nor release his identity publicly, the U.S. military officially assumed that Boyington was dead and awarded him a posthumous Congressional Medal of Honor in 1944.

After Boyington's capture, the Black Sheep Squadron continued with constant changes of officers and pilots and transfers to other units. The squadron's final wartime assignment was aboard the small aircraft carrier USS *Franklin*. The *Franklin* suffered severe casualties and damage when a Japanese dive-bomber landed a bomb on the flight deck full of armed and fueled U.S. aircraft ready for takeoff.

Postwar Honors

After the end of the war, Boyington was safely returned to U.S. military officials and sent to the mainland United States. The media had made him a popular war hero, and he went on a tour of appearances. On October 5, 1945, U.S. president Harry S. Truman personally awarded Boyington his Medal of Honor in a White House ceremony. The president's words were, "Congratulations, I would rather have this honor than be President of the United States."

The Black Sheep Squadron as a whole received the Presidential Unit Citation, which recognized its air-to-air combat missions from April 7, 1943, to January 6, 1944. The squadron was credited with 132 pilots, 160 downed Japanese airplanes, 70 airplanes lost, 28 pilots killed or permanently missing in action, 13 pilots wounded, and a casualty rate of 30 percent. Some of the original members later flew in the Korean War and in the 1948 Israeli War for Independence.

In his autobiography, Boyington mentions a happy reunion with twenty of the Black Sheep Squadron members in Oakland, California, soon after the war's end. Until his death in 1988, he kept in touch with many of his fliers, who, in interviews throughout the years, were generous with praise for their colonel, mainly because his teaching and experience helped them return home safely from the war. However, the fliers believed that the news media and the 1976-1978 television series *Baa Baa Black Sheep*, for which Boyington was a paid adviser, exaggerated the rowdy behavior of the unit as a whole. Boyington freely admitted in his autobiography his own troubles but provided very few examples of rowdy behavior by his men.

Alan Prescott Peterson

Bibliography

Boyington, "Pappy." *Baa Baa Black Sheep*. New York: G. P. Putnam's Sons, 1958. Boyington's personal, and not necessarily historically accurate, autobiography describes his memories of the exploits of the Black Sheep Squadron and especially his time as a prisoner of war.

Gamble, Bruce. *The Black Sheep Squadron*. Novato, Calif.: Presidio Press, 2000. A historian's account of the squadron, drawing from military records, archives, and the fliers' personal letters home during the war.

McCullough, David G., ed. *The American Heritage Picture History of World War II*. American Heritage, 1966. A picture book of World War II, containing photos and a summary of Boyington's and the Black Sheep Squadron's contribution to the Pacific war effort.

Walton, Frank E. *Once They Were Eagles: The Men of the Black Sheep Squadron*. Lexington: University of Kentucky Press, 1996. A historian's account of the squadron and its individual members.

See also: Bombers; Fighter pilots; Flying Tigers; Kamikaze missions; Marine pilots, U.S.; World War II

Blimps

Also known as: Airships, nonrigid or pressure-airships, dirigibles, balloon-dirigibles

Definition: A lighter-than-air, pressurized airship, comprising an elliptical, gas-filled bag, a means of propulsion, a means to control buoyancy and flight, and one or more gondolas to hold crew, passengers, the power unit, and cargo.

Significance: The nonrigid airship was the first form of controlled human flight, and the blimp was the last airship to be used in wartime.

Development

The early days of aviation witnessed a competition between two very different vehicles: the heavier-than-air airplane and the lighter-than-air airship. Although the airship initially prevailed, it would, by the 1930's, be largely replaced by the airplane. Airships, however, continue to perform functions that are beyond the capabilities of airplanes.

Airships evolved from the free, or hot-air, balloon, first launched in 1783 near Lyons, France, by Jacques-Étienne and Joseph-Michel Montgolfier. This balloon would be modified. Henry Cavendish, a British chemist, found that hydrogen gas was at least seven times lighter than air, and by 1785, French army engineer Jean Baptiste Marie-Meusnier designed a bag of an ellipsoidal shape. French inventor Henri Giffard took these notions, added mechanical propulsion and steering, and flew a dirigible-balloon, named from the Latin *dirigere*, "to steer," on September 24, 1852. This 143-foot-long airship, driven by a screw propeller rotated by a 3-horsepower steam engine, traveled at the speed of 10 miles per hour. It was the first successful flight of an airship. Thirty-one years later, the Tissandier brothers, Albert and Gaston, built an electrically powered, 37,000-cubic-foot airship. On August 9, 1884, Charles Renard and Arthur Krebs piloted the 66,000-cubic-foot, electrically driven *La France* for 5 miles, returning safety to the point of departure. A Brazilian aeronaut, Alberto Santos-Dumont, who "mused on the exploration of the aerial ocean," launched a series of fourteen airships in France before 1905. His airship *Number 6* made headlines when it successfully circled the Eiffel Tower. The eccentric Santos-Dumont popularized airships by parking them over the rooftops of his Parisian hosts, descending to join them for dinner.

Design

An airship has five crucial components: an elliptical bag filled with either hydrogen or helium and covered with a strong, light "envelope" (an outer skin initially made of cotton and rubber; today synthetic fabrics are used); a means of propulsion, using propellers and engines powered by fuels ranging from steam and electricity to gasoline; control of buoyancy attained by releasing ballast for ascent, gas for descent; flight control, with the pilot using vertically hinged rudders for steering, horizontally hinged elevators for lift; and one or more gondolas for crew, passengers, the power unit, and cargo.

There are three classes of airship. One is the nonrigid, or pressurized, airship. Without a metal frame, the bag collapses when the gas is released. During World War I, this type of airship was the most common in the Royal Navy and gave rise to the slang term "blimp," which took its initial "b" from "British Class B Airship," and "limp" from its nonrigid nature.

Another type of airship is the semirigid, in which, to maintain the form, gas pressure acts in conjunction with the longitudinal keel. A third type is the rigid airship, or zeppelin, named for the German count, Ferdinand von Zeppelin, who perfected it. With a skeleton, it retains its shape when deflated.

Use

The Germans stressed the rigid, the British the nonrigid type. During World War I, the Germans had some sixty-seven zeppelins flying a variety of missions. The British Navy favored blimps, deploying over two hundred of them for submarine and mine detection, aerial observation, coastal patrols, scouting, and escorting troop and merchant vessel convoys.

Following World War I, rigids were preferred. That popularity ended dramatically when Germany's pride, the *Hindenburg*, perished in fire at Lakehurst, New Jersey, in 1937. The United States, fortunately, had not abandoned blimps. By 1930, the Goodyear Tire and Rubber Company had a fleet of twelve blimps, used primarily for advertising. The only nation to make effective use of blimps in World War II, the United States had a fleet of 150 of them, serving in fifteen airship squadrons on three continents, patrolling three million square miles. The first nonrigid airship crossing of the Atlantic occurred from May 29 and June 1, 1944, when a U.S. Navy blimp squadron made the 3,145-mile flight from South Weymouth, Massachusetts, to Port Lyautey, French Morocco. Blimps proved effective in detecting German submarine wolfpacks. Not a single blimp-escorted convoy lost a ship. Only one blimp was downed by enemy fire.

During the Cold War, blimps were of value not only for coastal patrols, but also as an early-warning device against piloted bomber flights. In 1958, the U.S. Navy commissioned a series of four ZPG-3W airships, each 403 feet in length, 85 feet in diameter, with a capacity of 1,500,000 cubic feet. These were the largest blimps ever. When, after 1962, the piloted bomber gave way to the intercontinental ballistic missile, the value of blimps declined.

By the 1990's, however, there was a renewed interest in blimps. From a commercial standpoint, they could carry passengers and cargo cheaply and efficiently. Television networks used them for aerial views of sporting events. Advertising (as with the well-known Fuji and Goodyear blimps) was profitable. The recreation use of airships was appealing. Synthetic fibers, computer-aided design, and enhanced engineering led to such "super blimps" as the *Sentinel 5000* launched in 1997. It had a three-story pressurized gondola. Virtually impervious to weather (icing, snow, sleet, rain, fog, hail) and radar, it traveled in excess of 60 miles per hour. Blimps, because of their range, fuel efficiency, low cost of development and maintenance, capacity for in-flight refueling, and lack of negative environmental impact, proved attractive to both military and civilian agencies for a variety of surveillance work. The blimps promise to have a long and useful future.

C. George Fry

Bibliography

Botting, Douglas. *The Giant Airships*. Alexandria, Va.: Time-Life Books, 1981. A concise and profusely illustrated introduction to the history of lighter-than-air aviation.

Collier, Basil. *The Airship: A History*. New York: G. P. Putnam's Sons, 1974. A readable and reliable survey of the subject from its inception until the late twentieth century.

Dick, Harold G., and D. H. Robinson. *The Golden Age of the Great Passenger Airships*. Reprint. Washington, D.C.: Smithsonian Institution Press, 1992. A valuable analysis of the early decades of airship history by two skilled authors.

Horton, Edward. *The Age of the Airship*. Chicago: Regnery, 1973. Though somewhat dated, this remains a useful introduction to the subject for the beginning student.

See also: Buoyant aircraft; Dirigibles; Goodyear blimp; *Hindenburg*; Lighter-than-air craft; Montgolfier brothers; Reconnaissance; Alberto Santos-Dumont; World War I; World War II; Ferdinand von Zeppelin

Blue Angels

Also known as: The United States Navy Flight Demonstration Squadron

Date: Formed in 1945; first flight demonstration on June 15, 1946

Definition: A flight demonstration team organized to showcase naval aviation and serve as positive role models and goodwill ambassadors for the United States military.

Significance: The Blue Angels demonstrate the pinnacle of precision flying, representing the United States Navy to the civilian community, providing exciting entertainment for millions of spectators every year, and serving as a recruitment tool for the United States Navy and Marine Corps.

Formation and Development

At the end of World War II, Admiral Chester W. Nimitz, the Chief of United States Naval Operations, ordered the formation of a flight demonstration team to illustrate precision flying and maintain public interest in naval aviation. After several months of organization and practice, the first squadron demonstrated its initial public aerial performance on June 15, 1946, at the Southeastern Air Show and Exhibition at the Naval Air Station (NAS) at Craig Field in Jacksonville, Florida. They won the trophy for the most outstanding performance. The flight leader was Lieutenant Commander Roy "Butch" Voris. The other team members were Lieutenant Mel Cassidy (left wing), Lieutenant Maurice "Wick" Wickendoll (right wing), Lieutenant Al Taddeo (solo), and Lieutenant Gale Stouse (backup). The aircraft they flew was the Grumman F-6F Hellcat.

On August 25, 1946, the Blue Angels changed their aircraft to the Grumman F-8F Bearcat. By 1947, Lieutenant Commander Robert Clarke had become the flight leader. He introduced the diamond formation, which became the trademark of the Blue Angels. Near the end of the 1940's, the squadron was flying their first jet aircraft, the Grumman F9F-2 Panther.

With the outbreak of the Korean War, the Blue Angels were assigned to the aircraft carrier USS *Princeton* in 1950, forming the core of Fighter Squadron 101, which became known as "Satan's Kitten." They adopted a squadron insignia that portrayed a fiendish cat riding the devil's three-pronged fork and hurling lightning bolts at the enemy. In 1951, they were sent to the NAS in Corpus Christi, Texas, where they began flying the Grumman F9F-5 Panther. In October, 1951, a directive from the Chief of Naval Operations reactivated the Blue Angels to perform the same duties that they had performed prior to the war.

In 1954, the Blue Angels were assigned to their present home at the NAS at Sherman Field in Pensacola, Florida, where the crew began flying the newer, faster, swept-wing

Grumman F9F-8 Cougar. In 1957, the Blue Angels began flying the Grumman F-11 Tiger. By 1969, they were doing their aerial shows in a dual-engine jet, the McDonnell Douglas F-4J Phantom II.

New Focus

In December, 1974, the Blue Angels were reorganized as the United States Navy Flight Demonstration Squadron, with Tony Less as the commanding officer. Further changes included the addition of a number of support officers and a new aircraft, the McDonnell Douglas A-4F Skyhawk II. The mission of the squadron became focused on Navy recruiting. In celebration of their fortieth anniversary in 1986, the Blue Angels flew the sleek McDonnell Douglas F/A-18 Hornet. This aircraft 3was the first dual-role fighter and attack jet serving on the front lines of U.S. defense.

After a nineteen-year absence, the Blue Angels were deployed on a one-month European tour in 1992. Over a million people in Sweden, Finland, Russia, Romania, Bulgaria, Italy, Spain, and the United Kingdom witnessed their performances. In November, 1998, the first Blue Angel jet was landed on an aircraft carrier, the USS *Harry S. Truman*, by squadron Commander Patrick Driscoll.

The only Marine Corps aircraft that performs with the Blue Angels is the Hercules Transport C-30, nicknamed "Fat Albert." It is flown by an all-Marine crew consisting of three pilots and five enlisted personnel. In the course of a show season, Fat Albert is flown over 140,000 miles. It transports the necessary personnel and equipment that support the Blue Angels from one performance site to another.

Blue Angels Aircraft

Years of Use	*Manufacturer*	*Model*
1946	Grumman	F-6F Hellcat
1946-1949	Grumman	F-8F Bearcat
1949-1951	Grumman	F9F-2 Panther
1951-1954	Grumman	F9F-5 Panther
1954-1957	Grumman	F9F-8 Cougar
1957-1969	Grumman	F11F-1 Tiger
1969-1974	McDonnell Douglas	F-4J Phantom II
1974-1986	McDonnell Douglas	A-4F Skyhawk II
1986-present	McDonnell Douglas	F/A-18 Hornet

Source: Data taken from (www.chinfo.navy.mil/navpalib/aircraft/b-angels/blues.html), June 6, 2001.

Demonstrations

The flight demonstrations of the Blue Angels exhibit choreographed refinements of Navy-trained flying skills. Flight shows include graceful, aerobatic maneuvers of the four-plane diamond formation, in conjunction with the fast maneuvers of its two solo pilots, and the renowned six-jet delta formation. During the show season (April to December), the Blue Angels are stationed at Pensacola, Florida. During the other three months of the year, they are stationed for training at the NAS at El Centro, California.

At the beginning of a Blue Angels show, Fat Albert often demonstrates its jet-assisted takeoff capability. Eight solid-fuel rockets are attached to the sides of the aircraft. When they are ignited, Fat Albert climbs at a 45-degree angle to an altitude of 1,000 feet in a few seconds. Shortly thereafter, six Blue Angel Hornets engage their afterburners and climb into the sky to perform their maneuvers. Each Hornet is 56 feet in length, 15 feet high, with a wingspan of 40 feet, and the capability of reaching speeds well in excess of supersonic velocities.

Since the inception of the Blue Angels in 1946, there have been twenty-three pilots killed in air shows or training. Two Blue Angels were killed on October 28, 1999, in southern Georgia while trying to land during a training flight. The last fatality prior to that incident was on July 13, 1985, when one pilot died in a fireball crash after two planes collided during an air show.

During 2001, the Blue Angels performed in nearly seventy shows at thirty-six locations in the United States and Canada under the direction of Commander Robert A. Field. In 2000, they performed before more than 17 million fans. Since their first show in 1946, the Blue Angels have performed for more than 374 million spectators.

Alvin K. Benson

Bibliography

Bledsoe, Glen, and Karen E. Bledsoe. *The Blue Angels: The U.S. Navy Flight Demonstration Squadron*. Mankato, Minn.: Capstone Press, 2001. Excellent overview of the Blue Angels, their history, and aircraft; written for younger readers.

Van Steenwyk, Elizabeth. *From Barnstormers to Blue Angels*. New York: Franklin Watts, 1999. Discusses the air shows and aircraft of the Blue Angels and contains many photos.

Veronico, Nicholas A., and Marga R. Fritze. *Blue Angels: Fifty Years of Precision Flight.* Osceola, Wis.: Motorbooks International, 1996. The history of the Blue Angels over their first fifty years, discussing and showing pictures of the people, places, and aircraft.

See also: Aerobatics; Hornet; Military flight; Navy pilots, U.S.

Boarding procedures

Definition: Airline procedures that process passengers and allow them onto the correct aircraft for their destination.

Significance: Boarding procedures ensure that only ticketed passengers board the correct aircraft for their destination. Boarding procedures ensure on-time departures by taking place within a scheduled time period.

Boarding procedures were established to ensure that passengers are boarded onto the right airplane at the right time and are seated in assigned or available seats. To accomplish this task, a number of steps are taken before, during, and after passengers get onto their aircraft.

Boarding planning first considers the kind of flight that is being processed. Flight departures can be of two types: one in which an aircraft is coming from another location and proceeding on to its destination or one in which the departure is the flight's origination point. Operations departments determine whether the flight is on schedule, what gate will be assigned for the departure, and the expected number of passengers. Passengers are of three types: first, local passengers are those beginning their trip; second, connecting passengers are those arriving on other aircraft to continue their trip on the departing flight; and third, continuing passengers are those arriving and continuing onto the flight's destinations. If the flight is oversold, or overbooked, oversale procedures are initiated.

Keeping in mind that a departure may involve an arriving aircraft, gate agents report to the assigned gate typically thirty minutes before an aircraft's arrival or one hour before its departure. They prepare and post signs that indicate departure information such as the flight number, the destination, and the scheduled or adjusted departure time, if necessary. Adjusted departure times reflect any information that may or will, if known, affect the departure, such as weather, air traffic, maintenance, or crew matters.

When passenger counts are low, all of the boarding procedures can be performed by one person. When twenty-five or more passengers are expected, it is customary to have two gate agents. Three to four gate agents are needed for flights of larger aircraft in which two to three hundred passengers are expected.

Boarding responsibilities are divided into two functions, known by a variety of titles. The boarding agent, or coordinator, is responsible for all announcements, for the actual taking of tickets and boarding passes from passengers, and for all communication with the crew. The gate, or control, agent is responsible for checking passengers in if needed, for producing all needed reports, and for making the entries that calculate and finalize how many passengers are on board.

Within the hour before departure, passengers begin to arrive at the gate. Some need to be checked in and given their seat assignments. Most already have been checked in at a ticket counter or in their originating location if they are on a connecting flight.

Different aircraft have different boarding requirements and time frames that take into account the aircraft's size and the number of passengers. A full medium-sized aircraft may take as much time as a half-full large aircraft. Boarding may begin as much as one hour or as little as fifteen minutes before departure.

Preboarding

Boarding begins with a consultation and agreement with the flight crew that all is in order on board the aircraft. Boarding is managed and coordinated by announcements usually made through a public-address system. The first announcements identify the airline, the flight number, the destination, the departure time, and also include certain reminders regarding the size and the number of carry-on items allowed. Recognizing that certain passengers have special needs and that certain passengers enjoy the privileges of being preferred customers, the second announcement is called the preboarding announcement. Preboarding allows those with special needs or those with preferred privileges to board ahead of others. The third announcement begins the general boarding process. Row numbers, normally in sets of five, are called to board, starting with the back rows and progressing toward the front. Boarding from the rear rows to the front eliminates congestion on board the aircraft and allows passengers to proceed without interruption to their assigned seats.

After preboarding and while general boarding is conducted, other steps leading to final passenger and departure documentation take place. Almost every airline makes what is called a cutoff announcement twenty minutes prior to departure. Computer entries are then made releasing the advance seat assignments of passengers who have not already checked in. Other entries are then made to assign seats to standby passengers. Standby passengers are of two kinds, revenue and space available. Revenue standby passengers are passengers that were ticketed for earlier or later flights. Space-available passengers are passengers who are traveling on various kinds of passes and are boarded only if there are remaining available seats.

Final Boarding

At ten minutes prior to departure, the final boarding announcement is made. Passengers arriving at the departure gate after this announcement are late and may not be boarded. After the final boarding announcement is made, various reports and passenger counts are prepared and calculated and are given to the crew and to operations departments. At five minutes prior to departure, there is another consultation with the crew notifying them that all passengers who can be boarded have been boarded, and that the gate is prepared to close the door. The authority and direction to close the door comes from the captain. After the plane has left, the gate staff take several other steps, such as generating several other reports, documenting the actual time of departure and the exact number of passengers and crew, sending the now-used tickets to airline accounting departments, and communicating any relevant passenger information to the destination city.

Jim Oppermann

Bibliography

America West Airlines. *On-Time Performance Training.* Phoenix, Ariz.: America West Airlines, 2001. A handbook detailing America West's 2001 training initiative aimed at better coordinating and timing the steps taken before, during, and after boarding by gate agents, flight attendents, captains, and ramp service personnel.

Irrgang, Michael E. *Airline Operational Efficiency.* Washington D.C.: McGraw-Hill/Aviation Week, 2000. Describes the importance of timely loading of passengers, baggage, and cargo.

See also: Air carriers; Airline industry, U.S.; Airports; Baggage handling and regulations; Overbooking; Takeoff procedures; Ticketing

Boeing

Also known as: The Boeing Company

Definition: The world's largest builder of commercial aircraft.

Significance: Throughout its history, Boeing has been the world's largest commercial aircraft manufacturer, a major U.S. defense contractor, and an active participant in the U.S. spaceflight program. During the 1960's, Boeing lent the skills of 2,000 employees to the U.S. effort to land humans on the Moon.

Early History

In 1903, the same year that the Wright brothers completed their first flight at Kitty Hawk, North Carolina, William Boeing left Yale University's college of engineering for the West Coast. After accumulating a considerable amount of money trading in forest lands around Grays Harbor, Washington, Boeing moved to Seattle, Washington, in 1908.

Boeing had always been curious about air travel, which was in its infancy during his youth. In 1910, he attended the first American air meet in Los Angeles, California. He sought a ride on one of the airplanes shown at the meet, but not one of the dozen aviators participating would do him the favor. Little did the early pilots realize that they were refusing a man whose name would become synonymous with commercial aviation around the world.

In 1916, Boeing formed the Pacific Aero Products Company, which was renamed the Boeing Airplane Company the following year. He and G. Conrad Westerveldt developed the B & W seaplane. World War I brought Boeing lucrative contracts for Navy trainers and flying boats. However, by 1919, after the war's end, the company was on the brink of bankruptcy. Boeing scrambled to keep his workers busy making furniture, repairing Army planes, and building speedboats that would become popular with local bootleggers during Prohibition.

Boeing also earned money by pioneering airmail service. The Air Mail Act of 1925, also known as the Kelly Act, authorized the U.S. Post Office to contract with private carriers on designated routes. On September 15, 1926, Vern Gorst's Pacific Air Transport (PAT) delivered Seattle's first bag of domestic airmail to a Boeing airstrip. The following year, Boeing purchased PAT and introduced larger Model 80 and 80A trimotors. The cabins, carrying up to eighteen passengers, were attended by registered nurses who became the first flight attendants.

On February 1, 1929, William Boeing and Fred Reutschler, president of the Pratt & Whitney engine manufacturer, formed the United Aircraft & Transport Corporation (UATC). It quickly acquired other aircraft companies, including Stearman, which established Boeing's presence in Wichita, Kansas. In March, 1931, this carrier, a pioneer in commercial aviation, would be incorporated, along with numerous others, as United Air Lines.

The years immediately following the end of World War II were filled with changes for Boeing. After the military canceled its bomber orders, Boeing factories shut down, and 70,000 people lost their jobs. The same day the plants closed, attorney William M. Allen took over as company president. Allen promised to start hiring people back as soon as airlines ordered the Stratocruiser, a luxurious commercial airliner version of the company's four-engine C-97 troop transport first flown in 1944. The Stratocruiser did not provide Boeing's hoped-for financial windfall, however. Instead, Boeing earned substantial profits by adapting its C-97 air freighter as both a propeller-powered troop carrier and the KC-97, an aerial fuel tanker.

In the meantime, wind-tunnel data discovered in Germany as the war ended helped Boeing engineers design the country's first multiengine, swept-wing jet bomber, the XB-47. After World War II, Boeing also applied the technology of jet bombers to revolutionize civilian passenger airline travel.

As early as the 1940's, Boeing 707-120B's used to transport government officials had been given the call sign *Air Force One*. Boeing 707-320B airframes later were adapted specifically for use by the U.S. president, designated VC-137C, and officially called *Air Force One*. VC-137C's served as presidential aircraft until 1990, when they were replaced by two new aircraft using Boeing 747-200 airframes.

Events in Boeing History

1916: Boeing is first incorporated by William E. Boeing as the Pacific Aero Products Company to develop the B & W seaplane. The company is renamed the Boeing Airplane Company the following year.

1929: After building air mail and military aircraft throughout the 1920's, the company is merged into the United Aircraft and Transport Corporation (UATC), a group of aircraft manufacturers and airlines.

1934: Federal antitrust regulations require the splitting of UATC into three separate companies: the United Aircraft Company, the Boeing Airplane Company, and United Air Lines.

1935: Boeing's Flying Fortress (B-17) bomber, which later plays a crucial role in U.S. success during World War II, is first flown.

1942: Boeing's Superfortress (B-29) bomber, which also contributes greatly to the U.S. war effort, is first flown.

1952: Boeing's Stratofortress (B-52) bomber, which will remain the primary U.S. bomber for the next four decades, is first flown.

1957: Boeing's first jetliner, the 707, makes its first flight, entering service the following year. The company subsequently develops a series of jetliners that are enormously popular worldwide.

1960's: Boeing participates in the U.S. space program by designing and manufacturing Apollo and Saturn rockets and lunar orbiters.

1970: The first wide-body jumbojet, Boeing's 747, with twice the passenger capacity of any previous jet, enters service.

1980's: Boeing develops both the air-launched cruise missile and the MX intercontinental ballistic missile (ICBM).

1996: During a period of reorganization in the aerospace industry, Boeing purchases divisions of Rockwell International involved in aerospace and defense electronics.

1997: Boeing merges with the McDonnell Douglas Corporation.

Role in Space Exploration

In 1961, U.S. president John F. Kennedy committed the United States to landing a person on the Moon before the end of the decade. At that time, the far side of the Moon remained a mystery. Because Boeing president William Allen believed in the space program, he loaned 2,000 executives to the National Aeronautics and Space Administration (NASA) to coordinate activities. Boeing also provided overall systems integration for the entire Apollo project.

Boeing-built orbiters circled the Moon and sent photographs of the Moon's surface back to Earth, so NASA could select safe landing sites for the astronauts. Boeing also built the Lunar Roving Vehicle (LRV), which astronauts used to explore the Moon on the last three Apollo missions. Boeing often shared space-program construction responsibilities with other large aerospace companies. For the Saturn launch vehicles, for example, Boeing built the S-1C's first stage, North American Rockwell built the second, and McDonnell Douglas the third.

Despite the success of the space program, Boeing was buffeted during the mid-1960's by the loss of several crucial defense contracts. The company also launched its new 747 jumbojet on the eve of a depression in the airline industry. The company's workforce, once numbering more than 100,000, declined by more than 60,000 workers during the ensuing "Boeing Bust." A billboard erected in early April, 1971, teased, "Will the last person leaving Seattle turn out the lights."

Boeing's Influence

On August 1, 1997, Boeing and McDonnell Douglas merged and began operations as a single company with more than 220,000 employees. Phil Condit remained as chief operating officer and chairman of the new Boeing board of directors. Harry C. Stonecipher, formerly McDonnell Douglas president and chief executive officer, became president and chief operating officer of The Boeing Company.

Boeing announced on March 21, 2001, that its headquarters, employing about 1,000 people, would depart Seattle, the city that had been its corporate home for eighty-five years. However, many other Boeing manufacturing plants in the area would remain there. Dallas, Chicago, and Denver were listed as likely new Boeing headquarters locations before Chicago was chosen in early May. The surprise announcement came as a shock to many Seattle families that had sent several generations to work in Boeing's assembly plants in the Puget Sound area. Despite the loss of Boeing's headquarters, nearly 80,000 company jobs would remain in the state of Washington, but it was feared that some of those would leave as well. A few weeks after announcing the headquarters move, Boeing said it would move the assembly of its 757 jet fuselage to Wichita, Kansas, from Renton, a suburb of Seattle, transferring five hundred jobs.

Before the founding of Microsoft, Boeing was the Seattle area's signature, and singularly dominant, industry. Employees of the aircraft manufacturer influenced the city's traffic patterns, housing prices, and even department store sales, which coincided with Boeing's holiday bonuses. The famed World War II icon Rosie the Riveter was a Seattle-area Boeing assembly-line worker before her image was featured nationally on war posters.

Boeing is also a large-scale defense contractor as well as a commercial aircraft manufacturer. During 2001, the company competed with Lockheed Martin to build the Joint Strike Fighter. At an expected worth of about $300 billion, the defense contract was the most lucrative in history. Each of roughly 3,000 aircraft was expected to cost $25 to $30 million.

Boeing employees also have irrigated an eastern Oregon desert, managed housing projects for the federal Department of Housing and Urban Development, built a desalinization plant that converted sea water to fresh water for a resort in the Virgin Islands, and built voice scramblers for police departments. The Boeing Company also produced light-rail vehicles for the cities of Boston, Massachusetts, and San Francisco, California, introduced personal rapid transit in Morgantown, West Virginia, and built three gigantic wind turbines in the Columbia River Gorge.

Future Boeing Projects

For sale after 2007, Boeing planned to build a new 700-mile-per-hour Sonic Cruiser, which will reduce the current seven-hour transatlantic airline journey by one hour. Boeing also planned to increase aircraft speeds significantly with an entirely new engine technology using a mixture of conventional jet fuel—derived from oil, a fossil fuel—with clean-burning hydrogen. Prior to Boeing's new tests, the top speeds of commercial aircraft had been stagnant since 1970, when the record for the fastest civilian aircraft (1,600 miles per hour) was set by a Russian Tupolev Tu-144. Typical jet aircraft speeds (500 miles per hour) had not changed since the 1950's.

In 2001, Boeing unveiled a prototype superfast aircraft that could fly passengers between London and New York in forty minutes. In May, the Hyper-X, "a flying engine that looks like a surfboard with fins," designed jointly by Boeing and NASA, was tested over the Pacific Ocean 75 miles off Los Angeles.

In the engine test, the Hyper-X was bolted beneath the wing of a B-52 bomber. The B-52 released the "flying surfboard" at 20,000 feet, as a conventional booster rocket drove it to about 2,000 miles per hour. Revolutionary scramjets then cut in and, for ten seconds, the hypersonic plane reached a maximum speed of 5,000 miles per hour, making it the fastest aircraft in history.

Ordinary jet engines are propelled by blades that drag air into a chamber, compress it, mix it with jet fuel, and explode it out of the rear to create forward momentum. Scramjets have no blades, but depend on previously generated speeds to force air through an oval-shaped mouth into a copper chamber, where it mixes with hydrogen to produce a much more powerful explosion.

The Hyper-X can fly at speeds of up to 5,000 miles per hour, more than three times as fast as the next-fastest airliner, the thirty-year-old Concorde, which had become technologically obsolete by the year 2000. Other tests were foreseen with prototypes able to fly as fast as 7,000 miles per hour. Such vehicles could circumnavigate the earth in fewer than four hours. Boeing intended initially to design such aircraft for the U.S. military and then to build a bigger version for cargo operators. After all tests were completed, Boeing would build a version for commercial customers, such as British Airways, starting in 2016.

Boeing's hypersonic aircraft would be much smaller than the jumbojets that comprised parts of many airline fleets during the late twentieth century. The bigger planes lack the structural integrity required to withstand vastly accelerated speeds. The development of hypersonic aircraft also has been made possible by advances in the strength of manufactured metals. For structural reasons, the new airliner probably will have no windows. Passengers will be protected from a gravitational force of 6 g's by a highly pressurized cabin. The aircraft also will accelerate and decelerate slowly to lessen the effects of changing gravity. Such aircraft also will produce sonic booms as they accelerate, so routes will need to be configured to avoid large population areas at the point of transition to hypersonic flight.

Bruce E. Johansen

Bibliography

Bauer, Eugene E. *Boeing in Peace and War.* Enumclaw, Wash.: TABA, 1990. A concise summary of Boeing's history.

Bowers, Peter M. *Boeing Aircraft Since 1916*. Annapolis, Md.: Naval Institute Press, 1989. A good source for technical material on aircraft manufactured by Boeing.

Norris, Guy, and Mark Wagner. *Boeing*. Osceola, Wis.: MBI, 1998. A twentieth century history of Boeing.

Rodgers, Eugene. *Flying High: The Story of Boeing and the Rise of the Jetliner Industry.* New York: Atlantic Monthly Press, 1996. Boeing's development in the context of the aviation industry.

Serling, Robert J. *Legend and Legacy: The Story of Boeing and Its People*. New York: St. Martin's Press, 1992. An excellent history of Boeing for the general reader.

See also: *Air Force One*; Airplanes; Bombers; Hypersonic aircraft; Jumbojets; Manufacturers; McDonnell Douglas; National Aeronautics and Space Administration; Orbiting; Seaplanes; 707 plane family; Spaceflight; United Air Lines; X planes

Bombers

Definition: Military aircraft designed with the primary mission of dropping bombs.

Significance: Before the advent of effective missiles, bombers were the weapons used by air forces to attack enemy nations. Following World War I, bombing enthusiasts predicted that strategic bombardment could force an enemy to surrender and thus change the way nations would fight wars.

Bombers are generally classified by the type of bomb they deliver (a torpedo-bomber delivers a torpedo), their size (light, medium, or heavy), or their mission (fighter-bomber). While World War II saw a wide variety of bomb-dropping aircraft, the number of distinct bomber types produced has dwindled, as newer aircraft, such as fighters that are tasked with bombardment missions, perform multiple roles.

Development

The first military aircraft were used for reconnaissance. Only after trench warfare began did generals come to see airplanes as platforms that could carry ordnance beyond enemy lines to strike specific targets. In order to successfully bomb targets, bombers had to carry bomb loads that were heavy enough to inflict significant damage, to fly both high and quickly enough to bypass enemy defenses, and to deliver bombs accurately enough to hit the desired targets.

In order to build successful bombers, aircraft manufacturers had to overcome many technical difficulties that limited aircraft capabilities and performance. It is difficult, for example, to drop a bomb from a moving airplane so that it arrives on target, especially if the target is obscured or camouflaged. Bombs must be able to penetrate deeply enough or must carry enough explosive force to destroy the target. Bombers must be capable of flying high enough or quickly enough to avoid enemy fire or must be armored well enough to render enemy fire ineffective. Although defensive armor increases a bomber's odds of survival, its greater weight limits speed, range, and bomb load. Better accuracy is found at lower speeds and altitudes, where bombers are more vulnerable. A longer range allows bombers to hit a greater variety of targets but requires more fuel and, thus, a lighter bomb load. Finally, bombers require defensive armaments, or weapons, to avoid being shot down by fighters.

World War I

World War I saw experiments with almost every possible bomber mission. The first bombs were simply grenades tossed at enemy positions. Because these weapons were too light and inaccurate to cause serious damage, heavier bombs were designed. Early bombs that were released over the side of an airplane were often inaccurate. The first

Major Bombers of World War I

Name	*Date*	*Country*	*Speed (miles per hour)*	*Range (hours)*	*Ceiling (feet)*	*Number of Machine Guns*	*Bomb Load (pounds)*	*Wingspan (feet)*	*Weight (pounds)*
Airco D.H.4	1917	Britain	143	7	23,500	4	460	42.33	3,742
Blackburn Kangaroo	1918	Britain	100	8	10,500	2	930	74.83	8,017
Breuget Br.M.5	1917	France	88	5	14,110	2	661	29.67	4,235
Breuget B14B2	1915	France	110	3	19,030	2	661	47.08	3,892
Caproni Ca.30	1917	Italy	85	3.5	13,451	4	1,000	72.83	8,400
Caproni Ca.42	1918	Italy	78	4	13,451	8	3,197	98.08	14,793
Caudron R.11	1918	France	114	3	19,520	5	265	58.75	4,775
Friedicshafen G.III	1917	Germany	85	5	14,765	3	3,307	77.75	8,646
Gotha G.V	1917	Germany	88	6	21,325	3	1,102	77.75	8,745
Handley Page V/1500	1918	Britain	97	6	10,000	5	7,500	126.08	24,700
Short	1916	Britain	77	6	9,500	1	920	85.00	6,800
Sikorsky Ilva Mourometz V	1916	Russia	75	5	9,840	7	1,150	97.75	10,117
Vickers Vimy	1918	Britain	103	9	10,500	4	2,476	67.16	12,500
Voisin 5	1915	France	65	3.5	11,485	1	132	48.33	2,516
Zep Staaken R.VI	1917	Germany	80	8	12,467	7	4,000	138.41	25,269

bombers were observer aircraft converted for bombing missions. To inflict serious damage, many bombs were needed, so bombers began to fly in groups. Enemy fighters also forced bombers to fly together to mass their defensive firepower. By 1917, both sides had introduced specialized, bomb-carrying aircraft. The German Gotha G-IV, for example, was designed to attack enemy cities or port facilities far behind the battle lines. Fighters or observation planes were assigned to attack frontline headquarters or troop concentrations. Britain's Royal Navy Air Corps also experimented with launching planes from ships and, thus, introduced the aircraft carrier.

World War II

Although World War I bombers flew many varied missions, these operations were more ad-hoc responses to opportunities or threats than they were planned innovations. After the war, airmen began to develop theories about how aircraft could change the nature of war. The most famous of these was Italian general Giulio Douhet, who predicted that bombers could fly over battlefields to avoid costly ground battles. Bombers could attack specific cities, which Douhet termed a nation's "vital centers." Bombing would result in such damage and terror that citizens would force their governments to sue for peace. This terror bombing was justified as being more humane and less costly than the losses and suffering caused by a protracted war such as World War I. Essential to Douhet's theory was his claim that bombers would always perform as well as fighters, would carry more armament, and could thus always get though to their targets.

In the United States, General William "Billy" Mitchell, a strong advocate of air power, embraced Douhet's ideas, because they seemed to justify the creation of a separate branch of service. Mitchell believed that only a new and independent organization would be free from traditional preconceptions and could, thus, be innovative in utilizing new technologies.

Improved engines and metallurgy allowed aircraft designers to create weapons tailored to suit Douhet's predictions. Especially significant was the United States' adoption of the Norden bombsight, a complex instrument that was, in effect, an early analog calculator linked to an autopilot that determined the appropriate bomb release point using altitude, speed, and bomb type. Once the bomber was situated over the target, the bombsight automatically released the bombs at precisely the correct point, which could easily be missed by a pilot's human error.

Naval aviators perfected dive-bombing as another method to increase accuracy. A bomber would dive toward the target and release the bomb at the last possible instant before the aircraft pulled out of the dive, so that the bomb's trajectory was an extension of the dive. Dive-bombing was very accurate and was especially useful against maneuvering ships.

During World War II, many specialized types of bombers were used on missions that included strategic bombing, air superiority, interdiction, and tactical air support. In each case, bombers proved useful but neither as invulnerable nor as decisive as Douhet had predicted. Airframe and power plant developments resulted in fighters of great speed and heavy armament. Massed fighter attacks against bomber formations inflicted terrible damage on bombers that had to fly straight during their final bomb runs.

Technological advancements also increased the lethality of antiaircraft artillery. By 1945, antiaircraft defenses had been developed that used radar to determine the altitude and bearings of incoming aircraft, and proximity fuses ensured airbursts near the bombers. To avoid the more potent antiaircraft defenses, bombers were flown at higher altitudes that rendered them less accurate. During the war, more than one-half of the bombs dropped landed more than 1,000 feet from their targets.

Bomber enthusiasts failed to anticipate the effectiveness of what are now called passive defenses. Camouflage, decoy targets, smoke screens, and blackouts obscured targets and often rendered bombing raids ineffective, and civil defense measures reduced damage. Civilian morale proved more resilient than expected; terror attacks angered rather than intimidated civilians. Modern, organized national economies could thwart strategic bombing attacks by utilizing more civil defense personnel and replacement laborers or by dispersing factories to make them less vulnerable. These tactics made the concept of vital centers too vague to be useful. Throughout the war, airmen sought to destroy specific enemy industries, such as ball-bearing plants or petroleum refineries, which when destroyed, might cripple the enemy's ability to continue the war. Although these efforts inflicted terrible damage and chaos, enemy economies did not collapse.

Strategic bombardment also suffered from dispersal of effort. Anglo-American airmen such as U.S. general Henry Harley "Hap" Arnold faced continued demands for the diversion of bombers to other missions. Navies wanted long-range bombers for extended antisubmarine or reconnaissance missions. Theater commanders called for air assets to be used to interdict the movement of enemy troops and supplies, while battlefield commanders cried for tactical air support. After the war, air power theorists would claim that their lack of success was driven by these diversions.

The German experiences in World War I and the Spanish Civil War indicated that strategic bombardment alone would not bring victory. Many German airmen had been infantrymen in World War I and saw bombardment and observation as great force multipliers for ground attacks. As a result, German doctrine demanded heavy attacks on enemy air bases after which most of the bombers would support battlefield actions instead of attacking enemy cities. German bombers also tended to be of medium range and tended to use dive-bombing for accuracy. Early in the war, the Germans were successful with this strategy, but as the Allies introduced improved fighters and tactics, German bomber losses began to climb. Nevertheless, the German tactics, based upon a combination of mobile ground units and tactical air support, proved very potent. By 1943, the British and American air forces began to develop their own methods of tactical air support, which, by the summer of 1944, had proven to be a crucial factor in Germany's defeat.

World War II saw the invention of many technologies that would shape bombing missions and capabilities into the twenty-first century. Germany's emphasis on ground attacks led to improved communications between the bombers and ground forces. Specialized ground-attack bombers were created, and standard bombers such as the Stuka were modified for improved ground-attack lethality. To improve accuracy, the Germans introduced guided bombs, which were successful on numerous occasions. The V-1 rocket was a pilotless rocket-driven bomb that anticipated the American cruise missiles of the 1980's and 1990's. Germany fielded the world's first operational jet bomber when it introduced the Arado 234 Blitz Bomber.

Allied designers also introduced groundbreaking bomber technologies, inventing effective radar guidance systems to overcome the difficulty of finding targets. In order to stifle German antiaircraft defenses, the British invented what became known as chaff, thin strips of aluminum foil, which, when dropped in bundles, reflected aircraft-sized radar images that confused German gunners and night fighters. The Allies also introduced faster and more powerful bombers, including very-long-range strategic bombers, the most well known of which was the U.S. B-29 Superfortress.

Specialized bombs, such as the "Tallboy," which was designed to maximize penetration of hardened targets and used on German U-boat pens, were created. Ultimately, the greatest and most complex bomb used was the atomic bomb. Although some historians argue that the Japanese surrender was due to more than just the dropping of the atomic bombs, atomic weapons came closer than any other weapon to realizing Douhet's vision.

Postwar Developments

At the end of World War II, the United States investigated the effectiveness of bombing in a study called the United States Strategic Bombing Survey. According to the survey, bombing achieved a mixed record. Although strategic bombardment had not induced a complete collapse of the enemy, it had clearly played a major role in the Allied victories. Reflecting these results, postwar air forces became more balanced organizations. The U.S. Air Force, for example, divided bombing duties between Strategic Air Command (SAC), which operated long-range bombers armed with atomic bombs, and Tactical Air Command (TAC), which provided close air support for ground forces.

Throughout the second half of the twentieth century and into the twenty-first, bombers continued to play an active role in conflicts around the globe. New technologies made bombers increasingly powerful and more difficult to destroy. Some developments, such as jet engines and radar guidance systems, offered significant improvements over technologies introduced during World War II. Others, such as stealth technology, were entirely new. Stealth technology uses special aerodynamic shapes, radar-absorbing paint, and specially textured surfaces to render planes nearly invisible to radar.

The effectiveness of such technology was shown during the 1991 Persian Gulf War and the brief 1999 air campaign over Kosovo, where stealth aircraft penetrated enemy air defense zones. The effectiveness of bombers has also been enhanced by the development of powerful missile armaments and precision-guided munitions. In the 1982 Falkland Islands War, for example, French Exocet air-to-ground missiles sank a destroyer and damaged a number of British ships. In both the Gulf War and Kosovo, air-launched smart missiles guided by either infrared or radar inflicted considerable damage.

Both conventional bombers, such as the B-52, and stealth bombers can launch precision-guided munitions. The success of operations over Kosovo has resurrected the debate over the use of strategic bombardment as an alternative to ground warfare. While some claimed Kosovo was history's first successful independent air campaign, evidence indicates that bomb damage was not as extensive as claimed and that diplomatic factors were as decisive as bomb damage in Serbia's decision to surrender.

Kevin B. Reid

Bibliography

Corum, James S. *The Luftwaffe: Creating the Operational Air War, 1918-1940*. Lawrence: University Press of

A World War II B-17 bomber departs after successfully hitting its target. (Hulton Archive)

Kansas, 1997. Provides an insightful overview of how Germans developed their doctrine of tactical air support.

McFarland, Stephen L. *America's Pursuit of Precision Bombing, 1910-1945.* Washington, D.C.: Smithsonian Institution Press, 1995. Describes in detail the American efforts at achieving a war-winning strategic bombing campaign.

Sherman, Don. "The Secret Weapon." *Air & Space Smithsonian* 9, no. 6 (February/March, 1995). Describes the history of the Norden bombsight.

Wildenberg, Thomas. *Destined for Glory: Dive Bombing, Midway, and the Evolution of Carrier Airpower.* Annapolis, Md.: Naval Institute Press, 1998. An excellent history of America's development of dive-bombing techniques and the use of aircraft carriers to project air power across the oceans.

See also: Air Force, U.S.; Airplanes; *Enola Gay*; Fighter pilots; Flying Fortress; Gulf War; Korean War; Luftwaffe; Manufacturers; Military flight; Royal Air Force; Stealth bomber; Strategic Air Command; Stratofortress; Superfortress; Tactical Air Command; Vietnam War; World War I; World War II

Boomerangs

Definition: A curved, multiwinged projectile which, when properly thrown, returns near the original starting point.

Significance: Used as a toy and in sport, the boomerang's unique wing configuration generates forces in flight which return the projectile back to the original point from which it was thrown.

Evolution of the Boomerang

The boomerang originated in Australia and was used as a hunting tool by the Aborigines. Although the boomerang is often thought of as a weapon, it has primarily served as a recreational and sport toy.

The killer-stick, believed to be the predecessor of the boomerang, was used for both hunting and fighting. The killer-stick has a similar shape and shares many of the boomerang's properties with one important difference: the killer-stick does not return to the thrower. The stick was smoothed, sanded, and shaped to provide an airfoil cross-section like a wing and could be thrown fast, far, and with great accuracy. Like many other sports projectiles, such as the discus, the killer-stick was thrown with rotational spin stabilizing its flight path.

The boomerang, which is smaller, lighter and has a more pronounced separation of wings than the killer-stick, was not used to kill game, but to trap birds. An Aboriginal hunter would imitate a hawk's call and throw the boomerang over a flying bird flock. To evade the hawk, the flock would swoop down into the hunter's waiting nets.

Shape and Construction

The boomerang is composed of two connected wings. The point of connection is called the elbow. There is a front, or leading, wing and a rear, or trailing, wing. The elbow separates the two wings at an angle generally ranging between 105 and 110 degrees.

Each of the boomerang's wings has a traditional airfoil cross-sectional shape with a leading and trailing edge. As with any other flight vehicle, the leading edge strikes the air first and the air flows over the top and bottom of the wing, past the trailing edge.

Unlike a bird's or aircraft's wings, the two boomerang wings are not mirror images of one another. Thus, there are right-handed and left-handed boomerangs. When thrown vertically into the wind, the upper wing's leading edge is located on the inner concave portion of the boomerang. The lower wing's leading edge is on the outer convex portion. Air first strikes the upper leading edge. As the boomerang rotates, this allows the lower wing's leading edge to meet and strike the air.

Aerodynamic Forces and Stability

Common to all sports projectiles, the aerodynamic forces acting on the boomerang are lift, drag, and gravity, or the boomerang's own weight. The spin imparted to the boomerang stabilizes the flight path. When a boomerang is thrown correctly, these forces cause the boomerang to circle around and return.

As the boomerang flies through the air, each wing produces lift. Although the shape of the wing generates lift, the lifting force is not enough to sustain the boomerang's flight. A boomerang is thrown with a spin similar to that of a discus. Without spin, a boomerang will wobble and fall to the ground; the boomerang's flight is not stable. Airplanes and birds have tail configurations that provide stability, while the rotational spin of a boomerang stabilizes its flight and produces a curved flight path. Stabilizing effects of spinning also are observed in a toy top and a bicycle wheel.

The turning force produced is a result of the unequal airspeeds over the spinning wings. The wings of a stationary, spinning boomerang produce the same amount of lift. When launched with a forward velocity, the forward-moving wing experiences more lift than the retreating wing. The net result is a force which turns the boomerang.

As with anything flying through the air, a boomerang is subject to drag and its own weight. The drag slows the boomerang down, limiting the flight time. However, given enough spin and initial velocity, the boomerang might circle above the thrower's head a few times before landing.

Boomerang Throwing Technique

A boomerang is launched almost vertically, based on the speed of the wind. The boomerang incurs a continuous turn throughout the duration of its flight, which causes the boomerang to lay down as it turns. Thus, the boomerang returns to the thrower in a horizontal hover. If a boomerang were thrown horizontally, it would climb until the wings stalled and simply fall to the ground.

The boomerang is launched at an angle to the wind. The thrower faces the wind and turns approximately 50 degrees to the right or left, depending on whether the person is right handed or left handed. Thrown at the proper angle, the boomerang will return.

Modern Designs

Simple and sleek in design, the boomerang's unique motion utilizes complex aerodynamics and physics. Based on these same scientific principles, some modern boomerangs have advanced technical or artistic designs. Several wings may be joined at a centralized hub. Modern boomerangs may be constructed to resemble letters of the alphabet or birds, for example. Some boomerangs are constructed so that the wings' tips are slower, making the boomerang easier to catch. All boomerangs use the same basic aerodynamic and physical principles to return to the thrower at the end of their flight.

Jani Macari Pallis

Bibliography

Hess, Felix. "Aerodynamics of Boomerangs." *Scientific American* 219 (1968): 123-136. A classic and comprehensive technical work on the aerodynamics and basic science concepts related to the boomerang.

Mason, Bernard S. *Boomerangs: How to Make and Throw Them*. Mineola, N.Y.: Dover, 1974. Comprehensive information on the boomerang design, construction, and throwing techniques.

Ruhe, Benjamin, and Eric Darnell. *Boomerangs: How to Throw, Catch, and Make Them*. New York: Workman, 1985. A nicely illustrated book on the technical and athletic aspects of boomerangs.

Walker, Pearl. "Boomerangs! How to Make Them and Also How They Fly." *Scientific American* 240 (1979): 130-135. A technical overview of the construction and aerodynamics of boomerangs.

See also: Aerodynamics; Airplanes; Birds; Forces of flight; Wing designs

Richard Branson

Date: Born on July 18, 1950, in Shamley Green, Surrey, England

Definition: Founder and chairman of the Virgin group of companies that includes Virgin Atlantic Airways.

Significance: British entrepreneur and venture capitalist Branson's Virgin empire melds fun and business, reflecting his "people-first" policy. His upstart Virgin Atlantic Airways has succeeded despite competition from more established airlines.

Richard Charles Nicholas Branson is the driving force at the center of a web of more than 200 companies employing more than eight thousand people in twenty-six countries. His web of interest in travel, retail, hotels, consumer goods, financial services, computer games, radio, television, cinema, and publishing makes Branson a regular entry into *Forbes* magazine's lists of the richest people in the world.

In 1970, the twenty-year-old Branson founded Virgin Mail Order Records and shortly thereafter opened a record shop in Oxford Street, London. He established his own record label, Virgin Records, in 1973, building a recording studio in Oxfordshire, where the first Virgin artist, Mike Oldfield, recorded *Tubular Bells* (1973). Another Virgin act, the Rolling Stones, helped make Virgin Records one of the top six record companies in the world.

In 1984, Branson became the majority backer of an airline he renamed Virgin Atlantic Airways. An upstart in a fiercely competitive field of established carriers, Virgin Atlantic eventually became the second-largest British long-haul international airline, with a fleet of Boeing 747 and Airbus A340 aircraft flying routes to New York, Miami, Boston, Los Angeles, Orlando, San Francisco, Hong Kong, Athens, and Tokyo. The airline is founded on the concept of offering competitive, high-quality upper class and economy service. It holds many major awards, including several Airline of the Year awards from *Executive Travel Magazine*.

Since 1985, Branson has been involved in a number of world record-breaking attempts. In 1986 he and a teammate made the fastest ever-recorded crossing of the Atlantic Ocean in his powerboat *Virgin Atlantic Challenger II*. A year later, Branson crossed the Atlantic with Swedish aeronaut Per Lindstrand in the hot-air balloon *Virgin Atlantic Flyer*, which was not only the first hot-air balloon to cross the Atlantic but also the largest ever flown, at 2.3 million cubic feet. The two men crossed the Pacific Ocean in 1991. Branson has also been a member of three teams that made unsuccessful attempts at transglobal hot-air balloon flights during the late 1990's. On the third attempt, in December, 1998, the team traveled 8,200 miles (13,200 kilometers), becoming the first hot-air balloonists to cross the entire Asian continent.

A child of a revolutionary 1960's, Branson has forged a unique synthesis of the youth revolution's values and the needs of a modern business. He captivates the public and employees by the unexpected prospect of making the gray world of work come alive with fun, excitement, and challenge.

Lori Kaye

Bibliography

Branson, Richard. *Losing My Virginity: How I've Survived, Had Fun, and Made a Fortune Doing Business My Way*. London: Virgin, 1998. Branson's memoir about his business and aviation exploits.

Burger, William. "Up, Up, and Away." *Newsweek* 123, no. 24 (June 13, 1994). An article about Branson's balloon flights.

Conniff, Richard. "Balloon Challenge." *National Geographic* 192, no. 3 (September, 1997). A well-illustrated article about Branson's attempts at transglobal balloon flight.

See also: Air carriers; Balloons; Transatlantic flight; Transglobal flight; Virgin Atlantic

Wernher von Braun

Date: Born on March 23, 1912, in Wirsitz, Germany; died on June 16, 1977 in Alexandria, Virginia

Definition: German-American rocket engineer who developed the first practical space rockets and launchers and became known as the father of the space age.

Significance: Von Braun designed the first ballistic missile, led the research team that put the first American satellite into orbit, and designed the Saturn rockets that sent humans to the Moon.

As a young man, Wernher von Braun became intrigued with the possibilities for space exploration, joining the Verein für Raumschiffahrt (Society for Space Travel) in the spring of 1930. In 1932, he went to work for the German army to develop rockets and missiles. After earning his doctorate in physics in 1934, von Braun was appointed the director of Germany's military rocket development program. Under pressure from German chancellor Adolf Hitler, Wernher subjugated his dreams of space travel to Germany's demand for weapons. Operating at a secret laboratory along the Baltic Coast, von Braun and other German scientists built and tested the V-1 cruise missile and the V-2 ballistic missile. When the German war machine collapsed in 1945, von Braun hid all of the classified rocket documents in an abandoned mine in Germany's Harz Mountains.

On May 2, 1945, the German rocket team surrendered to American forces. Von Braun and his research team were transferred to Fort Bliss, Texas, along with the German rocket documents and approximately 150 captured V-2 missiles. At the U.S. Army's Redstone Arsenal in Huntsville, Alabama, during the 1950's, von Braun and his team built the Jupiter ballistic missile. Between 1952 and 1954, von Braun developed one of the first comprehensive space exploration programs in the world. He led the team that put the Explorer 1, the first American satellite, into orbit on January 31, 1958. After being transferred to the newly established National Aeronautics and Space Administration (NASA) in 1960, von Braun was given the mandate to build the giant Saturn rockets.

Von Braun was appointed the director of NASA's Marshall Space Flight Center in Huntsville, serving in that capacity from July, 1960, until February, 1970. In that position, he designed and oversaw the development of the Saturn I, Saturn IB, and Saturn V rockets. On July 16, 1969, a Saturn V launched the crew of Apollo 11 to their successful landing on the Moon.

In 1970, von Braun moved to Washington, D.C., to oversee the strategic planning effort of NASA. He was awarded the National Medal of Science by President Gerald Ford in early 1977. Von Braun was one of the world's first and foremost rocket engineers and a leading authority on space travel. His intense desire to expand man's knowledge through the exploration of space led to humans' setting foot on the Moon.

Alvin K. Benson

Wernher von Braun stands in front of the Saturn IB launch vehicle, designed by his team of expatriate German scientists, at the Kennedy Space Flight Center in 1968. (NASA)

Bibliography

Bergaust, Erik. *Wernher von Braun: The Authoritative and Definitive Biographical Profile of the Father of Modern Space Flight.* Washington, D.C.: National Space Institute, 1976. An authoritative, definitive biographical profile.

Lampton, Christopher. *Wernher von Braun.* New York: Watts, 1988. Traces the life and achievements of von Braun as the father of modern rocketry.

Piszkiewicz, Dennis. *The Nazi Rocketeers: Dreams of Space and Crimes of War.* Westport, Conn.: Praeger, 1995. An excellent account of the history of rocket development and the role von Braun played in it.

See also: Apollo Program; National Aeronautics and Space Administration; Rocket propulsion; Rockets; Saturn rockets; Spaceflight; World War II

British Airways

Also known as: British Overseas Airway Corporation (BOAC), British European Airways (BEA), British Caledonian Airways

Date: Founded in 1924

Definition: The United Kingdom's national airline since 1939.

Significance: British Airways is one of the world's largest international airlines. Along with Air France, it is one of only two airlines to fly the supersonic Concorde jet.

History

The origins of British Airways lie in the post-World War I era of civil aviation. On August 25, 1919, its forerunner company, Aircraft Transport and Travel (AT&T), began the world's first daily international scheduled air service flying between London and Paris. That initial flight, in a single-engine De Havilland DH-4A biplane carrying one passenger and a cargo of newspapers, Devonshire cream, and grouse, taking off from Hounslow Heath, made history. The first flight took two and one-half hours and was the inspiration for further growth of British companies, starting services to Paris and to Brussels. Instone, a shipping group, and Handley Page, an aircraft manufacturer, became pioneer air companies despite facing difficulties of few passengers, high fares, and the danger and unreliability of early air travel. For instance, it took one pilot two days to complete the two-hour flight to Paris, making thirty-three forced landings along the way. Pioneer British airlines faced undercutting from their Dutch and French competitors and suffered severe losses.

By 1924, Britain had four main airlines: Instone, Handley Page, Daimler Airways (a successor to AT&T, who succumbed to Dutch and French competition), and British Air Marine Navigation Company, all of which merged to form Imperial Airways. The following year, Imperial Airways was servicing Paris, Brussels, Basel, Cologne, and Zurich. Further service was introduced to Egypt, the Arabian Gulf, India, South Africa, Singapore, and West Africa. Cooperating with Qantas Empire Airways, which serviced Singapore and Australia, service between the United Kingdom and Australia was established by 1935. Smaller air transport companies also began business. In 1935, these smaller companies merged to form the original privately owned British Airways. By 1939, Imperial Airways and British Airways were nationalized to form British Overseas Airways Corporation (BOAC).

Postwar development of British Airways has been a substantial leap from its humble initial origins. Long-haul services were provided to many international routes. Continental European domestic flights were flown by a new airline, British European Airways (BEA). BOAC introduced services to New York in 1946, Japan in 1948, Chicago in 1954, and the West Coast of the United States in 1957. Domestic flights included service to Belfast, Edinburgh, Glasgow, and Manchester. By 1967, the government recommended that a holding board be responsible for two main airlines, BOAC and BEA. British Caledonian was born in 1970, when Caledonian Airways took over British United Airways. Two years later, BOAC and BEA combined under a British Airways board, and a separate airline emerged as British Airways in 1974. Defeating the odds against it, including severe financial challenges, in January, 1976, British Airways launched the world's first supersonic passenger service, simultaneously with Air France: the Concorde. In 1987, the British government began selling shares in British Airways and the company was completely privatized.

As of 2001, British Airways' fleet comprises seven Concordes, seventy-one Boeing 747's, thirty-eight Boeing 777's, twenty-one Boeing 767's, forty-eight Boeing 757's, and fifty-two Boeing 737's, as well as eighty-three Airbus A318's, A319's, and A320's in service or on order. British Airways is a member of the oneworld Alliance, along with American Airlines, Qantas, Cathay Pacific, Iberia, Finnair, Aer Lingus, and LanChile, and also has code-sharing agreements with Air Mauritius, America West, Crossair, Emirates, LOT Polish Airlines, and Malev.

Operational Structure

British Airways operates from two main bases, London's two primary airports, Heathrow (the world's largest international airport) and Gatwick. In the year 2000, forty-eight million people flew on 529,807 British Airways flights, or an average of eighty passengers checking in every minute around the clock, and a British Airways flight either taking off or landing safely every thirty seconds. British Airways maintains an enormous fleet capable of circling the globe, with flight crews trained to serve and protect business, royal, and vacationing tourists around the clock all year long. Maintaining a huge fleet of aircraft and maintenance support systems, the flagship of the British Airways fleet remains the Concorde; British Airways boasts more supersonic flying in one year than all of the world's air forces combined.

British Airways attributes its success to its behind-the-scenes activities, such as investor relations. On its World Wide Web site, the company provides comprehensive information regarding its financial performance, a presentation provided to the financial community, an online version of its own investor magazine, and even more information, including its history, in the British Airways Factbook. British Airways has also created partnerships and alliances with other airlines around the world to better serve its customers.

British Airways is involved in cargo service and tracking to support globalization, big business uniting the world. British Airways engineering services offer a wide range of technical and support services, individually tailored toward operational and financial efficiency of the business customer. British Airways also offers flight training in what are considered to be world-class facilities.

British Airways participates in Dreamflight, a British charity whose sole purpose is to transport seriously ill children on a holiday of a lifetime to Walt Disney World in Florida. British Airways provides information on the program through a Dreamflight Web site.

Ever mindful of its responsibility to protect the environment, British Airways maintains a program of corporate responsibility to maintain and improve the environment. In conjunction with this program, there is a community learning center in Waterside, Harmondsworth, promoting opportunities for young people and adults to develop knowledge and skills to enable them to grow in active, positive community participation. Furthermore, British Airways provides awards that encourage environmental awareness in the hospitality industry.

British Airways is today the world's biggest international airline, carrying more passengers from one country to another than any of its competitors. Because it is the world's longest-established airline, it bears the distinction as the industry leader.

Pamela M. Gross

Events in British Airways History

1939: Two British airlines, Imperial Airways and British Airways, are nationalized to form the British Overseas Airways Corporation (BOAC).

1946: A new British airline, British European Airways (BEA), is established to handle continental European and domestic British flights. BOAC introduces London-to-New York service.

1952: BOAC flies De Havilland Comet jets on service to Johannesburg, South Africa.

1954: After two crashes in one year, Comets are removed from service.

1958: BOAC makes the first transatlantic jet flights, between London and New York.

1974: British Airways is formed by the merger of BOAC and BEA.

1976: Cooperating with Air France, British Airways inaugurates supersonic travel on the Concorde.

1987: The British government privatizes British Airways through a public stock offering.

1993: British Airways enters into a partnership with USAir, which is dissolved four years later.

1999: British Airways joins the oneworld Alliance, a global network of airlines, also including American Airlines, Canadian Airlines, Cathay Pacific Airways, and Qantas.

2000: British Airways suspends its Concorde operations after an Air France Concorde crashes shortly after takeoff from Paris, killing all on board and four on the ground.

2001: British Airways modifies its Concorde fleet, restoring service later in the year.

Bibliography

Gregory, Martyn. *Dirty Tricks: British Airways' Secret War Against Virgin Atlantic*. Boston: Little, Brown, 1996. An exposé of British Airways' and American Airlines' attempts to prevent upstart Virgin Airways from competing with them.

Jackson, A. J. *Imperial Airways and the First British Airlines, 1919-1940*. Lavenham, Suffolk, England: T. Dalton, 1995. A history of the early British airlines that eventually merged to become British Airways.

Marriott, Leo. *ABC British Airways Book.* 2d ed. Plymouth, Mich.: Plymouth, 1998. A corporate history.

Reed, Arthur. *Airline: The Inside Story of British Airways.* London: BBC Books, 1990. A behind-the-scenes look at the airline.

See also: Air carriers; Concorde; Supersonic aircraft

Buoyant aircraft

Definition: Aircraft, such as hot-air balloons and dirigibles, or airships, that fly because they are filled with lighter-than-air gases.

Significance: Balloons and airships played important roles in the development of aviation, serving as reconnaissance, battle, and commercial aircraft. They continue to be used for recreation and advertisement and may see future use in coast watching, scientific study, and short-haul transportation of heavy goods.

Development

The first buoyant aircraft was the hot-air balloon, invented in 1782 by the French brothers Joseph-Michel and Jacques-Étienne Montgolfier, who found that closed bags of hot air rose and stayed aloft until the air inside them cooled. The basis for hot-air balloon flotation is that hot air is less dense, or weighs less, than the volume of air it displaces, so a hot-air balloon is a lighter-than-air aircraft. It soon became clear that hot-air balloons could be kept aloft by hanging under them baskets holding firepots. In this way, gondolas that also carry passengers were created.

Modern hot-air balloons differ from early prototypes only in the fabrics and heaters used. Flammable hydrogen and safer helium are other fill gases, with helium replacing hydrogen in all modern balloons that do not use hot air. Hot-air balloons cannot be steered. Hot-air balloonists rise into the sky by throwing out ballast and descend by letting gas out. Destinations and arrivals depend on the winds. Helium-filled balloons can remain aloft for longer periods and can travel farther than can hot-air balloons. However, their flight methodology and limitations do not change. Recreational balloonists prefer hot air, and commercial balloons use helium.

In 1852, France's Henri Giffard ended the problems of uncertain destination and arrival times by adding a 3-horsepower steam engine and propeller to a cigar-shaped, hydrogen-filled balloon, allowing for a maximum speed of 6 miles per hour. This nonrigid aircraft and all that followed it were named dirigibles, from the Latin word *dirigere*, which means "to steer." Although Giffard's invention was impractical for prolonged flight, it was innovative in its cigar shape that reduced wind resistance and increased maneuverability. All following dirigibles have been constructed with a similar shape.

Throughout the late nineteenth century, increasingly better-designed buoyant aircraft were invented and used. In 1883, French aeronaut Gaston Tissandier built the first workable airship with an electric motor. From 1898 to 1905, Alberto Santos-Dumont, a Brazilian living in Paris, set airship records in nonrigid, powered airships. All these airships, however, had engines that made them too heavy for commercial flight.

Airships

There are three main airship types: nonrigid, semirigid, and rigid. Nonrigids are essentially large, streamlined cylindrical balloons, nicknamed "blimps," supposedly for the sound made by a finger thumping into the side of the envelope, or gas bag. Nonrigids, such as the Goodyear blimps, get their shape from the gas within a single gas bag or envelope. The engines and a car or a gondola hang below the gas bag.

The design of nonrigid dirigibles simplifies cost and minimizes structural weight, which in turn reduces the net lifting capability. However, nonrigids are limited in size, because an unsupported gas bag may bend unpredictably under heavy loads or strong winds. In a worst-case scenario, a partially deflated gas bag may flop over the gondola or propellers. Conversely, the gas bag cannot be filled too tightly, lest it burst. The one gas bag is a single point of failure that could cause a crash, although the large size of dirigibles means that operations could continue for some hours, even with significant leaks. Another method to compensate for pressure loss in the gas bag is the use of an inner ballonet, which can be inflated with outside air.

Semirigid dirigibles have a keel on the bottom to support a larger gas bag, and the keel can hold the gondola and engines, at the cost of additional weight. The risks associated with a single gas bag also apply to semirigid dirigibles. The most famous semirigid was the Italian semirigid airship *Norge*, which in 1926 was used by explorer Roald Amundsen to make the first transpolar flight, from Spitsbergen Island to Alaska.

Rigid dirigibles have a framework to support an outer skin and individual gas bags. In rigid dirigibles, an individual gas bag can fail without damaging the aerodynamic integrity of the craft, and there are usually sufficient reserves among the other cells to maintain buoyancy. Those advantages cost additional weight. However, greater weight can be compensated for by greater size. There is theoretically no limit to a rigid's size. The German passenger airship *Hindenburg* had an LTA gas capacity of 7,000,000 cubic feet, and designs of twice that size have been proposed.

During World War I, the French army used semirigids for reconnaissance, coast patrol, and to find submarines. Near the end of the war, the British armed forces began developing rigid airships, in response to the expected large-scale availability of nonflammable helium gas. The 643-foot-long *R-34*, with a gas capacity of 2,000,000 cubic feet, was put into service in 1919 and made the first transatlantic airship flight, from Scotland to Mineola, New York, and back in eight days. Both the *R-34* and its sister ship, *R-38*, were destroyed in 1921.

Count Ferdinand von Zeppelin, a German military officer, had spent time as a military observer for the Union Army during the American Civil War. While in the United States, he took some balloon flights and became fascinated with balloons before returning to Germany to continue his military service in the Seven Weeks and Franco-Prussian Wars. After his retirement from the military in 1890, Zeppelin devoted all his efforts to the construction of rigid airships.

Zeppelin worked to allay one of the main problems in airship design, that of maintaining a cigar-shaped gas bag and avoiding partly deflated bags that cannot be steered. He made steerable, unchanging shapes by designing rigid, strong, light frames that were mated to gas bags. In 1900, Zeppelin's first airship, *LZ-1*, flew for seventeen minutes. In 1908, the 446-foot-long *LZ-4*, with a gas capacity of 500,000 cubic feet, flew for one half-day at 40 miles per hour.

Soon, Zeppelin's airships, widely known as zeppelins, made practical air transport available. In 1910, the first commercial airline and dirigible manufacturer, Deutsche Luftschiffahrts Aktien-Gesellschaft (Delag), was established. Delag's luxurious dirigibles interconnected many German cities, carrying 15,000 travelers per year. By 1914, when World War I began, Delag had in service twenty-seven dirigibles, including the *Sachsen* and the *Deutschland*, that had achieved airspeeds of over 50 miles per hour. In the next five years, 1,600 dirigible flights carried more than 35,000 passengers over 200,000 miles. Although Zeppelin died in 1917, before transatlantic airship flight was achieved, Delag carried on.

Military Uses

The first military buoyant aircraft were French hot-air/hydrogen tethered reconnaissance balloons, crewed by *aerostiers*. From 1794 to 1799, reports from these balloons aided French armies in battle. Similar reconnaissance balloons were used in the American Civil War, but true military aviation began only with the perfection of navigable airships in the late nineteenth century. Airships were the most formidable aircraft before World War I. They were made of fabric-covered metal frames and gas bags full of hydrogen. Airships were used by the Allies for antisubmarine patrols during World War I. The German zeppelins were the most functional airships, used during World War I to carry five 110-pound bombs and twenty 7-pound incendiaries at a time when military airplanes carried no weapons. More than one hundred zeppelin bombers were used by the Germans to bomb Paris and London. The German dirigibles were so effective that, after Germany lost the war, they were all confiscated. Further German production of zeppelins was prohibited until the late 1920's.

Commercial Uses

Airship construction continued in Europe and the United States throughout the 1920's and 1930's. The British dirigible *R-34* made a round-trip transatlantic crossing in July, 1919. In 1926, Amundsen flew the Italian semirigid airship *Norge* in his expedition to the North Pole. In the late 1920's after Germany was again allowed to produce airships, Delag built the *Graf Zeppelin*, an outstanding zeppelin of the time. The *Graf Zeppelin* had an approximate length of 800 feet and a gas capacity of almost 4,000,000 cubic feet. The luxurious airship began its nine-year service in 1928, ultimately making 600 flights, including 144 ocean crossings, covering 1,000,000 miles. It crossed the Atlantic to North and South America 139 times and made a complete trip around the world in three weeks at a cruising speed of 71 miles per hour, faster than comparable train or boat transportation.

In 1936, Germany began regular transatlantic passenger service with a new Delag zeppelin, the *Hindenburg*, also known as *LZ-129*. The *Hindenburg* cruised at a speed of 78 miles per hour, held 7,000,000 cubic feet of hydrogen, and carried fifty passengers in luxurious style. The airship had been rushed into production by German chancellor Adolf Hitler to show off the greatness of his Reich and was successful in a number of transatlantic flights. However, on May 6, 1937, while docking at Lakehurst, New Jersey, the *Hindenburg* caught fire and exploded, killing thirty-six people in one of the greatest air disasters of all time. After this event, the reputation of dirigibles as air carriers was irretrievably damaged.

Airship Disasters

Although the wreck of the *Hindenburg* was the most notorious airship disaster, it was not the only one. Others had preceded it. For example, a U.S. purchase, the semirigid *Roma*, 400 feet long, with a gas capacity of more than 1,000,000 cubic feet, was lost in 1922. The USS *Shenandoah*, a Navy airship, became the first zeppelin built in the United States in 1923. Filled with helium, the airship was 700 feet long, with a gas capacity of more than 2,000,000 cubic feet. After making a few long trips, the *Shenandoah* broke apart in a thunderstorm and was wrecked in 1925.

The U.S. Navy's *Los Angeles*, a German war reparation, was a bright spot in the airships' dark safety record. The 650-foot-long airship, with a gas capacity of 2,500,000 cubic feet, carried thirty passengers on each of 250 long flights between 1924 and 1932. However, the success of the *Los Angeles* did not outweigh the failures of other airships, which continued to prove problematic around the world.

In 1926, Italy's semirigid *Norge* flew Amundsen's expedition to Alaska but had to be dismantled. A polar flight two years later in the closely related *Italia* was disastrous. Only German zeppelins seemed to survive for long time periods.

Elsewhere, though, airship catastrophes continued to occur. In England, the dirigibles *R-100* and *R-101*, with lengths of approximately 700 feet and gas capacities of 5,000,000 cubic feet and accommodating one hundred passengers, had short terms of service. The *R-100*, christened in 1929, crashed in 1930 and burned to cinders, killing forty-six people. After this incident, the *R-101* was immediately scrapped, and Britain gave up dirigible construction.

The U.S. Navy built two dirigibles, the USS *Akron* and the USS *Macon*, each of which was 785 feet long, with a gas capacity of 6,500,000 cubic feet. Both ships could carry five scout planes and release or take them aboard in flight. Each ship survived for only about two years before crashing into the sea during storms. After the destruction of these two ships, the United States discontinued the building of rigid airships. Despite their varied uses, airships were virtually abandoned in the late 1930's, due to their high production cost, low speed, vulnerability to storms, a series of airship disasters, and advances in heavier-than-air craft. In 1938, American military blimps were put under Navy control for World War II scout, convoy, and antisubmarine work. In peacetime, small blimps have provided aerial television views of sports events and advertising.

Future Uses

After World War II, the U.S. Navy continued for approximately twenty years to develop blimps for antisubmarine, research, and early warning use. The largest, ZPG-2 type, 324 feet long, with a gas capacity of 0.9 million cubic feet, could remain aloft for a week without refueling. The Navy stopped almost all airship use in 1961. During the late 1980's, there was renewed Coast Guard and Navy interest in airship use for early warning and electronic antisubmarine warfare, although not much came of it.

However, enthusiasts attest that blimps can be used for many tasks unsuited to planes. For example, because airships can move very heavy, large objects over short distances much better than planes can, blimplike airships are sometimes used for short-haul and heavy-lift operations in remote areas. This use may continue to increase.

Another proposed use of airships is for intraurban passenger service and other short-distance transportation. Here, low airship speed is not a problem and ability for vertical and short takeoff and landing (V/STOL) is a major advantage over planes. Another advantage of airships over airplanes is their low noise production and fuel consumption.

One final modern use of airships could involve making rigid airships from modern materials and adding contemporary flight instruments and computers. Such airships, it is thought, would be useful for military surveillance of the oceans and coastal areas or for the scientific study of the environment. Thus, it seems probable that in the future many blimps and dirigibles will again take to the air.

Sanford S. Singer

Bibliography

Capelotti, Peter Joseph. *By Airship to the North Pole: An Archeology of Human Exploration.* New Brunswick, N.J.: Rutgers University Press, 1999. An account of the transportation of Amundsen's party by dirigible and their exploration of the North Pole.

Collier, Basil. *The Airship: A History.* New York: G. P. Putnam's Sons, 1984. A history of airships, their proponents, and the events fostering their development and leading to the near-cessation of their modern manufacture.

Hall, George, Baron Wolman, George Larson, and Neil Shakery. *Blimp.* New York: Van Nostrand Reinhold, 1981. A brief, well-designed book with useful information on the history of buoyant aircraft, blimp construction, airship operation and uses, and ideas for future uses, as well as plentiful, interesting illustrations.

Hayward, Keith. *The Military Utility of Airships*. London, England: Royal United Services Institute for Defence Studies, 1998. A brief publication covering many aspects of military airship use.

Toland, John. *Ships in the Sky: The Story of the Great Dirigibles*. New York: Henry Holt, 1957. A classic book describing the great events in the history of balloon and airship development and use.

Ventry, Lord, and Eugene M. Kolinski. *Airship Saga*. Poole, England: Blandford Press, 1982. A book covering aspects of airship history and development in airship-producing nations, with much detail on prospects for future airships. Included are good illustrations and a useful bibliography.

See also: Balloons; Dirigibles; Goodyear blimp; *Hindenburg*; Lighter-than-air craft; Military flight; Montgolfier brothers; Reconnaissance; Alberto Santos-Dumont; Transatlantic flight; Vertical takeoff and landing; World War I; World War II; Ferdinand von Zeppelin

Richard E. Byrd

Date: Born on October 25, 1888, in Winchester, Virginia; died on March 11, 1957, in Boston, Massachusetts

Definition: Naval aviator and premier twentieth-century polar explorer.

Significance: Byrd claimed to be first to fly over the North and South Poles and led five Antarctic expeditions that called public attention to that part of the world.

Born in Virginia, Richard E. Byrd was the son of Richard Evelyn Bird, a lawyer, and Eleanor Bolling Flood. After graduating from the United States Naval Academy in 1912, Byrd received an ensign's commission and made his first airplane flight in 1914. During World War I (1914-1918), he was forced into retirement by continuing problems with an injured ankle but was recalled to active duty and won his wings as a naval aviator in 1918.

After the war, Byrd worked on NC flying boats, pioneering seaplane landings. He helped create a government bureau of aeronautics. In 1925, he commanded a naval unit on an expedition to northern Greenland sponsored by, among others, John D. Rockefeller and Edsel Ford. During this expedition, he flew over Ellesmere Island and the interior regions of Greenland.

On May 9, 1926, Byrd and Floyd Bennett flew north from Kings Bay, Spitsbergen Island, in a Fokker monoplane, the *Josephine Ford*. The two journeyed for more than fifteen hours and recorded that they had passed over the North Pole during their flight. Although their claim would later be disputed, both men returned to the United States as heroes and were awarded the Congressional Medal of Honor. In June, 1927, less than a month after Charles A. Lindbergh's transatlantic flight, Byrd and three companions crossed the Atlantic in a Fokker trimotor, crash-landing in France after a forty-two-hour journey.

Byrd turned his attention toward the South Pole, gaining private funding for his five visits to Antarctica, which included a flight over the South Pole in November, 1929, aboard a Ford trimotor plane, the *Floyd Bennett*. The round-trip flight, with three other fliers, took nineteen hours and earned Byrd a promotion to rear admiral. It departed from a base named Little America that was constructed on the Ross Ice Shelf, a flat area of ice fronting the Ross Sea.

From 1933 to 1935, Byrd conducted a second expedition, extending the exploration of Antarctica. He continued his exploration of territory that he had previously named Marie Byrd Land, in honor of his wife. He caused some controversy when he spent the winter of 1934 alone in a station hut and had to be rescued, frostbitten and sick with carbon monoxide poisoning.

After being named as head of the U.S. Antarctic Service by President Franklin D. Roosevelt, Byrd led a third Antarctic expedition in 1939 and 1940. He established two bases on the continent and discovered Thurston Island. During World War II, Byrd served on the staff of the chief of naval operations.

At the end of World War II, Byrd led his fourth Antarctic expedition, named Operation Highjump, which mapped and photographed more than half a million square miles of the continent. This massive operation involved approximately five thousand men and thirteen ships, including an aircraft carrier. During this expedition Byrd flew over the South Pole for a second time on February 16, 1947.

Byrd's fifth and final expedition, Operation Deep Freeze, was an exploratory and scientific project conceived to coincide with the International Geophysical Year (IGY) activities involving thirteen nations. Traveling to his base aboard the icebreaker *Glacier*, Byrd made his final flight over the South Pole on January 8, 1956. Skilled at applying technology to his explorations, Byrd employed helicopters, seaplanes, and skibased airplanes. His expeditions to Antarctica claimed thousands of square miles for the United States.

Niles R. Holt

Bibliography

Byrd, Richard Evelyn. *To the Pole: The Diary and Notebook of Richard E. Byrd, 1925-1927*. Edited by Raimund E. Goerler. Columbus: Ohio State University Press, 1998. The publication of Byrd's diary and notebook, including the official navigational report of the 1926 expedition, was prompted by allegations in Richard Montague's *Oceans, Poles, and Airmen* that Byrd did not reach the North Pole in 1926. Rather than providing clear answers, however, Byrd's diary and notebook only extend the controversy.

DeLeeuw, Adele. *Richard E. Byrd: Adventurer to the Poles*. 1963. Reprint. New York: Chelsea House, 1992. An account of Byrd's career and achievements written for younger readers.

Montague, Richard. *Oceans, Poles, and Airmen: The First Flights over Wide Waters and Desolate Ice*. New York: Random House, 1971. Conceived as a tribute to 1920's and 1930's aviators, this book questions whether Byrd's plane actually reached the North Pole in 1926.

See also: History of human flight; Navy pilots, U.S.; Seaplanes; Transatlantic flight; World War I; World War II

C

Cargo aircraft

Definition: Cargo aircraft are aircraft dedicated to hauling freight rather than transporting passengers. Any airplane, regardless of size, is considered a freighter if carrying cargo rather than people is its primary use.

Significance: In the Western industrialized world, most food, newspapers, and mail is transported by air. Because airfreighters often have separate terminals, fly primarily at night, or utilize different airports than passenger traffic, it is easy to overlook how vital air freight has become to the modern economy. Air freight companies, such as Federal Express, DHL, and United Parcel Service, operate thousands of cargo aircraft daily, most of which go totally unnoticed by the general public. DHL Worldwide Express alone operates 358 aircraft out of thirty-six regional hubs around the world.

Cargo aircraft, also called freighters, transport fresh flowers from Africa, grapes from Chile, and many other items that in previous decades would have traveled by ship or truck, or not at all. While much of the commercial freight traffic in many countries still travels by land routes, cargo aircraft handle ever-increasing amounts of material. In the United States and Europe, for example, catalog companies and Internet merchants rely on freight companies such as Federal Express to deliver orders to customers quickly. Without air freighters, the overnight delivery of packages and documents over thousands of miles, which is now taken for granted, would be impossible.

Although the general public often thinks of cargo aircraft as hauling freight that is small, light, or perishable, air freighters are just as likely to be hauling oversize materials as they are to be transporting perishable cargoes that must reach customers quickly. Items too large in weight, length, width, or height to be shipped by truck or rail are now flown to their destinations on oversize aircraft such as the civilian Super Guppy and the U.S. military's C-5. Until recently, railway freight cars could not carry items exceeding 60 feet in length. The C-5, in contrast, has a cargo bay 143 long.

The Beginning of Cargo Flight

Using aircraft for hauling cargo was not an immediate priority in the aviation industry. For the first several decades of the twentieth century, aircraft remained relatively small. Following the Wright brothers' success in 1903, aviation pioneers focused first on increasing speed, distance, and number of passengers before contemplating using aircraft to carry freight. Any cargo hauled was highly specialized and lightweight, such as medical supplies. The United States Post Office began airmail service in 1918, but the service was limited almost entirely to letters. Writers were urged to use thin onionskin paper to reduce the weight of individual pieces of correspondence.

As the size and range of aircraft increased, however, aviation's potential for hauling freight became more apparent. In the mid-1920's, the U.S. military acknowledged the divergence of freighters from transport aircraft and began numbering the former with a "C" designation, to indicate cargo. The Douglas C-1 was the first airplane so designated, in 1925. The airplane, a single-engine biplane, had an enclosed passenger compartment that could transport six people. With the seats removed, it became a freighter. This remains typical of military freighters, as many aircraft bearing a "C" designation are used as troop transports as well as for carrying cargo.

On the civilian side, United Parcel Service (UPS), founded in Seattle in 1907 as a messenger firm, began shipping packages by air in 1929. For many years, however, the company restricted that service to the West Coast. In 1953, UPS expanded its air delivery system nationally. Companies such as UPS generally use conventional aircraft that have been converted for use as freighters. This is nothing new in aviation history. Many aircraft designed for transporting passengers have been pressed into service as freighters, while numerous passenger aircraft carry some freight in addition to their human cargo. The rear bulkhead in the main cabin of many aircraft is movable, which gives airlines the flexibility to increase or decrease the size of the passenger compartment and the baggage compartment behind it if necessary. Most commercial airlines, for example, have contracts with the U.S. Postal Service to transport mail. Ironically, one of the aircraft most closely associated in the public's mind with long-distance passenger transport, the Boeing 747, was designed originally to maximize its cargo capacity.

Passenger/Cargo Craft

When Boeing company aerospace engineers began planning the 747 in the early 1960's, industry analysts believed the future of air passenger service would lie in the area of supersonic transports (SSTs), such as the Concorde airliner then being developed in Europe. Although the 747 was initially designed as a jumbojet with a double-decker passenger compartment, the price of aviation fuel in the 1960's was so low—barely 10 cents per gallon—that many experts believed SSTs would be economical to operate despite their high fuel requirements and relatively small passenger cabins. The Concorde, for example, has a maximum passenger capacity of 144 persons, in comparison with the 747's high-density 624. The energy crisis of the 1970's combined with public concerns about negative side effects of SSTs, such as noise, proved the experts' predictions wrong. With the 747 being highly useful for either passenger transport or as a freighter, it is not surprising that by the twenty-first century, the SST had become a curiosity, while the Boeing 747 dominates international air traffic. Boeing ultimately chose not to use the double-deck concept for the passenger cabin of the 747. Instead, they placed the cockpit above the main cabin, giving 747's used as freighters an exceptionally roomy cargo compartment. When configured as a passenger plane, a 747 will seat passengers in rows of ten persons across. When configured as a freighter, two 8-foot wide cargo containers can be placed side by side.

Noted cargo aircraft over the years have included the Ford Tri-Motor and the Douglas DC-3. Both aircraft were developed primarily for use in transporting passengers, but were quickly pressed into service as freighters. The Ford Tri-Motor, introduced in 1926, was notable for its all-metal construction, an aviation first. One of the first aircraft designed to be inherently stable, the Tri-Motor could fly well on only two engines and maintain a level flight path with only one. The aircraft was manufactured for only seven years, from 1926 to 1933, with a total of 199 being built. The Tri-Motor was a rugged aircraft capable of surviving a great deal of rough use. As recently as 1998, a few Tri-Motors remained in service, including one being used by a sightseeing company in Ohio to fly daily tourist excursions.

The venerable DC-3 made its debut in 1935. Within a short time it gained a reputation for being virtually indestructible. It has been estimated that by 1944, over 90 percent of the aircraft being used by commercial airlines were DC-3's. The Douglas Aircraft Company built a total of 18,000 aircraft before discontinuing production. During World War II, the military version of the DC-3, the C-47, saw wide use in both the European and Pacific theaters. General Dwight D. Eisenhower was quoted as saying that the DC-3 was one of the four "weapons," which also included the jeep, the bazooka, and the atomic bomb, that helped the Allies win the war. Over one thousand DC-3's remained in service as of 200l, with the majority being used for transporting cargo.

Specialized Freighters

The DC-3 was typical of many freighters in that it was designed initially to serve as a passenger plane. In 2001, many commercial freighters are civilian aircraft, such as Boeing 737's or McDonnell Douglas DC-10's, that have exceeded the maximum number of hours allowed for use as a passenger plane, although a few specialized cargo aircraft have emerged in the civilian market. These specialized freighters include the Super Guppy, a modified Boeing Stratocruiser, developed by Aero Spacelines in the early 1960's. The diameter of the upper portion of the fuselage was increased, giving the aircraft the ability to transport oversized items such as sections of the Saturn rockets used in the U.S. space program. The rocket sections were too large in diameter to be transported via rail or truck. The resulting rather bloated profile earned the Super Guppy its nickname. The Super Guppy has a sideways-hinged nose for straight-in loading. Other freighters may be hinged so the nose swings up, or feature a large rear cargo door with a ramp. The military's C-5 Galaxy has both a hinged nose and a ramped rear cargo door.

Some military freighters do share an airframe design with a civilian equivalent, but the manufacturer modifies the aircraft at the factory with specialized cargo doors and other features. The C-131 used for many years by the U.S. Air Force was a cargo transport version of the Convair 340 used by civilian airlines. Military freighters, such as the C-5 Galaxy developed by Lockheed, are generally built specifically to be used as cargo aircraft. The C-5 is the United States' largest military freighter. The cargo hold of a C-5 is large enough to carry six Apache helicopters. The aircraft has a payload capacity of 270,000 pounds, which is less than that of its Russian counterpart, the Antonov AN-124, but still sufficient to transport two M-1 battle tanks weighing 135,000 pounds each. Fact sheets on the C-5 Galaxy point out that the aircraft's cargo bay, at 143 feet long, is a greater distance from end to end than that which Wilbur and Orville Wright covered in their first powered flight at Kitty Hawk. The C-5 is remarkable in its ability to land and take off from very short runways. The aircraft's landing gear has twenty-eight tires, giving it the high flotation necessary for landing on dirt runways. The landing gear are hydraulically hinged, allowing the plane to "kneel" to bring the level of the cargo bay down to truck-bed height, making loading and unloading easier and faster.

Just as U.S. aerospace engineers developed the C-5 to transport oversized military equipment such as tanks, engineers in the Soviet Union designed the Antonov AN-124. The AN-124 made its first public appearance outside the Soviet Union at the May, 1985, Paris air show. The aircraft has a maximum payload of 330,700 pounds. Like the C-5, the AN-124 is notable for its twenty-four-wheel landing system, which enables it to land on dirt runways and even hard-packed snow, despite being possibly the heaviest aircraft in the world.

Nancy Farm Mannikko

Bibliography

Boyne, Walter J. *The Leading Edge*. New York: Workman, 1987. The former director of the Smithsonian Institution's National Air and Space Museum discusses various innovations in aviation. Very accessible to the general reader and lavishly illustrated with spectacular color photographs.

Green, William, Gordon Swanborough, and John Mowinski. *Modern Commercial Aircraft*. New York: Portland House, 1987. Easy to understand explanations of aircraft design and technical data, provides concise descriptions of hundreds of aircraft, including many not well known in the United States.

Holder, Bill, and Scott Vadnais. *The C Planes: U.S. Cargo Aircraft 1925 to the Present*. Atglen, Pa.: Schiffer, 1996. A general history of military freighters which looks exclusively at developments in the United States.

Matricardi, Paolo. *The Concise History of Aviation: With Over 1,000 Scaled Profiles of Aircraft from 1903 to the Present*. New York: Crescent Books, 1984. Good overview of aviation history written from a European perspective. Illustrations of aircraft are excellent.

Scharschmidt, Oliver. *Cargo Airlines*. Osceola, Wis.: Motorbooks International, 1997. An overview of commercial air cargo.

See also: Airplanes; Airline industry, U.S.; Airmail delivery; Apache helicopter; Boeing; Commercial flight; DC plane family; Lockheed Martin; Manufacturers; Military flight; 707 plane family

Sir George Cayley

Date: Born on December 27, 1773, in Scarborough, Yorkshire, England; died on December 15, 1857, in Brompton, Yorkshire, England

Definition: Early nineteenth century theoretician and experimenter who laid the foundation for heavier-than-air vehicle design.

Significance: Cayley was the first to conceive and publish the modern idea of the airplane: namely, the concept that an airplane should consist of one or more fixed wings, a fuselage, and a tail. He also designed and flew model gliders complete with control surfaces, was the first scientist to perform detailed tests of lift and drag as a function of angle of attack, and championed powered airplanes during the age of balloons.

George Cayley was born in Yorkshire, England in 1773 to a wealthy landowning family of noble lineage. He was privately tutored, with emphases on the scientific and mechanical arts, at which he excelled.

Although his official duties revolved around the administration of Brompton Hall, his baronial estate, Cayley applied his inquisitive mind to a wide range of practical issues affecting early nineteenth century English society. He devised a hot-air engine in 1799, more than a decade before the engine built by the Reverend Robert Stirling, whose name has since been attached to the concept. He directed the Polytechnic Institution in Regent Street, London, from its inception in 1839 to showcase technical achievements. He contributed improvements to railway carriage safety during the fledgling years of the railroad industry. He carried out experiments on ballistics and airships. He also invented several versions of an articulated mechanical hand to replace a lost limb.

Despite these many and varied achievements, Cayley is chiefly commemorated as the father of aerial navigation, a title first bestowed on him by the unsuccessful airplane designer William S. Henson, in an 1846 letter to Cayley. The appellation has since been reinforced by the discovery and publication in 1933 of Cayley's experimental notebooks, which along with his published writings, convey a sound knowledge of the dynamics of heavier-than-air flight almost a century before the Wright brothers applied themselves to the topic.

As early as 1799, Cayley had engraved on a small disk the force balance on a rudimentary aircraft. By 1804, he was performing whirling-arm experiments on the lift and drag characteristics of thin plates at low angles of attack and had noted the benefits of camber. In 1809 and 1810, he published seminal articles in *Nicholson's Magazine* that examined power requirements, structural issues, and aerodynamics for airplanes that could carry human passengers, given a sufficiently large lifting surface and a suitably light prime mover, or propulsion system. His sketchbooks show model vehicles with a movable rudder and tail, a variable center of gravity, and a dihedral angle for lateral stability.

Late in life, Cayley designed a remarkable vehicle that was described in an 1843 article in *Mechanics' Magazine*. It featured airscrews for vertical liftoff that converted to flat planes for horizontal flight under the impetus of a propeller. In 1849, Cayley published the results of model glider experiments and by 1852 had designed a vehicle with a 500-square-foot lifting surface, weighing 300 pounds and capable of supporting a person on board. It appears that he did oversee some manned glider experiments on his estate in or around the same year. Active almost to the end of his life, Cayley died at home on December 15, 1857, just short of his eighty-fourth birthday.

David M. Rooney

Bibliography

Anderson, John D. *A History of Aerodynamics*. Cambridge, England: Cambridge University Press, 1997. Chapter 3 of this monumental work places Cayley's work within the context of theoretical and practical aerodynamics in the early 1800's.

Gibbs-Smith, Charles H. *Aviation: An Historical Survey from Its Origins to the End of the World War II*. 2d ed. London: Her Majesty's Stationery Office, 1985. A balanced survey of the historical development of aviation, including Cayley's work.

Pritchard, J. Laurence. *Sir George Cayley: The Inventor of the Aeroplane*. London: Max Parrish, 1962. The definitive scholarly biography by one of the world's great historians of aeronautics.

See also: Aerodynamics; Gliders; Heavier-than-air craft; History of human flight

Cessna Aircraft Company

Definition: Wichita, Kansas-based company specializing in the manufacturing of small planes.

Significance: Cessna became a market leader in the general aviation field before World War II. After the war, the company continued its success by branching out into the production of business jets, eventually selling more business jets than any other company.

The Early Years

Cessna Aircraft Company had its inception in 1911 when founder Clyde Cessna attended an air show in Oklahoma City. Aviation so enthused Cessna that he moved to New York and spent three weeks working on an aircraft assembly line to learn all he could about flight. After returning to Oklahoma, Cessna purchased a monoplane and began making demonstration flights. Cessna also hoped to use his new monoplane as a blueprint for his own planes. Despite having no formal engineering training, Cessna produced several airplanes over the next seven years, first on his farm west of Wichita and after 1916 in Wichita itself. The United States' entrance into World War I in 1917 stifled Cessna's hopes to build a successful flight training school, and he was unable to secure any government manufacturing contracts. Cessna left the aircraft business in 1918 and did not return for seven years. Cessna focused on farming until 1925, when two Wichita aircraft builders, Walter Beech and Lloyd Stearman, asked Cessna to join them in establishing a new company, TravelAir. Cessna and Stearman designed the TravelAir 5000, which won the 1927 Dole Race to Hawaii. Other TravelAir designs also became very popular. However, Cessna was not content at TravelAir, and he sold his stock in 1927 to start Cessna Aircraft Company.

The Depression

Cessna enjoyed success during the late 1920's. The United States was in love with aviation, particularly after Charles A. Lindbergh's transatlantic flight in 1927. Cessna did well with two production models, the AW and BW, as well as specialized racing planes, but the good times did not last long. The Great Depression crippled the American economy, and aircraft manufacturers went bankrupt as their market disappeared. Cessna closed its doors in 1931, and Clyde Cessna never returned to the company he founded, though he did produce several highly successful racing aircraft during the 1930's. Cessna's nephews Dwane and Dwight Wallace resurrected the company in 1934. The Wallace brothers immediately set to work on a new aircraft, the C-34. The C-34 was designed for efficiency and long-term service, hoping to attract businessmen looking for reliable and cost-effective air transportation. The new plane boasted a cruising speed of 143 miles per hour and a range of 550 miles, while consuming only five gallons of fuel per hour. The single-engine C-34 and its descendants became known popularly as Cessna's Airmaster line. The company also produced a twin-engine plane, the T-50, nicknamed the Bobcat. The first T-50 flew in 1939 and boasted such features as retractable landing gear, hydraulic brakes, and a 1,000-mile cruising range. Though not as luxurious as some of its competitors, the T-50 cost considerably less. During World War II, the T-50 saw extensive service as a trainer aircraft, redesignated as the AT-17.

Wartime Production

As World War II loomed, Cessna's established reputation made the company an important part of the nation's preparedness efforts. The firm received its first order from Canada, which ordered five hundred modified T-50's and two hundred extra engines. The U.S. government soon followed with its own orders. As a result of this new demand, Cessna's employment increased from 200 in 1940 to 1,500 by the spring of 1941. After the United States entered the war, Cessna diversified its production to meet wartime demands. The company built subassemblies for the massive Boeing B-29 bomber. Cessna also designed a cargo plane, the C-106, but the craft never went into production. Cessna built 6,111 planes for the U.S. war effort, including 750 CG-4A-CE gliders. Production lines ran twenty-four hours a day, seven days a week, beginning early in 1942. Employees could make use of their own club, with a gym and a lounge. The company was among only 3 percent of U.S. manufacturers to earn the prestigious Army-Navy "E" rating, awarded to manufacturers that met production goals, five times during the war.

Postwar Hopes

Aviation enthusiasts envisioned a postwar United States that would take to the air in great numbers. The idea of providing families with safe, affordable airplanes as an alternative to automobile or rail travel attracted many within the aircraft industry. Cessna's production capabilities and experience positioned the company to take advantage of this new opportunity in the general aviation sector. Cessna produced the Model 120 in 1946, at a price of $2,695. This would be as close as any U.S. aviation manufacturer would come to producing an aircraft suitable for the needs and budgets of American families. The 120 could cruise at 100 miles per hour and promised durability with less maintenance. The company also produced a more expensive, upgraded version known as the Cessna 140. Cessna promoted its new product line by establishing distributorships around the world. In December, 1946, Cessna produced almost as many small planes as all of its competitors combined. The company built on the success of the 120 and 140 by introducing three new models, the 190 and 195 in 1947, and the 170 in 1948. The 190 and 195 were designed for greater luxury, and the 170 could carry four people, as opposed to only two in the 120 and 140. Despite these new designs, Cessna struggled through the end of the 1940's. The United States' economy sagged as it readjusted to peacetime production, and aircraft companies had to find other ways to make profits. Cessna's Wichita competitor Beech investigated automobiles and prefabricated housing. Cessna looked to somewhat more mundane products including furniture, hydraulics, and aluminum lockers. The furniture-making enterprise proved the most profitable, particularly after the company won a contract from the Army Quartermaster Corps. After the Army contract expired, Cessna tried to market its furniture line through Chicago-based department store Marshall Field's. Finally, the demand for aircraft increased to the point that Cessna could eliminate its furniture operation in 1951. In the 1950's, Cessna continued expanding its offerings. In 1953, Cessna debuted the Model 180 and the 182 Skyline which promised greater performance, though at a greatly increased price of $13,000. The company also introduced the 172 Skyhawk in 1956 as an improved version of the 170. Cessna diversified its product line even further in 1954 by entering the luxury twin-engine market with the Model 310, which would be immortalized in the television program *Sky King* (1951-1952, 1956-1962). Cessna's 2,489 planes sold in 1958 made the company the world's largest private-plane manufacturer.

The Jet Age

Cessna continued its advancements in design with the 1961 production of the Model 336-337 Skymaster. The Skymaster boasted two engines, but rather than placing one on each wing, Cessna put one on the nose and the other between twin tails. This configuration eliminated the dangerous situation of unequal thrust in case of engine failure. The Skymaster won excellent reviews from aviation experts, but it did not attract buyers due to its unconventional appearance. In fact, the Skymaster failure did not seriously damage the company because Cessna and Dwane Wallace had already turned to the jet market. Originally, company officials hoped to design a high-end turboprop plane that would fit the market niche between small private planes and the expensive business jets then making their first appearance. Wallace soon saw that a turboprop model was not the solution. Responding to the success of Wichita-based Learjet, Cessna produced the Citation in 1972. In designing the new plane, Cessna stuck with its established formula of offering the customer a safe and efficient aircraft at a reasonable price. The Citation offered quiet and reliable performance with less maintenance and better fuel economy. The plane had room for comfortable accommodations for six people at about half the price of its competitors. The new jet did, however, have its drawbacks. Its cruising speed of 400 miles per hour was 150 miles per hour slower than the Learjet models, and the Citation had a somewhat strange appearance, with its straight wings and blunt nose. Critics felt that the Citation stood little chance of success, and early sales favored the competition. The company stood behind its design, and by 1978, the Citation was the world's fastest-selling corporate jet. The company expanded its offerings by following up the original Citation with the Citation I and II models. Cessna hoped to move into new market territory, however, and began development on a larger, faster, and more luxurious jet, the Citation III. The first prototype of the Citation III flew in 1979 and Cessna delivered the first production models in 1983. The Citation III cruised at 509 miles per hour and could fly 2,500 miles. The cost of the jet in 1984 was $6,120,000, but that did not deter customers who purchased fifty of the planes that year, far above the totals of Cessna's competition. The company redesigned the Citation III extensively in 1990, giving the plane a new avionics package, an improved interior, and changes in the airframe. Cessna completed the evolution of the original Citation line with the VI and VII models. The Citation VI was a lower-priced version of the Citation III, which Cessna delivered in 1991. The Citation VII was a larger and more powerful model, delivered in 1992.

Corporate Changes

The economic pressure of developing the Citation line stretched Cessna's resources. Customers complained that their planes did not function reliably and the company's maintenance and repair services did not always meet expectations. Cessna sent out more than one hundred service bulletins addressing problems, publicly highlighting the shortcomings of the early Citations. The company also faced stiff competition from other firms. These problems made it impossible for Cessna to remain an independent company. In 1985, General Dynamics acquired Cessna and promised to continue and improve the Citation line. In 1992, Textron purchased Cessna from General Dynamics for $600 million in cash. These changes in corporate structure did not change Cessna's focus, however. The company remained the world's largest seller of business jets through the end of the century.

Matthew G. McCoy

Bibliography

Philips, Edward H. *Cessna: A Master's Expression*. Eagan, Minn.: Flying Books, 1985. A useful starting point for the history of Cessna Aircraft.

Porter, Donald. *The Cessna Citations*. Blue Ridge Summit, Pa.: Tab Books, 1993. This book covers the development of Cessna's highly influential Citation business jets.

Rowe, Frank Joseph, and Craig Miner. *Borne on the South Wind: A Century of Aviation in Kansas*. Wichita: Wichita Eagle and Beacon, 1994. This book covers the development of aviation in the state of Kansas. It does not go into great depth, but it does offer important information about Cessna and its role in aviation, both worldwide and as a key component to the state's economy. It is also well illustrated.

Octave Chanute

Date: Born on February 18, 1832, in Paris, France; died on November 23, 1910, in Chicago, Illinois

Definition: An experimenter in glider flight and design, a communicator who helped early experimenters exchange information, and an author who inspired many aviation pioneers.

Significance: Chanute's experiments in gliding flight near Chicago, Illinois, in the late 1800's and his correspondence with other experimenters in the United States and Europe led him to publish many articles on flight. His 1894 book, *Progress in Flying Machines* inspired many of the first American aviators, including Wilbur and Orville Wright, to begin their flying experiments.

Octave Chanute was born in France but emigrated to the United States in 1938 and studied civil engineering through an apprenticeship program. Working with railroads, he became one of the most successful engineers in the United States and designed the first bridge across the Missouri River. Chanute became interested in the challenge of flight in the 1870's and corresponded extensively with aviation experimenters, such as Otto Lilienthal in Germany. During the 1890's, he designed, built, and tested his own gliders in hundreds of flights on the shore of Lake Michigan.

Although Chanute learned much from his own experiments, he is best remembered for the encouragement he gave to others and for his role as a communicator. He pub-

Octave Chanute's glider experiments of the 1890's helped pave the way for the Wright brothers' successes in powered flight. (Hulton Archive)

lished many articles on flight in engineering journals and magazines of the time, as well as a book, *Progress in Flying Machines*, in which he reviewed the work of past and then-current flight researchers. Through his personal correspondence, he led many aviation pioneers into the field. Wilbur and Orville Wright read Chanute's book and corresponded frequently with him as they worked on their glider and airplane designs. Chanute visited the Wrights at their camp at Kitty Hawk, North Carolina, and gave them advice and encouragement as they chased the goal of powered flight.

It was Chanute who, in Paris in 1903, first revealed to European aviation researchers that the Wright brothers were on the verge of success. He published detailed accounts of that success within a few weeks of their first flights. A couple of years later, in an apparent effort to motivate the Wrights toward a more public exhibition of their success, Chanute informed Orville and Wilbur that the Europeans were nearing success in building a flying machine. The Wrights, who had been somewhat secretive about their post-1903 experiments in hopes of selling profitable airplanes to the military, rejected his advice, responding that there was no one in the world, not even Chanute himself, who was capable of building a flying machine within the next ten years. The first successful European flight took place the following month.

James F. Marchman III

Bibliography

Anderson, John D., Jr. *A History of Aerodynamics*. Cambridge, England: Cambridge University Press, 1998. An excellent review of the work of all who have contributed to advances in aerodynamics.

Chanute, Octave. *Progress in Flying Machines*. Reprint. New York: Dover, 1998. A reprint of the classic work that inspired the Wright brothers and others.

Roseberry, C. R. *Glenn Curtiss: Pioneer of Flight*. Syracuse, N.Y.: Syracuse University Press, 1991. This excellent biography of Glenn Curtiss includes extensive reference to Chanute and his correspondence with the Wright brothers.

See also: Airplanes; Gliders; Heavier-than-air flight; History of human flight; Otto Lilienthal; Wright brothers

Jacqueline Cochran

Date: Born c. 1910, in Muscogee, Florida; died August 9, 1980, in Indio, California

Definition: Pioneer in aviation who paved the way for future female American pilots.

Significance: At the time of her death, Cochran held more speed, altitude, and distance records than any other pilot in aviation history. She was the first woman to break the sound barrier and the first living woman to be inducted into the American Aviation Hall of Fame.

Born near Pensacola, Florida, sometime in 1910 (although she claimed her birth date was May 11, 1912), Jacqueline "Jackie" Cochran spent her early years in poverty. Orphaned while still an infant, she was raised by a foster family of sawmill workers. Her formal education did not go beyond the second grade. During her teens, she moved to Alabama to work as a beautician and enrolled in a three-year nursing program. Although she completed her training, Cochran, fearing failure, did not take the written examination and instead became a doctor's assistant near the sawmills where she had been raised.

Depressed by the poverty of the sawmills, however, Cochran returned to work as a beautician. This work took her from Pensacola to Philadelphia, and finally to New York City, where she worked at the Saks Fifth Avenue department store. On a business trip in 1932, she met Floyd Odlum, a wealthy investor who enjoyed aviation. Following their conversation, Cochran enrolled in flight school. She quickly completed the courses, becoming one of only a few women to have a pilot's license. Thereafter, she received a commercial pilot's license, bought her own plane, and began competing in air races.

Cochran opened a beauty shop and started a cosmetic manufacturing business. In 1936, she married Floyd Odlum; the couple settled on a ranch near Indio, California. Marriage allowed Cochran to concentrate most of her time on flying. During the next year, she set three speed records and was awarded the Harmon trophy as the outstanding woman aviator of the year. In 1938, she won the Bendix Transcontinental Race by setting a new speed record, and the following year set an altitude record of 33,000 feet while winning the New York to Miami air race.

During the early 1940's, Cochran used her aeronautical skills to help the Allied war effort. In 1941, she became the first woman to ferry a B-17 bomber to Britain, and thereafter recruited other female pilots to continue ferrying operations for the military. As a result of this success, in 1943 the military appointed her as the head of the Women's Airforce Service Pilots (WASPs). Under her direction, over one thousand women completed important missions across the Atlantic. In 1945, the Army awarded Cochran the Distinguished Service Medal for her accomplishments as head of the WASPs. Following World War II, she worked as a reporter for *Liberty Magazine*. As a journalist she covered the Nuremberg war crimes trials, was the first American woman to enter Japan after the war, and interviewed Chinese leaders Chiang Kai-Shek and Mao Zedong.

Following the war, Cochran returned to the skies to set new records. In 1948, she set an altitude record of over 55,000 feet, and in 1953, she flew an F-86 Sabre jet faster than the speed of sound (Mach 1), becoming the first woman to break the sound barrier. Her aviation records continued, as she became the first female pilot to land a jet on an aircraft carrier and the first woman to pilot a jet across the Atlantic Ocean. In 1964, she logged a record-setting flight at 1,429 miles per hour, over twice the speed of sound (Mach 2) and the fastest for a female pilot.

In 1970, Cochran retired from the U.S. Air Force Reserve with the rank of colonel, and the next year she became the first living woman to be inducted into the American Aviation Hall of Fame. She died in 1980, leaving behind more than two hundred flying records. In addition to holding more records for altitude, speed, and distance than any pilot in history, her pioneering efforts made it possible for women to serve their country as pilots, astronauts, and military officers.

Aaron D. Purcell

Bibliography

Cochran, Jacqueline, and Maryann Bucknum Brinley. *Jackie Cochran: An Autobiography.* New York: Bantam, 1987. A detailed autobiography compiled by Brinley consisting of interviews, personal recollections, and photographs.

Cochran, Jacqueline, and Floyd Odlum. *The Stars at Noon*. Reprint. New York: Arno Press, 1980. A fascinating autobiography providing important descriptions of Cochran's early years and the challenges women faced during the development of aviation, especially during World War II.

Cole, Jean Hascall. *Women Pilots of World War II.* Salt Lake City: University of Utah Press, 1992. Excellent overview of the WASP program with personal interviews.

Lomax, Judy. *Women of the Air.* New York: Dodd, Mead, 1987. The chapter on Cochran provides an excellent overview of her contributions to flight.

See also: Air derbies; Flying Fortress; Military flight; Women and flight; Women's Airforce Service Pilots; World War II

Cockpit

Definition: The area within an aircraft from which pilots operate the aircraft's controls.

Significance: Cockpits provide a central point from which airplane performance can be commanded and monitored.

The term "cockpit" originated with the ancient sport of cockfighting. Early pilots had to control unstable airplanes through control levers positioned without regard to one control's effect on another. Pilots stayed busy, their motions reminiscent of the frenzy in the gaming floor's cockpit.

Although early airplanes accommodated pilots, they had no cockpits by modern definition. The Wright brothers' *Flyer* pilot lay prone, having controls in reach but little else. No flight instruments existed until about 1911. In his underpowered, box-kite-like *14-bis*, Alberto Santos-Dumont stood erect while becoming Europe's first airplane pilot. By their first decade, airplanes had evolved cockpits as effective yet inefficient workstations.

By World War I, fighter cockpits gave their seated pilots a control stick, a rudder bar and precious few instruments. Open cockpits were a hallmark of pre-1920 airplanes—rarely were cockpits enclosed. As enclosures became prominent in the 1920's, some pilots disliked them, wanting the wind on their faces to indicate slips or skids. By the 1930's, most airplanes featured enclosed cockpits, although efficient pilot motion stayed a low priority.

The layout of cockpits only slowly became logical, with their instruments and installations sometimes cumbersome. Lockheed's prewar Model 14 Hudson is an example of cockpit inefficiency; its Royal Air Force (RAF) version was a handful for its single pilot. In his 1972 memoir, H. A. Taylor recounted the difficulties of solo flight in the Hudson, beginning with starting the engine. It was a procedure that "was preferably done with three hands, each with more than the usual number of fingers and thumbs," and involved simultaneously pressing buttons for both the starter motor and booster coil while holding a spring-loaded, three-position switch that selected the engine to be started. Meanwhile, an engine-doping pump and a wobble-pump had to be worked, and as soon as the engine fired, the idle cut-off lever had to be released and the throttle manipulated while the booster button was continually pressed. The layout of these vital mechanisms added to the challenge: "The buttons, switches and doper were on a fore-and-aft electrical panel to the pilot's right; the wobble-pump handle was at the rear of the throttle pedestal; and the cut-off levers sprouted, among a dozen or more others, from the top of this pedestal."

Not all 1930's manufacturers spurned pilot efficiency. By the early 1930's, Germany's Junkers Aircraft built its Ju-52/3m, called "Tante Ju" ("Auntie Junkers") by her adoring crews. Its innovations included dual instruments, a series of mechanical devices to reduce distraction-induced pilot errors, and effective weatherproofing. Logic arranged its flight instruments, and the pilot and copilot could both reach the brake lever. By the climax of World War II, cockpit efficiency had become a manufacturing priority.

Modern Cockpits

Airplane cockpits range from the single-place, where the pilot is the sole occupant and performs all duties, to the multi-place, in which several crew members share duties such as flying, communicating, navigating, and systems monitoring. Cockpit designs demand unique considerations.

Accessibility means that the pilot's station must be easily reached upon entry and easily departed at flight's end. Restraints must counter turbulence, yet allow quick crew egress in emergencies. Once seated, pilots must be able to reach all of the flight and systems controls. The control sticks so favored by early designers provide an unencumbered view of the instrument panel, and fall to hand naturally. Control yokes, or wheels, create an automotive feel that comforts new aviators, but blocks pilot vision of parts of the instrument panel. Both amateur airplane builders and conglomerates, such as Europe's Airbus Industrie, have found value and pilot acceptance of side-sticks, joysticks mounted on the cockpit bulkhead, or side wall, where they can comfortably be reached by the pilot's hand. These controls can be reliably gripped, even in tense moments or in turbulence, when jolts and jostling fling a pilot's reaching hand from levers or dials.

From the 1920's through the 1950's, training airplanes tended to have tandem cockpits, in which the student and instructor sat on the airplane's centerline, one behind the other. Advantages included the students' ability to perform maneuvers in either direction with equal challenge, for their field of vision either way remained identical. Additionally, students tended to develop cockpit skills more quickly because their instructors remained essentially hidden. Disadvantages included the need for duplicate instrumentation and the instructors' inability to see nuances in student facial expressions. In the 1950's, as most trainer cockpits adopted side-by-side seating, designers strove for cockpit efficiency. Sometimes that goal is still unmet. Ten accidents occurring between 1972 and 1982 prompted development of what is known as cockpit resource management. Accidents underscored the need for physical changes in cockpits. Studies revealed surprising clues to the dangers induced by poor design.

By the twentieth century's close, newly produced airplanes had begun to incorporate cockpit ergonomics. Ergonomics considers the design of the human body, including its ranges of skeletal and muscular motion. Normal operation is the first consideration, but airplanes encounter strong turbulence, operate in daylight and darkness, and can climb in minutes from searing heat at the airport to subzero temperatures at altitude. Designers must consider these factors and more, plus incorporate characteristics to maximize crash survivability. Like the rest of the airplane, the cockpit is a compromise, for which designers cannot rely on tradition. Today's cockpit designers use recent and exhaustive studies to meet their goals. Despite its claustrophobic faults, the cockpit holds strong allure to millions. Depicting airplanes, artists usually focus on cockpits, for therein sits an airplane's humanity, and what many see as its ultimate office.

David R. Wilkerson

Bibliography

Caidin, Martin. *The Saga of Iron Annie.* Garden City, N.Y.: Doubleday, 1979. An account of one of the world's most famous airplanes and the travails of restoring an airliner-sized antique to flying condition.

Connor, C. W. *Proceedings of the Seventh Aerospace Behavioral Technology Conference: Operational Infor-*

The modern cockpit is laid out for optimum ease of operation by both pilot and copilot, unlike many early aircraft designs. (AP/Wide World Photos)

mation Transfer Technology: Are We Designing for the Human Operator? Warrendale, Pa.: Society of Automotive Engineers, 1989. A compilation of twenty-three highly technical assessments of aerospace issues, including cockpit design.

Satchell, P. M. *Cockpit Monitoring and Alerting Systems.* Aldershot, Hampshire, England: Ashgate, 1993. A technical work directed to aviation's production professionals.

Szurovy, Geza. *Wings of Yesteryear: The Golden Age of Private Aircraft.* Osceola, Wis.: MBI, 1998. A nostalgic review of 1920's- to 1940's-era light planes, lavishly illustrated with superb color photos, posters, and contemporary black-and-white photographs.

Taylor, H. A. "Flying the Harassing Hudson." *Air Enthusiast Magazine* (December, 1972): 292. A pilot's memoir of Royal Air Force Service during World War II.

See also: Airplanes; Flight control systems; Instrumentation; Manufacturers; Rudders; Alberto Santos-Dumont; Training and education; Wright *Flyer*

Bessie Coleman

Date: Born on January 26, 1893, in Atlanta, Texas; died on April 30, 1926, in Jacksonville, Florida

Definition: The first African American woman to fly an airplane and the first to earn an international pilot's license.

Significance: Coleman overcame racial barriers in the United States to achieve her dream of flying, becoming an inspiration to women and minorities.

Bessie Coleman, known to her fans as "Queen Bess," grew up caring for her thirteen younger siblings on a small farm near Waxahachie, Texas. Coleman's father left the family when Coleman was nine, and her mother supported the children by picking cotton and taking in laundry. Although at the top of her class, Coleman had to leave school at the end of eighth grade to work as a laundress. Finding domestic work humiliating, she left Texas in 1915 and went to live with her brother in Chicago, where she studied to be a manicurist.

During World War I (1914-1918), Coleman read accounts of brave aviators and decided that she wanted to become a pilot and open a flight school for African Americans. When no flight school in the United States would accept her as a student, she went to France. Coleman began her training at the École d'Aviation des Frères Caudron at at Le Crotoy in November, 1920. In June, 1921, she became the first black woman to receive a license from the Fédération Aéronautique Internationale. She returned to Chicago in September, 1921, hoping to further her flight training, but again no flight school in the United States would accept her. Undaunted, she returned to France in February, 1922, and completed an advanced course. She sailed for the United States in August, ready to begin her aviation career.

Coleman flew in her first air show, the first public flight in the United States by an African American, in New York on September 3, 1922, and went on to perform in air shows across the country. She was offered a role in a motion picture, but when she found that it required her to dress in ragged clothing, she refused, believing the role would perpetuate a negative image of African Americans. After this episode, several of her backers withdrew their support of her.

Coleman began a lecture tour hoping to inspire young African Americans but found that the meager sums she collected from her audiences would not allow her to achieve her dream of opening a flight school. The chewing-gum manufacturer Edwin W. Beeman helped her purchase a plane in which to fly in a celebration sponsored by Jacksonville's Negro Welfare League. On the day before the event, Coleman's mechanic, William D. Wills, took her up so she could survey the area. Coleman sat in the back of the plane without wearing her seatbelt, so she could boost herself up to see out of the cockpit. A wrench left in the airplane jammed the controls, causing the plane to flip and go into a tailspin. Coleman was flung from the plane and plunged 2,000 feet to her death.

Polly D. Steenhagen

Bibliography

Haskins, Jim. *Black Eagles: African Americans in Aviation.* New York: Scholastic, 1995. A superb history of many of the greatest African American pilots and astronauts.

Plantz, Connie. *Bessie Coleman: First Black Woman Pilot.* Berkeley Heights, N.J.: Enslow, 2001. A biography written for younger readers, with bibliographical references and an index.

Rich, Doris. *Queen Bess: Daredevil Aviator.* Washington, D.C.: Smithsonian Press, 1995. An excellent, in-depth biography of Coleman.

See also: Air shows; Safety issues; Training and education; Women and flight

Commercial flight

Definition: The transportation of passengers and freight by commercial airline companies.

Significance: Commercial flight has made possible a global economy, drastically reducing the amount of time and money that must be spent in transporting people and goods over long distances.

The emergence of relatively reliable and safe airplanes during World War I induced people to attempt the organization of an airline to operate those craft on a scheduled basis over a consistent route. The Deutsche Luftreederei began service from Berlin to Leipzig and Weimar on February 5, 1919, followed only three days later by the French Farman Company on the cross-channel crossing from Paris to London using a converted Goliath bomber of World War I provenance. In August, 1919, the first daily service was established on this route from Le Bourget Airfield in Paris to Hounslow in the United Kingdom. The oldest surviving airline, KLM, was organized in the Netherlands in 1919 and, jointly with a British company, began flying the route between Amsterdam and London the following year. Outside of Europe, the Queensland and Northern Territories Aerial Services (Qantas) was founded in 1920. This eventually became the Australian national airline.

Most of the airlines founded in the 1920's and 1930's were created at least in part to encourage the purchase of aircraft of domestic manufacture. However, the privately owned Swissair was the first European airline to purchase American aircraft. The intertwining of domestic aircraft manufacture and national airline operation was widely advocated as critical to national defense.

In the United States, airline pioneers were private operators, as were the aircraft builders, and there was no national policy concerning either operation. Throughout the 1920's there were no adequately financed airlines, and most lasted for only short periods before failing or merging. Given the large expanse of the United States, an airline with routes of national or even regional coverage was the exception. It was only in the late 1920's that any thought was given to the question of encouraging a domestic aircraft industry or the promotion of domestic airline companies.

International Transportation

In Europe, in particular, the colonial airline emerged as a factor in the overall evolution of commercial aviation. Britain, France, and the Netherlands all developed colonial airlines, with Belgium, Italy, and the United States joining the operation less extensively. Routes for national airlines were limited to destinations within a country or its possessions, except by agreement. The extensive colonial empires still in existence in the 1920's and 1930's became natural sites for extended airlines. Britain, for example, created Imperial Airways by first using bilateral agreements with other European countries to reach the Mediterranean, and once there, to project a continuation based on British colonies and protectorates in Malta, Cyprus, Palestine, Trans-Jordan, Iraq and the Persian Gulf protectorates, India, Burma, the Malay Protectorate, Australia, and New Zealand. China, Central Africa, and South Africa could be reached by other routes. Only the North Atlantic and the northern Pacific resisted a "British" national airline. France shaped a colonial airline from Provence across the Mediterranean to Algeria, the French Sahara, French Equatorial Africa, and Madagascar. Working out the landing rights between Belgium and France provided a route to the Belgian Congo. The Netherlands, through trades with Britain, shaped a colonial route for KLM to the Dutch East Indies (present-day Indonesia).

In the 1930's, these colonial routes were the main long-distance air routes available not only because a far-flung empire simplified the problem of securing landing rights but also because the operating stage, the maximum distance that might be flown without stopping to refuel, was then only about 500 miles. The Pacific and Atlantic oceans were the major water jumps that remained unconquered by civil aircraft in 1930. The American air routes showed the way to the solution. Pan American Airways was first organized to fly from Miami to Key West in Florida and to Havana, Cuba, and by the 1930's from Brownsville, Texas, to Mexico City and Panama. Pan American founder Juan Terry Trippe advocated the concept of the "chosen instrument" or the idea that international connections for the United States should be provided by a single American company flying only outside of the country. The American "empire" in this sense was Latin America, where American investment was extensive but political control was only indirect. Germany, which after World War I lost its colonies, similarly turned to South America, particularly Colombia, to shape an extensive system of air routes. In the American case, Pan American's ultimately extensive route structure in the Caribbean, on the east coast of South America, and in Central America provided experience in operating a long-distance international airline.

By the early 1930's, three airlines in particular were seeking to develop world-scale route patterns: Pan American, Imperial Airways, and KLM. Such a development called for a set of aircraft that were entirely new in concept from those that had been derived from the planes of World War I. Specifically, what was needed were seaplanes, which offered some advantages. They could fly stages of considerably greater length than could be flown with standard land planes because the sea-based plane enjoyed an almost infinite takeoff runway, a long stretch of water in a sheltered embayment. Several miles might be used at a time when a 1,000-foot runway was the norm. Long runways, either on land or on water, meant that planes could be quite large, use multiple engines, have large enough fuel tanks to fly an extended stage, and require less strength in the undercarriage.

The tradition of high-powered planes introduced between 1907 and 1909 by American aircraft builder Glen H. Curtiss continued. In addition to the Curtiss Company, Martin and Sikorsky each produced large, four-engine seaplanes with the potential for stages of more than 500 miles. Because of its size, the United States showed a concern for lengthening the stage even of land-based planes. When Pan American adopted the seaplane in the early 1930's, the Sikorsky S-42 flying boat had four engines that permitted it to fly to Buenos Aires, Argentina, by making a series of water crossings between Puerto Rico and the Rio de la Plata.

Airmail Service

After World War I, another factor contributed to airline development, namely the desire for an air service to speed up delivery of mail. Unlike Europe, where the nationalized airlines carried the mail, in the United States the Army Air Corps was assigned the job, with generally poor results. The problems of flying in a country the size of the United States were considerable. Particularly in the East, with the broad band of the Appalachian Mountains lying athwart the main routes, bad flying conditions were endemic and crashes were frequent. The introduction of aircraft beacons helped, but the low altitude at which most contemporary planes operated continued to plague service. Commercial flying began in earnest in 1925 when, under the Air Mail Act of 1925, also known as Kelly Act, the United States Post Office Department established contracts for carrying mail over assigned routes. Payments were made in return for the weight of mail carried. This practice often generated earnings that made the difference between marginal operation and flying at outright losses. Later, the method of airmail payments was revised. Instead of paying for the weight of mail carried, the Post Office paid instead for the space reserved for airmail, were it to be offered to the airline company to transport. The result was an incentive to the companies to increase the size of the planes that they normally flew.

Growth of the Aviation Industry

Competition for the airmail routes led to the formation of several large American aviation companies. William Boeing, who during World War I was a lumber producer in Seattle, Washington, had built planes from Sitka spruce, a wood with fibers of great tensile strength. Boeing bid on what came to be called the "Columbia Route" (New York City to California's San Francisco Bay Area), winning the western segment from Chicago to Oakland. Henry Ford, who for several years had been building a trimotor plane, secured the route linking Cleveland and Chicago. To serve the western section, Boeing experimented with new and larger planes built by the Boeing Aircraft Company, which in the following sixty years became the world's largest and most comprehensive civilian aircraft manufacturer. United Aircraft & Transport joined with National Air Transport (which later became United Air Lines) and others to create a second aviation company that secured the contract for the eastern segment of the Columbia Route, linking Chicago and New York City, and for the north-south route on the west coast from Vancouver, Canada, to Los Angeles. A further recipient of an airmail contract was the Aviation Corporation (North American and Curtiss aircraft builders), which became American Airlines. The General Motors Corporation held major ownership in Transcontinental Air Transport (TAT) as well as Eastern Transport on the north-south airmail route on the East Coast. With Pan American, which was assigned several foreign routes, these aviation companies constituted the "Big Five" airlines, which survived as the dominant U.S. carriers until the 1990's.

Improvements in Aircraft Operation

In the late 1920's, airlines were stymied by two problems: night flying and high-altitude flying. Both were still too dangerous for passenger transportation. In the United States, crossing the Appalachians was possible, as the operating ceiling of the planes exceeded the necessary 3,000 to 4,000 feet. In the Rocky Mountains and the western Coast Ranges, however, there were 8,000- to 10,000-foot passes. Continuous flight over a major part of the United States could not be accomplished during daylight hours.

In 1929, Transcontinental Air Transport and the Pennsylvania Railroad joined forces to solve, at least in part, these altitude and darkness problems. They organized a rail-plane route between New York City and Los Angeles. The "Airway Limited" departed New York's Pennsylvania Station at 6:05 P.M., using a Pullman sleeper to reach Port Columbus, Ohio, a new landing field outside of the Ohio capital. There, passengers boarded a Ford Trimotor plane at 8:15 A.M., which carried ten passengers to Waynoka, Oklahoma, by 6:24 P.M., in time to board a second Pullman sleeper on the Santa Fe Railway at 11:00 P.M. This was to arrive in Clovis, New Mexico, at 8:10 A.M., when the passengers boarded a second plane to fly to Los Angeles, and, for through passengers, on to San Francisco by 7:45 P.M. The route avoided most night flying and any mountains over 5,000 feet.

Such an arrangement demonstrated the need for planes better than the Ford Trimotor, the workhorse of American carriers in the late 1920's. By 1928, Ford had improved speed on his plane, from 100 miles per hour on the 1926 model to 120 miles per hour on the 1928 model, through the introduction of stronger radial engines that were coming into use in the United States. By 1929, the United States was building 5,500 aircraft, up from only 60 five years earlier. The Vega of 1927 had increased cruising speed up to 150 miles per hour.

In 1930, Boeing's Monomail demonstrated the virtues of all-metal planes with the installation of retractable landing gear. Most experts view the Boeing 247 of 1933 as the first modern commercial aircraft. It showed that twin-engine planes were safer than trimotors because they could be maneuvered more easily and might be flown on a single engine. So many of the planes were ordered that when Transcontinental and Western Airlines (TWA) sought to order some, Boeing declined. TWA turned to a smaller builder, the Douglas Company, and commissioned a similar plane as a trial. The prototype was the DC-1. In its developed form as the DC-2 and DC-3, it proved to be the most significant commercial plane ever constructed.

The DC Planes

The plane was first introduced as a prototype (the DC-1) in 1933 and put into production as the DC-2 (and in evolved form as the DC-3 in 1936). The first DC-2 was put in service on the Newark-Pittsburgh-Chicago run, after only 11 months development time. In an era when American engine builders were introducing new and more powerful engines at a regular and rapid rate, the Wright Engine Company had been able to substitute an improved and more economical engine by the time that quantity production began. American Airlines asked for a slight enlargement of the DC-2. When fitted out with seats, this enlargement held twenty-one passengers and was called a DC-3. As such, it was the first airliner to operate at a profit with a reasonable load factor. The DC-3 had a ceiling above 5,000 feet, could fly on only one engine, and with a stressed aluminum sheathing, was a strong plane with a retractable landing gear. In the ten years of its production life, the DC-3 became the unrivaled master airliner, carrying the majority of American traffic. It was found in the fleets of most of the world's airlines, was used for military cargo (as the C-47 in the United States and as the Dakota in Britain), and was constructed in a run of more than 13,000 planes. Undoubtedly the greatest contribution of the DC-3 was that it demonstrated with great clarity the feasibility of safe, reliable, affordable, and profitable flying. Flying was a curiosity when the DC-3 was first built but had become standard transportation by the time of its last manufacture.

Between 1927 and 1939, the smaller aircraft engine rapidly advanced in its technology. Before World War I, the Russian aeronautical engineer Igor Sikorsky had constructed a twelve-engine flying boat. In the progression from DC-1 through DC-3, knowledge secured from earlier expressions of a basic design was used to enlarge that design so as to gain size, speed, and economy. Certain general qualities were standardized. The typical DC plane had a squarely rounded fuselage, a low wing, a particular way of carrying engine pods, and other features that had become standard. For example, if a larger passenger load was sought, the fuselage would be lengthened rather than widened. A longer plane required no other changes than enlarging the engines. Engines could be made more powerful by turbocharging them (supercharging them using centrifugal blowers driven by exhaust gas turbines), enlarging the cylinders, and making other mechanical elaborations. American aircraft builders became very adept at securing more power to go faster, farther, or cheaper.

The Four-Engine Plane

During the 1930's, American airline operators increasingly sought ways of constructing and operating four-engine planes, recognizing that such aircraft could potentially fly above altitudes normally characterized by turbulence. Consequently, the Boeing Stratoliner was introduced in 1940. Equipped with a pressurized cabin and capable of flying at 14,000 feet at a speed of 200 miles per hour, the Stratoliner had just begun service during the second year of World War II. Development of this pioneering four-engine plane was taken over by the U.S. government for the duration of that conflict. The Stratoliner was the only commercial aircraft able to be flown directly from Newfoundland, Canada, to Northern Ireland during World War II. With its powerful supercharged engines, the Stratoliner could navigate not only above weather but also over mountains, rather than around them. Thus, routes could be chosen because they formed parts of great circles on the earth's surface and were therefore the shortest possible distances between two points.

A second four-engine plane was designed just before World War II when the general configuration of the DC-3 was transformed into a four-engine size. Unlike the Stratoliner, this was not a pressurized plane, so it represented the last phase of one line of advance rather than the beginning of a postwar design. The enlarged DC-4 was flown throughout the war, becoming the main transatlantic aircraft in the form of the United States Army's C-54 troop transport.

Postwar Developments

After World War II, air transportation was quickly restored to civilian life. The Stratoliner and the DC-4 began immediate service on the longer routes, even across the Atlantic and Pacific Oceans. Even more important was the introduction of a plane that for one decade became the principal competitor of the DC-4, the Lockheed Constellation. The rapid growth in the power produced by American aircraft engines encouraged TWA to turn to the Lockheed company in search of a plane that would add more than 100 miles per hour to the speed of the DC-3 (175 miles per hour) rather than the marginal 25-mile-per-hour increase of the DC-4. In addition, TWA engineers sought to lengthen the stage of planes so that a single-stop transcontinental flight was possible in either direction. When it entered service, the Constellation had an 80-mile-per-hour speed advantage over the DC-4. When the Super Constellation went into service in 1957, it weighed twice as much as its precursor, was considerably faster, and carried a much increased payload.

The very rapid growth of air traffic in the ten years after 1945 called for a number of different planes to deal with extended routes and enlarging markets. In large part, this expansion could transpire because there was a market for used aircraft. As airlines strove to fly faster and with lengthened stages, more people switched from trains or ships to planes. By 1953, the DC-7 was put in service, with a stage of up to 3,000 miles and a speed reaching 300 miles per hour. By 1957, the number of passengers crossing the Atlantic by air was greater than by sea. Once jet planes came into service at the end of the 1950's, flying the Atlantic accelerated to the point that little more than a decade of steamship service remained before the end of the Atlantic Ferry.

The Jet Era

The realization that planes of varying size and purpose could carry jet engines had a profound effect on commercial flight. It was anticipated that the jet would revolutionize the speed of air travel. What was rather unexpected was that it could sharply reduce its cost when provided by a jetliner large enough to carry an economical load. The Boeing 707 was so economical when it was placed in service by Pan American on October 26, 1958, that it played the role for commercial jets that the DC-3's had played for piston planes. When the fan jet was substituted for the simple jet engine, the family of Boeing jets earned a reputation for economical working just as the DC-6 had in the last generation of piston planes. Within a few years, Boeing had developed specialized jets for nearly the full range of commercial flying. The Boeing 727 became an intermediate-range jet carrying more than one hundred passengers, rivaling in size the largest piston planes. Later, the Boeing 737 became the workhorse of North American airlines. When it was discovered that the cost of operating jets was considerably less per passenger mile than the cost of operating even the best piston-engine planes, flying grew rapidly and became quite common over considerably longer distances. The Boeing Company began planning what came to be known as a jumbojet, the 747. When placed in service in 1970, the 747 was capable of carrying up to about five hundred passengers, but most models were fitted out for about four hundred, with substantial space allocated for baggage, mail, and freight.

Oliver Griffin

Bibliography

Baker, David. *Flight and Flying: A Chronology.* New York: Facts on File, 1994. A very comprehensive reference text on aviation history.

Heppenheimer, T. A. *Turbulent Skies: The History of Commercial Aviation.* New York: John Wiley & Sons, 1998. A comprehensive history of commercial aviation from the biplane era to the end of the twentieth century.

Jane's All the World's Aircraft, 2001-2002. New York: Franklin Watts, 2001. An exhaustive almanac of current civil and military aviation.

Van der Linden, F. Robert. *Airlines and Air Mail: The Post Office and the Birth of Commercial Aviation.* Lexington: University Press of Kentucky, 2002. The curator of air transportation at the Smithsonian Institution's National Air and Space Museum tells the story of the Post Office's influence in the development of American commercial aviation.

See also: Airline industry, U.S.; Airmail delivery; Airplanes; American Airlines; Boeing; Cargo aircraft; DC plane family; Jet engines; Jumbojets; KLM; Lockheed Martin; Manufacturers; Pan Am World Airways; Qantas; 707 plane family; Swissair; Transatlantic flight; Transcontinental flight; Turbojets and turbofans; United Air Lines

Communication

Definition: The practice of exchanging safety and operating information between aircraft in flight and ground stations.

Significance: Communication enables aviation to serve society more completely by expanding the conditions and geographical areas of its operations.

Because they were few, underpowered, and only slightly engaged in commerce, airplanes before 1914 needed no communications between themselves or with ground-based stations. As World War I progressed, airships and specially equipped airplanes carried Morse code radio equipment for military purposes. It was not until the 1930's, however, that civil aviation communications radio became a truly useful appliance. Fledgling airlines in the United States began to install radios aboard their airplanes and at their dispatch hubs to monitor each airliner's progress. This practice brought about the earliest, most rudimentary form of what has become the air traffic control (ATC) system. Early pilots considered radios an unwelcome intrusion in the cockpit, and some pilots refused to use them. Despite these protests, aviation communications provided undeniable benefits to safe and efficient operation, so the system expanded. Following World War II, aviation radios had become widespread in all but the smallest airplanes, as airspace around major cities became congested. By the 1960's, radios were familiar even in small airplanes. By the 1970's, air travel had become sufficiently pervasive that medium-sized and smaller cities attracted enough air traffic to make communications important to safety. The number of control towers rose accordingly, and radio communication frequencies soon became congested. Few pilots could realistically consider their airplanes as operating apart from the air traffic system, but standardization of communications procedures and phraseology lagged behind hardware technology.

International Standardization

Standard phraseology is essential for several reasons. Flying is increasingly an international venture, for even those pilots who never venture far from their home airports encounter fliers from other lands. At the end of World War II, industry leaders of various nations recognized aviation's international tendency and formed the International Civil Aviation Organization (ICAO). The ICAO established English as the standard aviation language; international aviation communication was and is to be conducted in English. Pilots from non-English-speaking countries must be able to read, write, and speak English sufficiently to use the aviation system, but at the beginning of the twenty-first century, reliably judging that ability in every corner of the industry was still uncertain. The twentieth century's worst aircraft accident, the Tenerife, Canary Islands, collision of two loaded Boeing 747's, hinged solely on unclear communications. Responding to these deficiencies, the ICAO's Proficiency Requirements In Common English Study Group (PRICESG/2) completed its second meeting and final report in May, 2001. The ICAO's goal is to implement an English language proficiency standard for aviation in the twenty-first century. That standard is to address pronunciation, stress and intonation, grammar and syntax, vocabulary, fluency, comprehension, and interaction. The group suggested a list of items to be included in ICAO guidance material. These included the full ICAO scale with a glossary of terminology, elaboration of each level, and examples; an English language competencies chart specifying language performance objectives appropriate to the air traffic controller and pilot work domain; an introduction to English language acquisition and learning theories and methodologies; a manual describing the characteristics and attributes of sound English language training programs; a discussion of the importance of "extended" English, relevant to a controller and pilot's ability to handle unusual aviation circumstances and emergencies; and approaches to testing English language speaking and listening proficiency.

Aviator's Alphabet

At the beginning of the twenty-first century, aviation was largely dependent on radio communications for both safety and efficiency. Air traffic control has developed from what was basically a trial-and-error experiment in the 1930's to an essential segment of the aviation industry. It works best when all participants understand the system and use it properly. Understanding is the most important commodity in pilot-controller communications. To establish a solid basis for understanding, in the early 1970's the Federal Aviation Administration (FAA) of the United

States established a pilot/controller glossary. In that glossary, words and phrases to be used in flight have specific meanings.

Aviation communication relies on these standardized meanings. The FAA calls this "phraseology," and sets forth these words, phrases, and their meanings in the Aeronautical Information Manual (AIM). The AIM divides its treatment of communications into a user-friendly general discussion, placing the pilot/controller glossary handily at the end of the book. The FAA also had to deal with the issue of letters and numbers spoken over aviation radios. Each nation registers its airplanes using letters and numbers or letters alone; these tail numbers establish an airplane's identity in radio communication.

To facilitate this, one segment of the AIM displays a phonetic alphabet wherein individual letters are pronounced as specific and familiar words. The AIM treats numbers just as thoughtfully, rendering easily confused numbers with distinct sounds. For example, in conversational use, the numbers "five" and "nine" can be impossible to distinguish in noisy environments or when accents blur them. Aviation pronounces "five" as "fife" and "nine" as "niner." Number sets such as "fifteen" and "fifty" are easily misheard even in the quiet of casual office conversation. Aviation addresses this by instructing pilots to, in most cases, speak each number separately. "Fifteen," therefore, becomes "wun fife" and a correctly speaking pilot or controller says "fifty" as "fife zero." On the other hand, the AIM instructs pilots and controllers to speak airliner call signs and airways in the more conversational format. Airway V12 would be spoken "viktah twelve." Airliner 523 (the assigned flight number, not the tail number) would be spoken "Airliner fifetwenty-tree."

Aviators accepted the phonetic alphabet as they did the radio: Some loved it, some ridiculed it. As aviation brought regions, states, and nations into ever closer contact, the existing hodgepodge of dialects and accents justified the FAA's wisdom in detailing even phonetic pronunciation. This practice bolsters understanding between pilots and controllers, making the aviation system far safer than it was before standardization had become a goal.

Pilot/Controller Glossary

Even pilots native to English-speaking countries may have widely diverging accents, and syntax differs from region to region in many countries. In the United States, after 1972 the FAA established a pilot/controller glossary in the AIM that put forth words and phrases that were largely compatible with those of the ICAO. These words had developed by trial and error since the 1930's, and the FAA found them both efficient and effective. Common words include "Affirmative" to answer a question "yes," while "negative" answers such a question with "no." Flight students soon learn that on the radio, monosyllabic words such as "yes" or "no" might not transmit over the radio. Within the United States alone, different regions say "yes" in fashions confusing to the inhabitants of other localities. A commonly misused aviation word, "Roger," means simply that the hearer has received all of the last transmission. It does not indicate compliance with an instruction, nor understanding of information. When pilots or controllers do not understand a transmission, they should ask the sender to "Say again." Because radio communications frequencies are usually very busy, the ATC system has words that en-

AIM Phonetic Alphabet

Letter	*Word*	*Pronunciation*
A	Alpha	al-fah
B	Bravo	brah-voh
C	Charlie	char-lee *or* shar-lee
D	Delta	dell-tah
E	Echo	eck-oh
F	Foxtrot	foks-trot
G	Golf	golf
H	Hotel	hoh-tel
I	India	in-dee-ah
J	Juliet	jew-lee-ett
K	Kilo	key-loh
L	Lima	lee-mah
M	Mike	mike
N	November	no-vem-ber
O	Oscar	oss-cah
P	Papa	pah-pah
Q	Quebec	key-beck
R	Romeo	row-me-oh
S	Sierra	see-air-rah
T	Tango	tang-go
U	Uniform	you-nee-form *or* oo-nee-form
V	Victor	vik-tah
W	Whiskey	wiss-key
X	X ray	ecks-ray
Y	Yankee	yang-key
Z	Zulu	zoo-loo

capsulate entire sentences into a single word, easily understood by anyone without regard to their first language, accent, or any impediment. One example would be "Wilco," which the AIM defines as meaning, "I have received all of your last transmission, I understand it, and I will comply with it."

Spoken altitudes, radio frequencies, and headings have traits that mesh with the basic rule of pronouncing numbers. Pilots in the United States speak altitudes as thousands and hundreds of feet. In aviation English, the phrase "Two thousand, five hundred" spoken alone only refers to altitude; any other subject would follow the numbers, such as "two thousand, five hundred RPM" if discussing engine or propeller speed, or "two thousand, five hundred miles" when discussing range. The AIM also admonishes U.S. pilots to address radio frequencies by speaking the numbers individually, and to use the word "point" to define tenths and hundreds of a frequency allocation. Internationally, non-U.S. pilots use the three-syllable word "decimal" instead of the single-syllable "point," which the Americans find clearer and more succinct. A common ground control frequency is spoken as "wun too wun point seven" (121.7). Controllers and pilots use good procedure when they speak aircraft headings (the direction in which the aircraft travels in a straight line) by enunciating each number separately. To head east, therefore, is spoken as "zero niner zero." This system, properly used, allows the person familiar with it the ability to understand a message because the more it uses specific, meaning-rich words or phrases, the less aviation is encumbered by ambiguous, nonstandard ones. The result is increased safety (saving lives and property) and efficiency (saving money and resources). For pilots and controllers, the pride of professionalism should be a third benefit.

Benefits of Standardized Communication

Not all pilots agree with the principle of standard phraseology. To teach standard phraseology takes time, and its benefits are not readily apparent with each use. Articles in aviation magazines occasionally have derided established phraseology, some authors belittling aviators who used it or instructors who taught it. Many of these too quickly embraced the AIM's allowance that, should a pilot's understanding of phraseology fail, he might simply speak conversational English. Others retorted that every pilot's public duty is to learn the system and be a fully functioning part of that system, which includes established communications standards.

Within the aviation community, as in most others, effective communication remains elusive. Yet while other industries tend to have codes or jargon for internal use, the decades have forged aviation's communications system into an English-based specialty language. As such, aviation-speak is inefficient for face-to-face conversation but very succinct for time-critical communications in a fluid environment. That fact and its implications are only just beginning to make inroads into the flight training environment. Flight schools still concentrate on teaching aerodynamics, airplane systems, maneuvers, regulations, weather, or myriad other subjects that at the time seem far more immediate than communications. Overall, the aviation industry continues to awaken to communications as a serious public safety issue.

David R. Wilkerson

Bibliography

Federal Aviation Administration. *Aeronautical Information Manual*. Washington, D.C.: U.S. Government Printing Office, 2001. A continuously updated handbook of operating procedures and technical information for pilots and controllers.

Federal Aviation Administration. *ATC Communications Phraseology Guide*. Oklahoma City: FAA Academy Air Traffic Division, 1995. A controller training text supplementing the Air Traffic Handbook 7110.65 for air traffic controller training. This manual subdivides phraseology into subsets of airport operations, en route phase, radar, nonradar, and other specific situations for controllers in training.

International Civil Aviation Organization. *Proficiency Requirements in Common English Study Group Final Report*. Montreal, Canada: Author, 2001. The second meeting of the PRICESG/2, Luxembourg, May 15-18, 2001, discussing improvement of international aviation communications.

Cushing, Steven. *Fatal Words, Communication Clashes, and Aircraft Crashes*. Chicago: University of Chicago Press, 1994. A detailed examination of the communication process and its role in aviation accidents.

Gardner, Bob. *Say Again, Please, Guide to Radio Communications*. Newcastle, Wash.: Aviation Supplies and Academics, 1996. A non-FAA publication providing communication guidance to pilots.

Kern, Tony. *Darker Shades of Blue: The Rogue Pilot*. New York: McGraw-Hill, 1999. A study of the motivations and attitudes of pilots who, for whatever reasons, stray from standards.

See also: Air traffic control; Airline industry, U.S.; Federal Aviation Administration; Pilots and copilots; Safety issues

Concorde

Also known as: Supersonic transport (SST)

Date: First flight in 1969; placed in service in 1976

Definition: Name assigned to an Anglo-French fleet of supersonic passenger transport airplanes.

Significance: The Concorde was the first supersonic aircraft used for regularly scheduled passenger service, built jointly by British and French aircraft manufacturers and later operated by two carriers, British Airways and Air France. The Concorde, which crosses the Atlantic in a scheduled time of three hours and fifty minutes, reduces both flight times and the effects of jet lag.

Supersonic Flight

The laws of physics are absolute and mysterious, as aviators in the 1940's discovered when their planes approached the speed of sound: about 760 miles per hour at sea level and about 660 miles per hour at 50,000 to 60,000 feet. As pilots accelerated toward these speeds, they found their planes shaking violently and running up against some sort of invisible wall, later referred to as the sound barrier.

When a vehicle achieves a speed exceeding the speed of sound, it is said to be traveling at Mach 1. At twice the speed of sound, it enters Mach 2. Mach numbers refer to the ratio of an aircraft's speed to the speed of sound at the altitude of the vehicle. Speeds from Mach 1 to Mach 5 are designated supersonic; speeds above Mach 5 are hypersonic. When a plane travels at exactly the speed of sound, its speed is described as transonic. Speeds below the speed of sound are considered subsonic.

During World War II (1939-1945), before U.S. Air Force test pilot Chuck Yeager first achieved supersonic speeds in the Bell X-1 rocket plane in 1947, numerous pilots unwittingly achieved such speeds during dives. Under such conditions, they could not control their vehicles, because shock waves built up around the controls, locking them in place and rendering them useless. Some pilots ejected under such circumstances; others died when their planes plowed into the earth at supersonic speeds.

Following World War II, U.S. Air Force designers sought to develop supersonic aircraft for the military. Engineers had to cope with the effects of the shock waves that occur as the sound barrier is being breached. They also needed to devise ways for aircraft to endure the extremely high temperatures generated by friction on the craft's outer surface, as speeds of Mach 1 and higher are achieved. Such heat-resistant metals as titanium were employed to replace the aluminum that covered the exteriors of most subsonic aircraft.

The work of these engineers and designers had broad implications for the commercial aircraft industry. By the 1970's, both the United States and the Soviet Union had planes, notably the Soviet MiG-25 Foxbat interceptor and the U.S. SR-71 spy plane, that could fly at speeds higher than Mach 3.

Supersonic and hypersonic aircraft create shock waves because of sudden changes in air pressure. Although people on the ground experience sonic booms when supersonic and hypersonic craft fly overhead, people within them do not, because the vehicles fly faster than the sound their planes create and remain well ahead of it. Because sonic booms are destructive and annoying, often shattering both windows and the nerves of people on Earth, most supersonic flights are routed over oceans. When supersonic commercial aircraft fly over land, they usually fly at subsonic speeds.

Commercial SSTs

The commercial aviation industry passed through several stages before the 1960's. Single-engine planes from the first decade of flight gave way to more powerful and safer dual-engine planes. As airmail routes expanded, the size of aircraft also expanded to assure larger payloads and accommodate passengers. As early as 1914, regular passenger service was available between St. Petersburg and Tampa, Florida, a distance of about 25 miles.

World War II brought about considerable advances in aviation, including the development of jet planes, which, after the war, gradually became used as commercial passenger vehicles. These planes flew faster than propeller planes, often cruising close to 600 miles per hour but still not approaching the sound barrier. The next major development in commercial passenger service was the supersonic transport plane that reduced the transatlantic crossing time from eight or nine hours to three and one-half hours or fewer.

Four major world powers, the United States, the Soviet Union, France, and Great Britain, began to consider developing supersonic commercial air transport. It was presumed that SSTs would carry passenger loads comparable to those carried by existing jet planes, and that SSTs would offer the same two or three classes of service (coach, business, and first classes) typically available on most long-distance subsonic planes, thereby making supersonic air travel economically feasible.

While the actual SST prototypes were being developed, it became evident that they could not comfortably carry

The Concorde 2 takes off on its maiden flight from Bristol on April 9, 1969. (Hulton Archive)

more than about one hundred passengers, although some configurations would permit a maximum capacity of 144. At fares averaging 20 percent more than those of full-fare first class, approximately eleven thousand dollars for a round trip, supersonic air travel attracted an elite class of transatlantic passengers. These fares, however, did not begin to cover the high cost of flying supersonic aircraft. The Concorde, which accommodates about one hundred passengers, requires three times more maintenance than does a 747, which accommodates about four hundred passengers. The Concorde also burns 50 percent more fuel.

During a quarter-century of supersonic air service, Concordes incurred huge deficits for Air France and British Overseas Airways Corporation (BOAC). Keeping the Concordes aloft, however, became a matter of national pride for the French. Britain's contractual agreements with France prevented its withdrawal from participation, although there was a public outcry from the British public and members of Parliament to do so.

The Anglo-French Alliance

In the 1950's, once it had become clear that supersonic transport was the next logical step in the development of passenger air transportation, the four major powers began to look into developing SSTs. The British hoped to join with the United States in developing such planes, but the Americans were cool to entering into such a partnership. Finally, in November of 1962, the British and French, realizing it made economic sense to merge forces in the development of this project, agreed to proceed with SST research and development.

These nations had their own reasons for wanting to proceed with the Concorde, which cruised at 60,000 feet with an average speed of 1,320 miles per hour over water. For the British, the project would keep design teams employed when the economy was lagging. It might also enhance British attempts to join the European Community. For the French, the project would result in enhancing the image of France's national aircraft industry.

The estimated cost of the Concorde project was between \$420 and \$480 million. By the time the first Concordes were aloft, however, the cost had reached more than ten times the earlier estimates. Nevertheless, the prototypes, 001 and 002, were ready in September, 1968.

The initial flight of 001 from Toulouse, France, occurred in March, 1969. The following month, 002 flew in

Concorde Problems Between 1979 and 2000

Between its first commercial flight in 1976 and the crash of Air France Flight 4590 in July, 2000, Concordes carried more than four million passengers over many millions of miles. Its safety record is impressive. Concordes log fewer than one thousand hours per year, whereas Boeing 747's log many times that number. Heavy overhauls on Concordes occur with three times the regularity of overhauls on other passenger planes.

Concordes registered the following major problems between 1979 and 2000:

1979-1981: Tires blow out four times on Concordes during takeoffs in this period.

May, 1985: A London-to-New York flight encounters engine problems and makes an emergency landing in Boston.

March, 1992: On a London-to-New York flight, a Concorde loses a section of its rudder but lands safely at John F. Kennedy International Airport.

February, 1997: On a London-to-New York flight, a Concorde develops engine trouble and lands at Halifax, Nova Scotia.

January, 2000: Two Concordes are forced to make emergency landings within twenty-four hours of each other, one for engine failure on landing, the other for the sounding of a fire alarm in the cockpit.

July 25, 2000: Air France Flight 4590 crashes just after takeoff from Charles de Gaulle Airport outside Paris, killing 113. All Concordes are consequently grounded pending investigation.

Bristol, England. Both prototypes were displayed at the Paris Air Show in June, 1969. By April, 1970, after various design changes, the production of sixteen Concordes was confirmed. BOAC ordered five. Air France ordered four.

Both Trans World Airlines (TWA) and Pan American had taken options to buy Concordes. However, as environmentalists began to rally against permitting these noisy planes to fly into the United States, both TWA and Pan American dropped their options. When the first commercial Concordes were launched in January, 1976, the British Concorde flew the London-to-Bahrain route, and the French Concorde flew from Paris to Rio via Dakar.

In February, 1976, the Concordes won their battle to fly into both New York's John F. Kennedy International Airport and Washington's Dulles International Airport. Regular New York-to-London and Washington-to-Paris service began, continuing until July, 2000, when, after Concorde's long accident-free history, Air France Flight 4590 to New York crashed after takeoff from Charles de Gaulle Airport outside Paris. The death toll was 113. All Concordes were grounded pending a thorough investigation and modification of the remaining aircraft. Air France resumed Concorde flights on November 7, 2001, and British Airways followed suit two days later.

American and Soviet SSTs

In December, 1966, the United States commissioned Boeing to build a swing-wing aircraft with General Electric engines capable of carrying three hundred passengers at a cruising speed of about 1,800 miles per hour, or Mach 2.7. In early 1971, this project, for which the U.S. Congress had appropriated about $425 million between September, 1966, and October, 1967, was well under way, with both presidential and congressional support. In March, 1971, however, the House of Representatives voted to discontinue all SST funding. The Senate lacked sufficient votes to pass an amendment to restore this funding.

The Soviet Union was far ahead of the other three national powers that were directly involved in the development of passenger SSTs. The Soviet Tupolev-144 (Tu-144) was shown at the Paris Air Show in June, 1965. The first prototype flew in December, 1968, exceeding Mach 1 for the first time the following June. In May, 1970, shown off at Moscow's Sheremetyevo Airport, it exceeded Mach 2 for the first time.

A production model of the Tu-144 crashed at the Paris Air Show in June, 1971, killing all on board. The crash was attributed not to mechanical problems but to other factors. The Soviets, however, did not give up. By December, 1975, scheduled freight and mail flights were instituted between Moscow and Kazakhstan. Regular Tu-144 passenger service began between Moscow and Kazakhstan in November, 1977, lasting until June of the following year, while modifications were made in the aircraft. The route reopened in June, 1979, and continued to function until August, 1984, when Aeroflot discontinued such service.

The present fleet of Concordes is expected to operate until 2007. Meanwhile, research is afoot to produce an entirely new generation of SSTs, most of them capable of carrying relatively large numbers of passengers at speeds approaching Mach 3.

R. Baird Shuman

Bibliography

Feldman, Elliot J. *Concorde and Dissent: Explaining High Technology Project Failures in Britain and France.*

New York: Cambridge University Press, 1985. A striking analysis of the tensions present in the Anglo-French Concorde partnership.

Knight, Geoffrey. *Concorde: The Inside Story.* London: Weidenfeld & Nicolson, 1976. An intriguing, behind-the-scenes account of the Concorde's development by the French and the British.

Moon, Howard. *Soviet SST: The Technopolitics of the Tupolev-144.* New York: Orion, 1989. The best book in print on the development of Soviet SSTs.

Owen, Kenneth. *Concorde and the Americans: International Politics of the Supersonic Transport.* Washington, D.C.: Smithsonian Institution Press, 1997. A riveting account of the politics involved in the development of the SST. Thorough and easily accessible for general readers.

Sobieczky, H., ed. *New Design Concepts for High Speed Air Transport.* New York: Springer, 1997. Chapters 1, 6, and 16 through 19 are especially relevant for those interested in SSTs. Clearly written, intelligently conceived.

See also: Accident investigation; Aeroflot; Air France; British Airways; Mach number; Sound barrier; Supersonic aircraft; Transatlantic flight; Andrei Nikolayevich Tupolev; X planes; Chuck Yeager

Continental Airlines

Definition: A large U.S. air carrier of passengers and cargo based in Houston, Texas.

Significance: Continental is the fifth largest airline in the United States, despite two bankruptcies.

From very humble beginnings, Continental has risen to become the fifth largest air carrier in the United States. Born in scandal, Continental survived the deregulation of the airline industry in 1978 only to face two bankruptcies that would leave the company's employees demoralized and its customers angry over poor service. In one of the industry's most successful turnarounds, a new management team came onboard in October, 1994, and returned the carrier to profitability and award-winning service.

Early Days

Walter T. Varney learned to fly in the U.S. Army during World War I. He learned about the airline industry flying mail under contract to the U.S. Postal Service. A scandal involving the postmaster general, Walter Folger Brown, resulted in President Franklin D. Roosevelt canceling all airmail contracts on February 9, 1934. The new postmaster general, James Farley, called for new bids on April 20, 1934. Varney's recently formed airline, the Southwestern Division of Varney Speed Lines, bid on one of these routes. The Pueblo, Colorado, to El Paso, Texas, route was not as long or as profitable as the routes awarded to the large U.S. carriers such as United Airlines, Trans World Airlines, American Airlines, and Eastern Air Lines, but Varney and his financial backer, Louis H. Mueller, planned to use the mail revenues to support their first air route from Denver to Pueblo, stopping in Las Vegas, Nevada, Sante Fe and Albuquerque, New Mexico, and finally, El Paso. Unfortunately, Varney and Mueller found that the sparsely populated areas of western Texas did not generate a high level of passenger traffic.

The Robert Six Years

On July 5, 1936, Robert F. Six paid Mueller $90,000 for a 40 percent share of the company now called Varney Air Transport. On Six's initiative, Varney Air Transport purchased the majority of the Wyoming Air Service network, moved the company's headquarters to Denver, and changed its name to Continental Airlines. Six was elected president of the airline on February 3, 1938. He saw the passage of the Civil Aeronautics Act (1938), which created the Civil Aeronautics Authority (CAA) and gave it authority to issue permanent route certificates, as an excellent opportunity for Continental to expand. In order to do this, the company needed aircraft. To raise the money, Six arranged for Continental's first stock offering in late 1938.

The United States' entrance into World War II in 1941 interrupted Six's plans for Continental. Six himself joined the Army Air Transport Command, where he devoted most of his time to administration and logistics planning. Meanwhile, Terry Drinkwater, who had joined Continental in 1938 as a legal expert, assumed the job of temporary president of the airline. Following the end of the war, Six returned to Continental determined to expand the airline. Unfortunately, the only major expansion during the early postwar years was to Houston, Texas. On July 15, 1949, Continental became one of the first carriers in the United States to offer a promotional low fare to expand its passenger traffic. This Skycoach service and a reputation for technical excellence earned Continental the respect of its fellow air carriers, but respect did not translate into the expansion that Six envisioned.

All of this changed with the adoption by the Civil Aeronautics Board, which had replaced the Civil Aeronautics

Authority, of a new concept, interchange service. In the spring of 1951, Continental signed an agreement with American Airlines and Braniff. American agreed to provide service in California to San Francisco, Los Angeles, and San Diego, as well as to Phoenix, Arizona. Continental provided service in Texas from El Paso to San Antonio. Braniff initially flew the San Antonio-to-Houston segment. When Braniff withdrew from the agreement, Continental assumed this route as well. On February 1, 1952, Continental signed a second interchange agreement with Mid-Continental Air Lines to serve St. Louis. A third interchange agreement was worked out with United Air Lines in September, 1953. Under this agreement, Continental would fly between Denver and Tulsa, Oklahoma, while United would continue the route on to Seattle, Washington, and Portland, Oregon. During the period of the interchange service agreements, only Delta became involved in more agreements than Continental.

On July 22, 1958, Continental was authorized to operate service from Dallas-Fort Worth to El Paso, Lubbock, Midland-Odessa, Amarillo, Abilene, Albuquerque, and Santa Fe. This award, combined with the 1953 acquisition of Pioneer, gave Continental a full range of service in the state of Texas. Following Continental's acquisition of its first true jet, the Boeing 707, Six was now ready for further expansion. Continental became the first U.S. airline to introduce economy class fares in December, 1961. They also introduced a system of progressive maintenance in which aircraft maintenance was broken down into self-contained work periods and then spread out at regularly scheduled times over the operating life of the aircraft. This system was later adopted by the airline industry as a whole.

During the 1960's, Continental experienced three more firsts. On August 3, 1961, a Continental flight from Los Angeles to Houston became the first U.S. jetliner to be hijacked. On May 22, 1962, Continental recorded its first fatal accident in twenty-five years when what was presumed to be a bomb exploded in the lavatory of the same 707 that had been hijacked the previous year. A happier event occurred in May, 1964, when Continental entered the international market with a contract from the Military Airlift Command to fly troops to the western Pacific. Continental created a wholly owned subsidiary, Continental Air Services, in September, 1975, which served Vietnam, Laos, and Thailand. The service closed in December, 1975, when the Communists took over the entire region. In another Pacific venture, Continental invested in the newly formed Air Micronesia, holding 31 percent of its shares. Air Micronesia was to provide local service to the islands of the central Pacific.

By the late 1970's, Continental had grown to become the tenth largest domestic carrier in the United States. Unfortunately, its size did not protect it from the results of airline deregulation following the passage of the Airline Deregulation Act in 1978. The Airline Deregulation Act freed carriers to enter new routes and charge fares based on "market considerations." In other words, airlines were free to select routes based on their judgment of its potential profitability and to charge competitive fares that would attract necessary customers. The result was an expansion of low-cost, no-frills service that seriously weakened the higher cost, prederegulation carriers such as Continental.

Frank Lorenzo Takes Over

In 1972, Frank Lorenzo purchased a debt-ridden regional carrier named Texas International. With deregulation, Texas International, which had introduced its Peanut Fares in 1977 as a way of appealing to cost-conscious leisure passengers, was set for new growth and prosperity. After the creation of Texas Air Corporation in 1980 as a holding company for Texas International, Lorenzo formed New York Air to compete in the northeast markets of New York and Washington. Lorenzo began his takeover bid for Continental in 1981. The takeover was bitterly opposed by the management and unions of Continental, who finally conceded on November 25, 1981, when Texas Air assumed control of 50.8 percent of the stock. The two airlines agreed to merge in July, 1982, and the headquarters was moved to Houston, Texas. Amid mounting losses, the machinists union went on strike on August 13, 1983. After failing to reach agreement with the striking union, Lorenzo and Continental declared bankruptcy on September 24, 1983, and ceased all domestic operations. The company furloughed two-thirds of its workforce and resumed operation three days later. The pilots and flight attendants joined the striking machinists on October 1, 1983; however, Continental continued flying a reduced schedule. The old labor contracts were declared void and employees were given a "take it or leave it" option to work at reduced salaries and forfeit all seniority rights. These cost savings helped Continental to emerge from bankruptcy on September 2, 1986.

Intent on expanding, Texas Air took over Eastern Air Lines and People Express in 1986. People Express and New York Air were merged into Continental on February 1, 1987. Eastern Air Lines continued to fly under its own name, filing for bankruptcy in March, 1989, and ceasing operations on January 18, 1991. Stressed by these acquisitions and plagued with service problems, Continental was again in financial difficulty. In the summer of that year, Frank Lorenzo sold all of his assets in the company. Continental filed for bankruptcy a second time in December, 1990.

The Road to Recovery

Continental again reduced service and began to sell assets to raise cash, including its Seattle to Tokyo and Australian routes. However, it was not until Air Canada and a group of private investors headed by David Bonderman, of Air Partners, agreed to invest $450 million in the company that Continental was able to emerge from bankruptcy in April, 1993. Unfortunately, by the fall of 1994, Continental was once again faced with a serious shortage of cash. Earlier that year, Gordon Bethune had left the Boeing Aircraft Company to become the president and chief operating officer of Continental. He assumed the duties of chief executive officer in October, 1994. Under his leadership, Continental began what was called the "Go Forward Plan." The plan called for Continental to review its flight schedule and eliminate money-losing routes, restructure its balance sheet, improve its service performance, and restore employee morale. The plan was a tremendous success. By July, 1995, Continental had posted the largest quarterly profit in its history. In January, 1997, it was named the Airline of the Year by the industry journal *Air Transport World*.

In 1998, Continental announced the beginning of an alliance with Northwest Airlines. The agreement called for code sharing on all Continental destinations from Cleveland, Newark, Houston, and Los Angeles and on all Northwest destinations from Detroit, Memphis, Minneapolis, and Tokyo. A code-sharing agreement occurs when two airlines offer a flight under the flight designation of a single carrier. Each carrier agrees to fly one segment of the larger route. For example, on a route from Houston to Tokyo, Continental would fly the segment from Houston to Minneapolis while Northwest would fly the segment from Minneapolis to Tokyo. In effect, code sharing allows an air carrier to offer flights to destinations that it does not serve directly by placing its passengers on the aircraft of another airline. They also agreed to consolidate ramp, cargo, and ticketing activity in selected cities. In a more controversial move, Northwest also purchased the stock interest of David Bonderman, giving the airline a 14 percent share of Continental stock. Because the deal would have given Northwest a 51 percent share of Continental's voting stock, a trust was established for this stock and Northwest agreed to retain only a veto over any potential merger of Continental with another carrier. It was announced in 1999 that Continental would join the Wings Alliance, whose major partners were Northwest and the Dutch carrier KLM.

Although Northwest agreed to sell its controlling interest back to Continental in an effort to settle a lawsuit filed by the U.S. government, the marketing aspects of their alliance continue and have helped Continental to expand to serve over 220 destinations worldwide. Continental also continues to rank as one of the best airlines in the United States in terms of on-time performance, baggage handling, and customer satisfaction, while posting some of the best financial performances in its

Events in Continental Airlines History

1934: Continental Airlines's predecessor, Varney Speed Lines, makes its first flight, from Pueblo, Colorado, to El Paso, Texas.

1937: The name of Varney Speed Lines is changed to Continental Airlines, and the new airline's headquarters are moved from El Paso to Denver, Colorado.

1941-1945: Continental modifies B-17 and B-29 bombers for the United States during World War II.

1953: Continental merges with Pioneer Airlines, expanding its services further into Texas and New Mexico.

1959: The airline makes its first jet flights, with a small fleet of Boeing 707's.

1963: The airline moves its headquarters to Los Angeles, California, and transports U.S. troops to Asia during the Vietnam War.

1969: Continental inaugurates service to Hawaii.

1970's: The airline undergoes a period of extensive expansion, including service to Australia and New Zealand.

1982: After deregulation of the airline industry, Continental experiences a period of turbulence, merging with Texas International.

1983: The airline reorganizes under Chapter 11 of the Federal Bankruptcy Code.

1986: Continental reemerges from Chapter 11 and, a year later, becomes the third-largest U.S. airline, with the consolidation of Frontier, New York Air, and People Express.

1987: The airline introduces its OnePass frequent flier program.

1993: Continental purchases ninety-two new Boeing 737, 757, 767, and 777 aircraft.

1999: Continental's jet fleet, with an average age of about seven years, becomes the youngest of those of the ten largest U.S. airlines.

history. Despite financial crises and bankruptcy, Continental has become the transcontinental and international carrier that Robert Six dreamed of creating.

Dawna L. Rhoades

Bibliography

Bethune, G. *From Worst to First: Behind the Scenes of Continental's Remarkable Comeback.* New York: John Wiley & Sons, 1999. A very readable book about the efforts of Continental's new management team to turn around a struggling, debt-ridden airline.

Davies, R. E. G. *Continental Airlines: The First Fifty Years, 1934-1984.* The Woodlands, Tex.: Pioneer Publications, 1984. An interesting look at the early history of Continental. The book includes early photos as well as route maps of Continental's expansion.

Jones, G. *The Big Six Airlines.* Osceola, Wis.: Motorbooks International, 2001. An excellent pictorial history of the six largest U.S. carriers.

Serling, R. J. *Maverick: The Story of Robert Six and Continental Airlines.* Garden City, N.J.: Doubleday, 1974. An excellent account of the Six years, a period that spanned forty-six years of Continental history.

See also: Air carriers; Airline Deregulation Act; Airmail delivery; Boeing; KLM; Mergers; Northwest Airlines

Corporate and private jets

Definition: Jets owned wholly or in part by corporations or private individuals.

Significance: Corporate and private jets provide corporations and individuals greater freedom and luxury in planning their air flights and have other significant business advantages over commercial flight.

In the general scheme of air travel, corporate and private jets are considered general aviation. Having a private jet is a status symbol of some magnitude. In Africa, for example, Swaziland's King Mswati III became a member of the jet-owning club in 1999. In southern Africa alone, the presidents of Namibia, Zimbabwe, and Botswana each have their own private jets. In common with other world leaders and corporate executives, these leaders note that having one's own jet saves time and money.

Corporate and private jets have become more comfortable and safe over time and come in various degrees of luxury and comfort. Some companies tout that they have their own private jets. Other corporations and individuals buy shares in private jets, depending on how much time they wish to use the jet.

Jet Manufacture and Ownership

The major manufacturers of corporate jets are Cessna, Piper, and Beech, each striving to provide businesses with speed, comfort, and safety. In overall safety, corporate jets, if not all private jets, have an equivalent rank with the airlines.

A new trend in corporate jet ownership is fractional ownership of business jets, a time-sharing application. Richard Santulli of Executive Jet is one of the leaders in this trend. Shared ownership provides the comfort and convenience of owning a plane with the economy of timeshare. An eight-passenger Raytheon Hawker, for example, sells for $12.4 million plus the cost for personnel and servicing. To make its purchase economical, a business would have to fly four hundred or more hours.

Time-sharing allows eight customers to buy a single plane and for each to use it for one hundred hours of flying. Executive Jet guarantees a plane to be ready with six hours of notice. The company has about six hundred jets, giving it 40 percent of the world's business jet market. FlexJet and TravelAir offer similar plans.

There is also a flourishing market in used planes. The maintenance of private planes tends to be well done, because of both federal standards and the general culture of business regarding its planes.

Whether conventional business jet or time-sharing jet, planes are being designed to operate in many different terrains while still offering passenger comfort. Thus, the Cessna Citation Excel has more headroom and other amenities than did earlier jets, but it is also able to land and take off on sod, dirt, and other difficult runways. It also has greater range and speed than earlier business jets.

Manufacturers are engaged in a continual effort to improve corporate and private jets. For example, Cessna Aircraft, based in Wichita, Kansas, which has sold more than three thousand Citations in more than thirty years, will upgrade to a larger Model 680 Sovereign to increase its market share of the super-midsize business market. There are other efforts to expand the private jet market. Eclipse hopes to make the world of the private jet affordable for business and first class customers through building a plane with a smaller, more efficient engine. It has had a number of design breakthroughs, including a digital avionics system. In most private planes, avionics are largely analog. This system requires that each gauge on the instrument panel has its own box of electronics and wiring, adding a great deal of weight. The Eclipse system will combine all the display instruments in one system. Eclipse is also counting on automation to cut its costs. Laser welding is one of its options, as is the use of robotic painters.

Safety Concerns and Issues

The 1999 crash that killed golf champion Payne Stewart has, however, led to serious questions regarding the safety of private and corporate jets. The National Transportation Safety Board (NTSB) has determined that the accident was most likely caused by the flight crew's incapacitation following the loss of cabin pressure. A flight data recorder, which could have prevented the accident, was not required on the twenty-five-year-old plane.

Between 1990 and 2000, the number of business jets in the United States virtually doubled. That means that in 2001, there were about seven thousand small jets flying for either charter companies or private businesses. About half of the passengers on private jets are middle management personnel. It is important to note that federal regulations for commercial airlines do not always apply to private and corporate jets. Business jets have had safety records that stand up well in comparison with commercial airlines. However, with the growth of fractional ownership, there is growing concern that this type of corporate-owned plane operates under less stringent rules than do charter flights. Business spokespeople, nevertheless, argue that business has an excellent culture of training and maintenance.

Practical Aspects of Private Jets

There are a number of advantages to corporate airlines or private jets in any form. Passengers of private jets are able to land at the airport of their choice, many of which are not serviced by scheduled commercial airlines. Moreover, these flights do not require connecting flights and can avoid clogged airports. Business planes offer the convenience of flexible scheduling, so that travelers are able to leave on their own schedules and often can return the same day. All this convenience, moreover, can take place out of the public's sight. No time is wasted, since travelers choose virtually all details of the schedule and destination, and can change flight plans as needed. These private corporation planes tend to be under rigid safety inspections.

Privacy is another reason for using these jets. Confidential meetings can be held on the plane, and work can be pursued. These private aircraft also send a message to one's business guests of their importance.

The luxury of private jets extends to their entertainment centers as well as other features. Pacific Systems, for example, has long supplied entertainment and communication systems for top-level corporate jets, planes that carry heads of state, and even personalized 747's for the extremely wealthy. The company is introducing its IntelliJet touch-screen cabin management system. It is an application of Pentium processors and touch-sensitive screens, putting all possible cabin systems that control its environment and entertainment at the touch of the jet's owner. Pacific Systems will give the buyer gold- or titanium-plated switches, a karaoke system, and state-of-the-art music and video systems that can be independently operated by each passenger.

The trade-off for convenience is money. These jets are quite expensive. A small six-seat Cessna or Learjet sells for about $5 million. A Falcon 900, Canadair Challenger 601, or a Gulfstream IV will cost $23 million or more. The new long-range Gulfstream V costs about $35 million, fully fitted, but has a nonstop range of 6,500 nautical miles.

In addition to the initial cost for the plane, operation expenses are heavy. There are pilot salaries, hangar fees, mechanics' fees, fuel, airport charges, mandatory Federal Aviation Administration inspections, and more. For example, a Gulfstream IV costs around $3,000 an hour to operate. A midsize Cessna Citation V costs about $1,500 per hour.

Advantages of Private and Corporate Jets

Athletes are often dependent on private jets for transportation to their various engagements. Not even the death of Payne Stewart, who crashed on a private jet en route to a professional golf tournament, could shake the confidence of players in their use. Golf champion Tiger Woods, for example, said that private jets have become the only sensible form of transportation for celebrity golfers, whether they seek privacy or convenience in their travel. Both Arnold Palmer and Jack Nicklaus not only use private jets but also own them. Moreover, Palmer is a licensed pilot, and Nicklaus hires pilots as part of his own company.

The main benefit of a private or corporate jet is savings in time. Private and corporate jets give those who use them control of their schedules. The private or corporate traveler can avoid the usual delays that plague commercial travelers. For example, Bill Cosby in his Gulfstream IV, Arnold Palmer in his self-piloted Cessna Citation VII, Disney's Michael Eisner, and many other CEOs can save time and aggravation through the use of private and corporate jets. Business executives with sufficient money say that the planes are worth the expense. Jack Nicklaus, for example, states that he could not run his current business, much less his still-flourishing golf career, without the use of his private jet. As a prolific golf-course designer, Nicklaus travels throughout North America and the rest of the world in his Gulfstream II *Air Bea*. Traveling without the constraints of commercial airline schedules allows him to double the amount of work he can accomplish without exhaustion.

Motion picture stars often own private jets. Arnold Schwarzenegger has a G-III. John Travolta has a G-II. In addition to the status, comfort, and convenience of private jets, movie stars own them because of the security they provide. However, the movie-star image is not one that the private and corporate jet companies wish to convey. They prefer to emphasize the convenience and businesslike nature of the private plane. They point out that a great deal of business can be conducted on these planes. These planes transport midlevel executives from place to place, not simply the CEOs. Xerox is but one corporation that flies its executives from small airports such as Westchester County, New York, to Rochester for business. The planes are also ideal for bringing customers to the company. Sales personnel often start their sales pitches on the planes.

Frank A. Salamone

Bibliography

"Mystery Learjet Crash Puts Spotlight on Corporate Jet Aircraft Safety." *Airline Industry Information*, October 28, 1999. This article discusses the issue of safety concerns in private aircraft.

Minard, Lawrence. "The Highfliers." *Forbes*, November 15, 1999, 121-125. A glimpse into the world of corporate and private jet owners.

Phillips, Almarin, and Thomas Phillips. *Biz Jets: Technology and Market Structure in the Corporate Jet Aircraft Industry*. New York: Kluwer, 1994. Primarily an economic analysis of corporate jet industry history.

Schonfeld, Erick. "The Little (Jet) Engine That Could: With a Revolutionary 85-pound Engine and $60 Million in Backing, Vern Raburn Wants to Turn the World of Private Air Travel Upside Down." *Fortune*, July 24, 2000, 132. Describes the development of a smaller, more economical and affordable jet.

Szurovy, Geza. *Cessna Citation Jets*. Osceola, Wis.: Motorbooks International, 2000. A color guide to the history of one of the best-selling private jets.

See also: Airplanes; Beechcraft; Cessna Aircraft Company; Flight data recorders; Piper aircraft; Safety issues

Crewed spaceflight

Date: Beginning April 12, 1961
Definition: Any flight that carries humans into space.
Significance: Crewed spaceflight has allowed humans to explore the solar system and to react to discoveries and problems encountered during space missions.

Development

The flight of humans into space had been the dream of science-fiction writers and explorers for more than a century before the first crewed spaceflight. Explorers regarded the first human step into space as the beginning of a new age of exploration, in which humans eventually explored the Moon, Mars, the asteroids, and all of the solar system.

Humans were preceded into space by robotic spacecraft containing instruments to monitor the earth and space environment and to explore the solar system. These uncrewed spacecraft frequently suffered from a lack of intelligence, or, the ability to adapt to unforseen circumstances. Thus, after a new discovery was made, it was frequently necessary to design a follow-on spacecraft, with new equipment intended to help scientists to understand more fully the measurements from the previous mission. Crewed spaceflight was expected to allow human intelligence and ingenuity to respond and adapt to discoveries or problems encountered during flight.

Neither the drive to explore the unknown nor the need for human intelligence in space provided the motivation for the vast expenditure of government funds needed to send humans into space. The real motivation was the intense political competition between the capitalist and communist political systems during the Cold War. The political leaders of both the United States and the Soviet Union saw success in space exploration as a way to demonstrate to their own citizens and to the rest of the world the superiority of one political system over the other.

The First Humans in Space

Following the launch of the world's first artificial Earth satellite, Sputnik 1, on October 4, 1957, both the United States and the Soviet Union began serious efforts to launch humans into space. The U.S. effort was called Project Mercury, and the Soviet program was called Vostok.

The formal selection process for the Mercury astronauts began in January, 1959, when the National Aeronautics and Space Administration (NASA) chose 110 military

U.S. Crewed Space Missions, 1961-1972

May 5, 1961:	Mercury Redstone 3	December 4, 1965:	Gemini VII	March 3, 1969:	Apollo 9
July 21, 1961:	Mercury Redstone 4	December 15, 1965:	Gemini VI-A	May 18, 1969:	Apollo 10
February 20, 1962:	Mercury Atlas 6	March 16, 1966:	Gemini VIII	July 16, 1969:	Apollo 11
May 24, 1962:	Mercury Atlas 7	June 3, 1966:	Gemini IX-A	November 14, 1969:	Apollo 12
October 3, 1962:	Mercury Atlas 8	July 18, 1966:	Gemini X	April 11, 1970:	Apollo 13
May 15, 1963:	Mercury Atlas 9	September 12, 1966:	Gemini XI	January 31, 1971:	Apollo 14
March 23, 1965:	Gemini 3	November 11, 1966:	Gemini XII	July 26, 1971:	Apollo 15
June 3, 1965:	Gemini IV	October 11, 1968:	Apollo 7	April 16, 1972:	Apollo 16
August 21, 1965:	Gemini V	December 21, 1968:	Apollo 8	December 7, 1972:	Apollo 17

Source: Data taken from (www.nssdc.gsfc.nasa.gov/planetary/chrono_astronaut.html), June 4,2001.

test pilots from 508 candidates submitted by the Department of Defense. The 110 candidates were carefully screened for physical fitness, experience, skill, and size, to accommodate the small size of the Mercury spacecraft. On April 27, 1959, NASA announced the names of the seven astronauts chosen for Project Mercury: Navy lieutenant M. Scott Carpenter; Air Force captain Leroy Cooper; Marine lieutenant colonel John Glenn, Jr.; Air Force captain Virgil "Gus" Grissom; Navy lieutenant commander Walter Schirra; Navy lieutenant commander Alan Shepard; and Air Force captain Donald "Deke" Slayton. The Soviet Union was, at the same time, conducting its own selection process, although the Soviet selection effort received less publicity than did the U.S. effort. On March 14, 1960, the Soviet Union selected a group of twelve cosmonauts: Pavel Belyayev, Valeri Bykovsky, Yuri Gagarin, Viktor Gorbatko, Yevgeny Khrunov, Vladimir Komarov, Alexei Leonov, Andrian Nikolayev, Pavel Popovich, Georgi Shomin, Gherman Titov, and Boris Volynov.

The Soviet Union's second artificial satellite, Sputnik 2, launched on November 3, 1957, demonstrated the technology that was required for humans to fly in space. Sputnik 2 carried an 11-pound dog named Laika into orbit. Monitors on Sputnik 2 demonstrated that the interior temperature could be maintained in a range suitable for human survival and that a suitable atmosphere could be maintained using a system of reactive chemicals that give off oxygen and another chemical system to absorb exhaled carbon dioxide.

During the spring of 1960, engineers in both the United States and the Soviet Union were working to place the first human into space. The timetable for the U.S. program was highly publicized, whereas little information about the Soviet Union's effort was released. A series of uncrewed test vehicles preceded the crewed flights. The first flight of the Vostok capsule, on May 15, 1960, failed after the capsule did not reenter Earth's atmosphere. The second Vostok flight, in August, 1960, carried two dogs who were recovered successfully after reentry. The third flight, in December, 1960, suffered a failure to reenter the atmosphere, and the dog carried on board was killed. In March, 1961, the final Vostok test flight successfully carried a dog into orbit. The United States made successful suborbital flights of the Mercury spacecraft in December 19, 1960, and January 31, 1961. The January, 1961, flight carried a chimpanzee named Ham, the first primate to fly into space, on a 15-minute spaceflight to an altitude of over 100 miles.

At 9:07 A.M. Moscow time on April 12, 1961, the era of human spaceflight began with the launching of Vostok 1, carrying cosmonaut Yuri Gagarin, into orbit from a launch pad at the Baikonur Cosmodrome in Kazakhstan. Gagarin completed one Earth orbit and landed near Smelovka, on the Volga River, 108 minutes after liftoff.

Only three weeks later, the first crewed flight of the U.S. Mercury project was launched. At 9:34 A.M. on May 5, 1961, a Redstone rocket carried astronaut Alan Shepard from the Cape Canaveral Air Force Station. Five minutes after liftoff, Shepard's spacecraft, *Freedom 7*, reached its peak altitude of 107 miles above Earth's surface. Shepard's flight lasted 15 minutes and 22 seconds and carried him 290 miles over the Atlantic Ocean.

On May 25, 1961, fewer than three weeks after Shepard's successful suborbital spaceflight, President John F. Kennedy publicly committed the United States to the goal of landing a man on the Moon and returning him safely to the earth before the end of the 1960's. Recognizing that Shepard's suborbital flight did not match either the orbital flight conducted by Gagarin or the Soviet Union's leadership in space exploration, Kennedy established a more long-term goal for the space race.

On August 6, 1961, the Soviet Union launched cosmonaut Gherman Titov, the backup pilot for Gagarin's flight, on a seventeen-orbit spaceflight lasting 25 hours and 18 minutes. During his flight, Titov suffered severe space sickness in response to weightlessness, demonstrating that human flight into space would present physiological problems.

The United States matched Gagarin's orbital flight on February 20, 1961, when astronaut Glenn was launched into space. Glenn completed three orbits before returning to Earth after 4 hours and 55 minutes. During the flight, the ground controllers received a signal that the heatshield, which proctected Glenn and his *Friendship 7* capsule from burning up during the extreme heat of reentry, had come loose from the spacecraft. Ground controllers instructed Glenn to override the planned separation of the retrorocket package from the spacecraft and to use the straps that held the retrorockets to the spacecraft to hold the heatshield in place. Although the difficulty turned out only to be a faulty sensor, Glenn's intervention could have been vital to the successful completion of the mission if the heatshield had indeed come loose.

The Soviet Union followed its Vostok program with the Soyuz program, using a spacecraft that could carry up to three cosmonauts on spaceflights lasting several days. The nation also developed a series of space stations, called Salyuts, which were visited by cosmonauts carried aloft in the Soyuz spacecraft.

The U.S. Mercury project was followed by the Gemini Program, in which two astronauts flew in a single spacecraft. In the Gemini missions, astronauts demonstrated orbital rendezvous and spacewalking techniques that would be required for lunar landings. The Gemini Program was followed by the Apollo Program, which had the goal of landing astronauts on the Moon. A total of twelve Apollo astronauts walked on the Moon, beginning with astronauts Neil Armstrong and Edwin "Buzz" Aldrin, who landed on the Moon on July 20, 1969. Leftover Apollo hardware was used to launch the first U.S. space station, called Skylab, and to perform the first joint U.S.-Soviet spaceflight, Apollo-Soyuz.

On April 23, 1967, cosmonaut Vladimir Komarov became the first human to die in space. Komarov was piloting the first test flight of Soyuz 1. On the spacecraft's eighteenth orbit, its maneuvering system began to malfunction, and Komarov attempted to make a landing. However, he could not control the spacecraft, which became entangled in the cords of its parachute and hit the ground at more than 200 miles per hour, killing its pilot.

In June, 1971, three cosmonauts, George Dobrovolsky, Vladislav Volkov, and Viktor Patsayev, were killed when the Soyuz 11 spacecraft returned to Earth after a twenty-three-day stay at the Salyut 1 space station. A valve, designed to open after the spacecraft reentered the atmosphere, opened while the Soyuz was still in space, allowing the spacecraft's air to escape and suffocating the crew.

The United States has also suffered human losses in its development of spaceflight. On January 27, 1967, during a preflight Apollo test, a fire swept rapidly through the Apollo Command Module, killing all three astronauts participating in the test, Roger Chaffee, Virgil "Gus" Grissom, and Edward White. After the fire, NASA officials

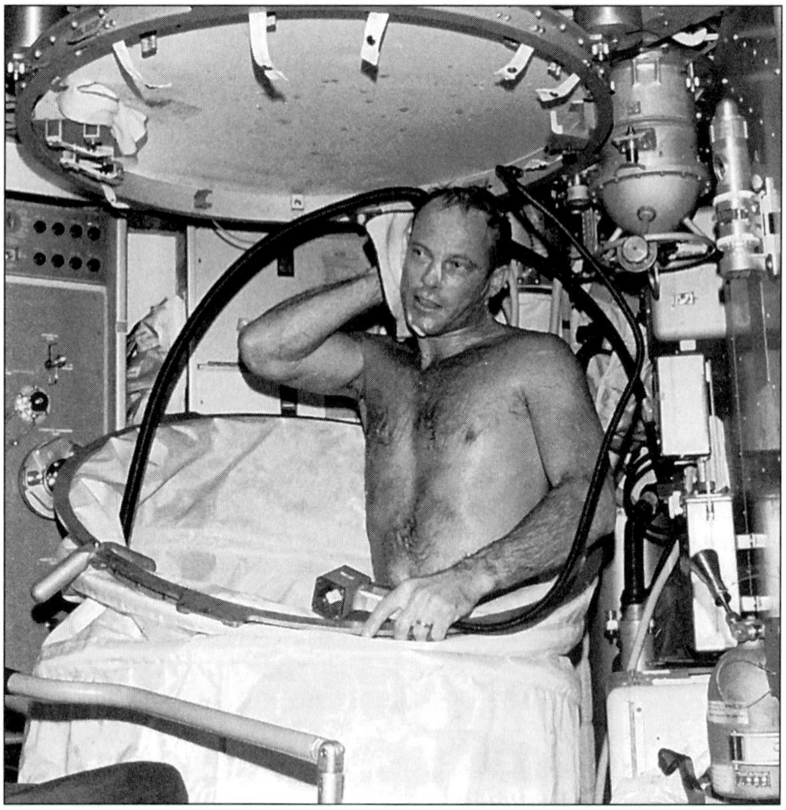

By the time Skylab 3 was launched in 1973, astronauts were able to dispense with their heavy spacesuits once in orbit and could even take showers while in space. (NASA)

designated the test as Apollo 1, honoring the crew. An extensive investigation revealed numerous design flaws, and manned launchings were postponed for more than a year while an extensive redesign was conducted.

On January 28, 1986, the space shuttle *Challenger* exploded shortly after liftoff. It carried a crew of seven astronauts: Francis Scobee, the commander; Michael Smith, the pilot; Judith Resnik, Ellison Onizuka, and Ronald McNair, all mission specialists; Gregory Jarvis, a payload specialist; and America's first schoolteacher in space, Sharon Christa McAuliffe. Below-freezing temperatures had hardened the O-ring seals between the segments of the solid-fueled rocket boosters causing one joint in the right solid rocket booster to develop a leak. After liftoff, hot gases cut through metal on the shuttle, and seventy-three seconds into the flight, the *Challenger* disintegrated, and all seven astronauts were killed.

More recent developments have been more positive. On February 20, 1986, the Soviet Union launched the Mir Space Station, which would be almost continuously occupied by a succession of crews for fifteen years before being deorbited on February 23, 2001. In 1999, NASA, working with a group of international partners including Canada, Japan, Russia, and the European Space Agency (ESA), began construction of the International Space Station (ISS). By 2001, after forty years of human exploration of space, more than 400 astronauts and cosmonauts had flown into space.

Risks Versus Benefits

Because space is not a natural environment for humans, sophisticated life-support systems are required to maintain atmospheric composition, temperature, and other features within a range suitable for human survival. The failure of any critical system can result in death for the crew.

Crewed spaceflight is also inherently more costly than robotic spaceflight, because the crew and its life-support systems must be carried into orbit at a cost of about $10,000 per pound. During the 1970's, NASA officials had decided that the space shuttle would carry all future payloads into space, even simple satellites that required no human intervention. NASA began to phase out rockets, such as the Delta, that had been used to launch uncrewed satellites. A highly focused debate developed over the next two decades over the relative merits of crewed versus uncrewed spaceflight. Critics of crewed spaceflight specifically targeted NASA's planned space station as a high-cost project whose scientific return would be less than if an equivalent amount of money were spent on robotic spacecraft. Following the *Challenger* accident, NASA officials decided that the crewed space shuttle fleet would be used only to launch satellites that required human intervention, and a fleet of new booster rockets was developed to launch robotic satellites.

However, in crewed spaceflight, humans can accomplish many tasks that may not be performed by robotic spacecraft. For example, when thrusters on his Gemini spacecraft began firing, causing the craft to rotate rapidly, astronaut Neil Armstrong was able to regain control of the spacecraft and return it safely to Earth. After an explosion on board the Apollo 13 mission to the Moon, astronauts modified the spacecraft's air-purification system so that they could survive the return to Earth. Cosmonauts overcame damage from an onboard fire and the leaks and damage caused by the impact of a resupply spacecraft, keeping the Mir Space Station operational. Space shuttle astronauts have repaired the Hubble Space Telescope, salvaged the improperly orbiting Westar-VI and Palapa-B2 satellites, and assembled large structures, including the International Space Station.

George J. Flynn

Bibliography

Catchpole, John. *Project Mercury: NASA's First Manned Space Programme*. London: Springer-Praxis, 2001. A extensive account of Project Mercury, including its history, accomplishments, and personalities.

Olberg, James E. *Red Star in Orbit*. New York: Random House, 1980. A comprehensive account, drawn mainly from Soviet media reports, of the Soviet space program, including the Vostok series of crewed spacecraft and the development of the ICBM that served as the Vostok launch vehicle.

Stoiko, Michael. *Soviet Rocketry*. New York: Holt, Rinehart and Winston, 1970. Provides an exhaustive discussion of the design, flight, and accomplishments of the first Vostok satellites and describes the development of the R-7 ICBM that launched the early Vostok satellites.

Yenne, Bill. *The Astronauts: The First Twenty-five Years of Manned Space Flight*. New York: Exter, 1986. A comprehensive account of all crewed space missions from Yuri Gagarin's flight in 1961 through the Challenger accident in 1986, with extensive coverage of the flights of both the United States and the Soviet Union.

See also: Apollo Program; Neil Armstrong; Astronauts and cosmonauts; Yuri Gagarin; John Glenn; Gemini Program; Mercury project; National Aeronautics and Space Administration; Russian space program; Alan Shepard; Space shuttle

Crop dusting

Also known as: Agricultural aviation

Definition: The aerial application of dusts, granules, sprays, and other materials for agricultural purposes.

Significance: Crop dusting allows the rapid, even dispersal of pesticides, fertilizers, seeds, and other materials over wide areas without the compaction of soil or crushing of vegetation commonly caused by tractors and other heavy farm machinery.

Early History and Development

The first patent relating to the use of aircraft in agriculture was granted to Alfred Zimmermann in 1911 by the Imperial Patent Office in Berlin, Germany, for an invention allowing the aerial application of lime water to control moth damage to pine forests. In practice, neither Zimmermann's invention nor other sporadic efforts succeeded until two Americans, C. R. Neillie and J. S. Houser, controlled *Catalpa sphinx* caterpillars by dusting them with lead arsenate from a Curtiss JN-6H Jenny biplane near Troy, Ohio, on August 31, 1921.

Types of Aircraft

Agricultural aircraft are designed or equipped to enhance visibility in all directions, to avoid pilot exposure to chemicals with special ventilation, to reduce pilot fatigue, to ensure pilot safety, to protect the aircraft and equipment from corrosive chemical mixtures, and to ensure high performance at slow speeds with heavy loads. The types of aircraft employed in crop dusting include specially designed agricultural monoplanes and biplanes, ex-military and ex-civil aircraft, and helicopters. Specially designed agricultural aircraft, such as the Cessna Ag-Truck, Turbo Thrush, and Ag-Cat and the Skyfarmer T-300A, tend to be more expensive than nonspecialized aircraft that can be adapted to agricultural use. Both ex-military aircraft, such as the Grumman Avenger and the Boeing Stearman, and ex-civil aircraft such as the Douglas DC-6, DHC Beaver, Piper Aztec, Russian Antonov AN-2M, and Pilatus Turbo Porter, have been used as agricultural aircraft.

The use of helicopters in crop dusting has increased worldwide because of helicopters' many advantages over fixed-wing aircraft, including their superior efficiency as sprayers due to their greater downwash at slower speeds, their ability to land almost anywhere without the need for an airstrip, their greater maneuverability, and their superior visibility. Helicopters' advantages may be outweighed by their disadvantages, however. Helicopters are far more expensive to purchase, and their operation is much more affected by changes in temperature and humidity than that of other types of aircraft. In addition, helicopters have more moving parts than other aircraft, requiring more maintenance, and they also have smaller centers of gravity, necessitating more careful loading.

Crop-Dusting Equipment

During the 1920's, hand-cranked and horse-drawn ground dusters had been employed with some success against boll weevil infestations devastating the cotton crops of the American South, but application rates were too slow for control. Early crop-dusting airplanes were equipped with sheet metal containers, called hoppers, mounted behind the pilot in the rear seat. An assistant balancing behind the hopper was required to turn the hopper's feeder crank, causing calcium arsenate to be discharged through a tube in the bottom of the fuselage. Dusting from aircraft proved successful for pest control over large areas.

Experience taught crop dusters that the pesticide dusts had to be constantly agitated and sifted in order for them to be dispersed evenly, leading to hopper designs incorporating rotating paddles or sweeping, windshield-wiper-like blades. Hoppers centrally mounted under the fuselage and equipped with pilot-operated levers eliminated the need for assistants. Tanks suitable for carrying liquid applications, accurate gauges to measure loads, and efficient filters and pumps evolved through trial and error.

Dispersal equipment also evolved, the most popular being the boom and nozzle for spraying liquids and the spreader for dispersing dry materials. The boom, a long, rigid pipe usually attached behind or below the wing of an airplane, supports movable nozzles that can be variably spaced along the boom and rotated to point in any direction to vary the spray pattern. In addition, the size and shape of the orifices on the nozzle can be changed like those on a showerhead to vary droplet size and spray intensity. The most frequently used spreader is suspended beneath the hopper and the fuselage so that the airstream can help blow the hopper contents out through the opening gate of the hopper to be deflected by the spreader. Some spreaders are fan- or wing-shaped, whereas others have divided outlets or rotating disks to disperse materials.

Application Materials and Methods

Applications of insecticides, fumigants, herbicides, fungicides, and defoliants are most often associated with crop dusting, but many other materials are commonly applied, including fertilizers, trace elements, and poison bait dropped over wide areas for the control of animal pests such as rabbits and mice. Even seeds for crops such as rice, grass, and vetch are sown from aircraft.

The usual practice in crop dusting is to fly back and forth in straight, parallel lines across the field being treated. If the area is too steep or irregular, the flight lines should follow the contours of the land. Coverage should begin downwind, so that the aircraft can make each swath without passing through an already-sprayed area. Flags are often posted at each end of a field to mark the spacing for each swath.

Turnarounds are the most likely maneuvers to cause accidents and must be made carefully, to prevent both crashes and the accidental dispersal of chemicals over adjoining areas. Maintaining a constant speed and height is important for even distribution. Speed and height of the aircraft are determined by atmospheric conditions and by the material to be dispersed. Controlling drift is necessary to prevent the exposure of humans, livestock, water sources, adjacent crops and pasture, and structures to contamination by hazardous materials. Drift is influenced by many factors, including weather conditions, particle size, specific gravity, evaporation rate, height of release, horizontal and vertical air movement, and aerodynamic effects of the aircraft. These factors lead to concerns about the potential use of crop dusters to disperse hazardous biological or chemical material in a terrorist attack. These planes were grounded temporarily following the hijackings of September 11, 2001.

Sue Tarjan

Bibliography

Anderson, Mabry I. *Low and Slow: An Insider's History of Agricultural Aviation*. San Francisco: California Farmer, 1986. Entertaining, informative, and well-illustrated account of the history of agricultural aviation in the United States by a professional crop duster.

Quantick, H. R. *Aviation in Crop Protection, Pollution, and Insect Control*. London: Collins Professional and Technical Books, 1985. Well-written, encyclopedic coverage of all aspects of agricultural aviation.

United States Animal and Plant Health Inspection Service. *Aerial Application of Agricultural Chemicals*. Washington, D.C.: U.S. Department of Agriculture, 1976. A practical manual containing clearly presented technical information and simple, accurate illustrations that enhance comprehension.

See also: Airplanes; Biplanes; Helicopters; Jennys; Pilots and copilots

Glenn H. Curtiss

Date: Born on May 21, 1878, in Hammondsport, New York; died on July 23, 1930, in Hammondsport, New York

Definition: The most prolific aeronautical inventor and manufacturer of airplanes and airplane engines in the United States well into the 1920's.

Significance: Curtiss developed ailerons for flight control, designed the first American amphibious airplanes, and built the first airplane to cross the Atlantic Ocean. He was also the first U.S. licensed pilot and the first to make a public flight.

Born in 1878, Glenn Hammond Curtiss, a champion bicycle racer, developed gasoline engines to power his bicycles, and he set international speed records on bicycles powered by engines with up to eight cylinders. It was Curtiss's engine expertise that led him to join Alexander Graham Bell's Aerial Experiment Association (AEA) in 1907 as its director of experiments. In the AEA, Curtiss quickly became instrumental in the design of a series of successful airplanes. The Curtiss White Wing became the first American airplane to take off on wheels instead of skids and the first to use ailerons for roll control in turns. His June Bug made the first "public" flight that was filmed and witnessed by the press in 1908. This flight would win Curtiss the *Scientific American* prize. The French press proclaimed Curtiss the "Champion Aviator of the World" after he set new speed records in winning the Gordon Bennett trophy in France in 1909.

Curtiss was sued by Wilbur and Orville Wright, who claimed that his use of ailerons violated their patents for controlling the roll of an airplane. Although many experts believed that ailerons were different from the Wrights' use of wing-warping, the courts were harder to convince. Repeated lawsuits kept Curtiss tied up in the courts for years, until the government intervened in the national interest, as the country entered World War I.

Curtiss further angered the Wrights when, at the request of the Smithsonian Institution, he agreed to prove that Samuel Pierpont Langley's aerodrome, an uncrewed flying machine driven by a gasoline-fueled, steam-powered engine which had crashed in the Potomac River twice in attempting the first airplane flight, was actually capable of flight. Curtiss made significant modifications to Langley's design, and when it flew, the Smithsonian proclaimed the aerodrome to be the first heavier-than-air craft capable of flight. The Wrights never forgave Curtiss for trying to usurp their rightful claim as the first to fly.

Curtiss established flying schools throughout the country and contracted to train Navy and Army aviators as he continued to develop newer airplanes. His Curtiss JN-4, or Jenny, a trainer aircraft, was the best American-designed plane to come out of World War I, and surplus Jennys became the airplane of choice for hundreds of aspiring pilots after that war.

A prolific designer of land-and-water-based airplanes for the Navy, Curtiss designed the NC-4, the first airplane to fly across the Atlantic, in 1919, and a series of racing planes that set world speed records in the early 1920's. His OX series of airplane engines were dominant in the U.S. market. Curtiss died in Hammondsport, New York, on July 23, 1930, of a pulmonary embolism suffered after a bout with acute appendicitis.

James F. Marchman III

Bibliography

Bilstein, Roger. *Flight in America*. Baltimore: Johns Hopkins University Press, 1987. Chapter 1 offers a thorough review of the history of aeronautics and space technology in the United States.

Christy, Joe. *American Aviation: An Illustrated History*. Blue Ridge Summit, Pa.: Tab Books, 1987. Chapters 1 through 3 provide an excellent overview of U.S. history in aviation and space. Profusely illustrated with historic photographs.

Roseberry, C. R. *Glenn Curtiss: Pioneer of Flight*. Syracuse, N.Y.: Syracuse University Press, 1991. The definitive biography of Glenn Curtiss with particular detail given to his many inventions and his court battles with the Wright brothers and others.

See also: Airplanes; Jennys; Heavier-than-air craft; History of human flight; Samuel Pierpont Langley; Transatlantic flight; World War I; Wright brothers

D

DC plane family

Date: Beginning in 1933

Definition: The most widely used passenger airplane series between the 1930's and the 1970's.

Significance: From the 1930's through modern times, the DC series of planes made air travel possible for most Americans by introducing such innovations as sleeper cabins, nonstop coast-to-coast flights, pressurized cabins, and ever-expanding fuselages to allow for more passengers per plane and to allow airlines to fly profitably.

The Beginning of the DC Series

The greatest contributor to the expansion of domestic and international air travel was the family of planes known as the Douglas Commercials or DC series. Built by the Douglas Aircraft Company, the DC's became the dominant brand of commercial passenger plane starting in the 1930's, and later served the needs of the American military beginning in World War II. The first DC model, the DC-1, was built in 1933. Capable of carrying twelve passengers, the two-propeller plane could travel coast-to-coast in a little over eleven hours. The DC-1 took passenger comfort into account in comparison to its main rivals. To combat the noise from the propeller-driven plane, the company used carpeted floors, sound-absorbing fabric, and rubber supports for the seats. The only DC-1 built was purchased by TWA, which saw the plane as the one that would allow it to compete with the more established air carriers. Within a year of the DC-1 rolling off the assembly line, the Douglas company built the DC-2, also for use in passenger flight by TWA. Known as the Sky Chief, the DC-2 could carry fourteen passengers, and in terms of physical size it had 2 feet more space in the fuselage and nearly 6 feet more in the wingspan. While it had a limited range of 1,000 miles, the DC-2 proved to be a workhorse, with 134 produced between 1934 and 1937. The third of the line was appropriately known as the DC-3 and was first flown as a passenger plane in 1935. This was the best known and the most popular of the DC series and is frequently called the greatest cargo plane ever built. American Airlines was the first to use the craft, after seeing its competitors tie up the other aircraft manufacturers with large orders of other passenger plane models. The airline sought a plane that would allow passengers to rest during the lengthy flight from New York to Los Angeles. The DC-3 had fourteen seats that folded into sleeping berths for passengers. The plane could carry fourteen passengers on an extended coast-to-coast flight or use all of the seats to fit twenty-eight passengers per flight for a shorter trip. The DC-3's larger capacity, its sleeping berths, and its nearly 1,500-mile range provided a boon for the passenger airline business and more importantly for the Douglas Aircraft Company. By the 1940's, approximately 90 percent of all passenger planes flying in the United States were either DC-2's or DC-3's. Some 455 DC-3's were built for commercial use, but the start of World War II saw a surge in the need for military transports that the DC-3 also filled. Over 10,000 DC-3's were produced for the military to carry both men and matériel to the European and Asian war zones. Even after the war and the end of production in the 1940's, the DC-3 continued to influence the passenger and freight airline markets and it continued to be flown in both capacities at the turn of the century.

From DC-4 to DC-8

The highly popular and profitable DC-3 was followed by a less successful version, the DC-4. Nearly twice the size of its predecessor, the DC-4 could carry up to forty-two passengers, but its size made maintenance and flight expensive, relegating the DC-4 to use almost exclusively as a military transport. In this role, the DC-4 was known as the C-54 Skymaster. The DC-4's were used mainly to fly supply missions across the North Atlantic. The four-engine plane proved to be reliable in this task and was used as a cargo carrier for civilian purposes at the end of the war.

In 1939, the DC-5 made its first flight. However, only five DC-5 aircraft, with seven more as R-3D military transports, were ultimately built.

The next in the series, the DC-6, was best known as the first regular aircraft to make around-the-world flights. Flying for the first time in 1946, the DC-6 was used by American, United, and Pan American airlines. Featuring the first pressurized cabin in the DC series, the DC-6 was able to fly at 20,000 feet while keeping passengers comfortable within the fuselage. The new DC-6 was a considerable improvement over its predecessors, carrying 102

A DC-3 in flight in 1959. (Library of Congress)

passengers and traveling at a speed of 308 miles per hour, a full 90 miles per hour faster than the DC-4. The DC-6 became the workhorse for the airlines in their extended international flights. In 1951, the DC-6B, with modifications of the original DC-6, became first official presidential airplane. Known as the *Independence*, it was first used by President Harry S. Truman to allow him to travel quickly across the country or around the world. The DC-6B was also adapted for use as a cargo carrier in the Korean War. Over seven hundred of them were built for military and civilian use, and by end of the century, scores continued to be used.

The DC-7 proved to be the last propeller-driven plane in the DC series. It represented the greatest increase in range among the models, with each plane able to fly 5,135 miles. By increasing the distance it could fly, the DC-7 became the first passenger plane to fly nonstop from New York to Los Angeles. Because the DC-7 did not have to stop for refueling, the flying time of the trip was reduced. This reduced flying time increased profits and lowered the ticket price for the flight, while the shorter flying time made a cross-country trip less burdensome for most people. The DC-7 was also known as the Seven Seas because its long range allowed for flights around the world. The DC-7 was introduced in 1953 and it could carry 110 passengers, a small improvement over the DC-6. There were 338 of the planes built and a few continued to operate a half-century later.

The DC-8, introduced in 1959, was the first jet-powered plane of the DC series. The four jet engines allowed the plane to reach speeds exceeding 600 miles per hour. The DC-8 became the first commercial jet to break the sound barrier. Along with its speed, the DC-8 had an expanded fuselage that doubled the passenger load to 260. While the plane had a slightly shorter range—4,500 miles—than its predecessor, its passenger capacity and freight-hauling abilities made it one of the largest commercial planes at that time. Over 550 of the planes were built, with more than 350 continuing to fly through the 1990's. Three different models of the DC-8's were built: the DC-8-61, the DC-8-62 and the DC-8-63.

The Modern DC's

The DC-9 has the distinction of having the largest number of commercial airplanes produced of any of the DC series.

Some 976 planes were built, of five different types, each one extending the fuselage and allowing for more passengers. The DC-9-10 was the smallest version, carrying only ninety passengers and used primarily for shorter range flights. The DC-9-20 also had a smaller fuselage, carrying fewer than one hundred passengers while utilizing larger engines to create greater thrust and carry larger payloads. The DC-9-30 added 15 feet to the fuselage and carried 115 passengers. The plane was specifically designed for rapid takeoff, allowing it to be used on smaller air fields. This made the DC-9-30 the most frequently used of all the aircraft. The DC-9-40 added another 6 feet to the fuselage and expanded passenger cargo to 125. The DC-9-50 was the largest plane in the family, with 8 more feet of fuselage beyond the DC-9 40, a passenger capacity of 139, and more space for cargo. Each of the DC-9's was introduced in the 1960's and many continued to fly both passengers and cargo at the turn of the century.

The DC-10 was the last of the series to be produced. While many of the features of the series would be found in its successor, the MD, the merger of Douglas Aircraft with McDonnell Aircraft led to the end of the name DC. The first model DC-10 flew in August, 1971. The DC-10-30 and the DC-10-40 were both extended-flight airplanes, with ranges of 5,900 and 5,800 miles, respectively. Three other types of DC-10's were used, mainly for carrying freight. The DC-10 Convertible was able to carry passengers or freight, though it was mainly a cargo carrier. The DC-10-15 resembled the original DC-10 but had a longer range of approximately 6,000 miles. The last of the DC-10's was the 30F. It was used exclusively as a freight carrier and became one of the standard planes for package delivery companies. The 30F was renamed the KC-10 cargo plane for the U.S. Air Force. When DC-10 production was halted in 1989, approximately 380 planes were flying commercially, while 60 more were being used as cargo carriers for the Air Force. Yet even with this commercial success, the DC-10 had a mixed safety record. A 1974 crash near Paris killed 346 people and was blamed on a cargo door blowing open in flight. Similar problems were discovered in other DC-10's. In a six-month period in 1979, some five hundred people died in three DC-10 crashes. This was attributed to structural fatigue, with one crash caused by a pylon collapsing in flight. In July, 1989, in Sioux City, Iowa, the most spectacular crash occurred, when a DC-10's hydraulic system failed. Over one hundred people died, although more than twice that many survived. These safety problems gave the DC-10 a bad reputation but it continues to fly in many airline fleets.

The DC series ended with the DC-10. In 1967, the Douglas Company, suffering from severe financial losses caused by problems in the production of DC-8's and DC-9's, merged with the McDonnell Corporation to form McDonnell Douglas. The next series of DC planes were renamed the MD series, and when McDonnell Douglas merged with Boeing in 1997, the planes took on the 700 family name associated with that company.

The Legacy of the DC's

The DC series of planes may have been the most important of all families of passenger carriers. With their start in the 1930's, the DC series helped make air travel affordable for the individual and profitable for many airlines. The DC planes also established such innovations as nonstop flights across the United States, larger fuselages to carry ten times the passengers of the original DC models, and a dependability that sees many DC's flying local routes to smaller airports and others longer routes across countries or continents. While the DC line ended with the DC-10 and the original company that developed the model was merged into oblivion, the plane series continues to strike the imaginations of both those who study passenger airlines and those who fly them.

Douglas Clouatre

Bibliography

Badrocke, Mike, and Bill Sunston. *The Illustrated History of McDonnell Douglas Aircraft from Cloudster to Boeing*. Oxford, England: Osprey, 1999. A colorful, well-illustrated book describing the history of the McDonnell and Douglas airplane companies, their merger, and how their planes revolutionized air travel.

Endres, Günter. *McDonnell Douglas DC-10*. Osceola, Wis.: Motorbooks International, 1998. A primer on the DC-10, with illustrations and an in-depth discussion of its flying capabilities, its many features, and its uses in airlines across the world.

Francillon, Rene. *McDonnell Douglas Aircraft Since 1920*. Annapolis, Md.: Naval Institute Press, 1990. Discusses the civilian and military aircraft developed by both companies prior to their merger and after their combination.

Graves, Clinton H. *Jetliners*. Osceola, Wis.: Motorbooks International, 1993. A wide-ranging book with illustrations of many of the major McDonnell and Douglas aircraft used for civilian and military purposes.

Norris, Guy, and Mark Wagner. *Douglas Jetliners*. Osceola, Wis.: Motorbooks International, 1999. Focuses on the Douglas passenger planes with special emphasis on the DC family and its development and capabilities.

Singfield, Tom. *Classic Airliners*. Leicester, England: Midland, 2000. An introduction to many of the original planes used during the early years of the airline industry, including the DC-3 and other Douglas planes.

Waddington, Terry. *McDonnell Douglas DC-9*. Osceola, Wis.: Motorbooks International, 1998. Focuses on one of the best known of the Douglas planes with pictures of the exterior and interior and an in-depth discussion of its capabilities.

_______. *McDonnell Douglas DC-10*. Osceola, Wis.: Motorbooks International, 2000. Examines the last of the DC models, providing details on its upgrades over its predecessors and its continued use.

See also: Airplanes; Boeing; Cargo aircraft; Commercial flight; Jet engines; Manufacturers; McDonnell Douglas; MD plane family; Military flight; 707 plane family; Trans World Airlines; Transatlantic flight; Transcontinental flight; Turboprops; World War II

Delta Air Lines

Definition: The world's first crop-dusting company, which became one of the world's foremost passenger airlines.

Significance: One of the world's most successful airlines, the history of Delta Air Lines spans the aviation era. From the beginnings of commercial flying, Delta has been a presence in aviation, first as the original aerial crop dusting company and later growing to become a leader in passenger carriers.

Crop-Dusting Beginnings

As an official incorporated entity, Delta Air Lines dates from 1945, when what was then the Delta Air Corporation changed its name. The organization actually dates back some twenty years earlier to the first aerial crop-dusting company. In the early years of the twentieth century, boll weevil depredations on cotton crops forced the Bureau of Entomology of the United States Department of Agriculture to establish its Delta Laboratory in Tallulah, Louisiana, to research methods of controlling the pest. Dr. Bert R. Coad directed the research, often assisted by Collett Everman Woolman, a district agent of the extension department of the Louisiana State University. Powdered arsenates were effective against the pest but an efficient, broad-scale delivery method was required. Coad decided to try aerial dusting from airplanes. With surplus Curtiss Jenny airplanes acquired from the Army, he and Woolman began to perfect aerial crop-dusting procedures.

In 1923, mechanical problems with his airplane forced George Post, an executive of the airplane maker Huff Daland Manufacturing Company, to land in Tallulah. Excited by the prospects for crop dusting, Post convinced his company to form a new division, the Huff Daland Dusters. Huff Daland then began building the first airplanes specifically designed for crop dusting.

The forerunner of Delta Air Lines, Huff Daland Dusters began operations in Macon, Georgia, in 1924 but cotton farming in the area was insufficient to support activities and, at Dr. Coad's suggestion, operations were shifted to Monroe, Louisiana, for the next year.

Woolman then joined the company as vice president and field manager. Because crop dusting is seasonal, the company soon sought ways of generating off-season revenues. Their first effort was to continue crop dusting through the winter, shifting operations to Peru, where the seasons are the reverse of those in the northern hemisphere. Additionally, Woolman acquired airmail service rights for a 1,500-mile route between Peru and Ecuador.

The company's operational base at the time was the Mississippi Delta. Accordingly, the word "delta" appeared in the company name for the first time when ownership changed in 1928. Huff Daland sold the division to a group of Monroe businessmen. Woolman remained as vice president and general manager of Delta Air Service. The new company continued crop-dusting operations under that name until 1966.

A political revolution in Peru forced the closing of crop dusting and airmail operations there in 1928. The planes were sold to what later became Peruvian Airlines; the airmail route went to Pan American Grace. Delta Air Service used the money to purchase three five-passenger TravelAir monoplanes and, on June 17, 1929, began 90-mile-per-hour passenger service on a route from Dallas, Texas, to Jackson, Mississippi, with stops in Shreveport and Monroe, Louisiana. Strictly a passenger operation without any associated airmail contracts or revenue, this was an ambitious step into what was to be an increasingly important service.

Airline Regulation

The election of President Herbert Hoover brought important changes to the air industry. Hoover's postmaster general, Walter Folger Brown, was determined to use the awarding of airmail contracts to improve and streamline what he saw as a chaotic air carrier structure. On April 29, 1930, Congress passed the Air Mail Act of 1930, also known as the McNary-Watres Act, empowering Brown to do just that. Lacking night flying experience, Delta did not fit Brown's requirements and consequently lost its mail contracts in 1930. Delta was back to being merely a crop-dusting operation. In response, Dr. Coad was brought on as chief entomologist and dusting operations expanded. Passenger planes and routes went to Aviation Corporation (AVCO), a holding company whose most important asset was American Airlines, but crop-dusting rights and equipment were retained by Delta Air Services, which was then reincarnated as Delta Air Corporation.

Franklin D. Roosevelt won the 1933 presidential election, sweeping Hoover and his postmaster general out of office. Within months, the new administration cancelled all airmail contracts. After a disastrous attempt to use the Army Air Corps to fly the mail, a call was placed for new bids. In the bidding, Delta acquired the mail route from Charleston, South Carolina, to Dallas and Fort Worth, Texas, with stops along the way in Atlanta and Birmingham. Delta purchased trimotor Stinson-T planes, which could carry seven passengers and mail at 100 miles per hour, and resumed airmail operations on July 4, 1934.

The Great Depression of the 1930's was a time of often painfully slow development and consolidation for the American airline industry. Airlines made little or no profit. The Civil Aeronautics Authority was created to regulate and control the airlines and airline routes. Despite recurring difficulties, Delta gradually improved its routes and position. Delta also experienced its first passenger fatalities in 1935 when the propeller of a Delta Stinson-A broke in flight. The resulting crash in a cotton field near Gilmer, Texas, killed two passengers and the crew of two. In the next year, mechanical failure of another Delta Stinson-A seriously injured a veteran Delta test pilot. The pilot, Charles H. Dolson, eventually recovered and returned to work. He was a key figure in unionizing Delta pilots, bringing them into the Air Line Pilots Association in 1935, and he later replaced C. E. Woolman as president of Delta.

The Delta complex of routes increasingly passed through Atlanta, making it a natural home for the organization. Accordingly, when leases in Monroe came due for renewal in 1941, the Board of Directors moved Delta headquarters from Monroe to Atlanta.

With the entry of the United States into World War II, key personnel were lost to the war effort and vital matériel and supplies were in short supply. Nevertheless, the airline forged ahead. Air routes were added and the workforce increased. Assets, working capital, and passenger totals climbed. Delta undertook a major aircraft modification program for the military. Under a two-year contract, Delta personnel prepared bombers for conditions in the Pacific and European theaters and installed long-range fuel tanks on P-51 Mustangs.

As the war ended in 1945, Delta renamed itself Delta Air Lines. At this point, Delta had flown more than 300 million passenger miles in ten years without a passenger or crew fatality. In a major victory for Delta that year, the Civil Aeronautics Board (CAB), instituted in 1940, awarded it the lucrative Chicago-to-Miami route. Company fortunes continued to rise. By 1946, Delta had carried more than one million passengers.

Delta lost a number of key personnel in a horrifying and bizarre accident in 1947. On April 22 of that year, a Delta C-47 carrying seven major Delta executives and piloted by Delta's operations chief was approaching the runway at Muscogee County Airport in Columbus, Georgia, when a BT-13 flown by an experienced Civil Air Patrol pilot landed on the C-47. No one survived the ensuing crash and fire.

The Delta Fleet

Aircraft Type	*Total Owned and Leased*	*Average Age*
Boeing 727-200	74	22.4
Boeing 737-200	54	16.1
Boeing 737-300	26	14.1
Boeing 737-800	43	0.9
Boeing 757-200	120	9.5
Boeing 767-200	15	17.6
Boeing 767-300	85	10.9
Boeing 767-400	15	0.2
Boeing 777-200	7	1.3
Lockheed 1011-100	4	19.7
Lockheed 1011-250	4	18.1
MD-11	15	6.9
MD-88	120	10.5
MD-90	16	5.1
ATR-72	19	6.5
EMB-120	57	10.6
CRJ-100/200	155	2.8
Total	829	9.6

Source: Data taken from (www.delta.com/inside/investors/corp_info/fleet/index.jsp), June 5, 2001.

Mergers and Expanding Routes

Another benefactor of the 1934 bidding was Carleton Putnam's Pacific Seaboard Air Lines. Flying passenger routes in California without airmail contracts, the line was struggling until, like Delta, it captured an airmail contract. The new route, from Chicago to New Orleans, shifted its operations to the Mississippi Valley and inspired a change of name to Chicago and Southern Air Lines (C&S). With Delta's newly acquired Chicago-to-Miami route, C&S came quickly into Delta's purview and, in 1953, the two airlines merged. The new company went by the name Delta-C&S for about two years before reverting to Delta. The move significantly expanded Delta's range of routes and enhanced its competitive position.

A month after the merger, a Delta DC-3 flying from Dallas to Shreveport encountered a thunderstorm near Marshall, Texas. Unaccountably breaking Delta regulations, the pilot attempted to pass through the storm. Seventeen passengers and the crew of three perished when the plane went down. One passenger survived when her seat detached and landed upright.

Important improvements to company operations were made in the 1950's. In 1955, Delta pioneered the hub-and-spoke system. Shortly thereafter, weather-avoidance radar was installed in the noses of all Delta aircraft. Delta entered the jet age in 1959, becoming the first airline to introduce the new passenger jet aircraft, the DC-8. The greater maintenance needs of jet aircraft thrust Delta into a vast effort to upgrade training, inspection, and maintenance procedures. Air traffic control improvements were also introduced at this time, when Congress created a new safety regulatory agency, the Federal Aviation Agency, later called the Federal Aviation Administration (FAA). The agency was to develop and manage an air traffic control system to maintain safe separation distances between all commercial aircraft through all phases of flight.

Tragedy struck again when one of Delta's new Convair-880 jets crashed and burned during a 1960 training flight. It seemed to take off normally but almost immediately rose steeply, then banked left and right and crashed, killing the four crew members. The cause of the accident was never determined.

The 1960's were a time of intense competition for new routes and milestone changes. Bert Coad, still managing the crop-dusting division, died in 1966 and the division was soon closed down. Then, in September of that year, C. E. Woolman died following surgery for an aneurysm. He had only recently given up his position as president and general manager in order to become the chief executive officer.

Throughout the decade, Delta lost many route wrangles with the CAB. Although these efforts were a drain on energies and resources, company assets, revenues, and net income rose impressively. Delta's market share also rose steadily. The CAB did award Delta a few lucrative routes, such as the Dallas-Fort Worth-to-Phoenix route. Delta also gained access to the major North Carolina airports.

The route system continued to expand in the 1970's, both through CAB awards and also through the 1972 merger with long-ailing Northeast Airlines. The Atlanta-to-London route was established in 1978. Together with its New England and Northeast routes, Delta's route system now encompassed the entire eastern United States, with side routes to western cities as well as to the Caribbean and London. However, the merger with Northeast also brought equipment problems. In the 1970's, Delta had a dozen different types of planes in service, creating enormous maintenance headaches. The Northeast planes had not been made to Delta's specifications. A five-member crew and eighty-three passengers were all killed when one of the DC-9-31's obtained from Northeast hit a seawall while attempting to land in Boston in a thick fog on July 31, 1973. No single cause was identified, but a number of errors and failures of both the crew and ground control apparently combined to create the disaster. This accident followed the May 30, 1972 loss of the four crew members during the takeoff of a DC-9 training flight in Fort Worth caused by turbulence from the previous landing.

Deregulation

Deregulation of the air industry in 1978 created a host of opportunities and risks. Delta greatly extended its western routes in a merger with ailing Western Airlines in 1987. Then, in the greatest transfer of flights in airline history, Delta gained many additional transatlantic routes with the 1991 purchase of the routes of bankrupt Pan American. The former crop-dusting outfit had become a giant of global passenger flight. By 1998, Delta was the most-flown airline in the world, with 105 million passengers that year.

Delta suffered its most deadly accident on August 2, 1985. A Lockheed Tri-Star attempting to land at Dallas in a thunderstorm was caught in a microburst downdraft and landed more than a mile short of the runway, striking a car and killing the occupant. On board, 8 crew members and 126 passengers were killed.

Almost exactly three years later, on August 11, 1988, a Delta Boeing 727 crashed trying to take off from Dallas, killing two crew members and twelve passengers. Improper procedures and failures of discipline were cited as causes.

John A. Cramer

Bibliography

Davies, Ronald E. G. *Delta: An Airline and Its Aircraft*. Miami: Paladwr Press, 1990. A short but nicely illustrated history by a major airline historian.

Davis, Sidney F. *Delta Air Lines: Debunking the Myth*. Atlanta: Peachtree, 1988. A critical but constructive personal account of Delta during deregulation by a former Delta vice president.

Newton, Wesley Phillips, and W. David Lewis. *Delta: The History of an Airline*. Athens: University of Georgia Press, 1979. A major history of Delta up to deregulation by two technology historians.

See also: Accident investigation; Air carriers; Airline Deregulation Act; Airline industry, U.S.; Airmail delivery; Crop dusting; Jennys; Mergers; Safety issues

Dirigibles

Definition: Aircraft that float because of lighter-than-air gas and that have power sufficient to direct their course of flight.

Significance: Dirigibles were the leading edge of aviation from the 1850's until they were supplanted by airplanes and helicopters. Dirigible aviation developed many techniques that were later adopted for airplanes. In the twenty-first century, dirigibles may serve a number of niche functions, such as telecommunications repeaters, high-altitude science platforms, and heavy cargo transporters.

Nature and Use

Dirigibles, like balloons, are often referred to as lighter-than-air (LTA) craft, in contrast with airplanes and helicopters, which are heavier-than-air (HTA) craft. The term "dirigible" is a shortened form of "dirigible balloon," meaning directable, or steerable, balloon. Buoyancy is the key to dirigible flight. The ancient mathematician Archimedes stated that a body immersed in a fluid is buoyed up by a force equal to the weight of the displaced fluid. For dirigibles, two LTA gases, hydrogen and helium, are combined to provide the buoyancy that lifts the dirigible and any payload.

Typically, hydrogen lifts 60 pounds per 1,000 cubic feet. Helium lifts 14 percent less (53 rather than 60 pounds) per 1,000 cubic feet. Helium has a major safety advantage over hydrogen, in that it does not burn, whereas hydrogen can ignite explosively.

Unfortunately, helium did not become available until the 1920's, and even then, the U.S. government, which controlled most of the world's supply, was slow to allow exports. Consequently, dirigibles manufactured outside the United States flew using highly flammable hydrogen, which caused many catastrophic fires. A third LTA gas, hot air, has only one-third the amount of lift of hydrogen, meaning the propulsion unit must have proportionately more thrust.

Another dirigible concern is that the density and pressure of the surrounding air decreases with altitude. Hence, there is less lift available per unit volume, so the craft must be larger to carry a given payload to higher altitudes. Consequently, dirigibles with heavy payloads tend to be limited to low altitudes of a few thousand feet. For higher altitudes, designers can compensate for decreased lift per unit volume by using lighter payloads, such as remotely controlled instruments instead of people.

Types

Dirigibles are divided into three categories: nonrigid, semirigid, and rigid. Nonrigids are essentially large, streamlined cylindrical balloons, nicknamed "blimps," supposedly for the sound made by a finger thumping into the side of the envelope, or gas bag. Nonrigids, such as the Goodyear blimps, get their shape from the gas within a single gas bag or envelope. The engines and a car or a gondola hang below the gas bag.

The design of nonrigid dirigibles simplifies cost and minimizes structural weight, which in turn reduces the net lifting capability. However, nonrigids are limited in size, because an unsupported gas bag may bend unpredictably under heavy loads or strong winds. In a worst-case scenario, a partially deflated gas bag may flop over the gondola or propellers. Conversely, the gas bag cannot be filled too tightly, lest it burst. The one gas bag is a single point of failure that could cause a crash, although the large size of dirigibles means that operations could continue for some hours, even with significant leaks. Another method to compensate for pressure loss in the gas bag is the use of an inner ballonet, which can be inflated with outside air.

Semirigid dirigibles have a keel on the bottom to support a larger gas bag, and the keel can hold the gondola and engines, at the cost of additional weight. The risks associated with a single gas bag also apply to semirigid dirigibles. The most famous semirigid was the airship *Norge*, which made the first transpolar flight from Spitsbergen Island to Alaska.

Rigid dirigibles have a framework to support an outer skin and individual gas bags. Although the term "zeppelin" is sometimes used to describe any rigid dirigible, the name legally applies only to the type of craft manufactured by the Luftschiffbau Zeppelin company of Germany.

In rigid dirigibles, an individual gas bag can fail without damaging the aerodynamic integrity of the craft, and there are usually sufficient reserves among the other cells to maintain buoyancy. Those advantages cost additional weight. However, greater weight can be compensated for by greater size. There is theoretically no limit to a rigid's size. The German passenger airship *Hindenburg* had an LTA gas capacity of 7,000,000 cubic feet, and designs of twice that size have been proposed.

History

Beginning in the 1790's, balloons made true humankind's dream of the possibility to drift like clouds. Like clouds, however, balloons drifted wherever the wind blew. Thus, inventors realized their craft must be directable as well as lighter than air. The key to this directional ability was generating sufficient power while remaining light enough to fly. Repeated attempts in the first half of the nineteenth century showed that human power was insufficient against even slight winds. A number of inventors flew models powered by springs or clockworks during that time, but none of the models' mechanisms could be sufficiently scaled up to power a craft carrying a person.

Henri Giffard of France had the first partial success on September 24, 1852, with a dirigible powered by a steam engine. The engine, advanced for its time, produced 3 horsepower and weighed as much as two large men. Giffard's aerial steamer, as it was known, launched from the Paris Hippodrome and hissed sedately to a landing 17 miles away. In a later flight, Giffard circled around Paris. However, because his craft's top speed was only 6 miles per hour, it was not steerable against even a breeze.

Fortunately for dirigible designers, the development of metallurgy and power plants advanced in the second half of the nineteenth century. In 1886, an electrolytic process was invented for producing aluminum inexpensively enough so that it could be used to replace the heavier steel in dirigible support structures. In 1876, German engineer Nikolaus August Otto began marketing a four-stroke, internal combustion engine yielding more power per unit weight than the external-combustion steam engines. In 1885, another German engineer, Gottlieb Daimler, patented significant improvements to the internal combustion engine and offered it for use in dirigibles.

On November 12, 1897, an airship built by Austrian David Schwartz sported a 10-horsepower Daimler motor. Before it was ready to launch, a gust of wind pulled the craft loose from its moorings and toward nearby buildings. The pilot panicked and valved out so much gas that he crashed on the field. Despite its misfortune, this ship represented the first rigid dirigible, with a solid structure and a thin aluminum skin around the gas bag.

By this time, both airships and balloons had developed a bad reputation. Fortunately, public relations assistance and superb flying skill arrived from Brazil in the form of wealthy experimenter Alberto Santos-Dumont, who took a single-cylinder engine from each of two tricycle automobiles to make a single, 66-pound, two-cylinder engine delivering 3.5 horsepower, roughly five times the power-to-weight ratio that had been available to Giffard. Santos-Dumont launched this eighty-two-foot nonrigid craft with 64,000 cubic feet of gas volume, along with himself and a basket.

On September 20, 1898, Santos-Dumont began flying around Paris in his airship, usually flying low enough to greet people on the streets. As both his flying skills and his dirigibles became progressively more advanced over the next several months, he aroused tremendous public interest, especially because he commuted around Paris in his compact dirigibles, mooring his craft above the spots to which he traveled.

Santos-Dumont engendered so much interest in flying that a prize was offered to the aviator who could fly a seven-mile course to the Eiffel Tower and back within in thirty minutes. After several heroic attempts, Santos-Dumont won the prize. He became a global celebrity and inspired many others built nonrigid airships.

In Germany, Count Ferdinand von Zeppelin built a large rigid dirigible, *Luftschiff Zeppelin Number 1*, or *LZ-1*, which was 420 feet long and 42 feet in diameter with a gas volume of 400,000 cubic feet, sixty times greater than that of Santos-Dumont's model number 1. *LZ-1*, which first flew in July, 1900, had seventeen separate gas cells held together by an aluminum framework and covered with fabric.

However, the two 15-horsepower engines gave *LZ-1* a top speed of only 16 miles per hour, still insufficient to fly against moderate winds. Zeppelin raised more money to build *LZ-2* and *LZ-3*, both of which had two 65-horsepower engines. *LZ-2* was destroyed at its mooring by winds, but successful flights of *LZ-3* led the German government to offer payment for a still-larger craft, if it could stay aloft for twenty-four hours.

On August 4, 1908, the *LZ-4* began a majestic tour from its home base on the Swiss border, heading north along the Rhine River. People along the way cheered the giant airship. *LZ-4* flew for eleven hours, as far as Mainz, Germany,

Notable Airship Disasters

August 24, 1921: The U.S. *R-38* airship, built for high altitudes, maneuvers hard at a low altitude, breaks in half, and explodes.

December 21, 1923: The French airship *Dixmude* explodes in flight during a thunderstorm, killing fifty crew members.

September 3, 1925: The USS *Shenandoah* breaks apart in thunderstorm, killing fourteen of the forty-three crew members.

May 25, 1928: The Italian airship *Italia* loses buoyancy and crashes attempting to reach the North Pole.

October 5, 1930: The British *R-101* airship crashes and burns, killing forty-eight of the fifty-four total passengers and crew.

April 3, 1933: The USS *Akron* is driven into the sea by downdrafts in a storm, killing seventy-three of the seventy-six crew members.

February 12, 1935: After the fin of the USS *Macon* is ripped off in storm, the airship loses buoyancy and crashes at sea, killing two of the eighty-three crew members.

May 6, 1937: The German airship *Hindenburg* explodes while docking at Lakehurst, New Jersey, killing thirty-six of the ninety-eight passengers and crew.

and had begun its return when one engine failed. Rather than press on in the dark with only one engine, Zeppelin set *LZ-4* down near the town of Echterdingen. That night a storm pulled the craft loose and destroyed it.

Yet Zeppelin's story continued, as envelopes of cash began arriving from all over Germany. The so-called Miracle of Echterdingen supplied his company with more money than the German government had offered. The count continued to build, and by 1910, Zeppelin dirigibles had begun carrying sightseeing passengers and mail. By 1914, a number of Zeppelin dirigibles were in regular service.

That year, World War I began. At first, the dirigibles dominated the skies, and the competing airplanes posed no threat to them. In 1915, German rigid dirigibles conducted the first long-range bombing attacks against targets in Great Britain, with little effective resistance from airplanes. However, faster and larger airplanes were soon able to catch the dirigibles, which proved to be large, slow targets. A single incendiary round of fire passing through a hydrogen gas cell could transform an airship into a fireball. In order to escape the airplanes, the Germans piloted their dirigibles to higher altitudes, where at 20,000 feet, water froze in the crews' canteens. The airplanes, however, were improved enough reach the dirigibles. By the end of the war, large airplanes had replaced rigids for long-distance bombing. The only dirigibles successful throughout the war were two hundred nonrigids the British used to guard convoys against submarines.

The long flights made by rigid dirigibles during the war suggested that dirigibles might be used for intercontinental passenger service, or even as flying warships. Continued research was conducted by four countries: France, Great Britain, the United States, and Germany.

France had a number of smaller nonrigids, as well as one large rigid taken from Germany as part of its war reparations. The airship was renamed the *Dixmude* and flew for several years, making a record-breaking flight over the North African desert. After the airship exploded in flight during a storm in 1923, France abandoned large dirigibles.

During World War I, the British had built an R (for rigid) series of dirigibles, which the British continued to develop after the war by reverse engineering from a captured German dirigible. On July 2, 1919, the *R-34* left England, and, four days later, it had completed the first east-to-west aerial crossing of the Atlantic Ocean.

In 1924, the British government started two competing programs to build dirigible airliners. The *R-100*, built with private funding, was known as the capitalist ship, and it flew well on a demonstration flight to Canada and back.

The *R-101*, built by the government, was known as the socialist ship and was heavy with safety features. To increase lift, the builders cut the ship in half and inserted an additional gas bag. They also loosened wire netting around the gas cells, so they could be expanded. Unfortunately, this adjustment allowed the cells to rub against the framework, causing many small leaks. Because officials wanted to use the airship for a prescheduled demonstration flight to India, the major changes were not flight-tested. The *R-101* launched from England on October 4, 1930, and early the following morning, it crashed into a hillside and exploded 40 miles northwest of Paris; forty-eight of the fifty-four people aboard died. As a result of this accident, Great Britain abandoned passenger airships and even scrapped the successful *R-100*.

In the 1920's and 1930's, the U.S. government operated four rigids as military ships intended for long-range reconnaissance. Two of the airships, the USS *Akron* and the USS *Macon*, actually carried their own fighter planes for defense. Because the United States held most of the world's helium supply and used helium for its LTA gas, none of these craft exploded. However, three were lost in storms, and the United States abandoned the giant rigids after the last, the *Macon*, broke up in a storm and went into the sea off Point Sur, California, on February 12, 1935.

The Luftschiffbau Zeppelin company of Germany, with its experience in building more than one hundred rigids and its thorough design details, had the best safety record of any dirigible manufacturing company. For several years after World War I, Germany was forbidden by the Treaty of Versailles from possessing dirigibles larger than 1,000,000 cubic feet. However, in 1922, the U.S. Navy placed an order for a dirigible, which was named the *Los Angeles*. Zeppelin's brilliant manager, Hugo Eckener, flew the craft to the United States. After the size limit on German dirigibles was lifted in 1925, Eckener organized construction of the *Graf Zeppelin*. Beginning in 1928, the *Graf Zeppelin* circled the world, flew regularly to Brazil and North America, made an Arctic expedition, and traversed one million miles before being retired.

The last and greatest rigid was the *Hindenburg*, launched in 1936. The *Hindenburg* was 803 feet long and 135 feet in diameter. Its 7,000,000 cubic feet of gas allowed it to carry fifty passengers and sixty crew in absolute luxury at a speed of 84 miles per hour for a range of 11,000 miles. The *Hindenburg* and the older *Graf Zeppelin*, represented great profits for Luftschiffbau Zeppelin and good propaganda for Germany's Nazi regime.

Then disaster struck. Although Luftschiffbau Zeppelin was negotiating with the U.S. government for helium, it still employed hydrogen in its airships. As the *Hindenburg* was docking at Lakehurst, New Jersey, on May 6, 1937, several crew members noticed a small fire in one gas cell. Within one minute, the craft had exploded into a ball of fire and lay on the ground, a smoldering wreckage. Although many theories proposed causes such as lightening, leaking gas, and anti-Nazi sabotage, filmed footage of the event convinced the public that large dirigibles were unsafe.

After the *Hindenburg* disaster, only nonrigids remained, and they played a major role in the antisubmarine warfare of World War II. However, they were retired in the 1950's, after it became clear that helicopters provided the same hovering capability with greater dash capability and easier storage. In the last third of the twentieth century, the few working nonrigid dirigibles were limited to use as advertising billboards and as vehicles for television cameras providing overhead views of sporting events.

Economics and Prospects

Although dirigibles at the beginning of the twenty-first century enjoyed a small resurgence in several niche markets, they will probably never recover their primacy in aviation for five major reasons.

The first reason is the massive investment cost of building and developing dirigibles. Several factors make dirigibles more efficient as their size increases. However, the increase in size increases the cost of design and building. Large size also reduces the number of units made, so dirigibles have less chance for lower costs and improved designs than do HTA craft, which are typically made by the hundreds or thousands.

Second, hangar costs are high. Dirigibles are kept inflated because their helium lifting gas is expensive and would require too much time and effort to pump back into tanks. However, inflated dirigibles can easily be swept off their parking areas by winds. Consequently, dirigibles must be housed in their own special hangars instead of being parked on runways as airplanes are.

Third, dirigibles are vulnerable to bad weather, which limits their performance. The giant buoyant structures can be seized by gusts of wind on takeoffs and landings and are more vulnerable than airplanes to icing. Zeppelin passenger flights were not scheduled in winter. Dirigibles are so large that winds may pull them in different directions while they are in flight, destroying them. The USS *Shenandoah*, *Akron*, and *Macon* were all destroyed in this way. Moreover, unless they are specially designed for high altitude, dirigibles cannot readily climb above storms as jet-propelled airplanes can.

Fourth, because dirigibles' great size causes more drag per unit mass of cargo, dirigibles are significantly slower than their HTA competition. They can at best obtain one-half the speed of propeller-driven planes and one-fifth that of jets. Thus, a jet with one-fifth of the cargo capacity of a dirigible can deliver the same cumulative mass of cargo. For the passenger market, shorter flight times are crucial.

Still, dirigibles have potential for certain markets because they can run quietly and smoothly, linger for long periods, carry heavy and awkwardly large payloads, and land without runways. These advantages have been increased by lighter and more fireproof materials. The number of advertising dirigibles increased steadily beginning in the 1980's. At the start of the twenty-first century, the present-day Luftschifftechnik Zeppelin company marketed sightseeing semirigids one-third the size of the *Hindenburg*. A German-American company called CargoLifter designed a cargo-carrying rigid larger than the *Hindenburg*.

Meanwhile, an entirely new concept was being developed: the use of dirigibles in the lower stratosphere as high-altitude platforms. Such platforms could serve many functions of communications satellites and astronomical satellites at a fraction of the cost of spacecraft.

Roger V. Carlson

Bibliography

Botting, Douglas. *The Giant Airships*. Alexandria, Va.: Time-Life, 1981. An exhaustive but readable history and technical description of the earliest dirigible attempts through the destruction of the *Hindenburg*.

Cross, Wilbur. *Disaster at the Pole*. New York: Lyons Press, 2000. An historical account of the airship *Italia*'s disastrous mission of scientific research at the North Pole and the political backlash in Italy against the expedition's commander and dirigibles in general.

Hogenlocher, Klaus G. "A Zeppelin for the Twenty-first Century." *Scientific American* 281, no. 5 (November, 1999): 104-109. A detailed description of the technical innovations of the "new technology" Zeppelin airships of the 1990's.

Kunzig, Robert. "Dirigibles on the Rise." *Discover* 21, no. 11 (November, 2000): 92-99. A description of the new dirigible enterprises being developed at the end of the twentieth century, including new passenger craft and heavy cargo lifters.

See also: Balloons; Blimps; Buoyant aircraft; *Hindenburg*; Icing; Lighter-than-air craft; Military flight; Reconnaissance; Alberto Santos-Dumont; Weather conditions; Ferdinand von Zeppelin

Dogfights

Definition: Aerial battles between two or more aircraft.

Significance: Since World War I, air-to-air combat and the establishment of air superiority has been one of the most crucial components of success in modern warfare.

History

World War I. By the beginning of World War I in August, 1914, many military strategists had already predicted the possibility of combat between aircraft. At the time, military aviation on all sides was limited to a few hundred rudimentary aircraft that were expected to perform reconnaissance missions, artillery spotting, and courier duties. The low performance of available aircraft at the time made the carrying of effective weapons initially pointless, because their added weight made the aircraft incapable of climbing to altitude or of overtaking any opposing aircraft. Early in the war, there existed a camaraderie of the air. Pilots treated each other with a restrained civility, often saluting or waving at enemy pilots in passing. Piloting an aircraft was akin to membership in an elite gentlemen's club.

As the value of aerial observation became apparent to ground force commanders, it soon became necessary to disrupt the enemy's reconnaissance activity in order to wage successful land and sea campaigns. In short order, both pilots and observers began attacking enemy aircraft with rifles, revolvers, semiautomatic pistols, and steel-dart flechettes in attempts to down opposing fliers. As the possibility of being shot out of the sky while on a mission became a real threat, aggressive pilots and resourceful ground crews soon initiated rapid development in both aircraft and aircraft missions in World War I. The three technological developments most noteworthy in the early intensification of aerial combat include the design and production of more powerful engines and robust machines; the installation of lightweight machine guns, synchronized to fire through the propeller arc of single-engine aircraft; and long production runs of mass-produced, standardized aircraft that made possible the institution of formation tactics. As soon as more powerful machines were available, flexible machine-gun mounts were fitted to either the sides or the upper wing surface of the aircraft. This positioning was necessary because the sides, rear, or above the propeller arc were the only safe directions in which to shoot without possibly destroying the front-mounted tractor drive propeller. These early aircraft could not be pointed so both pilot and aircraft were in alignment with the targeted enemy, making for dangerous flying circumstances during an aerial battle. After several experimental attempts, the forward-firing synchronized machine gun was designed and fitted to the cowl of high-performance single-seat scout aircraft. The mission of these aircraft was primarily offensive, and they were employed to destroy enemy reconnaissance and bomber aircraft. These were the first true fighter aircraft. In an effort to protect airplanes on reconnaissance and bombing missions, groups of fighter planes began flying as escorts. Flying out to meet the enemy's reconnaissance, bombers, and escorts was called interception. When fighter escort aircraft encountered fighter interceptors, an aerial melee, which became known as the dogfight, resulted. The sole purpose of the dogfight was to destroy as many enemy aircraft as possible before they could return the favor.

World War II. World War II saw the most prolific application of interceptor and escort strategies. Air-to-air

The use of airplanes by the military in World War I moved combat from the ground to the sky. Here, British SE-5's are locked in a dogfight with German Fokker D-7's in 1915. (Hulton Archive)

Messerschmitt Bf-109, and the Focke-Wulf Fw 190. MiG-15 and F-86 Sabre jet fighters were used in the Korean War. MiG-21 and F-4 fighters were used in the the Vietnam War. MiG-23, F-15, F-16, F-18, and Mirage fighters were used in wars in the Middle East during the last half of the twentieth century.

Tactics

The duel between fighter aircraft to gain control of the skies above a battle theater has become a necessary command strategy. Control of the skies means unfettered access for one's own reconnaissance and bombers to the exclusion of the enemy's. The basic rules of air-to-air combat established during World War I have not changed since. Air-to-air combat, from its very inception, remains exclusively individualist. Early air warfare tactics were essentially individual in nature, evolved by pilots to reflect their own experiences and personalities and altered to suit the circumstances and the aircraft and its armament. Despite advances in technology, this warrior tradition remains in place.

During World War I, pilots learned that the key to success and survival in a dogfight was to gain surprise and get off the first shot. A protracted aerial dogfight, in which the advantage hinges on pilot skill, higher maneuverability, tighter turning radius, munitions, and greater speed, is not the optimum scenario. Drawn-out dogfights typically end in stalemate or random losses due to some unforeseen circumstance. The primary rule of all air-to-air combat is to take the enemy by surprise. Nearly all aerial kills are the result of the surprise attack, in which the attacking pilot obtains a favorable position, usually high and to the rear, and fires the initial attack. The victim usually never sees the attacker. The average aircraft-to-aircraft aerial duel takes less than ninety seconds.

An effective fighter pilot must not only be skilled but also must be able to apply those skills quickly under the intensity and pressure of a life-and-death struggle that takes place on a three-dimensional battlefield at incredible speeds. A dogfight is not a planned mission. Once the duel begins, all operational order is gone. One of the most common tactics in dogfighting is to force the enemy into elabo-

combat and superior dogfighting aircraft swung the balance of power and ultimate air supremacy toward the Allied forces. Aerial duels during the Battle of Britain (1940), the Allied daylight bombing raids on Germany (1942-1945), the Pacific Island campaign (1942-1945), and operations on the Russian front (1941-1944) established the doctrine of air supremacy as the key to victory in modern conventional warfare.

Fighter Planes. Some of the most recognizable and renowned aircraft in the history of aviation have been fighter planes. Many well-known aircraft were designed specifically for the air-to-air mission. World War I fighters included the Fokker Dr-I triplane, the Sopwith Camel, the Spad XIII, and the Albatros D-III. World War II fighters included the Spitfire, the Hurricane, the P-51 Mustang, the P-38 Lightning, the Corsair, the Mitsubishi Zero, the rate maneuvers that deplete the enemy craft's fuel supply and force the enemy to break off the engagement, at which point the enemy becomes exposed and vulnerable to follow-up attacks. Interceptor pilots defending air space have an advantage in that they require less fuel. Defending interceptors can linger in their air space longer, and, because they are closer to their bases, they can land, refuel, rearm, and return to battle if necessary.

In modern warfare, weaponry and personnel are likely to be somewhat evenly matched. It has been known since World War I that excellence in fighter aircraft design is more important than greater speed and that maneuverability and weapons technology are the keys to successful fighter design. Often, however, the outcome of air-to-air combat is influenced by factors other than aircraft performance and firepower, such as the pilot's skill and morale, the tactical situation or mission, the weather, the balance of forces in the air, and intelligence data. Yet, to win a dogfight, the pilot must be equipped with an aircraft capable of keeping up with the enemy and must be trained to use the aircraft to its maximum potential. Superior aircraft coupled with inferior pilots is no match for skilled pilots in similar aircraft. Historically, about 5 percent of combat pilots account for more than 50 percent of all downed enemy aircraft during a conflict. Putting as many skilled pilots as possible into a battle theater is the most efficient way to gain air superiority.

From the beginning of air-to-air combat, spotting the enemy first, acquiring position, and firing the first shot have been the keys to success and survival. Although pilot skill remains an important factor, modern dogfighting is a matter of teamwork and applied technology. With the advent and application of long-range detection systems, weapons, and communications, pilots can detect, coordinate, and attack opposing aircraft from greater distances. In modern air warfare, the side with the superior detection systems usually gets the superior position and manages to fire the first shot. Early detection also allows for quicker adaptation to fluid battlefield conditions. Modern improvements in aircraft armament and sighting allow pilots to reach out and touch the enemy at greater distances and with a greater measure of success. A modern 30-millimeter cannon is highly accurate to 800 meters, compared with the 100 meters of an 8-millimeter machine gun of World War I. Modern air-to-air missiles have kill ranges of up to 200 kilometers and are highly reliable at ranges of 10 to 50 kilometers. Because of these long-range munitions, most modern dogfights often take place beyond the visual range of the combatants.

Randall L. Milstein

Bibliography

Cooksley, P. G. *Air Warfare*. London: Arms and Armour Press, 1997. A well-illustrated basic book covering weapons, bases, personalities, tactics, and events in the history of air warfare, with a bias toward British aviation history.

Gunston, B., et al. *Fighter Missions*. New York: Orion Books, 1988. A beautifully illustrated and informative book outlining the modern doctrines of air combat.

Guttman, J. *Fighting First: Fighter Aircraft Combat Debuts from 1914 to 1944*. London: Cassell, 2000. A volume covering the important aircraft and fliers from World War I through World War II and recounting the most famous air battles of both wars.

Park, E. *Fighters: The World's Great Aces and Their Planes*. Charlottesville, Va.: Thomasson-Grant, 1990. A beautifully illustrated, large-format book that outlines the exploits and histories of the most famous combat aircraft and renowned combat pilots.

See also: Battle of Britain; Black Sheep Squadron; Bombers; Fighter pilots; Fokker aircraft; Korean War; Luftwaffe; Messerschmitt aircraft; Military flight; Reconnaissance; Manfred von Richthofen; Eddie Rickenbacker; Sopwith Camels; Spitfire; Triplanes; Vietnam War; World War I; World War II

Jimmy Doolittle

Date: Born on December 14, 1896, in Alameda, California; died on September 27, 1993, in Pebble Beach, California

Definition: A pilot and pioneer of military aviation and instrument flying.

Significance: As a member of the U.S. Army Air Service in 1922, Doolittle made the first transcontinental flight in less than twenty-four hours. He was most noted for leading the first air raid over Japan during World War II.

James Harold "Jimmy" Doolittle was born in California but spent much of his youth in Alaska. He left the University of California in 1917 to enlist in the U.S. Army Reserve and was assigned to the Signal Corps. During World War I (1914-1918), he served as an aviator and flight instructor. Commissioned as a first lieutenant in 1920, he spent much of the following decade in the development of military aviation.

During this period, Doolittle combined his interest in aviation with the sport of flying. He took part in numerous races, winning a number of trophies. In September, 1922, he carried out the first transcontinental flight from Florida to California, a distance of more than 2,100 miles, in fewer than twenty-four hours. The purpose of the flight was to support the growing role of the U.S. Army Air Service in the nation's defenses. At the same time, the flight brought Doolittle to national prominence.

In 1930, Doolittle resigned from the Army to work for the Shell Petroleum Company. He continued to race, setting a world speed record in 1932. In 1940, Doolittle rejoined the Army Air Corps with a rank of major. On April 18, 1942, as a lieutenant colonel, he led a force of sixteen B-25 bombers from the USS *Hornet*, hitting targets in Japan more than 800 miles across the Pacific. Although the targeted cities, Tokyo, Yokohama, Kobe, and Nagoya, received negligible damage, the raid shattered the impenetrable image of the Japanese islands. Most of the planes and their seventy-five fliers crash-landed in China. Doolittle was awarded the Congressional Medal of Honor for his action.

During the war, Doolittle rose to the rank of lieutenant general, commanding the Twelfth Air Force in North Africa and the Fifteenth Air Force elsewhere in the region. In 1944, Doolittle assumed command of the Eighth Air Force, directing bombing of Germany until the end of the war. From 1948 to 1958, Doolittle served on both the National Advisory Committee for Aeronautics and and the President's Science Advisory Committee. He became director of Space Technology Laboratoriesfollowing his retirement from the Air Force in 1959.

Richard Adler

Bibliography

Doolittle, James, and Carroll Glines. *I Could Never Be So Lucky Again*. New York: Bantam, 1991. Doolittle's autobiography, covering his extensive career in the air service, with emphasis on the Tokyo raid.

Glines, Carroll. *The Doolittle Raid*. Atglen, Pa.: Schiffer, 1999. Among the most detailed and most recent of numerous books on the subject.

Schultz, Duane. *The Doolittle Raid*. New York: St. Martin's Press, 1988. Contains excellent material on Doolittle, with emphasis on his famous raid and its aftermath.

See also: Air Force, U.S.; Bombers; National Advisory Committee for Aeronautics; Transcontinental flight; World War I; World War II

Doppler radar

Definition: A device for determining the radial velocity of an object by measuring the frequency change in the echo of a radio wave reflected from the object.

Significance: Doppler radar may be used to determine the speed of a distant object along the line of sight, or when installed in a moving aircraft, used to determine the true speed relative to the ground. Doppler radar is also used to map the internal structures of severe storm systems posing a potential hazard to nearby aircraft.

Radar, an acronym derived from "radio detection and ranging," uses radio waves of constant frequency reflected from a target to determine its position, distance, direction, and speed. When radio waves emitted from a transmitting antenna are interrupted by a solid object, a portion of the energy, called an echo, is reflected back toward the transmitter, which can also function as a receiver. The distance to the object is determined by the time required for the radio wave to travel from the transmitter to the object and back to the receiver. Since radio waves, like light, travel at the constant speed of 186,000 miles per second, the measured time delay is proportional to the distance. If the transmitter were operating continuously, however, reflected signals would also be continuous, making it impossible to disassociate the emitted signal from the returning signals of different objects. The emitted signal is therefore labeled by emitting the signal in short, high-powered pulses rather than continuously. During the short interval between pulses, the transmitter is operated as a receiver. If the transmitter is not emitting when the return signal arrives, the received signal can be associated with a specific transmitted pulse.

Doppler Effect

The Doppler effect, discovered by Austrian physicist Christian Johann Doppler in 1842, is the change in the observed frequency of a wave due to relative motion between the observer and the wave source. When observer and source approach one another, the emitted frequency of the waves is measured to be higher due to the velocity of approach; the greater the relative speed, the greater the frequency shift. When the source and observer are receding from each other, the emitted frequency is measured to be lower, in direct proportion to the velocity of recession. The effect is particularly noticeable for sound waves; when an ambulance speeds past, the pitch (frequency) of its siren drops noticeably.

Although the Doppler effect has been well-known for sound and light since the mid-nineteenth century, radar systems utilizing this effect were not developed until after World War II. By detecting the frequency shift in the reflected signal caused by an object having a component of velocity toward or away from the observer, the object's speed along the line of sight can be calculated from the Doppler equation. Unlike the pulsed systems, Doppler radar is a continuous wave system. A single antenna can be utilized because the reflected signal from a moving object returns at a different frequency, hence the outgoing and incoming signals are not confused.

Applications

Radar was developed by the military for its own use and still finds its major applications in the military arena. It is used to detect aircraft, missiles, artillery projectiles, ships, land vehicles, and satellites. Civilian applications include the surveillance of aircraft and weather in the vicinity of airports. Air route surveillance tracks aircraft between airports up to 200 miles away. Radar is also used as a surface detector at airports to give the controller the location and movement of ground-based vehicles within the airport.

A Doppler navigator is a simple continuous-wave system used to determine a plane's ground speed. The plane's radar has an antenna that directs a beam forward and down toward the ground at a 45-degree angle to the direction of flight. The plane's velocity can then be determined from the relative radial motion. A radar altimeter gives the height of an airplane above the ground by reflecting signals straight down.

The National Weather Service (NWS) and the Federal Aviation Administration (FAA) now deploy a network of Doppler radar systems to monitor potential weather hazards to aircraft. By measuring the radial velocity of precipitation in conjunction with the strength of reflected signals, the severity of storms up to 250 miles away can be accurately gauged. The intensity of the echoes from raindrops and ice particles reveals the type of approaching storm and enables forecasters to predict when violent storms will reach specific regions. Doppler radar can also be used to pinpoint hazardous wind conditions such as downbursts, strong blasts of air associated with storm systems and a major cause of aircraft accidents.

Since Doppler radar is the only remote sensing instrument that can detect and measure the radial velocity of wind inside areas veiled by clouds, it is also used to probe the internal motions and structure of tornadoes or other potentially hazardous weather systems. In addition to aiding researchers better to comprehend the dynamics and life cycles of severe storms, this unique capability provides improved early warnings of impending weather hazardous to human communities.

George R. Plitnik

Hearing the Doppler Effect

It is very easy to hear the Doppler effect at the auto races. Race cars are noisy and emit a jumble of sounds dominated by the roar of their engines. When a race car approaches a listener (or a microphone), the pitch of the sounds it emits is higher. As the car passes, the pitch drops noticeably lower. When a car, or an airplane, approaches a radar set, the reflected radar waves will be Doppler-shifted. This means that when the echo reaches the receiving antenna, the wavelength will be shorter, and the frequency will be higher than the radar waves originally sent out by the transmitter. Because the amount of the Doppler shift depends on the target's speed, this speed can be calculated. If the target is moving away from the radar set, or even if the set itself is on a moving aircraft, the target's speed can still be found from the Doppler effect.

Bibliography

Blake, Bernard, ed. *Jane's Radar and Electronic Warfare Systems*. 6th ed. London: Jane's Publishing, 1994. Complete description of the various types of radar systems and their military applications.

Doviak, R. J., and Dusan Zrnic. *Doppler Radar and Weather Observations*. 2d ed. New York: Academic Press, 1993. A comprehensive summary which introduces basic theory enhanced with numerous observations and measurements not available in other texts. Although the presentation is often technical, there is a wealth of information accessible to the general reader.

Hitzeroth, Deborah. *Radar: The Silent Detector.* Murray Hill, N.J.: Lucent Books, 1990. A clear and concise treatment of the principles and practice of radar and its applications.

Skolnik, Merrill, ed. *Radar Handbook*. 2d ed. New York: McGraw-Hill, 1990. A complete handbook detailing all aspects of radar, including a wealth of technical information.

Strong, W., and G. R. Plitnik. *Music, Speech, Audio*. Provo, Utah: Soundprint, 1992. An easy-to-read introduction to the science of acoustics, containing a complete explanation of the physics of the Doppler effect in descriptive terms easily understood by the general reader.

See also: Air traffic control; Airports; Federal Aviation Administration; Instrumentation; Military flight; Radar; Weather conditions; World War II

Dresden, Germany, bombing

Date: February 13-14, 1945

Definition: A series of British and American air raids during World War II that nearly destroyed the scenic and historic German city of Dresden.

Significance: The controversial Allied bombing of Dresden raised the issues of whether the mass bombing of an urban area was a legitimate military strategy and whether the bombing of Dresden, in particular, was more of a political than a military action.

Air Campaign

With the World War II nearing an end, the British air ministry in January, 1945, devised a plan called Thunderclap, an air offensive that was to be directed at Berlin and population centers in eastern Germany. The Allies' major justification of this campaign was that it would add to the growing chaos in Germany created by the rapid westward advance of Russian troops and thus make it more difficult for the German army to summon reinforcements and armaments to meet the Russian advance. The attack also was intended to crush German morale. The Russians had been pressuring the British and Americans to conduct such an offensive in order to paralyze German communications. It was an outgrowth of a grand Allied strategy initiated in 1943, which called for combined operations to crush the German war machine. Specifically, the plan called for bombardment by the Allies from the air, by Allied ground operations from the west, and by Russian ground operations from the east.

Dresden was officially designated a military target for several reasons. First, it was considered a primary communications center in the Berlin-Leipzig-Dresden railway complex. Second, it was an important industrial and manufacturing center directly associated with the production of aircraft components and other military items, including poison gas, antiaircraft guns, and small arms. Third, it was believed that a raid would devastate the area, curtailing communications within the city and disrupting the normal civilian life upon which the city's larger communications activities and manufacturing enterprises depended. In addition, it was theorized a widespread assault that included bombing strikes against the city's industrial plants, which were interspersed throughout the region, would be construed as part of the overall pattern of the raid. However, many historians have argued that Dresden had, from a military perspective, virtually no great strategic importance. The city had little heavy industry and for this reason had been spared earlier bombings, except for a small raid by the Eighth U.S. Army Air Force in October, 1944.

Noted for its magnificent architecture and its manufacture of fine china, Dresden long had been considered one of the most beautiful cities in Europe. Its streets were adorned with statuary and other art, much of which dated from the seventeenth century. Among the city's residents, there was a mistaken notion that the city's grandeur protected it from an all-out attack. Until February, 1945, Dresden suffered from only those problems confronting most other German cities of the time: the loss of men in action and the economic hardships resulting from the war. Residents further believed that a nonstrategic city with a large number of military hospitals, POW compounds, and refugees would not face the same attacks that other cities had. Consequently, most of the German air defense and flak batteries that would otherwise have been stationed in Dresden had been relocated to areas where it was assumed they were more needed.

Raids

On February 13, the British Royal Air Force (RAF) Bomber Command dispatched 796 Lancaster bombers and 9 Mosquitoes from the United Kingdom. The planes attacked Dresden in two waves, three hours apart, dropping first high-explosive bombs and then tons of incendiaries that precipitated a mammoth firestorm. The high-explosive bombs were intended to demolish roofs and windows, leaving the interiors of buildings vulnerable to the second wave of bombers that followed with the incendiaries. Soon, rising columns of intense heat merged into a single conflagration that sucked up oxygen and burned it, creating hurricane-force winds and temperatures of up to 1,000 degrees Fahrenheit. On the following day, U.S. B-17 bombers, in a third-wave strike against the city, contributed to the damage. Target sector markings on RAF photographs confirmed that the attack on the heart of the city was carried out as planned. Military areas situated north of the city, including factories and freight stations, received minimal damage, though part of an American P-51 Mustang fighter escort was ordered to strafe traffic on the roads around Dresden to heighten the chaos. Only eight Allied planes were shot down during the assault.

At the time of the attacks, Dresden was virtually defenseless, because all remaining German fighter planes assigned to the area had been grounded for lack of fuel. In addition, the inhabitants of the city were mostly women and children who recently had fled the Russian offensive in the east. Indicative of the town's feeling of relative security was the fact that it had never put into operation the civil defense precautions for lessening the effects of potential firestorms that had been taken in other cities.

The total devastation wrought on the city in the twenty-four-hour period was unprecedented in its suddenness and totality. The city was nearly extinguished in a single blow. For a long period of time, residents remained in a state of collective shock so intense that the shock itself nearly numbed their grief. However, their daily struggle for survival amid the ruins instinctively became their immediate concern. Helplessness and despair set in only later.

The firestorm created by the bombardment was impossible to extinguish, because the incessant assault made it extremely difficult for the fire service to utilize either the river or canals. The fires burned for seven days. Although a final death toll in the bombardment has never been established, estimates have run between 25,000 and 135,000. The hospitals that survived the raids were overwhelmed. Corpses were loaded onto farm carts for burial in mass graves or stacked in huge pyres, as distraught men and women roamed the desolate streets seeking any traces of relatives or friends. Survivors later related macabre scenes, such as a bus filled with dead soldiers, all sitting perfectly in their seats. In another area, some corpses were discovered dressed in costumes for a February 13 pre-Lenten carnival that some residents had been celebrating. The raid completely leveled the center of the city, including the cathedral, known as the Frauenkirche, or Church of Our Lady. Overall, almost 12,000 buildings were turned into a wasteland of rubble and smoldering corpses. Most of the deaths on the ground were caused by suffocation and carbon monoxide poisoning, as residents huddled in their cellars waiting for the fires from the first attack to be extinguished. The area of destruction was approximately three times that of the area damaged in London during the more than two months of sustained German bombing in the Battle of Britain (1940). The firebombing inflicted on Dresden left almost nothing standing and most of what did remain was bulldozed for safety reasons.

Aftermath

From its inception, the Allies' strategic bombing campaign in Europe had generated controversy. Despite its general horrors, the air offensive failed to break German morale and, until the last months of the war, did not decisively impact German industrial production. Although the bombing contributed to the ultimate Allied victory over the German forces, it clearly did not in itself bring about victory.

Much of the speculation on the motive for the attacks has centered on Air Vice Marshal Sir Arthur "Bomber" Harris, commander of the RAF Bomber Command. Harris was the principal proponent of nighttime area strategic bombing carried out by formations of heavy bombers attacking large targets, such as cities, which were nearly impossible to miss at night. He believed that long-term bombing was the most effective way of destroying an enemy's industrial centers and demoralizing its population. He also made it clear that he opposed the redeployment of bombers to other theaters of aerial operations. In the spring of 1944, he strongly objected to the temporary interruption of the strategic bombing campaign by Allied leaders who planned to relocate the bombers to France to help in the assault on rail lines and bridges in preparation for the Normandy invasion.

The Harris strategy was an outgrowth of the original doctrine of strategic heavy bombing that had emerged in Europe at the end of World War I and was adopted in part as a justification for separating air operations from the British Army. Prior to the bombing of Dresden, the cities of Lübeck, Rostock, Pforzheim, Hamburg, Hildesheim, Cologne, Magdeburg, Mainz, and Würzburg had been subjected to tactical bombardments, where the maximum number of bombs had been dropped within as large a target area as possible. Theoretically, strategic bombing would accomplish several goals, including the crippling of German industry and the undermining of the Nazi war effort. It was also felt that because the RAF was poorly trained in night bombing, a practice it was not expected to perform, it lacked efficient bombs and navigational aids to conduct precision bombing. However, even the daylight attacks proved wildly inaccurate and achieved little of strategic value. Photographic evidence revealed that only a small percentage of bombs was being dropped anywhere near designated targets. Consequently, the RAF adopted the policy of "area bombing," the less-discriminate bombing of entire cities and towns. During the course of the European air campaign, more than 500,000 German citizens and more than 55,000 RAF airmen fell victim to the strategic bombing campaign.

The Allied bombing of Dresden, Germany, on February 13-14, 1945, completely destroyed one of the most beautiful cities in Europe. (Hulton Archive)

Despite the criticism that followed the Dresden raid, Harris remained unrepentant, citing in his defense British prime minister Winston Churchill's approval of the raid. In the opinion of many military historians, Churchill only agreed to the bombardment because the RAF demonstrated such a great lack of accuracy during air operations. Although they were subjected to searchlights, heavy flak, and night fighter attacks, the RAF bombers and their crews were spared the vengeance of marauding Luftwaffe day fighters, who took a heavy toll on the Americans.

Following the war, Dresden became a part of East Germany, whose authorities continued the city's ruin by bulldozing vast areas of burned-out buildings and replacing them with Soviet-style, high-rise apartment buildings. With the fall of Communism, efforts to reconstruct the city received a fresh boost, as programs were developed to construct new housing and restore old landmarks.

Although some have viewed the Dresden bombing as cruel and senseless because it targeted civilians while German capitulation was so near, others have justified the raids by arguing that Germany started the war and carried out terror bombings on British cities. In the latter view, the real responsibility for the Dresden bombing rests with Nazi leaders. Although debate continues over the ultimate responsibility for the terror, there is little doubt that the bombing of Dresden represented a milestone in the annals of modern warfare.

William H. Hoffman

Bibliography

Clayton, Anthony, and Alan Russell, eds. *Dresden: A City Reborn*. New York: Berg, 2001. A comprehensive introductory history of Dresden, including the seventeenth century baroque period, events surrounding the World War II bombardment, and the postwar reconstruction efforts.

Irving, David. *The Destruction of Dresden*. New York: Holt, Rinehart and Winston, 1963. A detailed analysis by a British historian of the planning that went into the air strikes against Dresden.

McKee, Alexander. *Dresden, 1945: The Devil's Tinderbox*. New York: E. P. Dutton, 1982. A critical account of the Dresden raids, based in part on a study of official records and interviews with survivors and airmen who participated in the bombings.

Vonnegut, Kurt. *Slaughterhouse Five*. New York: Delacorte, 1969. Vonnegut, who witnessed the raids as a prisoner of war, recreates the scene of the terror in his critically acclaimed novel.

See also: Air Force, U.S.; Bombers; Flying Fortress; Luftwaffe; Royal Air Force; World War II

Hugh L. Dryden

Date: Born on July 2, 1898, in Pocomoke City, Maryland; died on December 2, 1965, in Washington, D.C.

Definition: American aerodynamicist who conducted pioneering research in high-speed aerodynamics and coined the word "transonic" to mean "at or near the speed of sound."

Significance: Dryden studied the phenomenon of compressibility associated with fluid motion near the speed of sound and designed wind tunnels to test the aerodynamic problems of transonic flight. His innovative use of a hot-wire anemometer to measure accurately the levels of turbulence in wind tunnels enhanced scientific understanding of turbulence and the boundary layer as they affect flight.

Born in 1898, Hugh Latimer Dryden finished high school at the age of fourteen and received a scholarship to Johns Hopkins University in Baltimore, Maryland. There he studied physics and mathematics and completed his undergraduate work in three years. In 1919, at the age of twenty, Dryden completed his doctorate in applied physics, with a dissertation entitled "Air Forces on Circular Cylinders."

Dryden then accepted a full-time leadership position in the aerodynamic division of the National Bureau of Standards (NBS). In 1934, he was named chief of the mechanics and sound division. In 1940, he was asked to develop a radar-guided aerodynamic missile head at the Office of Scientific Research and Development (OSRD). He received the Medal of Freedom in 1946 for his work for the U.S. Army Air Force.

After World War II (1939-1945), Dryden was appointed assistant director and then associate director at NBS. He was named the director of aeronautical research of the National Advisory Committee for Aeronautics (NACA) in 1947 and the director in 1949. When the National Aeronautics and Space Administration (NASA) was created after the Soviet Union's launch of Sputnik 1 in 1957, Dryden became deputy administrator of the new organization, remaining in that position until his death.

The bulk of Dryden's scientific research was performed prior to his post at OSRD. His ground-breaking research in wind-tunnel design allowed scientists to predict the effects of turbulence on aircraft performance. Using a hot-wire anemometer to measure rapid air fluctuations, he and a colleague found that turbulence could indeed account for the drag on aircraft. Dryden's wind-tunnel experiments enabled engineers to gain a clearer understanding of laminar and turbulent flows in the boundary layer. Dryden's work also led him to redesign wind tunnels in order to reduce the effects of turbulence and drag. More efficient wind tunnels led to the design of more effective aircraft. Dryden's experimental work on the boundary layer and laminar flow validated the earlier theoretical work of Ludwig Prandtl.

In 1938, Dryden became the first American to deliver the Wilbur Wright Lecture sponsored by the Institute of Aeronautical Sciences in England, with an address entitled, "Turbulence and the Boundary Layer." Dryden received the Daniel Guggenheim Medal in 1950 for his contributions to aeronautics. In 1976, the NASA Flight Research Center in California was renamed the NASA Hugh L. Dryden Flight Research Center. Although Dryden was diagnosed with cancer in 1960, he continued working and lecturing for the next five years before finally succumbing to the disease on December 2, 1965.

Said Elghobashi

Bibliography

Gorn, Michael H. *Hugh L. Dryden's Career in Aviation and Space*. Washington, D.C.: NASA History Office,

1996. Valuable information about Dryden's life and career.

Smith, Richard K. *The Hugh L. Dryden Papers, 1898-1965*. Baltimore: Milton S. Eisenhower Library, Johns Hopkins University, 1974. A good source of information regarding Dryden's life and work.

Thomas, Shirley. *Men of Space*. 8 vols. Philadelphia: Chilton, 1960-1968. A good overview of the U.S. space program, with reference to Dryden's work.

See also: Aerodynamics; Forces of flight; High-speed flight; National Advisory Committee for Aeronautics; National Aeronautics and Space Administration; Ludwig Prandtl; Wind tunnels; World War II

E

Eagle

Also known as: F-15, F-15A, F-15B, F-15C, F-15D, F-15DJ, F-15E, F-15J

Date: First flight on July 27, 1972

Definition: The dominant U.S. tactical fighter jet since 1972.

Significance: The F-15 Eagle has served since the mid-1970's as the primary jet fighter in the U.S. arsenal, and it remains superior to any other operational fighter in the world. Several countries allied with the United States also utilize the F-15, which enjoys an outstanding combat record and has evolved into several air-superiority and ground-attack variants.

Development

Conceived during the Vietnam War (1961-1975) by engineers at McDonnell Douglas, the F-15 Eagle was designed as an air-superiority fighter capable of defeating all types of enemy aircraft at close or long range under any weather conditions. The first F-15A flew in 1972, with a two-seat training version, the F-15B, following one year later. After extensive flight testing, the U.S. Air Force in 1976 took delivery of the first Eagle slated for a combat squadron.

From its beginnings, the F-15 demonstrated its superiority over contemporary fighters, combining high speed with spectacular maneuverability via a high thrust-to-weight ratio and a low wing loading, or ratio of weight to wing area. The F-15 had more power and wing area in relation to its weight than any other fighter in the world, and it performed accordingly.

Equipped with two Pratt & Whitney F100-100 engines capable of 23,930 pounds of thrust each, the F-15 could reach an altitude of 50,000 feet from a stationary start in just 2.5 minutes and could attain a maximum speed of 1,650 miles per hour.

Power and maneuverability were, however, only two of the F-15's advantages. The F-15 also included a revolutionary heads-up display (HUD), which allowed pilots to monitor critical flight data, such as speed, course, and altitude, without taking their eyes off the sky. It boasted an inertial navigation system that allowed pilots to fly anywhere in the world with unerring accuracy, an advanced computer, and extensive electronic countermeasures for defense against enemy radar and missiles. With its Hughes APG-63 pulse-Doppler radar and a service ceiling of 65,000 feet, the Eagle could fly well above most potential enemies, could detect enemies from a long range at either high or low altitude, and, if necessary, could destroy enemies at minimum risk. Early F-15 models carried four AIM-7 Sparrow and four AIM-9 Sidewinder missiles as well as a 20-millimeter cannon. Later versions were equipped with up to eight AIM-120A advanced-medium-range air-to-air missiles (AMRAAMs) in place of the Sparrows.

In response to an increasing threat from the Soviet Union, the U.S. Air Force decided to improve upon the F-15's capabilities and field the F-15C and F-15D training models in 1979. These new Eagles carried more internal fuel, as well as exterior conformal fuel tanks, for a greater range of more than 3,400 miles without refueling. An increased maximum takeoff weight of 68,000 pounds allowed F-15 pilots to carry more weapons than ever before.

The seemingly elastic ability of the F-15 to adopt more capabilities led the Air Force to stretch the design even further when, in 1982, it fielded the two-seat F-15E dual-role fighter. Designed for air-to-air and ground-attack missions, the E model featured a larger HUD, a more powerful central computer, a forward-looking infrared (FLIR) camera, and Maverick missiles for use against ground targets at long range. It boasted the Hughes APG-70 radar for use in either ground-attack or air-superiority mode, and automatic terrain-following avionics and could perform long-range missions at night or in bad weather.

These improvements were showcased in 1990 and 1991, when the Air Force deployed F-15C, F-15D, and F-15E models to Saudi Arabia during Operations Desert Shield and Desert Storm. In combat against Iraq, Eagle pilots destroyed thirty-six enemy aircraft in air-to-air engagements without losing a single plane, and F-15E models were used in night attacks on Scud missile sites and against artillery positions with great effect. Since 1991, F-15's have patrolled the no-fly zone over northern and southern Iraq, escorted cargo planes during Operation Provide Comfort

F-15 Eagle and F-15E Strike Eagle Characteristics

	F-15C/D Eagle	*F-15E Strike Eagle*
Primary Function	Tactical fighter	Air-to-ground attack aircraft
Builder	McDonnell Douglas Corporation	McDonnell Douglas Corporation
Unit Cost	$34.3 million	$31.1 million
Power Plant	Two Pratt & Whitney F100-PW-220 or 229 turbofan engines with afterburners	Two Pratt & Whitney F100-PW-220 or 229 turbofan engines with afterburners
Thrust per engine (pounds)	23,450	25,000-29,000
Length (feet)	63.8	63.8
Height (feet)	18.5	18.5
Maximum Takeoff Weight (pounds)	68,000	81,000
Wingspan (feet)	42.8	42.8
Ceiling (feet)	65,000	50,000 (service), 35,000 (combat)
Speed	1,875 miles per hour (Mach 2.5 plus)	1,875 miles per hour (Mach 2.5 plus)
Crew	F-15C, 1; F-15D, 2	2
Armament	One internally mounted M61A1 20-millimeter six-barrel cannon with 940 rounds of ammunition; four each AIM-9L/M Sidewinder and AIM-7F/M Sparrow air-to-air missiles, or eight AIM-120 AMRAAMs, carried externally	One 20-millimeter multibarrel gun mounted internally with 500 rounds of ammunition, four each AIM-9L/M Sidewinder and AIM-7F/M Sparrow air-to-air missiles, or eight AIM-120 AMRAAMs, any air-to-surface nuclear or conventional weapon in the U.S. Air Force inventory
Inventory	Active force, 396; Reserve, 0; ANG, 126	Active force, 217; Reserve, 0; ANG, 0
Date Deployed	July, 1972	April, 1988

Source: Data taken from (www.af.mil/news/factsheets/F_15_Eagle.html) and (www.af.mil/news/factsheets/F_15E_Strike_Eagle.html), June 6, 2001.

in Turkey, flown combat missions in support of NATO operations in Bosnia and Kosovo, and performed many other assignments around the world.

F-15's form an important part of the air forces of several countries allied to the United States, including Japan (which uses specially designed F-15DJ and F-15J models), Saudi Arabia, and Israel. They have been flown with distinction in Israeli combat operations against both Syria and Iraq, and will remain an integral component of fighter squadrons around the world until at least the mid-twenty-first century. By then, the 396 Eagles in active service with the U.S. Air Force will have been transferred to Reserve or Air National Guard squadrons, having been replaced by the F-22 Raptor. Raptor pilots will count themselves fortunate if they find themselves in aircraft that prove to be as dominant over the next thirty years as the F-15 has been since 1972.

Lance Janda

Bibliography

Foster, Peter R. *F-15 Eagle*. London: Ian Allen, 1998. A good basic overview of the F-15 and its history, with excellent photographs.

Jenkins, Dennis R. *McDonnell Douglas F-15 Eagle*. North Branch, Minn.: Specialty Press, 1997. This book is aimed at a general audience, with more emphasis on technical detail.

Verlinden, Francois. *Lock on No. 22: McDonnell Douglas F-15 Strike Eagle*. Lier, Belgium: Verlinden Publications, 1993. The best overall work on the F-15.

See also: Air Force, U.S.; Fighter pilots; Gulf War; Instrumentation; Military flight; Raptor; Vietnam War

Amelia Earhart

Date: Born on July 24, 1897, in Atchison, Kansas; died on July 2, 1937, near Howland Island in the Pacific Ocean

Definition: The most famous female aviator in the United States, known for her record-setting nonstop flights in the 1920's and 1930's.

Significance: Earhart demonstrated that a woman could withstand the rigors of long-distance solo flights just as well as a man. As the first woman to fly across the Atlantic Ocean, she gave popular lectures on her aviation adventures. While attempting to fly around the world in 1937, she and her navigator, Fred Noonan, disappeared; their bodies were never found.

Early Life

Amelia Mary Earhart and her sister Muriel never stayed long in one place as they grew up, because their father continually moved around the United States to find work. The girls spent some time living with their grandparents. Nevertheless, they had a good education and grew up to love books and music. While living in Toronto, Earhart befriended a Royal Flying Corps officer who took her to see his planes and airfield. This experience sparked her lifelong love of aviation.

In Los Angeles, Earhart began to take flying lessons and bought her first plane, financed partly by her parents and partly by money she had earned driving a truck. After returning to the Boston area in 1925, she went to work teaching English, first at the University of Massachusetts and later at Denison House, a social settlement. She continued her hobby of flying and became known among the local aviators.

Setting Records

In April, 1928, Earhart was selected by the publisher George Palmer Putnam to be a passenger on a flight that would make her the first woman to cross the Atlantic by air. The flight was sponsored by Amy Phipps Guest, an American flying enthusiast living in London who had bought the explorer Admiral Richard Byrd's Fokker Trimotor plane the *Friendship*. Unable to make the trip herself, Guest had asked Putnam to find a young woman to represent her in the promotion of women in aviation. Putnam saw qualities in Earhart that he hoped would make her an appealing icon of American womanhood.

Amelia Earhart was famous for her record-setting nonstop flights throughout the 1920's and 1930's. (Library of Congress)

Delayed by bad weather for several days, the *Friendship*'s transatlantic flight began on June 17, 1928, at Trepassy Bay, Newfoundland, when pilot Wilmer Stutz lifted off with Earhart and navigator Lou Gordon. Hampered by fog and a dead radio, they flew for twenty hours and forty minutes before landing their pontoon plane on a river near Burry Port, Wales.

Earhart became an overnight celebrity, even though she had not done the actual flying. She attracted crowds of admirers in Southampton and London and was the guest of honor at parties, where she spoke with Winston Churchill and danced with Edward, prince of Wales. After sailing back to the United States on an ocean liner, she found more fame and opportunity. She received product endorsement offers and writing assignments for *McCall's* and *Cosmopolitan*. Putnam immediately organized a lecture tour for her and rushed her account of the flight, *Twenty Hours Forty Minutes* (1928), into print.

In 1932, Earhart, now married to Putnam, made a solo flight from Harbor Grace, Newfoundland, to Culmore, Ireland, landing in a pasture on May 22 after fourteen hours and forty-five minutes in the air. As the first woman to make a solo flight across the Atlantic, she was showered with honors such as the Distinguished Flying Cross from the Congress of the United States, an award from the French Legion of Honor, and a medal from the National Geographic Society. Although Earhart made other solo flights from Hawaii to California, from Los Angeles to Mexico City, and from Mexico City to Newark, New Jersey, and also set various speed records, none of these achievements attracted as much attention as her 1932 flight. She remained active on the lecture circuit but yearned for one more spectacular flight. In 1935, the trustees of Purdue University purchased a twin-engine Lockheed Electra for her, and she began planning an around-the-world flight.

The Last Flight

After much fund-raising and organizational effort, the flight began on June 1, 1937, when Earhart and her navigator, Fred Noonan, left Miami and headed south, where they would follow the equator eastward to Africa and Asia. A month of flying brought them to Lae, New Guinea, on June 29. Departing on July 1 for tiny Howland Island, in the middle of the Pacific. Despite assistance from the U.S. Coast Guard cutter *Itasca*, Earhart became lost in the clouds and was unable to make sufficient use of radio signals to find her way. She was last heard from at 8:45 A.M., July 2, 1937, when she radioed that she was lost and running low on fuel. An extensive search mounted

Amelia Earhart's Final Flight, 1937

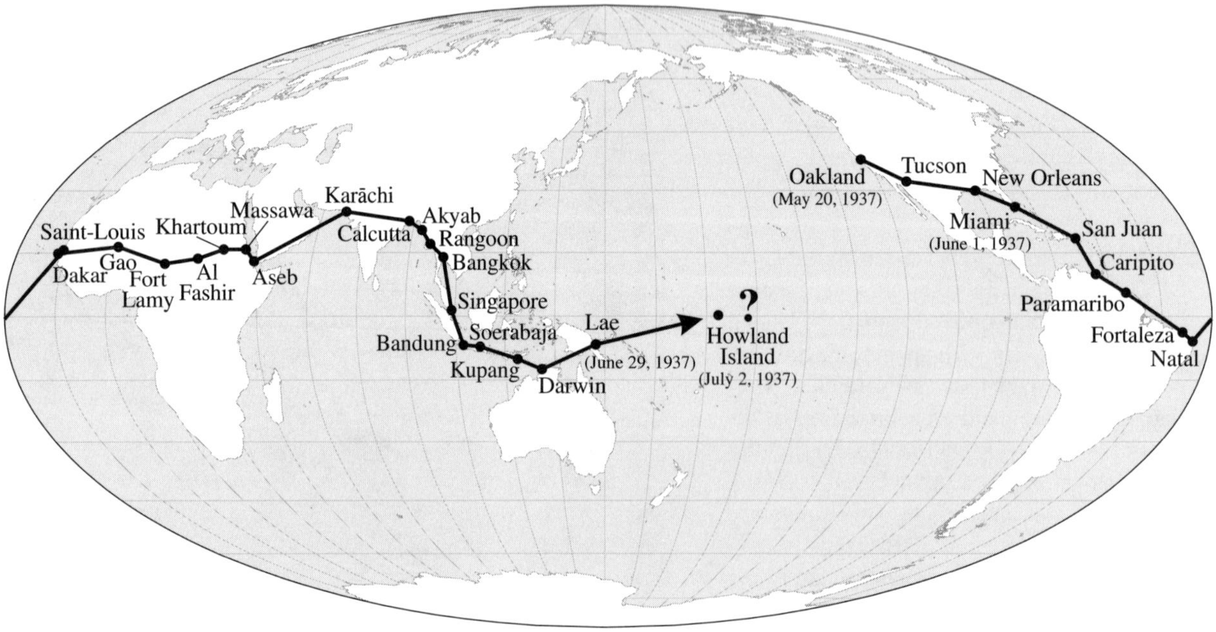

by the U.S. Navy failed to find any trace of her plane or its occupants.

Earhart's fate became one of the century's greatest mysteries, resulting in many unusual claims and theories, mostly unsubstantiated by physical evidence. The most compelling of these is that she turned back from Howland Island, crashed in the Marshall Islands, then under unfriendly Japanese control, and perished of her injuries, either immediately or after languishing in a Japanese prison.

John R. Phillips

Bibliography

Earhart, Amelia. *The Fun of It: Random Records of My Own Flying and of Women in Aviation*. New York: Harcourt Brace, 1932. The most complete autobiographical account of Earhart's life and the most interesting of her books.

_______. *Last Flight*. New York: Harcourt Brace, 1937. Earhart's last writings, compiled by her husband.

_______. *Twenty Hours Forty Minutes*. New York: Putnam, 1928. An account of the flight of the *Friendship*, based on Earhart's log book and her personal thoughts during the flight.

Goldstein, Donald M., and Katherine V. Dillon. *Amelia: The Centennial Biography of an Aviation Pioneer.* Washington, D.C.: Brassey's, 1997. A comprehensive biography illustrated with thirty photographs, some never before published.

Loomis, Vincent, and Jeffrey Ethell. *Amelia Earhart: The Final Story.* New York: Random House, 1985. Brings to light evidence from Japanese and other previously untapped sources in an attempt to explain Earhart's disappearance.

See also: Ninety-nines; Record flights; Transatlantic flight; Transglobal flight; Women and flight

EgyptAir

Also known as: Air Egypt

Definition: The national airline and flag carrier of Egypt, as well as its largest airline.

Significance: EgyptAir is one of the oldest and largest airlines of the Middle East, running comprehensive domestic air service and international services throughout the region, as well as to Europe, North America, Africa, Australia, and the Far East.

History and Fleet

EgyptAir is a major airline, established in 1932 and headquartered in Cairo, Egypt. EgyptAir has owned and operated many different kinds of civil aircraft and was the first airline in the Middle East to operate jetliners. Beginning in 1980, EgyptAir embarked on a modernization and growth plan, developing it from an airline using only seven Boeing 707's and seven 737's to a large, market-oriented, self-financing airline. EgyptAir is a fully state-owned company but is duly vested with its own legislation as a deregulated autonomous organization.

Starting in the 1980's, Egypt Air carried out several dedicated marketing surveys of international and local markets to evaluate its total domestic and international traffic volume. Accordingly, EgyptAir planned its network to maximize its traffic and scope, reaching main cities in all five continents. To carry out its marketing plan, EgyptAir purchased a mix of modern aircraft and then developed an autonomous infrastructure to support and serve its fleet in order to operate in a safe and efficient manner.

As a result of its market research, EgyptAir was reinvented in a stepped-pyramid form. The base of the pyramid was EgyptAir's most profitable markets. In September, 1980, EgyptAir purchased eight Airbus A300-B4 wide-body, medium-range aircraft to cover international markets in Europe and the Middle East, at a total cost of $382.8 million. This became the basis of EgyptAir's process of gradual development. The economic operation of these aircraft enabled EgyptAir to fulfill its financial obligations to the European banks that had financed the purchase of the aircraft. This was a significant step toward establishing the airline's financial credibility, which can be very challenging for an airline from an underdeveloped region of the world such as Egypt.

To cover the domestic market, especially in Sinai, EgyptAir purchased three Dutch Fokker F-27 aircraft. In order to develop long-haul markets, EgyptAir purchased three Boeing 767-200 ER's. To increase its capabilities to serve its long-haul markets and to cope with the needs of the fast-growing passenger and cargo markets, EgyptAir purchased two Boeing 747-300 combis (mixed passenger and cargo planes) followed by two 767-300 ER aircraft. Following that, EgyptAir sought to renew its medium-range, wide-body fleet and purchased nine Airbus A300-600's. At the same time, EgyptAir replaced its medium-range, medium-capacity aircraft with seven Airbus A320-200's, making the airline one of the world's first operators of this high-technology, fly-by-wire (completely computer-controlled) aircraft. In order to link its local tourist market directly with international

The EgyptAir Fleet

Type of Aircraft	*Number of Aircraft*	*Seats (Three-Class Congfiguration)*	*Delivery Date*
Boeing 747-300	2	450	1988
Boeing 777-200	3	319	1997
Boeing 767-300	2	217	1989
Boeing 737-500	5	104	1991-1992
Boeing 737-200	2	130	Scheduled for replacement
Boeing 737-300	1	121	in 2002 by 3 Airbus 318's
Airbus 340-200	3	260	1996-1997
Airbus 300-600	7	253	1990-1991
Airbus 320-200	7	144	1991-1993
Airbus 321-200	4	185	1997

Total Available Seats: 7,997 seats

Source: Data taken from (www.egyptair.com.eg/docs/inside/fleet.htm), June 5, 2001.

destinations, EgyptAir purchased five Boeing 737-500's to replace its aging 737-200's, which had served the airline for more than twenty years. To fulfill the needs of its markets in Japan and North America, EgyptAir purchased three Airbus A340-200's and three Boeing 777-200 ultra-long-range aircraft for nonstop flights. EgyptAir was one of the first operators in the region using these two types of aircraft. It then purchased four Airbus A321-200's to serve charter operations in the development of tourism. EgyptAir financed a contract to purchase three Airbus A318-200's to be delivered early 2002. The process of fleet renewal resulted in a young fleet whose average age is similar to that of most major international airlines.

Corporate Infrastructure and Plans

EgyptAir has invested heavily in constructing an engineering complex to be outfitted with the most modern technical support for its fleet. Its in-flight services complex has a production capacity of twenty-five thousand meals per day. EgyptAir has also established a ground services complex to carry out its loading and off-loading services, as well as the transportation of crew members and employees. It has also established an in-house computer center to control all EgyptAir activities, linking with international reservations systems such as Galileo, Amadeus, SABRE, and World Span.

EgyptAir embarked on a plan of renewing and modernizing its sales offices in Egypt and internationally, as well as modernizing its training center to provide EgyptAir with highly qualified crew members, technicians, and personnel in all other business-related fields. EgyptAir's corporate headquarters were designed, constructed, and equipped with the latest technology and can accommodate 4,500 staff. Finally, as part of its comprehensive modernization plan, EgyptAir established duty-free shops at all Egyptian international airports.

Integrating its activities in the model used by most major world carriers, EgyptAir owns shares in many tourism companies and hotel chains, such as Cairo Airport Mövenpick, Tut Amon, and Nefertari in Aswan and Abu Simbel, and Taba Hilton resorts in Sinai. EgyptAir also owns shares in many charter companies, such as Shorouk Air and Air Cairo. Such investments have increased EgyptAir's assets tenfold since the implementation of the airline's modernization and expansion plan in 1980. EgyptAir's network has also expanded to reach major cities and capitals in all five continents.

The Crash of Flight 990

On the whole, EgyptAir's safety record is good, with a few exceptions. Its most notorious accident was the crash of EgyptAir Flight 990 on October 31, 1999. The New York-to-Cairo flight crashed into the ocean between Long Island, New York, and Martha's Vineyard, Massachusetts, with the loss of all 217 people aboard. The official investigation concluded that the copilot had committed suicide by deliberately diving the plane into the water, although EgyptAir vehemently denied that the copilot had ever shown any signs of mental instability. The presence of a number of Egyptian military officers on the flight suggested the possibility of a terrorist attack. The incident was the focus of much conspiratorial theorizing due to the fact that it occurred in the same general area as the July 19, 1996, crash of TWA Flight 800 and the July 16, 1999, crash of John F. Kennedy, Jr.'s Piper Saratoga.

Triantafyllos G. Flouris

Bibliography

Groenewege, Adrianus D. *The Compendium of International Civil Aviation*. 2d ed. Geneva, Switzerland:

International Air Transport Association, 1999. A comprehensive directory of the major players in international civil aviation, with insightful and detailed articles.

Weimer, Kent J., ed. *Aviation Week and Space Technology: World Aviation Directory.* New York: McGraw-Hill, 2000. An excellent introductory guide on all global companies involved in the aviation business. Provides a basic introduction to the essential information on each company.

See also: Air carriers; Airports; Accident investigation

El Al

Also known as: El Al Israel Airlines

Definition: The national airline and flagship air carrier of Israel and a major international carrier.

Significance: Since its establishment in 1949 as the national airline of Israel, El Al has grown into an international carrier with a reputation for safety and customer service. The International Air Transport Association (IATA) has ranked El Al on several occasions as one of the world's most efficient carriers.

History

El Al Israel Airlines, in Hebrew *El Al Netive Awir Le-yisra'el*, is the Israel national airline and flagship carrier. It was founded in November, 1948, by the government of Israel, shortly after the establishment of the new state. El Al is headquartered in Tel Aviv, which is also the airline's hub. It was El Al's inaugural flight that brought the country's first president, Chaim Weizman, to Israel from Geneva, Switzerland. El Al flew its first commercial scheduled flights to Rome and Paris in July, 1949. By the 1980's, the airline was flying routes from Tel Aviv to many of the major cities of Europe, as well as to Asia, Africa, and North and South America.

In 1981 and 1982, the Israeli government considered liquidation of the airline due to financial difficulties exacerbated by a government ban on Sabbath flights, but ultimately decided against it. The airline has had a history of setting aviation records: In June, 1961, El Al set a new world record for longest nonstop commercial flight on its first nonstop flight from New York to Tel Aviv using a Boeing 707, covering 5,760 statute miles in 9 hours and 33 minutes. In May, 1988, El Al set yet a new first, operating its longest nonstop flight in history: from Los Angeles to Tel Aviv, a 7,000 statute mile trip, covered in thirteen hours and forty-one minutes. On May 24, 1991, an El Al Boeing 747 airlifted a record-breaking 1,087 passengers, Ethio-

The El Al Fleet

Type of Craft	*Total in Service*	*Range (miles)*	*Engines*	*Cruising Speed (miles per hour)*	*Length*	*Wingspan*
Boeing 737-700	2	2,500	C.F.M.	524	110 feet	112 feet
Boeing 747-400	4	6,400	4 Pratt & Whitney	575	231 feet, 11 inches	213 feet
Boeing 747-200	6	4,500	4 Pratt & Whitney	570	231 feet, 11 inches	195 feet, 9 inches
Boeing 767	6	5,400	2 Pratt & Whitney	542	159 feet, 2 inches	156 feet
Boeing 737-800	3	2,500	C.F.M.	542	129 feet, 3 inches	112 feet
Boeing 757	7	3,300	2 Rolls Royce	542	155 feet, 3 inches	124 feet, 10 inches

Source: Data taken from (www.elal.com/glance/fleet/index.htm), June 5, 2001.

pian Jews flying from Addis Ababa to Israel as part of Operation Solomon.

El Al operates a fleet of Boeing 747's, Boeing 767's, Boeing 757's, and Airbus A320's. In 2001, El Al introduced three new, technologically advanced, long-range Boeing 777-200ER models into its fleet. The new aircraft, powered by twin Rolls Royce Trent 895 engines capable of developing 95,000 pounds of thrust, can fly 5,561 miles nonstop with a full payload of passengers and cargo. Among the new features and amenities these aircraft offer to El Al passengers (6 in first-class, 47 in business, and 245 in tourist class) are the most comfortable seats available and fully digital electronic entertainment, with provisions for personal computers and interactive features. Their acquisition was part of the company's overall strategy to expand service to business-class customers, toward which goal El-Al also invested $15 million in upgrading its other long-range airliners to match the standards of the 777.

Organization

El Al has described itself as "a unique combination of amenities and advantages earning international appeal and making El Al a preferred global gateway to every corner of the world." Teshet, a subsidiary company wholly owned by El Al, operates travel agencies, catering facilities, and hotels in Israel, as well as airports in the United States. Teshet operates two kosher catering companies: Tamam, based in Israel, and Borenstein in New York. Teshet holds interests in Maman, handles cargo at Ben Gurion International Airport in Tel Aviv, manages the Laromme Hotel Chain, and is the Israel representative of the Howard Johnson Hotel Group, Alamo Rent-A-Car, Air Nevada, British Midland, and international technical aviation companies such as Pratt & Whitney and United Technologies. Another subsidiary, Arkia Inland Airlines, founded in 1950 and owned 50 percent by El Al, provides scheduled domestic air services within Israel and serves as El Al's feeder carrier. El Al is the sole owner of yet another subsidiary airline, Sun D'or International Airlines. Sun D'or, operating charter flights between Israel and Europe, helps develop new routes and tourist markets for El Al.

Triantafyllos G. Flouris

Bibliography

Groenewege, Adrianus D. *The Compendium of International Civil Aviation*. 2d ed. Geneva, Switzerland: International Air Transport Association, 1999. A comprehensive directory of the major players in international civil aviation, with insightful and detailed articles.

Weimer, Kent J. ed. *Aviation Week and Space Technology: World Aviation Directory*. New York: McGraw-Hill, 2000. An excellent introductory guide on all global companies involved in the aviation business. The information is basic but essential as a first introduction to each company.

See also: Air carriers; Airports

Emergency procedures

Definition: Established steps in memorized, paper, or electronic format used in critical situations to aid the memory of those involved in a crisis.

Significance: Emergency situations place human beings under greater than normal amounts of stress, making them less able to solve problems appropriately. Clearly understood procedures enable people to act appropriately under such circumstances.

Introduction and Overview

For as long as there have been aircraft, there have been unexpected events that have necessitated methodical procedures in order to help ensure the safest possible outcome for both aircraft and crew. In aviation, emergencies are defined as situations in which immediate action by those involved is required in order to ensure the safety of a flight. In general, humans are ill equipped to deal consistently and effectively with emergencies.

A detailed set of guidelines or procedures for people to follow in the event of an emergency often helps to positively impact the emergency situation. These procedures have evolved from the relatively simple memorized procedures used by pilots of early aircraft to the relatively complex procedures used by a flight crew to deal with anomalies aboard large aircraft, such as the Boeing 777 and 747 heavy-transport aircraft.

Emergency procedures range from small-aircraft checklists for dealing the accidental opening of a cabin door during flight to large commercial airports' detailed emergency plans for dealing with an incoming aircraft that has been rendered virtually uncontrollable. In the first case, the procedure may involve only the pilot of the aircraft, whereas, in the second scenario, emergency procedures would typically involve many people in several different organizations all engaging in a highly coordinated and rehearsed plan of action in order to effectively deal with the situation.

Procedures in the Aircraft

In the case of a small aircraft, it is recommended that pilots carry a set of emergency procedures checklists readily available to them in the event of an emergency. These checklists may be in paper or electronic format. Emergency procedures cover a variety of topics dealing with engine failures, in-flight fires, electrical failures, flight control malfunctions and others.

Emergency procedures checklists will often be color coded with red and white or red and black in order to command attention. These are then readily distinguishable from other procedures checklists. Often, certain emergency procedures are considered to be time-critical in order to effect a safe outcome. Such procedures, generally limited to three to five items, are called immediate-action items and are usually committed to memory. An instance in which such a procedure would be necessary is the failure of one engine of a twin- or single-engine aircraft immediately after takeoff. An event such as this leaves little time for a pilot to pull out a checklist and go over it line by line. Once the immediate-action items are attended to, the pilot can then methodically go through the remaining items.

Generally, the more complex the aircraft is, the more involved are the emergency procedures. In larger transport aircraft, more than one pilot is available to assist during crisis situations, and the delegation of responsibility at such times rests upon the pilot in command. In an emergency situation involving a multicrew aircraft, generally one pilot continues to fly and maintain control of the aircraft while the other pilot (or two) are freed up to focus on the emergency procedures.

Electronic Aids

In modern aircraft with electronic flight instrumentation there are often systems on board the aircraft that will assist the flight crew in diagnosing a problem and will provide the appropriate checklist on what is called a multifunction display (MFD) on the flight deck. This display highlights the appropriate checklist items and forces the crew to acknowledge each checklist item before proceeding to the next item. The MFD is a helpful feature during periods of high stress on the flight deck, because it makes it more difficult to forget or omit critical items, as can happen with paper checklists.

Larger aircraft, such as the Boeing 757 and 767, are equipped with an engine information and crew alerting system (EICAS), which immediately brings a fault diagnosis to the attention of the flight crew. The crew must then execute the pertinent emergency checklist in its entirety in a challenge-response format, in which the pilot who is not flying issues a challenge such as "throttle closed," to which the appropriate crew member response is "throttle closed." In this manner, each crew member is made aware of the checklist item and its completion status.

Cabin Safety

Emergency procedures also exist for the cabin crew, or flight attendants, and for passengers. All passengers are required by the Federal Aviation Regulations (FARs) to be briefed on these procedures by the cabin crew prior to flight. Research has shown that those passengers who listen to the preflight emergency briefing information are much more likely to survive an air accident than those who do not.

Airline cabin crew members are required to attend annual recurrent emergency procedures training. This training consists of a review of basic emergency and evacuation procedures for the particular aircraft the crew members fly. Most major airlines have or have access to aircraft cabin simulators, which can simulate an aircraft accident. In these scenarios, the cabin fills with a harmless smoke agent and the lights go down, as they would in an accident. The crew members are then evaluated on the accuracy and timeliness of their actions in getting people to safety.

Survival

In order to survive an air accident, the crew and passengers must be able to do three things successfully. First, they must survive the impact of the crash, if applicable. Second, they must evacuate the aircraft safely in a timely manner, especially in the event of a fire. Third, if the accident occurs away from an airport, they must survive the post-accident environmental conditions until they are rescued or until safety is reached. The first two items are often largely dependent upon how much attention was paid to the preflight safety briefing, whereas the third item depends upon previous training. Many organizations around the country specialize in postcrash survival training. A survival course varies from one day to one week or more and will cover a variety of subjects, including land navigation, rescue-signaling techniques, shelter construction, and food- and water-gathering techniques. Flights over remote areas are likely to carry emergency survival kits on board, which contain a variety of survival gear including signaling devices, drinking water, high-energy food, and first aid kits. Flights over water are required to carry flotation devices and life rafts, and training in their proper use is imperative. Most life rafts have an on-board emergency radio-locator transmitter and a visual strobe light to aid in aerial location, as well as sun shelters, sunscreen, water, and other survival gear.

Airport Emergencies

The majority of aircraft accidents happen on the premises of an airport. If an aircraft accident occurs on or in the immediate vicinity of an airport, the occupants have a much greater chance of surviving the post-incident conditions, because all publicly certified airports have emergency action plans. There are approximately 5,400 public-use airports in the United States. Of these, approximately 670 are certified under Part 139 of Title 14 of the Code of Federal Regulations as certificated airports. The remaining airports, classified as non-certificated or general aviation airports, handle a relatively low volume of mostly light aircraft. If an airport has scheduled passenger or cargo airline service, it is required to be a certificated airport.

Certificated airports are rated according to classes A, B, C, D, or E, in accordance with the types of aircraft they serve. These classes, among other things, determine the amount and type of airport firefighting and rescue services (ARFF) the airports have. In the event of an aircraft accident, the emergency operations plan for the airport will go into effect. This plan is contained in, and is a preapproved part of, an airport's certification manual. The plan is generally put into effect by an airport's control tower, which will immediately notify the ARFF unit in the event of an accident. Other calls will be made in accordance with the emergency plan to such outside agencies as local law enforcement, ambulance services, hospitals, and coordinating fire and rescue departments in order to put them either on alert or into action, in accordance with the preapproved plan.

Every three years, certificated airports are required to conduct a live-fire training exercise under simulated accident conditions. In this exercise, the emergency action plan is put into effect, and all agencies react just as they would in an actual emergency. In this rehearsal, usually an aircraft is towed into a simulated crash position on the airport and costumed victims are situated in and around the aircraft, as they might be in an accident. Simulated crash victims are then extracted and treated, and fires are extinguished, allowing everyone involved a chance to identify areas of needed improvement in the plan. Each year, these same airports are required to go through the motions of the emergency plan in a "table-top" scenario, in which phone numbers are verified, calls are made according to the plan, and treating-unit capabilities are updated to ensure the currency of the plan.

Emergency Procedures in Action

On July 19, 1989, a series of events occurred in which well-rehearsed emergency procedures helped save many lives in an aviation event that might have been considered unsurvivable. On that day, United Air Lines Flight 232, a DC-10 wide-body transport aircraft, was flying passengers from Denver to Chicago when the airplane's center engine disintegrated, causing the aircraft to lose all available hydraulic fluid. Because the flight controls on this type of aircraft are hydraulically operated, the aircraft was rendered virtually uncontrollable. Because the crew members had been trained in cockpit leadership resources (CLR), which taught them to utilize all available resources in order to save an aircraft under adverse circumstances, they were able to regain some control of the aircraft using differential thrust from the two remaining engines on each wing, even though there was no specific procedure for such an occurrence. This solution was made possible with the help of a passenger who happened to be a DC-10 pilot for United Air Lines. The cabin crew alerted the flight crew of this passenger's availability, an event which might not have occurred had the crew not been trained in CLR. The aircraft subsequently crashed under limited control, broke apart, and then erupted in flames at the Sioux City, Iowa, airport. More than one-half of the occupants survived the ordeal, due, in large part, to the extraordinary efforts of the crew and to the fact that, very recently prior to the accident, the Sioux City airport had conducted its emergency action live-fire drill. This practice drill had enabled a much faster accident reaction time for rescuers.

A common theme throughout all aviation emergency procedures, whether they are on board an aircraft in flight or on the ground at an airport, is the importance of having structured, well-rehearsed and well-coordinated plans of action to follow. With these in place, human beings are much better able to perform under adverse circumstances, ensuring the minimum loss of life and damage to property.

R. Kurt Barnhart

Bibliography

Brown, Gregory N., and Mark J. Holt. "Emergency and Abnormal Procedures." In *Turbine Pilot's Flight Manual*. Ames: Iowa State University Press, 1995. An explanation of emergency and abnormal procedures from a pilot's perspective and an excellent introduction to larger aircraft.

Dee, Emily. *Souls on Board: Responses to the United Flight 232 Tragedy*. Siox City, Iowa: Loess Hills Press, 1990. A collection of stories from survivors of the runway tragedy in which many lives were saved by the effective application of emergency procedures.

Wild, Thomas W. *Transport Category Aircraft Systems*. Casper, Wyo.: IAP, 1990. An in-depth examination of the various systems on board transport aircraft. Good illustrations and explanations make for a very informative text for those new to large aircraft. Information on the EICAS is presented throughout the text.

See also: Accident investigation; Air carriers; Airline industry, U.S.; Airport security; Flight attendants; Hijacking; Landing procedures; Maintenance; Runway collisions; Safety issues; Takeoff procedures; Taxiing procedures; Training and education

Enola Gay

Definition: The B-29 bomber that dropped the first atomic bomb to be used against a civilian population, on the Japanese city of Hiroshima.

Significance: The *Enola Gay* dropped the first of the two atomic bombs ever used in combat, on the Japanese city of Hiroshima, effectively ending World War II.

The B-29 Superfortress

Lieutenant Colonel Paul Warfield Tibbets, Jr., was twenty-nine years old when he first heard of the atomic bomb. At the time, he was one of the U.S. Air Force's best bomber pilots, having flown twenty-five B-17 missions, including the first raid against occupied Europe and the first mission in support of the North African invasion. As a test pilot, he helped bring the B-29 Superfortress into service. In September, 1944, he was briefed on the Manhattan Project, the code name of the project to build the atomic bomb. He was given the assignment to organize, equip, and train a unit to drop atomic bombs on Germany and Japan. Tibbets was told that if the bombs worked well, they might end the war.

The B-29 Superfortress was the first intercontinental bomber. Powered by four 2,200-horsepower engines, it stood three stories tall. With a 141-foot wingspan and a 99-foot fuselage, it filled half of a football field. It was armed with up to twelve 0.50-inch machine guns mounted in four turrets and the tail, and a 20-millimeter cannon mounted in the tail. To save weight, Tibbets requisitioned fifteen B-29's without protective armor, turrets, or guns, except for the 20-millimeter tail cannon. The B-29 had a pressurized cabin and could cruise above 30,000 feet, beyond the reach of antiaircraft fire and most enemy fighters.

Tibbets took command of the 393d B-29 bombardment squadron at the Wendover Army Air Base on the Utah-Nevada border. He had the squadron pilots train to drop a single large bomb from 30,000 feet and to perform a strange evasive maneuver in which they turned the planes 150 degrees while diving to gain speed. The squadron was told its mission would have an important effect on the war, but were given no other details. On December 17, 1944, orders were issued activating the 509th Composite Group, which eventually included more than 1,500 enlisted men and 200 officers. In the spring of 1945, the group began to move quietly to Tinian Island in the Marianas Islands.

The Hiroshima Bombing

On August 5, 1945, President Harry S. Truman gave his approval to use the atomic bomb on Japan. The bomber crews were briefed by William S. "Deke" Parsons of the Los Alamos National Laboratories. Honoring his mother, who had encouraged him to join the Air Force, Tibbets had the name *Enola Gay* painted on the nose of his selected B-29. With Tibbets as commander and pilot, the *Enola Gay* took off from Tinian at 2:30 A.M. on August 6, en route to Hiroshima. At first Tibbets flew at less than 5,000 feet, so that Parsons and his assistant could enter the unpressurized and unheated bomb bay to insert the cordite explosive that would propel the uranium-235 slug into awaiting rings of uranium-235 to form a supercritical mass. At 7:30 A.M. Parsons returned to the bomb bay and armed "Little Boy," the nickname for the 9,700-pound bomb.

During the 45-minute climb to the 31,000-foot bombing altitude, the weather plane flying ahead of the *Enola Gay* reported favorable conditions over Hiroshima. The plane was flying at a speed of 328 miles per hour as the bombardier, Major Thomas W. Ferebee, took control of the plane for the bombing run. Finding his aiming point, he let Little Boy fall away from the *Enola Gay.* Tibbets threw the bomber into its 150-degree escape maneuver, so that they were 11.5 miles away 43 seconds later, when Little Boy exploded 1,900 feet above the ground. After a blinding flash of light, Hiroshima was hidden beneath a huge, boiling cloud, that was simultaneously incredible and terrible. The bomb's estimated yield was 12,500 tons of TNT, a common high explosive. Captain Theodore J. Van Kirk, the *Enola Gay*'s navigator, later admitted to thinking as he watched the destruction, "Thank God the war is over and I don't have to get shot at any more. I can go home." The *Enola Gay* then returned to Tinian and landed at 2:58 P.M.

The Smithsonian Exhibition

As part of a fifty-year commemoration, the forward section of the *Enola Gay*'s fuselage was placed on display in the National Air and Space Museum of the Smithsonian Institution from June, 1995, to May, 1998. Initial plans for the display produced a hurricane of controversy. Objecting to a bare display of military hardware as a glorification of war, museum directors sought a larger context for the *Enola Gay* exhibit. One suggested theme was the dark side of air power and the inhumanity of nuclear weapons. An early script said that for most Americans, the war against Japan was a war of vengeance, but for most Japanese it was a war to defend their unique culture against the imposition of Western imperialism. Veterans' groups, along with many others, were outraged at what they considered to be a very biased treatment that glossed over the aggression and atrocities of Japanese warfare, focusing instead on the victims of the two atomic bombs. After Congress stepped in and threatened to cut off funds, the exhibit was restricted to the forward section of the *Enola Gay*, a plaque explaining its mission, and a video of the flight crew's training and experiences.

Charles W. Rogers

Bibliography

Laurence, William L. *Men and Atoms*. New York: Simon & Schuster, 1959. The author, the science editor for the *New York Times* attached to the Manhattan Project describes his experiences flying in one of the observation planes during the bombing of Nagasaki, along with the events surrounding the flight of the *Enola Gay*.

Newman, Robert P. *Truman and the Hiroshima Cult*. East Lansing: Michigan State University Press, 1995. The author answers the questions of those who initially sought to revise history with the *Enola Gay* display.

Rhodes, Richard. *The Making of the Atomic Bomb*. New York: Simon & Schuster, 1986. A comprehensive, richly detailed, accurate, and very readable account of

The Enola Gay, *the B-29 Superfortress that dropped the first atomic bomb on Hiroshima on August 6, 1945, returns to base.* (AP/Wide World Photos)

the politics, history, and science of the atomic bomb. The flight of the *Enola Gay* is described in the chapter entitled "Tongues of Fire."

See also: Air Force, U.S.; Bombers; Kamikaze missions; Military flight; Pearl Harbor, Hawaii, bombing; Superfortress; World War II

Evolution of animal flight

Definition: The process by which certain animals have biologically adapted to engage in three modes of flight: active gliding, passive gliding, and true or powered flight.

Significance: The air has provided a valuable ecological niche for many living things. The advantages conferred by flight are great: escape from predators, migration to suitable climates, and a means to search for food and mates.

Animal Adaptation to Flight

Insect flight evolved in the mid to late Carboniferous period, about 325 million years ago. The flight of certain reptiles, the first vertebrates to fly, dates from the Triassic period, between 230 and 195 million years ago. The first primitive birds evolved during the Jurassic period, 195 to 135 million years ago.

Throughout evolutionary history, flight has been a crucial characteristic of many insects, reptiles, birds, and bats, and more than one-half of the animal species now living can fly. Most of these contemporary species are insects, but about nine thousand species of birds and nine hundred bat species are at home in the air. The largest animal capable of flight was an extinct reptile, *Quetzalcoatlus northropi*, which weighed over 140 pounds (63 kilograms). The smallest flying creature, the chalcid wasp (*Encarsia formosa*), weighs about 0.0001 ounces (0.025 milligrams). Within the range of these extremes, the variety of flying animals is immense.

Flying animals that succeeded in surviving for long periods of time had to solve basic problems. Their survival in a particular ecological niche depended on achieving an optimum size, bone structure, wingspan, and cardiovascular system. Through trial-and-error methods, some flying animals survived and prospered while many others became extinct. Over the long years of animal evolution, animals have chanced upon a wide spectrum of solutions to the problems of flight. Some fliers developed long wings, others stubby ones. Some flapped their wings vigorously, others used gliding and soaring techniques. Gliding utilizes gravity by either launching toward a target (directed or active gliding) or relying on wind for motion (passive gliding or parachuting). Soaring is sustained gliding, taking advantage of rising columns of air. True or powered flight consists of using muscles to take off, fly, and land. The details of the evolution of animal flight are poorly understood, because most flying animals left no trace in the fossil record. How did primitive insects, reptiles, and birds solve the problems of flight? Did flying vertebrates evolve from running, jumping, or perching animals? Did early fliers have rigid or flexible wings? Some of these questions have been answered, others have been only partially answered, but many others remain to be satisfactorily answered.

Devices Developed for Flight

To enhance their ability to spend a significant part of their lives in the air, flying animals developed some devices that were similar to those now used by humans in constructing aircraft, but others were unlike the mechanisms found in flying machines. For example, the Wright brothers were helped in devising a wing-warping mechanism to control their *Flyer* by observing the flight of birds. When a jetliner is landing, the pilot lowers thc landing gear and wing flaps to reduce speed in preparation for touchdown, just as birds extend their legs and lower their tails to increase drag when they are landing. On the other hand, aircraft designers have been unable to successfully copy the complex structures of feathers and the intricate ways that birds twist and flap their feathered wings.

Animal flight involves lift, thrust, and control, and such animals as insects, reptiles, birds, and bats have used a variety of methods in achieving flight, some of which are specific to a species, while others are shared across classes. For example, the largest insects have wings similar to those of the smallest hummingbirds, but animals such as flying squirrels lack the ability to flap their "wings" and so are confined to gliding. Scientists have studied the theoretical limits for such variables as size, wingspan, bone mass, and muscular strength in trying to understand flying animals.

Theories of the Evolution of Animal Flight

To explain how animals developed the devices that allowed them to fly, scientists have proposed various theories. Two classical models, dating from the nineteenth century, are the arboreal and cursorial theories. These theories try to explain how limbs developed into wings, how bones became lightened, and how feathers evolved from scales. Some scientists speculate that elemental wings first sprouted as lengthwise ridges along the sides of vertebrates. These ridges were constructed from available materials such as fur, skin, or scales. According to the arboreal theory, these protowings adapted the animals for life in the trees, especially for leaping from limb to limb. Selective pressures from predators subsequently changed these winglets into wings. The cursorial theory, whose name derives from the Latin *cursor,* "a runner," explains the origin of flight differently. To evade enemies, bipedal vertebrates habitually combined running jumps with short glides. Their increasingly winglike forelimbs helped them to generate the lift and thrust to spend time in the air instead of in the jaws of predators.

Both these theories have been criticized, but because of the deficiency of relevant fossils, it has been impossible to eliminate either one. Even when relevant fossils exist, they fail to resolve all problems. For example, *Archaeopteryx lithographica*, a fossil that had a reptilian skeleton and feathered wings, created a sensation when it was discovered in Germany just two years after Charles Darwin had published *On the Origin of Species by Means of Natural Selection* (1859).

This 150-million-year-old protobird was acclaimed as the "missing link" between reptiles and birds. As more *Archaeopteryx* fossils were found, a debate ensued over whether it was a birdlike reptile or a reptilelike bird. It resembled birds because it had feathers and a wishbone, but it resembled dinosaurian reptiles because it had teeth in its jaws, three clawed fingers on its wings, and a lizardlike tail. Some argued that *Archaeopteryx* was cold-blooded and flightless; others that it was warm-blooded and a flier.

Theories of the origin of animal flight have encountered what evolutionist Stephen Jay Gould has called the "5-percent-of-a-wing problem." Aerodynamic analysis revealed that flying animals needed to have wings that were long, light, flexible, and strong, but critics of Darwin's theory pointed out that protowings, which were short, dense, rigid, and weak, would have been a hindrance rather than a help. However, Darwinists responded that this all-or-nothing argument that wings are worthless until well developed ignores the possibility that winglets could confer such benefits as aiding animals in escaping from their enemies. Furthermore, rudimentary feathers could have served as heat insulators. Some scientists have called these winglets and protofeathers "preadaptations," and some writers have called these primordial creatures "hopeful monsters."

The Evolution of Insect Flight

Flying creatures developed independently several times during the course of evolution, but the fossil record reveals that the first to fly were insects. Because of their small size, insects need little energy to launch themselves into the air, but due to an inadequate fossil record, scientists have been unable to develop incontestable interpretations of the evolution of insect flight. Fossils of flying insects in amber have not yet been discovered from the Paleozoic era (600 to 230 million years ago), and the sparse evidence scientists possess of the number, variety, structures, and ecological niches of flying insects provides only meager clues for constructing explanations of how insect flight originated. Paleontological evidence does exist for giant dragonflies (*Protodonata*) during the Carboniferous and Permian periods (between 300 and 225 million years ago). Some scientists think that these insects were aerial predators. *Meganeura*, the largest known flying insect, dates from this period, and it had a wingspan of 2.4 feet (73 centimeters) and weighed just under 8 ounces (200 grams). Evolution certainly played a role in the creation of these giant insects, and the mechanism may have been an escalating competition between predator and prey.

Some paleontologists have used later fossils and aerodynamic studies of contemporary insects to theorize about early flying-insect evolution. Many scientists believe that insect wings evolved only once and that all fliers derive from a wingless ancestor, but how these wings evolved has been passionately debated. Some scientists hold that winged insects originated on land, where jumping due to a startle reflex in response to predatory attacks may have been the selective force that promoted crude flying. Other scientists argue that insect flight began in an aquatic environment, where winglets allowed insects to walk on water since flapping their winglets kept them from breaking through the surface tension of the water.

Explanations of the history of insect flight are complicated by the great structural and functional diversity of insects. Flying insects now range from large butterflies who oscillate their wings about five times per second to tiny midges who beat their wings a thousand times per second. As impressive as contemporary insect diversity is, even more remarkable are the varieties found in Mesozoic fossils (245 to 65 million years ago). Paleontologists have estimated that about twenty-eight insect orders existed prior to the great Permian extinctions that occurred about 245 million years ago. These insects had many kinds of wings, and they differed in their ability to use them. Enhanced maneuverability in the air also facilitated sexual selection. For example, in some species, males hovered, explored, and then engaged in high-speed chases to capture females. In short, flight has been a key element in the survival and proliferation of an increasing variety of insect species.

Pterosaur Flight Evolution

The evolution of flying vertebrates presents paleontologists with different problems from those of flying-insect evolution. For example, adult vertebrates generally lack exoskeletons and weigh much more than insects. Because of these differences, scientists have had to modify their theories of insect evolution to explain how vertebrate fliers such as pterosaurs, birds, and bats evolved. A commonly held theory has pterosaurs (flying reptiles) and birds evolving from thecodonts, the direct ancestors of the dinosaurs. Most thecodonts were small reptiles that walked on their two hind legs, as do many birds.

Flying reptiles have been found fully developed in the Lower Jurassic (195 to 135 million years ago), but they had a much longer history, even though paleontologists have failed to find any intermediate forms between thecodonts and pterosaurs. The most distinctive trait of a flying saurian was its membranous wings buttressed by greatly elongated fourth fingers. Their three other fingers bore claws that may have allowed them to cling to rocks or tree limbs, from which they either hung head down (like bats) or perched head up (like birds). Some paleontologists believe that pterosaurs descended from a small arboreal reptile that spent its life in trees where, like modern flying squirrels, it used flaps of skin attached to its limbs to facilitate its glides and brake its falls. Others believe that pterosaurs evolved from bipedal reptiles that ran along the ground, perhaps spreading their upper limbs for balance. Through gradual growth, these forelimbs evolved into wings.

All these ideas are highly speculative, and most paleontologists agree that the question of pterosaur origins remains open. However, sufficient pterosaur fossils have been found for scientists to conclude that these "dragons of the air" were one of evolution's early success stories. Unhampered by enemies in the air, pterosaurs diversified into over 120 known species. The sparrow-sized *Pterodactylus elegans*, the smallest pterosaur, had a ten-inch wingspan, whereas the largest, *Quetzalcoatlus*, known from a fossil in the Big Bend region of Texas, had a wingspan of nearly 40 feet (12.2 meters). The small pterosaurs fed chiefly on insects, while the large pteranodons preyed primarily on fish. Though often described as flying reptiles, the pterosaurs were unlike modern reptiles in several respects. They were most likely warm-blooded and had a hairlike surface and highly developed brains. Pterosaurs lived and thrived for 150 million years, but all of them became extinct in a very short period of time 65 million years ago, at the end of the Cretaceous. Just as the origin of pterosaurs is disputed, so too is their extinction. Some paleontologists claim that pterosaurs, with their fragile bodies, had become so specialized that they were unable to adapt to changes in the Cretaceous climate. Others blame a large asteroid that slammed into Earth, making the demise of the pterosaurs part of the calamity that wiped out the dinosaurs. However, critics of the catastrophic theory point out that pterosaur species had been dwindling for millions of years prior to the cataclysm, suggesting that other factors may have contributed to their fate.

From Pterosaurs to Birds

Paleontologists have found no transitional forms between pterosaurs and the first birds, and most believe that birds and pterosaurs evolved separately from thecodonts rather than that birds evolved directly from pterosaurs. All modern theories of bird evolution have been influenced by a crow-sized creature that perished in a shallow lagoon 150 million years ago and whose bones and feather imprints were preserved in lithographic limestone found in 1861. Scientists naturally focused on this *Archaeopteryx lithographica* in their studies on the descent of birds. Strangely, during the nineteenth century, *Archaeopteryx* was not accepted as the missing link between reptiles and birds by most scientists nor by the public. Influenced by religious views of the fixity of species, the public viewed reptiles and birds as unchanging and unchangeable forms. In contrast, scientists had multiple interpretations of *Archaeopteryx*. For Richard Owen, the scientist who coined the name "dinosaur," *Archaeopteryx* was the earliest bird, which was a transmuted form of a long-tailed pterosaur. On the other hand, Thomas Henry Huxley, traditionally described as "Darwin's bulldog" because he eloquently defended natural selection, saw this fossil as proof that birds had evolved from dinosaurs.

The first and most influential book on avian evolution, *The Origin of Birds*, was written by a Dane, Gerhard Heilmann, in 1926. He argued for a thecondontian ancestry of birds, and his theory was supported by most textbooks and scholarly works on avian origins for the next fifty years. However, toward the end of the twentieth century, this classic theory came under attack. Some researchers interpreted thecodonts as a heterogeneous assemblage rather than a well-defined group. Others emphasized that birds were descendants of dinosaurs who were warm-blooded land-dwellers, and the first fliers, with their feathered wings, originated "from the ground up." This cursorial theory of avian flight denied dinosaurs any life in the trees, and thus directly contradicted the arboreal theory championed by Heilmann, who had reconstructed in detail the evolution of birds from tree-dwelling to flying animals.

Besides questions of whether birds evolved "from the ground up" or "from the trees down," problems arose over the origin of feathers, the most beautiful and well-known adaptation in evolutionary history. The central difficulty with the explanation of feathers' evolution from reptilian scales is that a feathered airfoil had to meet stringent aerodynamic criteria to function as a manipulable wing for controlled flight. Some have proposed that feathers arose as netlike devices to catch insects, but nets must be pervious to air whereas airfoils need to be impervious to air. Furthermore, feathers are present in bird tails, where they could scarcely serve fly-catching functions. Other scientists speculate that feathers initially developed for temperature regulation, a preadaptation later used for flight. The difficulty with this thermoregulatory theory is that the microarchitecture of feathers, with their numerous filaments (barbs) and interlocking fringes (barbules), is so well adapted to flight that some scientists proposed that avian feathers evolved directly for flight. However birds originated, they rapidly diversified and colonized a variety of environments. The cataclysm that annihilated the pterosaurs at the end of the Cretaceous created ecological voids that birds increasingly occupied in an extraordinarily explosive evolutionary diversification.

Mammalian Flight

The final flying vertebrate to evolve was the bat. The only mammals to have developed true flight, bats have existed since the start of the Cenozoic era, 65 million years ago. As with pterosaurs and birds, transitional forms of bats have not been found in the fossil record. The wings of the earliest fossilized bat, from about 50 million years ago, were as completely developed as those in modern species. This absence of incipient-winged ancestors has not prevented scientists from speculating about bat evolution. Some paleontologists interpret bat evolution as a succession of small mammals whose fingers gradually lengthened as wings and the specialized muscles necessary to power them developed, but this theory encountered the criticism that these protowings, which restricted normal hand movements, would be disadvantageous. It was argued that a creature would not sacrifice usable hands for half-developed wings. Other paleontologists use bats as an example of quantum evolution, believing that wings and other biomachinery needed to support flight developed in an evolutionary spurt. In this theory, bats arose from arboreal insectivores who developed a membrane that stretched between their fore- and hind-appendages. These protowings were initially used for gliding from tree to tree. Gliding evolved into flying, because wings allowed bats to occupy the niche of flying insect eaters. However, at that time the niche of daylight insect feeding was occupied by birds, forcing primitive bats into becoming nocturnal insect eaters. Like birds, bats quickly diversified and are now represented by over 850 species. They have been particularly successful in tropical regions, where more bat species exist than all other mammals combined.

Paleontological Conclusions

Paleontologists have been able to sketch an outline for the evolution of animal flight, but the details of this picture have yet to be worked out in any completely satisfactory way. New fossil discoveries continue to expand and deepen scientific knowledge of flying-animal evolution, and computers have come to the aid of paleontologists, who are able to model prehistoric flying creatures by making use of detailed aerodynamic and physiological data. Thus scientists have new evidence to determine whether certain vanished creatures really flew. These studies have enhanced the appreciation of scientists and their students for the great diversity of flying insects, pterosaurs, birds, and bats that have flourished through evolutionary time. Once these animals conquered the air, new worlds were opened up to them, from deserts to mountains, from arctic tundra to tropical forests. The impressive distances covered by migratory birds and butterflies show the great energies these flying animals are willing to expend to traverse vast distances in their search for suitable environments in which to feed and reproduce. It took 300 million years to create the variety of flying creatures that live in the modern world, but it took *Homo sapiens* a much shorter time to achieve heavier-than-air powered flight. However, humans, who profited from their observations of birds in solving the problems of flight, have yet to solve the myriad of puzzles posed by the evolution of animal flight.

Robert J. Paradowski

Bibliography

Dudley, Robert. *The Biomechanics of Insect Flight: Form, Function, Evolution*. Princeton, N.J.: Princeton University Press, 2000. The first detailed study of how insects actually fly and how they evolved into fliers, this well-researched book also contains an analysis of the roles that natural and sexual selection played in insect evolution.

Feduccia, Alan. *The Origin and Evolution of Birds*. New Haven, Conn.: Yale University Press, 1996. A heavily illustrated volume treating the origin and early evolution of birds and avian flight and the later evolution of a great diversity of highly developed birds, including raptorial and flightless birds.

Templin, R. J. "The Spectrum of Animal Flight: Insects to Pterosaurs." *Progress in Aerospace Sciences* 36 (2000): 393-436. A comprehensive scientific paper summarizing data on the flight characteristics of many kinds of winged animals, it uses flight simulation to explore the characteristics of hypothetical proto-fliers, favoring the "trees-down" rather than the "ground-up" theory of vertebrate flight origins.

See also: Aerodynamics; Animal flight; Bats; Birds; Forces of flight; Insects

Experimental aircraft

Definition: Aircraft that is still in the experimental testing phase of development, or craft built by amateurs from kits or plans.

Significance: Experimental aircraft are important for the progress of the aviation industry. Flight test of an experimental aircraft validates expected performance criteria, tests structures, evaluates handling characteristics, and more. Scientific research is dependent on flight test results from experimental aircraft. A new airplane as such is experimental, but a proven and certified aircraft is experimental when mounted with new, untried applications requiring testing.

Types of Experimental Aircraft

A distinction must be made between true experimental aircraft and amateur-built airplanes constructed from kits or plans. The former category contains aircraft involved in research and design at costs reaching into the millions, while the latter group contains aircraft constructed by individuals for fun, at a substantially smaller cost. An experimental military project or airline endeavor may reach hundreds of millions of dollars before the production run gains approval. The homebuilt or kit plane project, on the other hand, is notably less expensive. Single airplanes built for personal use have flown at a cost well below $10,000. Another difference between the two groups of experimental aircraft includes the time spent in flight test. The large company projects may spend more than a year in flight test while a homebuilt or kit plane undergoes a basic forty-hour testing period.

The Design Process

Regardless of the group to which the airplane belongs, every aircraft flying today began in the much the same manner. Industry discovers a need for a particular design or a mission requirement. After the creation of a new concept, pilots, engineers, and mechanics discuss and research ideas for the production of the new craft.

The next step is sketching the design idea. More than a few new ideas begin as pencil drawings on napkins in restaurants over lunchtime discussions. From the first idea sketches, aeronautical engineers further refine the drawing by use of computer-aided design software. During the initial design period, engineers exchange ideas and make concessions and compromises that are eventually lofted into the drawing process. Modelers next construct a model of the new aircraft.

Models of the new airplane are necessary for many reasons. Wind-tunnel testing requires models of different sizes. Models enable designers to visualize proportional sizing. Problems not visible on a drawing board may vividly stand out in three dimensions.

After studying the models in depth, technicians and engineers develop mock-ups of various sections and components of the new aircraft. These mock-ups allow others to test the airplane and offer their opinions to the designers regarding positive and negative aspects of the new craft. Mock-ups also allow pilots, engineers, and mechanics to spot problems before production of the aircraft. The earlier design changes can be made, the more economically they can be incorporated into the production schedule.

The Prototype and First Flight

At the completion of the design process, construction of the prototype begins. Construction of one or two copies of the new craft is required for testing. This is an expensive proposition; until the craft goes into mass production, per unit cost of the craft can be phenomenal. The purpose of the prototype is further research and development. Changes follow in quick succession as shortcomings become evident and better construction methods become available. As the first prototypes are readied for flight, flight profiles are developed for the initial flight and the flight test program to follow. The most exciting event at an aircraft plant is the first flight of a new design, an event anticipated by everyone in the company.

Typically, first flights do not last more than about three-quarters of an hour. The only concern the company and the pilots have with the aircraft is whether it will fly. Initial flight testing does not try to test the edge of the flight envelope. Pilots, engineers, and managers are not interested in how fast or high the airplane is capable of flying on the first flight; they only want to see it fly and see it handle the way it is expected. Test crews will address other questions regarding the performance envelope later.

As test pilots put the aircraft through its paces, they keep meticulous data on every aspect of each flight. The company uses the data to refine the flying qualities of the prototype and suggest changes in the production run.

After completion of the flight test program, the new airplane will finally reach acceptance. If it is an airliner, company officials from the airline will either accept or reject the new craft. Military acceptance is somewhat different, in that the airplane flies through a more intense flight test program. Additionally, flight tests of military aircraft involve weapons systems compatibility.

Homebuilt Experimental Aircraft

The other type of experimental airplane is the burgeoning kit plane and homebuilt industry. In 2001, approximately twenty-two thousand homebuilt aircraft constructed by amateur builder-pilots were flying in the United States.

One organization dedicated to homebuilt enthusiasts is the Experimental Aircraft Association (EAA). The national headquarters of the EAA is located in Oshkosh, Wisconsin. There are local chapters of the EAA throughout the United States that allow homebuilding proponents to exchange technical information and discuss problems or other building concerns. Members also enjoy the camaraderie of other members, along with the encouragement offered at monthly or biweekly meetings.

Advantages and Disadvantages of the Homebuilt

There are many reasons pilots choose to build their own aircraft. One reason is performance. Since the 1950's, aircraft manufacturers have done relatively little in the way of increasing or improving aircraft performance. For example, a popular older model production aircraft flies at 100 knots, burns 7.5 gallons of fuel per hour, and can travel up to about 500 nautical miles before refueling. The latest model of the same aircraft today can fly about 106 knots, burns approximately 8 gallons of fuel per hour, and has a range of about 500 nautical miles. There has been virtually no change over the years in its performance.

However, ingenious amateur aeronautical engineers and pilots have produced airplanes using the same engines in newly designed airframes. These homebuilts are capable of 175 knots, with ranges beyond 1,000 nautical miles, using the same power plant, producing the same power at the same fuel flow. The backyard engineers have managed to attain much greater speeds over longer distances for the same amount of fuel.

While an increase in performance is advantageous, there are limitations to the use of a homebuilt airplane. For instance, homebuilts or kit planes are restricted from use in commercial operations. They also require a large amount of time for construction. Some pilots complain about the handling qualities of the smaller airframes. However, in many instances the advantages far outweigh the disadvantages.

One very important advantage is that of knowing the aircraft. As the owner-builder of the aircraft, the pilot is intimately familiar with all the systems of the airplane. Another benefit to constructing an airplane is that as the manufacturer of the craft, the builder is qualified to perform all maintenance and inspections on the airplane according to the Federal Aviation Regulations (FARs). In other words, each year, when it is time to inspect the airplane, the builder of the craft can save hundreds if not thousands of dollars in shop fees.

Another advantage homebuilt owners have over production aircraft owners is that they can build the airplane at their own pace, according to their own budget. A homebuilder can spend $600 one month, and if funds are lacking the next month, the building process can slow down while the builder spends only $50, or nothing at all. For the pilot who has purchased an aircraft with a loan, however, the bank will want payment each month.

Building Time

The time required to build a kit plane varies. With a fast-build option, a builder can have a plane airborne in less than a year. On the other hand, some builder-pilots have dragged out a project for twenty-five years. Average build times, depending on the make and model of the airplane, is about 2,000 to 3,000 hours of work. Working part time, this equates to two to four years.

After the airplane is finished, it is time for the first test flight. For this important first flight, many homebuilders opt to hire a professional test pilot familiar with their design. This is a smart choice for pilots who have allowed their flying skills to degrade during the construction process. Many would like to fly their homebuilt on the first flight, but this is a case where vanity must defer to common sense.

Following the initial flight, the homebuilder is free to fly the airplane through a test program. During this time, the airplane will be restricted to one geographical area for forty flight hours. After the airplane is proven through this test period, the restrictions are lifted. Now the owner-builder-pilot is free to use the airplane for personal use as any other airplane.

Some experimental aircraft begin as homebuilt inventions to test new designs, such as Fred Weick's homebuilt 1934 W-1A, which experimented with the use of tricycle landing gear. (NASA)

Types of Homebuilts

While the production market is limited, homebuilts offer a wide selection of airplanes to the potential builder, running the gamut from very simple single-seat ultralights to highly sophisticated, six-place family airplanes. A pilot desiring a production four-place family airplane that can cruise faster than 145 knots will spend more than $400,000, plus ongoing maintenance and upkeep. On the homebuilt market, however, one can choose from among many relatively inexpensive four-seat, high-speed, long-range airplanes. Similar savings can be made in homebuilding seaplanes and amphibious aircraft.

After choosing the group of airplanes from which to select, a pilot might research the safety aspects of particular designs. The next decision is the airplane's appearance, which is a matter of personal preference. Not only are there a wide variety of kits available for homebuilt plane types, but builders who want more than what is available on the market also can design a new craft incorporating all the desired attributes.

Most homebuilts and kit planes are smaller in size and weigh less than manufactured planes. Coupled with proportionally larger engines, this tends to increase the performance of the design. Many builders opt for two-place designs that provide opportunities for a great deal of compromise. Most pilots find themselves flying alone or with only one other person. When they have a need to carry more, they rent larger airplanes.

Construction Techniques and Materials

An exciting aspect of homebuilding is the selection of construction materials. Construction techniques vary with different airplanes. A popular airplane that uses construction techniques of the 1930's and 1940's is the Pitts Special. The fuselage of the Pitts is constructed of steel tube and wood formers covered with doped fabric. The wings are constructed of wood spars and ribs and covered in fabric. Other homebuilts constructed completely of wood are beautiful examples of artistic creation. Some use woodworking techniques that date back to World War I.

Conventional construction of modern light airplanes is sheet aluminum riveted on formed aluminum structures. In the early 1970's, homebuilders began experimenting with foam, epoxy, and other composite materials. Many of the new materials, such as Kevlar fabric and carbon graphite, are lighter than steel and provide greater tensile strength. In addition to being stronger and lighter, some of the new composite materials are easier to work with and enable the builder to form the compound curves of aeronautical structures more easily than when working with conventional materials.

Engines

With the use of Stirling engines or engines that use other alternative fuels, pilots may be able to fly farther and faster than ever imagined. Modern engines are one reason homebuilts are capable of such great speeds. The fact that the aircraft is experimental allows owner-pilot-builders to select the engine of their choice. Selection of the engine can be just as varied as selection of the aircraft itself.

The modern certified aircraft engine is a costly item. A new Lycoming or Continental aircraft engine can easily exceed costs of $20,000. Propellers and other accessories on the engine can drive that cost up another $5,000 or $6,000. Although such high prices may be discouraging, the homebuilder has many engine options.

The engine on a homebuilt does not have to be a certified aircraft engine. Because the airplane does not adhere to the specifications defined by the Federal Aviation Administration, builders can use any power plant they find suitable for their design. Indeed, many homebuilts are flying using engines from Volkswagen cars, chainsaws, snowmobiles, or outboard motorboats.

An automotive V-8 engine powers one of the most popular homebuilts, the Lancair IVP. The IVP can carry four passengers in pressurized comfort at altitudes above 25,000 feet, at speeds greater than 330 knots. Another example for those who doubt the use of automotive engines is that of the Volkswagen engine. Properly adapted for aerial use, this little engine has powered airplanes as fast as 230 miles per hour while getting more than 80 miles to the gallon.

While some doubt the validity of homebuilt and kitplane flying, this class of airplane is here to stay. For individuals with some technical background and the ability to work with their hands, homebuilding is a way to acquire an airplane inexpensively. The rewards they reap flying their own creations are many; chief among them the cost savings the homebuilder will realize over the years of flying the airplane. Saving money, however, is only a part of the compensation. The greatest reward is watching people admire the airplane. Most homebuilders are very pleased to hear the comments others make regarding the craftsmanship and work invested in the airplane.

Joseph F. Clark III

Bibliography

Dwiggins, Don. *Build Your Own Sport Plane: With Homebuilt Aircraft Directory.* New York: Hawthorne Books, 1979. An informative book explaining the process of selecting and building a personal flying sportcraft.

Jablonski, Edward. *Flying Fortress: The Illustrated Biography of the B-17's and the Men Who Flew Them.* Garden City, N.Y.: Doubleday, 1965. This books aptly describes the design process of World War II's B-17 Flying Fortress. Jablonski takes the airplane from design idea to flight tests to production, and then tells the stories of the men who flew the airplane in the war.

Stinton, Darrol. *The Design of the Aeroplane: Which Describes Common-sense Mechanics of Design as They Affect the Flying Qualities of Aeroplanes Needing Only One Pilot.* New York: Van Nostrand Reinhold, 1983. Outstanding text relating aircraft design to flying qualities. This manual, written in a technical format, has a great deal of mathematical explanation.

Van Sickle, Neil D. *Van Sickle's Modern Airmanship.* New York: McGraw-Hill, 1999. This volume is a more technical work providing readers everything they might desire to learn regarding flying and the aviation industry. It is very extensive, covering all aspects of the business.

See also: Airplanes; Manufacturers; Military flight; Model airplanes; Test pilots; Testing; X planes

F

Federal Aviation Administration

Date: Formed on October 15, 1966, with the creation of the Department of Transportation

Definition: The U.S. government organization primarily responsible for overseeing aviation safety, air traffic control and navigation, federal funding for airport and airway facilities, and civil aviation security.

Significance: The Federal Aviation Administration is responsible for issues of aviation safety regulation, inspection, examination, certification, and issuance of licenses. The FAA oversees pilots, aircraft, airports, airlines, air traffic control and navigation, aircraft and parts manufacturing, repair, civil aviation security, and even commercial space transportation—blasting private satellites into space.

Functions and Structure

The Federal Aviation Administration (FAA) has three main areas of responsibility: air traffic control and navigation; civil aviation safety regulation, certification of airlines and aircraft, and licensing of pilots, mechanics, and other aviation personnel; and civil (as opposed to criminal) aviation security regulation and enforcement to safeguard airports, airplanes, and personnel and passengers from terrorism and other criminal threats to aviation. To accomplish these functions, the FAA maintains a headquarters in Washington, D.C., nine regional offices, and hundreds of other offices in the United States and worldwide. The FAA has two major research centers in Oklahoma and New Jersey.

The FAA employs almost 50,000 employees, approximately 35,500 of whom perform air traffic services. The job of regulating, inspecting, and licensing airlines, aircraft, pilots, and mechanics is performed by a regulation and certification workforce numbering approximately 6,000. More than 1,000 personnel work in the area of civil aviation security, and the remaining personnel work predominantly in administration, in research, or even in the overseeing the safety of commercial space launches to put satellites into orbit for telecommunications companies or other businesses.

The FAA is headed by an administrator who serves a five-year term under the U.S. secretary of transportation. A deputy administrator and several associate and assistant administrators oversee the different areas of FAA responsibility.

Although the size of the FAA workforce may seem extraordinary, it is appropriate to the role of aviation in the United States. Each year, air traffic controllers must handle approximately forty-five million flights, and FAA airport towers log fifty million operations. As of 2000, there were 19,281 airports in the United States, 3,953 of which were public-use airports with paved runways. There were 651 FAA-certificated airports in the United States, serving air carrier operations with aircraft seating more than thirty passengers. As of December 31, 2000, there were 635,472 active U.S. pilot certificates: Whether they were students in small propeller planes or airline captains commanding jetliners carrying hundreds of passengers, more than one-half million people held licenses permitting them to fly.

The FAA expends eleven billion dollars yearly in the performance of its functions. Much of that total is paid by the traveling public through excise taxes added to the price of airline tickets. These taxes, totaling almost twelve billion dollars annually, go into a national aviation trust fund to pay for improvements to airports and airways.

Origins

Soon after Jean-François Pilâtre de Rozier and the marquis François d'Arlandes completed the first untethered balloon flight on November 21, 1783, the first effort at aviation regulation was made. In April, 1784, a French police ordinance required permits for balloon flights over Paris.

Early laws were hardly conducive to aviation. Roman law proclaimed that whoever owned land also owned the sky above that land. Early property law provided that if one owned the land on the surface, then one owned it to the center of the earth and to the heights of the sky. This legal concept did not prove especially troubling until air travel became possible.

At the time of Wilbur and Orville Wright's success on December 17, 1903, the prevailing legal concepts made the dominion of the skies somewhat like that of the ocean: Both were considered to belong to all people but not to any one person. By the end of the World War I, that theory had been replaced by the realization that a nation's skies were the key not only to its defense but also to its prosperity. Thus, each nation's skies became protected airspace. Treaties were drafted to keep nations' aircraft from entering other nations' airspace and to regulate the economics and safety of international aviation.

Regulatory Framework

In 1919, the world powers met in Paris to devise a plan for the implementation of an international regulatory framework to carry out civil aviation in a peaceful, safe, and efficient manner. The sovereignty of each nation's airspace was recognized, and the group proposed minimum standards for certification and safety regulation as well as general rules for air traffic control. Each nation would be required to adopt regulations to certify its airlines, aircraft, and pilots and to oversee the safety of its operations. Although the United States sent representatives to attend the Paris conference, it did not adopt the convention's agreements.

The United States would become a signatory to later air commerce and navigation treaties, and eventually the international oversight of aviation would be governed by the International Civil Aviation Organization, a part of the United Nations. To this day, the international regulatory plan depends on each nation having aviation safety laws and a government agency to enforce them. Although the FAA would eventually fulfill that role for the United States, it was still decades in the making.

The U.S. Air Mail Service

The U.S. Post Office was the beneficiary of the first U.S. aviation regulation. In 1920, after numerous airmail accidents, the head of the U.S. Air Mail Service set about to improve the situation. Pilots were required to complete 500 hours of flight training, pass an examination, and undergo a physical to establish medical fitness. Orville Wright assisted the effort to qualify and license the nation's pilots, personally signing some of the earliest U.S. pilot's licenses.

Air mail was privatized with the Air Mail Act of 1925, known as the Kelly Act. The routes were put up for bid, and wealthy American industrialists, such as Henry Ford, William Rockefeller, Cornelius Vanderbilt, and Marshall Field, garnered the first contracts.

On May 20, 1926, the Air Commerce Act was passed at the urging of the aviation industry, after the aviation industrialists realized aviation could not reach its significant commercial potential without the federal government providing safety regulation. It is unsurprising, then, that the job of aviation safety was given to the U.S. Department of Commerce. The secretary of commerce was charged with promoting air commerce, enforcing air traffic rules, licensing pilots and planes, certificating aircraft, establishing airways, maintaining aids to air navigation, and generally working to improve aviation's dismal safety record. With that, the seeds of the future FAA were planted, and there was much to be done. There were only 6,000 passengers willing to brave the airlines in 1926.

Forerunners of the FAA

By 1933, the nation's system of 18,000 miles of airways with 1,500 beacon towers and 263 landing fields was finished. Aerial navigation was very much a ground-based enterprise, with a cross-country system of ground beacons, small towers with a flashing rotating light and two course lights. In addition, ninety radio navigation stations had been built to provide aural and visual guidance to pilots. In 1930, the first radio-equipped air traffic control tower was built in Cleveland, Ohio, with twenty more to follow by 1935. In 1935, the cities of Chicago and Newark set up air traffic control systems to control their flights. The Bureau of Air Commerce was formed in 1934 within the Department of Commerce, and, two years later, it took over the responsibility of air traffic control.

By the 1930's, the airlines, in the throes of destructive price-cutting competition spurred by mail-contract bidding, were themselves clamoring for federal regulation. The airlines wanted to upgrade their fleets with the new, sleek, metal marvels of the aviation world: the Douglas DC-3 aircraft. To afford these airplanes, the airlines needed to be spared from cutthroat price wars. The airlines' solution was federal economic regulation. By having the federal government regulate not only air traffic control, safety, and certification, but also airline profits, they would be protected from huge losses caused by destructive competition, and they could afford to buy the marvelous new DC-3's.

Thus, in 1938, the Civil Aeronautics Act created the Civil Aeronautics Authority to regulate safety and economics. In 1940, the authority was split into the Civil Aeronautics Board (CAB), which had the powers of safety regulation, accident investigation, and economic regulation and also established airline fares and routes, and the Civil Aeronautics Administration, which was responsible for air traffic control, pilot and aircraft certification, safety enforcement, and airway development. The airline plan worked. Americans loved the DC-3, which remains the most successful transport plane ever. Almost 11,000 were built in the United States and at least as many were manufactured overseas. By 1941, there were three million U.S. airline passengers, who looked to the U.S. government to protect their safety.

FAA Regions

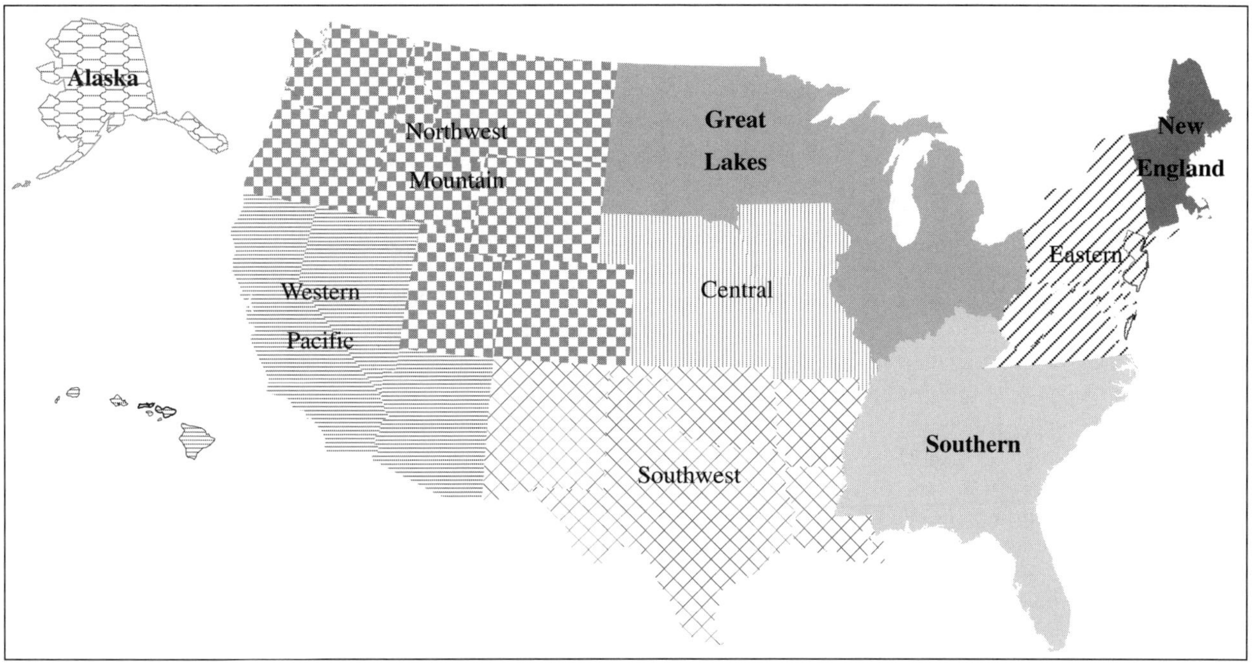

Source: Federal Aviation Administration.

During World War II, both military and civil aviation changed dramatically. Newly developed radar technology was applied to air traffic control. In 1944 alone, the United States produced 96,318 airplanes. Aviation was credited by many historians with winning the war.

During the Cold War, the federal government provided money for airports and instrument landing systems. Equipment, such as pressurized airplanes, airborne weather radar, and autopilots, were dramatically improved, and passenger comforts were increased. By 1956, U.S. airline passengers outnumbered rail passengers, a trend that was never reversed.

However, during this time, several tragedies shook the confidence of the flying public. The world's first jet commercial airliner, the British De Havilland Comet commenced passenger jet service in May, 1952, but its success was short lived. Of the nine Comets in commercial passenger service, three seemingly came apart in midflight. In 1956, a United Air Lines flight collided in midair above the Grand Canyon with a Trans World Airlines (TWA) flight, killing 128. After it was discovered that an air traffic controller had seen the planes' collision course on his radar and had failed to warn the pilots, the reputation of the CAA was tarnished. After two more U.S. midair collisions, between U.S. Air Force planes and civilian airliners, it was recognized that changes were required.

Federal Aviation Agency

In 1958, the new, independent Federal Aviation Agency was created and took over the CAA's functions. It also took from the CAB the job of promulgating safety regulation and coordinating military and civilian air traffic control. The CAB retained the responsibilities of economic regulation and accident investigation, but not for long. A midair collision on December 18, 1960, between a United Air Lines and a TWA flight in the skies above New York killed 135 people, including 8 on the ground, and intensified the public demand for improvements.

The 1960's brought the radar-based air traffic control (ATC) system, with its banks of green screens that enabled controllers to monitor the nation's airports and airways. In 1966, the Department of Transportation (DOT) was formed to coordinate the regulation of all modes of transportation within one department. The Federal Aviation Agency, now operating under the DOT, became the Federal Aviation Administration. The accident investigation function was removed from the CAB's jurisdiction, and an independent accident investigation organization, called the National Transportation Safety Board (NTSB), was established. Although the FAA may assist in aircraft accident investigation, the NTSB remains primarily responsible.

Reacting to Tragedies

Unfortunately, in the coming decades, air tragedies continued to direct the course of the FAA. Aircraft hijackings in the 1960's caused the FAA to institute security regulations and requirements, followed years later with more-stringent requirements following more deadly and catastrophic aircraft crimes, such as the terrorist bombing of Pan American Flight 103 in 1988. Plastic explosives were hidden in a personal tape recorder, which was loaded in Frankfurt, Germany, into the baggage compartment of the doomed plane. There was no passenger on board the flight to accompany the baggage.

Major domestic security changes were ordered after a tragic episode in 1987, in which a fired Pacific Southwest Airlines (PSA) employee boarded the airliner with his old employee badge and, after takeoff, shot his former boss, a passenger. The killer then shot the aircraft's pilots, and the plane plunged to earth, killing all on board.

The most shocking tragedy occurred on September 11, 2001, when teams of terrorists hijacked four commercial jets. Two 767's out of Boston's Logan Airport, American Airlines Flight 11 and United Air Lines Flight 175, were intentionally crashed into the Twin Towers of the World Trade Center in New York City, causing both buildings to collapse. The third plane, American Flight 77, a 757 out of Washington's Dulles Airport, was crashed into the Pentagon in Washington, D.C. The fourth plane, United Flight 93 out of Newark, New Jersey, crashed in a field near Pittsburgh, Pennsylvania, when passengers stormed the cockpit of the 757. In total, more than 5,500 people, from the planes and on the ground, were killed. The hijackers smuggled box cutters on board the aircraft, gained access to the cockpits, either killed or incapacitated the flight crews, switched off the transponders, and took over the controls. Two of the terrorists, Islamic fundamentalists associated with Osama bin Laden's al-Qaeda network, were on a Federal Bureau of Investigation (FBI) watchlist but were allowed to purchase tickets.

Eventually, the NTSB would discover the trend that the most frequent among many causes of accidents was the failure of the FAA to act to avert catastrophe.

Airline Deregulation

In 1978, the FAA faced another problem, when the airline economic deregulation unwittingly dealt airline safety an insidious blow. Bowing to intense political pressure, the federal government hastily freed airlines from almost all economic regulation. The debate over the wisdom of deregulation has continued ever since. The CAB was abolished, and with the elimination of economic regulation, a substantial part of airline regulation disappeared. The government no longer regulated the routes that airlines could fly or how frequently they could fly them. The airlines could set fares and invent combinations of arbitrary fare restrictions. The airlines could price their tickets below the cost of buying, flying, and maintaining the planes. As the U.S. airline fleet aged, cash-strapped carriers delayed maintenance, cut corners on safety, and, in some cases, even falsified maintenance records.

The FAA, however, did not change the way it policed the airlines. In the years following deregulation, dozens of upstart carriers entered the airline business. Many of these companies operated with meager financing, old planes, little experience, and low-paid employees. They planned to meet vital functions, such as maintenance and safety, by contracting with the lowest bidders. Such airlines came to be called virtual airlines, and almost all of them went bankrupt or otherwise ceased to exist within a few years.

One such carrier, ValuJet, caused the biggest FAA crisis in history, but also caused the FAA to increase by 267 percent its remedial action. On May 11, 1996, a ValuJet flight crashed into the Florida Everglades, killing all 110 on board. The American public's faith in the FAA was shaken when both the FAA administrator and the DOT secretary of transportation stood at the crash site and, before any investigation, on national television pronounced the airline safe. The airline was not safe, however, and days later, ValuJet was grounded for safety violations. A document produced from within the FAA showed that FAA inspectors had recommended grounding ValuJet months before the crash.

Congressional and Senate hearings probed the problems within the FAA, and the FAA admitted to Congress that with the advent of virtual airlines, its abiiity to inspect and oversee the airlines had been significantly hampered. The dual mission of the FAA, both to promote aviation and to regulate safety, inherited decades before from its predecessors within the Commerce Department, was an obvious inherent conflict. The FAA was tasked by Congress to set about improving its own safety record, as well as that of the U.S. airlines.

Air Traffic Control and Crowded Skies

By the 1980's, the nationwide air traffic control system of banks of blinking green screens and paper strips tracking thousands of planes across the skies had become antiquated. Congressional hearings examined the problem of demand for air travel exceeding the ability of the old air traffic control system to handle the traffic. The FAA's initial efforts to replace the system had failed by the mid-1990's. The first replacement program was woefully behind schedule and well over budget. The FAA, ordered by Congress to start over, began phasing in over many years parts of the new air traffic control replacement system. The overall completion date was targeted for 2015, with most of the system projected to be finished by 2008. The completed system allows for completely computerized and automated air traffic control aided by global positioning system (GPS) satellites. The aircraft itself will be able to communicate with the air traffic control system. Should something happen to the pilots, a verbal or electronic command from the aircraft's home base or air traffic control can tell it to return to its home airport or to another designated airport.

Under the new system, pilots will finally be able to legally and safely choose paths across the sky, without bonfires or beacons, without cumbersome air routes dictated by green blinking radar screens and strips of paper, and without needless tragedy that imperils pilots, passengers, and even pedestrians on the ground below.

Mary Fackler Schiavo

Bibliography

Nader, Ralph, and Wesley Smith. *Collision Course*. Blue Ridge Summit, Pa.: TAB Books, 1994. A frank discussion of the problems of aviation regulation and ideas about how to fix them.

Nance, John. *Blind Trust*. New York: Morrow, 1986. A commercial airline pilot's view of the problems caused by the deregulation of the airline industry and the FAA's role in aviation safety enforcement.

Schiavo, Mary. *Flying Blind, Flying Safe*. New York: Avon, 1998. The personal story of the DOT inspector general's investigation of the FAA, the surprising problems she found, and the federal government's even more surprising reaction.

See also: Accident investigation; Air carriers; Air traffic control; Airline Deregulation Act; Airline industry, U.S.; Airports; Commercial flight; Manufacturers; National Transportation Safety Board; Pilots and copilots; Safety issues; Training and education

Fighter pilots

Definition: The pilots of tactical jet aircraft used for defensive posturing and offensive attacks. The men and women who fly tactical aircraft are usually the best pilots available in terms of both talent and training.

Significance: Since World War I, fighter pilots have been a major component of modern warfare. They usually fly the most technologically advanced aircraft of their day, and are trained to exceptionally high standards. The fighters may be land based or deployed aboard aircraft carriers.

The Beginning of Aerial Combat

At the beginning of World War I, airplanes were very scarce and primarily used for surveillance. Both the English and the Germans had aircraft designed to be observation platforms, and that was the extent of their use. Few people seriously regarded airplanes as practical war machines; early military officers, like much of the public, thought of airplanes as expensive toys, frivolous and of little practical use. They were noisy, breezy, hard to communicate within, and dangerous. Typically, their engines would quit at any time without notice, and flight crews would be lucky to survive.

In the early days of the war, it was very rare for a pilot to come across another aircraft in flight. It was even more unusual to come across an enemy aircraft. Eventually, however, that is exactly what happened. When the two pilots realized that they were flying alongside one another, after the cursory waves to each other, they proceeded to fly along a little further. Then one of them realized that the other truly was the enemy and decided that action was required. Reaching into his tunic, he pulled out his revolver, carefully aimed it at the other aircraft, and squeezed the trigger. The other pilot decided to turn and run. Thus, aerial warfare was born.

From those very humble beginnings in World War I, the job and title of fighter pilot has become synonymous with heroism. From airplanes that flew no faster than 80 miles per hour to jets capable of more than 1,800 miles per hour, the duty and challenge of guiding these machines has been one sought after by many.

Training

The fighter pilot must be capable of multitasking, maintaining extreme situational awareness, and working in a very demanding and hostile environment, all the while being capable of using his airplane and its weapon systems to their limits. Fighter pilots are typically young men or women in their late twenties or early thirties. They are college graduates, and some possess graduate degrees.

Typically, they have wanted to be fighter pilots all their lives. They maintain a driving desire to attain any goal they set their minds to achieving. They have probably the most refined single-minded focus of any group of modern professionals.

Fighter pilots begin their careers with maintaining good grades throughout high school and college. The choice of the major course of study in college is not as important as aptitude and attitude. Of course, a technical degree will serve an aspiring career military aviator better than a nontechnical degree.

Fighter pilots must be commissioned as officers in the military services. To obtain a commission, a candidate will take part in the Reserve Officers' Training Corps (ROTC) while in college or Officer Candidate School (OCS) after graduating from school. Another route to a commission is through appointment and graduation from one of the nation's military academies.

After graduation and commissioning, the newly appointed second lieutenant or naval ensign passes through flight training, a process that takes from two to three years. During the time spent in flight training, new pilots learn all the basics of flying, including formation flying, instrument flying, and finally, aerial warfare.

Fighter Aircraft of the U.S. Navy and Air Force

In 2001, U.S. Air Force pilots could expect to fly the F-16 Fighting Falcon or the F-15 Strike Eagle. Another attack airplane in the Air Force inventory is the A-10 Warthog, the tank killer of the Gulf War. The F-16 and F-15 are true fighters, while the A-10 is an attack aircraft.

The pilots flying the F-15 and F-16 can reach speeds well in excess of 1,000 miles per hour, while the mission of the A-10 does not require such speeds. The A-10's strength is in "tank-busting" and ordnance delivery. The A-10's straight-wing design gives it the advantage of more wing area, so it can carry a bigger bomb load than the F-15 or F-16. The integrated gun is also larger, with 30 millimeters on the A-10 versus 20 millimeters on the other aircraft.

While the mission of the A-10 is ground attack, the F-15 and F-16 can perform both missions of ground attack and aerial warfare. They have the capability of attacking ground targets, but they excel in the air-to-air arena.

Navy fighter pilots fly the F-14 Tomcat and the F/A-18 Hornet. Each airplane, carrier-based and used for air-to-air fighting and ground attack, can fly faster than 1,000 miles per hour. While the F-14 came off Grumman's design table as a fighter aircraft, the F/A-18 was a multimission aircraft from the start. The F/A-18 is intended eventually to take on all roles of fleet defense and ground attack, with the F-14 being phased out.

Unlike Air Force aircraft, there are special considerations for naval aircraft. They have to be able to take the abuse of being aboard ships and dealing with the very harsh saltwater environment. While all tactical jets are built to survive extreme wear, the navy aircraft are constructed a little more sturdily to survive the catapult shots for takeoff and the hard landings required to land aboard a floating runway.

From their floating airfields, F-14 pilots and their counterparts in the backseat, the radio intercept officers, defend the fleet against aerial assault. Depending on the location of the fleet, the threat environment, and the rules of engagement, the F-14's may or may not become involved in traditional dogfights. Once Navy fighter pilots complete their missions, they must return to the ship. While the Air Force pilots have the luxury of landing on long runways, the naval aviator is faced with the daunting task of landing a multiton fighter on a pitching carrier deck.

Air Combat Maneuvering

Dogfighting, the term applied to airplanes engaged in aerial combat against one another, has come a very long way from the time the first British and German pilots shot at each other from their observation aircraft. From World War I to World War II, aerial combat was refined into a lethal art. Pilots learned maneuvers that would best allow them to get behind their enemies and bring their weapons to bear. In World War I, a dogfight lasted minutes. By contrast, a modern dogfight may take less than one turn to complete and is over in seconds.

If the fight degenerates into a turning fight, fighter pilots put their aircraft through basic air combat maneuvering, which includes maneuvers such as the rolling scissors, flat scissors, the high yo-yo and low yo-yo, and the Lufberry circle, to name a few. These are the maneuvers a pilot would use to gain an advantage over an enemy for a close-in gun shot or an intermediate missile shot.

In contrast to the fighters of the earlier days, today's fighter pilots have the capability for what is known as beyond visual range (BVR) shots. Depending on the theater, the threat environment, and the rules of engagement, fighter pilots may have the option of shooting down enemy aircraft without ever seeing them. This is dependent on the certainty that any aircraft coming from a certain sector is the enemy.

It is a risky endeavor and has resulted at times in Allied losses due to friendly fire. Everyone operating in the particular area has to be operating by the same rules. If not, there is the possibility of mistakes with terrible results. As a result, fighter pilots tend to be absolutely sure of themselves. There is always a chance that the consequence of any mistake may be fatal to themselves or someone else. Consequently, fighter pilots tend to be on the cautious side, only acting when certain that their knowledge and their actions are positively, literally, and completely correct.

Joseph F. Clark III

Bibliography

Gandt, Robert L. *Bogeys and Bandits: The Making of a Fighter Pilot*. New York: Viking, 1997. A well-written narrative account of Hornet pilot training at NAS Cecil Field, Florida.

Rosenkranz, Keith. *Vipers in the Storm: Diary of a Gulf War Fighter Pilot*. New York: McGraw-Hill, 1999. A personal account of the author's participation in the Gulf War as an F-16 fighter pilot.

Wolfe, Tom. *The Right Stuff*. New York: Farrar, Straus and Giroux, 1979. Excellent account of test pilots at Edwards Air Force Base and the first Mercury pilots.

Yeager, Chuck, and Leo Janos. *Yeager: An Autobiography*. New York: Bantam Books, 1985. An excellent read on the life and times of Chuck Yeager, fighter pilot, ace, and test pilot who broke the sound barrier. Illustrates through the accounts of Yeager and others involved in tactical aviation and test flight what it is like in the cockpits of fighters.

See also: Air Force, U.S.; Aircraft carriers; Airplanes; Dogfights; Eagle; Gulf War; Hornet; Kamikaze missions; Korean War; Luftwaffe; Marine pilots, U.S.; Navy pilots, U.S.; Pilots and copilots; Manfred von Richthofen; Eddie Rickenbacker; Royal Air Force; Spitfire; Tomcat; Training and education; Vietnam War; World War I; World War II

F-16C/D Fighting Falcon Characteristics

Primary Function: Multirole fighter
Builder: Lockheed Martin Corporation
Power Plant: One Pratt & Whitney F100-PW-200/220/229 or General Electric F110-GE-100/129
Thrust: 27,000 pounds
Length: 49 feet, 5 inches
Height: 16 feet
Wingspan: 32 feet, 8 inches
Speed: 1,500 miles per hour (Mach 2 at altitude)
Ceiling: Above 50,000 feet
Maximum Takeoff Weight: 37,500 pounds
Range: 1,740 nautical miles
Armament: One M-61A1 20-millimeter multibarrel cannon with 500 rounds; external stations can carry up to 6 air-to-air missiles, conventional air-to-air and air-to-surface munitions, and electronic countermeasure pods
Unit Cost: $34.3 million
Crew: F-16C, 1; F-16D, 1-2
Date Deployed: January, 1979

Source: Data taken from (www.af.mil/news/factsheets/F_16-Fighting_Falcon.html), June 6, 2001.

Fighting Falcon

Also known as: F-16, single-seat F-16A and F-16C; two-seat F-16B and F-16D; F-16XL delta wing

Date: First flight on January 20, 1974

Definition: A highly maneuverable, lightweight, air-to-air and air-to-ground attack fighter first developed in the United States and adopted by many world nations.

Significance: Proven in combat and in numerous world aircraft competitions, the F-16 Fighting Falcon is considered by experts to be the best multirole, cost-effective fighter aircraft ever made. Its technical evolution from the 1980's through the beginning of the twenty-first century has kept it ahead of all other contemporary potential threat aircraft.

Specifications

The F-16 is a single-seat air superiority and multirole fighter. Its wingspan is 32 feet, 10 inches, and its length 49 feet, 6 inches. Its power plant is one 25,000-pound thrust Pratt & Whitney F100-PW-200 (3) turbofan. When empty, it weighs 15,580 pounds; it weighs 35,400 pounds at gross capacity. The F-16's range is 2,000 miles and its service ceiling is 50,000 feet. The plane can reach a top speed of Mach 2, twice the speed of sound. For armament, it carries one 20-millimeter multibarrel rotary cannon plus two wingtip mounted sidewinder missiles, as well as seven external pylons for fuel tanks and other selected air-to-air and air-to-ground weapons.

Development

The F-16 was first planned and designed in the late 1960's

by Pierre Spray, a civilian working in the office of the assistant secretary of defense, and John Boyd, an Air Force major and flight instructor, along with Harry Hillaker of the General Dynamics Corporation. The project received further support from the 1972 Lightweight Fighter Prototype program sponsored by the U.S. Air Force. The plane was conceived to be smaller, lighter, faster, more maneuverable, and less expensive than the U.S. Navy's F-14 Tomcat and the F-15 Eagle of the U.S. Air Force. Another idea was to sell the plane to allied nations worldwide in order to increase production and lower the cost of the plane.

In August, 1972, the U.S. Air Force appointed General Dynamics Corporation and Northrop, another United States corporation, to build concept prototypes, one of which would be selected for production. On December 13, 1973, the first two General Dynamics prototypes, called YF-16, were ready to be tested and evaluated by government-appointed test pilots Phil Ostricher and Neal Anderson. On January 13, 1975, U.S. Air Force secretary John L. McLucas announced the YF-16 as the winner of the lightweight fighter competition and in February of that year, a North American Treaty Organization (NATO) consortium offered $5.16 million each for the production of two thousand F-16's for the United States and NATO allies. In December, 1976, the first test flight of the single-seat F-16A occurred and in January, 1979, the first military operational F-16's were delivered to the 388th Tactical Fighter Wing at Hill Air Force Base in Utah. The U.S. Air Force special demonstration team, the Thunderbirds, adopted the F-16 in November, 1982, and the U.S. Air Force increased its planned F-16 purchase total to 3,047 in February, 1986.

Technical Highlights

One technical innovation of the F-16 was the fly-by-wire electronic computer-assisted steering system, which was less vulnerable to damage from attack. This system had faster and more precise steering and maneuver capabilities compared to the old hydraulic systems.

The F-16 also had a new advanced radar system with look-down capability to track small high-speed objects below the airplane at treetop level, such as ground-to-air antiaircraft missiles.

Many experts consider the F-16 Fighting Falcon to be the best multirole, cost-effective fighter aircraft ever made. (U.S. Department of Defense)

Another innovation was the heads-up display (HUD), which projected data from the control panel onto the windshield, enabling pilots to keep their eyes on the sky and more quickly evaluate the data.

Combat

On Sunday, June 7, 1981, eight F-16's of the Israeli air force successfully bombed a nuclear reactor power plant located 11 miles southeast of Baghdad, Iraq, without suffering any losses of human life or airplanes. The mission is noteworthy in that the F-16's low-altitude flying eluded radar detection and the bombs dropped were precisely on target, thereby completely destroying the entire power plant facility.

During that same month, Israel launched a full-scale land and air campaign that removed the Palestine Liberation Organization (PLO) from occupied territories in Lebanon. On a single day, June 9, 1981, a total of ninety Israeli F-15 Eagles, F-16 Falcons, and Kfirs (Israeli-built versions of the French Mirage V) shot down about thirty of sixty Syrian MiG warplanes and bombed sixty SA-6 SAM ground-to-air missile sites. During air combat over the next two days, an additional fifty Syrian fighter planes were shot down, and by the end of the month, the Syrian total aircraft losses were numbered at eighty-five. During the entire month of June, 1981, not one single Israeli airplane was shot down.

In the Gulf War, 249 F-16's were deployed against Iraq during the forty-three-day air war in January and February, 1991. They flew the most missions or sorties of any coalition aircraft against ground targets, including parked Iraqi aircraft, airfields, Scud missile sites, and production facilities manufacturing weapons and chemicals.

Following the Gulf War, F-16's from several nations played a key role in enforcement of the United Nations sanctions and no-fly zones over Iraq and Bosnia. On December 27, 1992, the first United States Air Force F-16 to score an air-to-air victory occurred when an Iraqi MiG-25 was shot down by an F-16 over Iraq. In 1994, F-16's shot down three Serbian jets over Bosnia.

Alan Prescott Peterson

Bibliography

Drendel, Lou. *Viper F-16*. Carrollton, Tex.: Squadron/Signal, 1992. A comprehensive, well-illustrated factual review of the F-16 and its development, including personal essays of U.S. pilots who flew the F-16 in Operation Desert Storm.

Walker, Bryce. *Fighting Jets*. Alexandria, Va.: Time-Life Books, 1983. A historical and technical survey of jet airplanes from their development in the 1940's to the F-14, F-15, and F-16, with many illustrations and battle accounts.

Yenne, Bill. *The History of the U.S. Air Force*. New York: Bison Books, 1984. A study of civilian and military aircraft development.

See also: Air Force, U.S.; Eagle; Fighter pilots; Gulf War; Military flight; Testing; Tomcat

Firefighting aircraft

Also known as: Airtankers

Date: First aircraft used for firefighting in 1919; first airtankers produced in 1946

Definition: Aircraft, both planes and helicopters, modified to carry and release water or other flame-retardant liquids for the extinguishing of large fires, usually in remote areas.

Significance: The use of firefighting aircraft allows government agencies and civilian corporations to provide firefighting services primarily in large areas that are densely forested.

History

The first use of an airplane to combat a wildland fire occurred in California during 1919. Modifications to private aircraft continued for several decades but remained limited and experimental. In 1946, Glenn L. Martin, founder of the Martin Aircraft Company, designed an aircraft known as the Mars, which he initially envisioned as a long-range mission bomber with heavy lift capabilities. The four Mars planes, named the *Marianas*, *Phillippine*, *Hawaii*, and *Caroline*, transported troops and cargo between the islands of the South Pacific from 1946 to 1959. After setting world records for flight duration and airlift ability and logging more than eighty-seven thousand accident-free hours while in service with the U.S. Navy, the airplanes were retired. In 1959, after a series of catastrophic forest fires, a group of timber companies formed the Forest Industries Flying Tankers and purchased the four Mars planes, modifying them to serve as water bombers. In 1961, the *Marianas* crashed on a firefighting mission and in 1962 a hurricane destroyed the *Caroline* on land. The other two aircraft, the *Phillippine* and the *Hawaii*, were still in service at the beginning of the twenty-first century.

The Mars aircraft carries a crew of four, including the captain, first officer, and two flight engineers. Within ten minutes of receiving an emergency call, the planes are airborne and provide the initial attack on large fires, followed by repeated drops every fifteen minutes during sustained operations. Once the aircraft reaches the water source, the pilot begins the intake procedure by maintaining a constant speed of 60 to 70 knots while the scoops are turned to the down position. Water injected at the rate of 1 ton per second requires the flight engineer to continually advance the throttle to maintain the proper speed. When the tank is full the scoops are raised and the flight engineer takes off just as in a normal takeoff from land. Once the plane is back in flight, a foam concentrate is injected into the 7,200 gallons of water, where it remains inert until dropped. As the water falls, the tumbling action causes the foam to expand, transforming the water into a fire-retardant 4 percent solution. The aircraft, equipped with four Wright Cyclone R3350-24WA engines, measures 120 feet in length, has a 60,000-pound water/foam load limit, and can fly for over 5 hours before landing. When dropped from a height of 150 to 200 feet, the foam covers an area of 3 to 4 acres. Two additional aircraft assist the Mars planes in their firefighting efforts. The Grumman G21A Goose spots the fires and guides the Mars pilots into the bombing pattern. Responsible for coordinating efforts with the fire boss on the ground, the pilot of the Goose determines altitude and drop height and develops an exit plan. All of the pilots are trained on both the Mars and the Goose aircraft, enabling them to predict in-flight and drop requirements more accurately. Since 1974, Bell 206L-1 LongRange helicopters have also been used for smaller fires caused by lightning strikes. Each helicopter is equipped with a Bambi Bucket with a capacity of 140 gallons. The helicopters are also used to evacuate people and provide tactical support. While the Forest Industries Flying Tankers concentrate on large fires located on property owned by Weyerhaeuser Company Limited and TimberWest Forest Limited, the two lumber companies that own Forest Industries, they also contract out their services on occasion.

Private Aerial Firefighting Companies

Since the early 1960's, a number of other companies have offered similar services using a variety of aircraft developed for firefighting missions. Based primarily in heavily forested areas, the firefighting enterprises include several multiengine airtanker companies. Aero Flite, of Kingman, Arizona, was founded in 1963 and operates one C-54E, one C-54G, one M-18B Dromader, one Aztec, and one Piper Cherokee 6 within the continental United States and Alaska. Aero Union Corporation of Chico, California, the largest contractor of firefighting services in the United States, manufactures aerial firefighting aircraft and systems with a patented constant flow drop system used on Lockheed C-130's, Lockheed L-188's, and P3's, as well as on the Lockheed P-2V aircraft. The company also provides aircraft and personnel for firefighting missions in cooperation with the U.S. Department of Agriculture and numerous other federal agencies. Aero operates a fleet of six P3's, four SP-2H's, and two C-64's and contracts for services in California, Idaho, Oregon, Arizona, and Utah. The company manufactures retardant aerial delivery systems (RADS), helicopter-borne aerial firefighting systems, auxiliary fuel systems, aerial spray systems, aerial refueling tanks, modular airborne firefighting systems, bulk fuel transport tanks, 1080 refueling store systems, and automated cargo handling systems.

ARDCO of Tucson, Arizona, formed in 1976, has a fleet of three DC-4's and covers areas in Nevada, Oregon, and California, providing firefighting services for the U.S. Department of Agriculture and the Forest Service. Butler Aircraft Company of Redmond, Oregon, founded in 1946, introduced the B-17 into aerial firefighting and also uses two DC-7's, one DC-6, and one C-130 in its fleet providing services for the U.S. Department of Agriculture and Forest Service in Washington, Oregon, Arkansas, Tennessee, Alaska, Michigan, and Minnesota. Hawkins & Powers Aviation, out of Greybull, Wyoming, has specialized in aerial firefighting and agricultural spraying since 1958. Its fleet includes five PB-4Y-2's, fourteen P-2V's, seven C-130A's, six C-97's, and one P2-T, as well as a fleet of heli-tankers that includes two Bell UH-2B's, two Bell 206L3's, two Bell 206 BIII's, two Hughes 500D's, and two Hiller 12E's. The company provides services in Alaska, Australia, Washington, Oregon, Minnesota, Idaho, Nevada, Colorado, New Mexico, Arizona, Montana, Utah, Tennessee, North Carolina, Florida, Pennsylvania, New York, California, Oklahoma, Texas, and South Dakota.

Hirth Air Tankers of Buffalo, Wyoming, was formed in 1987 and operates two PV-2 airtankers, two PV-2 sprayers, and one Grumman American G164B-600 Ag-Cat single-engine airtanker. Neptune, of Missoula, Montana, was founded in 1993 and operates nine Lockheed P2-V Neptunes throughout the United States. International Air Response of Chandler, Arizona, founded in 1965, operates three C-130's, one DC-7 airtanker, and two DC-7 sprayers throughout the continental United States, Alaska, Spain, and France. T.B.M., of Tulare, California, founded in June, 1957, has provided aerial firefighting services since 1959.

Its fleet includes two C-130's, one C-54, one DC-6, one DC-7, and one SP-2H leased from Aero Union.

The two major companies that provide heli-tanker contract services are Erickson Air-Crane Company of Central Point, Oregon, and Heavy Lift Helicopters of Clovis, California. Single-engine airtanker companies such as Downstown Aero of Vineland, New Jersey, and Queen Bee Air Specialties of Rigby, Idaho, also provide services on a contractual basis.

Many of these companies use vintage World War II aircraft modified with retardant aerial delivery systems (RADS). One of the most common systems is used on the Lockheed P-3 Orion and L-188 Electra aircraft. Manufactured by Aero Union Corporation, the RADS II system is a constant-flow belly tank using a computer-controlled door system that allows the crew to select the appropriate flow rate. Equipped with a 3,000-gallon tank, the computerized system maintains a constant flow with a uniform drop rate with no overlapping or gaps. Used for initial attacks on grasslands or heavy timber fires, the advanced system prevents possible fire burn-through.

One of the more common heli-tankers is the S-70A/UH-60L Firehawk. The Firehawk has a one-thousand-gallon water tank, a 30-gallon foam tank, a 1,000-gallon-per-minute snorkel, and computer-controlled doors. The snorkel allows the heli-tanker to fill the water tank in one minute. The most recent addition to the fleet of heli-tankers is the S-64 Skycrane Fire Fighting Tank, with its 2,000-gallon water tank and 60-gallon foam tank. The Skycrane's snorkel is also capable of filling the water tank in one minute, even though its capacity is double that of the Firehawk.

Military Firefighting

Although several private companies across the country provide firefighting services, sometimes the U.S. military is called upon to assist when resources are limited and the extent of the fire warrants additional aircraft. Instead of building planes specifically for firefighting missions, the military transforms the Lockheed C-130 Hercules cargo planes into firefighting aircraft by adding systems such as the Modular Airborne Fire Fighting System (MAFFS).

Installation of the MAFFS system can be completed within one hour and contains a 3,000-gallon tank that can be filled in fifteen minutes and fully discharged in five seconds. Releasing the retardant alternately from a series of tanks, the MAFFS system allows pilots to maintain complete control over the aircraft during the drop, as opposed to other systems that require in-flight compensation as the nose of the aircraft pulls up and gravity pulls the retardant out of the rear of the plane. Installation of the MAFFS system is a two-step process. The initial installation of the lower tank is permanent and adds approximately 500 pounds to the aircraft. The upper tank attaches to the floor line and is removed after the firefighting mission is over, allowing the C-130 to be used once again for cargo and troop transport. The fire retardant released from the MAFFS system is a chemical called phos chek. In addition to reducing the combustion of plants and trees even after it loses its moisture, phos chek also contains a fertilizer that enhances regrowth of burned areas and helps prevent further soil erosion.

Mission and Operation

The mission of the firefighting aircraft is to extinguish forest and wildland fires in cooperation with federal, state, and local agencies. In the United States, the federal agencies responsible for wildfire firefighting are the Forest Service, the Bureau of Land Management, the National Park Service, the Bureau of Indian Affairs, and the Fish and Wildlife Service. Private airtanker companies work together with these agencies to control and manage fires. During the first six months of 1999, more than 1,097,400 acres of wildland had burned. Two of the worst fires occurred in Florida, where eight hundred firefighters fought the blaze for several weeks. When a wildland fire is reported, local crews respond first with an initial attack. This is usually provided by private companies, and 98 percent of fires are put out at this stage. If additional resources are required, one of the eleven nationwide coordinating centers are notified. Geographic area coordinating centers are located in Albuquerque, New Mexico; Riverside, California; Salt Lake City, Utah; Fairbanks, Alaska; Portland, Oregon; Broomfield, Colorado; Missoula, Montana; Milwaukee, Wisconsin; Reno, Nevada; Atlanta, Georgia; and Redding, California. The coordinating center then notifies the National Interagency Coordination Center in Boise, Idaho, where the National Wildfire Coordinating Group assumes control over the efforts. This group has extensive resources available nationwide, including 70 twenty-person Hotshot crews trained to handle complex firefighting situations; 409 individual smoke jumpers with 19 support aircraft for initial attacks in the western portion of the United States; 58 contract airtankers; a large transport for crew and equipment; 11 lead planes for tactical operations; 23 helicopters; 3 aircraft outfitted with infrared scanners capable of mapping the fires; 17 incident management teams trained specifically for handling complex situations; 447 twenty-person geographically located teams that can be called upon if necessary; 50 communication kits; 21 contractors that provide catering and shower facilities for the firefighting teams; and 11 warehouses full of firefighting equipment and supplies.

Cynthia Clark Northrup

Bibliography

Fuller, Margaret. *Forest Fires: An Introduction to Wildland Fire Behavior, Management, Firefighting, and Prevention*. New York: John Wiley & Sons, 1991. Discusses wildland fires, their environmental impact, and several well-known wildfires, and provides detailed information on firefighting behavior, management, and operations, including the use of firefighting aircraft.

Lowe, Joseph, et al. *Wildland Firefighting Practices*. Albany, N.Y.: Delmar Thomson Learning, 2001. Describes the process of firefighting in wildland areas and provides illustrations to help the reader understand topics such as burn-through. The author, a firefighter himself, includes information on ground operations and the use of fixed-wing aircraft and helicopters.

Perry, Donald G. *Wildland Firefighting: Fire Behavior, Tactics, and Command*. Bellflower, Calif.: Fire Publications, 1990. A reference for the tactical operations involved in fighting large fires in remote areas. Focuses on the command structure but provides a good deal of information on the use of firefighting aircraft.

See also: Airplanes; Bell Aircraft; Helicopters; Lockheed Martin; McDonnell Douglas; Manufacturers; Weather conditions

Flight attendants

Also known as: Stewards, stewardesses, cabin boys, sky girls or boys, air hostesses

Definition: Flight crewmembers who are responsible for aircraft passenger safety and comfort.

Significance: The training, duties, and employment requirements of flight attendants have changed greatly since the earliest passenger flights. Flight attendants are men and women of all ages, races, and family situations. According to the Bureau of Labor Statistics, there were 99,000 flight attendants in the United States in 1998. Flight attendants are highly trained, safety conscious, and skilled professionals serving masses of people traveling to all corners of the earth. A specific number of flight attendants are required by government regulations to be on all commercial airline flights, but they can also be found on private, corporate, and charter flights.

Background

Stewards provided the first in-flight service in the elegant dining rooms and private compartments of European airships between 1910 and 1937. In contrast to this luxurious and quiet mode of travel, the first paying airplane passengers in the United States traveled with the mail and fended for themselves.

In 1922, Daimler Airways of Britain hired the world's first airplane stewards, known as cabin boys. Selected on the basis of small stature and light weight, cabin boys provided general assistance and reassurance to passengers brave enough to fly the early flying machines but did not serve refreshments.

Stout Air Services of Detroit, which eventually became part of United Air Lines, hired male aerial couriers to serve aboard Ford trimotors between Detroit and Grand Rapids, Michigan, in 1926. Later, Transcontinental Air Transport, which eventually became Trans World Airlines (TWA), also hired male couriers on its trimotors. They were the sons of the airlines' investors and were promised that these jobs would launch their aviation careers.

Several airlines around the world introduced in-flight service in 1928. Western Air Express, forerunner of Western Airlines, began using stewards on its Los Angeles-to-San Francisco run to serve box lunches from a Los Angeles restaurant. Lufthansa of Germany hired a professional waiter to serve lunch on board its Junkers aircraft between Berlin and Vienna. Mexicana hired stewards on its Ford trimotors. Air Union, a forerunner of Air France, also employed stewards.

In 1929, Pan American Airways introduced the cabin boy type of attendant to serve on board Sikorsky flying boats and Fokker aircraft between Miami, the Caribbean, and South America. Food service was essential on flights to remote areas because there might not be food available for passengers along the way.

The crew often would bring several days' worth of food on the flight. The galley (kitchen) was in the tail of the flying boat. The cabin boys had to be small and nimble enough to crawl back to the galley, prepare the food, and serve it to the passengers. In May, 1930, New England and Western Airways was the first to employ African American attendants, Pullman railroad porters who had already been trained for on-board service.

On May 30, 1930, a more surprising and lasting innovation was brought into the airline cabin. Boeing Air Trans-

From the earliest days of commercial air travel, flight attendants have been dressed in distinctive uniforms to make them easily identifiable to passengers and to emphasize the brand image of the airline. (Hulton Archive)

port hired eight women to be the world's first airline stewardesses. Ellen Church, a registered nurse, is credited with convincing the company that would become United Air Lines to hire women cabin crewmembers. Pilots did not immediately welcome the women, and the pilots' wives were even less enthusiastic, but the flying public responded well.

Stewardesses went to work for Delta Air Lines in 1940, Continental Airlines in 1941, and Pan American Airways in 1944. The airlines had little choice but to hire women during World War II because all the able-bodied men were drafted into military service. By 1950, the stewardess was an integral part of air transportation. Her glamorous job was sought after by young women and idolized by little girls, but sometimes denigrated as being a glorified waitress.

In the mid-twentieth century, stewardessing was one of the few occupations open to women that provided adventure. Stewardesses personified the airlines for which they worked, were the main connection the public had with the airline, and were often used for promotional campaigns and advertising. On some airlines, stewardesses wore ethnic costumes as their uniforms. In the 1960's, miniskirts, hot pants, and sexist advertising overshadowed, but did not diminish, their safety responsibilities.

Qualifications

Body size and appearance dominated the hiring qualifications for the first in-flight personnel because of the constraints of the aircrafts themselves. Cabin ceilings were low and aisles were narrow in the early airplanes. Even though jet aircraft provided more space, stewardesses were still expected to be slim, have beautiful legs, and wear girdles. Weight checks were prevalent.

The original eight stewardesses were registered nurses. This was a requirement until nurses were needed in the military during World War II. Stewardesses could not be married, were age twenty-five or less, weighed no more than 115 pounds, and could be no taller than five feet four inches. Stewardesses had to retire at age thirty-two or when they got married.

The Civil Rights Act of 1964 promised equality for the cabin attendants, but it took a 1967 court case brought against Braniff Airlines by a stewardess to break down the marriage barrier. Until then, men who worked as stewards or pursers could be married and have children, but stewardesses had to remain single.

It took more litigation in the 1970's before women with children could work as flight attendants—the new non-gender-specific job title. These changes, plus the removal of the maximum age cap, encouraged men and women to make cabin service a career. Flight attendants can work into their seventies or when they are pregnant, as long as they can perform their duties. Some stay on the job for fifty years.

Qualifications vary among the airlines and change as equipment, routes, job markets, and regulations change. Flight attendants must be in good health and pass a physical examination and drug test. They must have good eyesight, hearing, and communication skills and be able to handle stress, irregular working hours, and being away from home. Airlines generally require flight attendants to be at least five feet two inches tall in order to reach the overhead compartments. Weight should be proportionate to height.

The minimum starting age for flight attendants is between eighteen and twenty-one. Some new hires start much later, choosing cabin service as an exciting second career after working as doctors, lawyers, teachers, or in any other profession. A high school diploma is required, some college education and experience working with people is preferred. Speaking two or more languages is a plus, especially for international airlines. Flight attendants must be able to work well in teams. The ideal flight attendant is well groomed, friendly, resourceful, and confident.

Duties and Working Conditions

A flight attendant's most important responsibility is to provide for the passengers' safety. This begins with checking safety equipment, securing carry-on items, and making safety announcements about exits and emergency equipment such as oxygen masks and life vests. Flight attendants enforce Federal Aviation Administration (FAA) requirements such as fastening seat belts, use of electronic equipment, and no smoking.

Flight attendants help passengers stay safe when flying through turbulent air and take charge when dealing with an emergency landing or other problems. They handle medical situations, unruly passengers, and children flying alone. Flight attendants stand for long periods of time and must be able to push and pull carts that weigh from 150 to 250 pounds.

Food and beverage service on airlines has changed dramatically over the years. After elegant dining on airships came box lunches and coffee from thermoses served on airplanes. Fried chicken became a staple for most early airline meals. Food service improved as galleys were installed on larger aircraft. European airlines had always provided alcoholic beverages, but U.S. airlines did not begin to offer liquor or wine until 1950.

During the 1960's and 1970's, when the price of airline tickets was regulated by the Civil Aeronautics Board (CAB), airlines competed with each other through the in-flight service they provided. Jet aircraft made more elaborate meals possible and introduced two to three different classes of seating and service. The wide-body jets, especially the Boeing 747, dramatically changed in-flight service and flight attendant duties. Cocktails could be served in VIP lounges on an upper deck and meals could be cooked in a galley in the aircraft's belly and brought up to the main level by elevator.

The Airline Deregulation Act of 1978 gave the airlines greater economic freedom and drastically affected every aspect of air transportation, including in-flight service. As price became the major ticket selling point, the labor-intensive, service-oriented industry struggled to cope with the new rules. Job security was replaced with uncertainty and sometimes layoffs as established airlines filed for bankruptcy and new airlines came and went throughout the 1980's and 1990's.

Most airlines only provided the minimum number of flight attendants required by FAA regulations—one per fifty seats—and all the seats were usually filled. Food service was reduced and often eliminated for economy passengers. The snack bags which passengers grabbed before boarding were less ample than the box lunches of the 1920's.

Flight Attendants in the Twenty-first Century

Full-time flight attendants work 75 to 85 hours per month in the air and another 75 to 85 on the ground. Flight delays can extend these hours. The airlines provide hotel accommodations and allowances for meal expenses when attendants are away from their home base. If flight attendants do not live at their assigned base, they must commute to the base on their own time to start their work assignments.

Most airlines use computerized bidding systems for work assignments. The most senior employees get the first choice of assignments, which are made one month at a time. Flight attendants on reserve status must be ready to work when they are called. Starting salaries averaged $15,000 annually in 2001, with experienced flight attendants earning about $25,000 and some as much as $50,000. Experienced flight attendants may become pursers or supervisors or become involved in training and recruiting.

Flight attendants work in a confined, moving environment at high altitudes, breathing recycled air. Their work is strenuous and they are susceptible to back injuries. While doing mundane work with a smile, they must be ready to handle disruptive passengers, medical problems, aircraft emergencies, hijacking, and terrorist situations.

Flight attendants have worked to improve the environment for themselves and their passengers. Their efforts led to U.S. airlines banning smoking on domestic flights in 1990 and on international flights in 1997. Through their professional associations, they work for improved safety features and regulations. The Association of Flight Attendants is the world's largest flight attendant union, with about fifty thousand members at almost thirty airlines.

The benefits of this occupation are enormous for those who like to travel and work with people. Flight attendants travel while they are working and may be able to sightsee on their layovers. They receive free airline travel passes for themselves and family members and discounts on other travel accommodations. Flight attendants enjoy the satisfaction of helping people, and many have heroically saved lives.

Training

Only a small percentage of applicants are selected for training and not all of those pass the four to seven weeks of rigorous, intensive training and testing at the airline's training center. Trainees learn aviation terminology, company policy and operations, FAA regulations, and all about the airplanes on which they may fly.

Trainees must thoroughly understand safety procedures and equipment, demonstrate that they can perform such duties as putting out fires, evacuating aircraft, and helping passengers survive a ditching at sea. Trainees learn to give first aid and cardiopulmonary resuscitation (CPR) and to use a defibrillator (AED). As new equipment is developed, it is added to the training. Flight attendants are trained to be alert to potential medical and security problems.

Safety and emergency procedures are the most important aspects of flight attendant training, but attendants also learn how to project their airline's image through their appearance and the service they provide. They learn to prepare and serve elegant meals and beverages for first class, as well as fast service and clean-up on short hops. They are schooled in personal grooming, weight control, and how to wear their uniforms.

Upon successful completion of initial training, flight attendants are assigned a base. They spend a certain length of time on probation and on reserve status. They may have to relocate. Qualifications are maintained by participating in a minimum of twelve hours of recurrent training every year. Safety related topics are emphasized. Flight attendants sometimes train with the pilots and other personnel to improve teamwork and communication.

Ursula Malluvius Davidson

Bibliography

Association of Flight Attendants. (www.afanet.org) A good source for current developments and issues affecting flight attendants.

Bock, Becky S. *Welcome Aboard! Your Career as a Flight Attendant*. Englewood, Colo.: Cage Consulting, 1998. Information for the aspiring flight attendant.

McLaughlin, Helen E. *Footsteps in the Sky*. Denver, Colo.: State of the Art, 1994. A comprehensive history of the flight attendant occupation in the United States with many photographs and personal stories written by flight attendants representing many airlines.

U.S. Department of Labor, Bureau of Labor Statistics. *Occupational Outlook Handbook: Flight Attendants*. (stats .bls.gov/oco/ocos171.htm) Up-to-date information on the duties and opportunities in the field.

See also: Air carriers; Airline industry, U.S.; Commercial flight; Emergency procedures; Food service; Training and eduction

Flight control systems

Definition: Electric, mechanical, and hydraulic systems that help to move an aircraft while flying.

Significance: Flight control systems allow pilots to adjust the speed, attitude, and direction of an aircraft.

Early History

The early experimenters and inventors who preceded Orville and Wilbur Wright, who made their first flight in 1903, did not fully appreciate the necessity for positive control of the machine. Prior to 1903, the prevailing ideas about aircraft control were that the airplane must have some kind of inherent stability and that the pilot's only function was to make small directional changes. Much of inventors' efforts prior to beginning of the nineteenth century were directed at obtaining a lightweight engine. The Wright brothers realized that a proper engine was a necessary ingredient in mechanical flight. However, they appreciated the importance of control and the fact that the pilot must be an active participant in the control of the airplane. By 1909, a control system had evolved consisting of ailerons, a rudder, and an elevator, which, in its essentials, remains in use today.

Types of Controls

Modern aerodynamic flight control systems, as opposed to engine controls, are essentially the same for all airplanes. Flight controls can be separated into two categories: primary and secondary controls. The primary controls change the angles that the airplane makes relative to the ground. The secondary controls are flaps that control the lift of the airplane, especially at low speeds, and tabs that reduce or eliminate the forces the pilot must exert on the controls in the cockpit.

All controls, whether primary or secondary, have three important subdivisions. The first are external moveable surfaces on the airplane, such as the rudder, aileron, and elevator. The second are the cockpit controls, which are moved by the pilot to change the direction of the airplane. The third are the links between the cockpit controls and the external surfaces of the airplane. These connections might be cables, electrical-conducting wires, electrical motors and computers, hydraulic lines, and hydroid motors.

Primary Controls

There are three categories of primary controls. Category A refers to the three hinged panels that are rotated about their hinge line to change the angular attitude of the airplane.

Category B controls are those which the pilot moves to change the direction of the aircraft and, to a limited extent, the speed of the aircraft, particularly the descent rate. These controls consist of the stick or wheel, which is moved to pitch and roll the airplane, and the rudder, which is moved to yaw the airplane. These controls have not changed significantly since 1915, during the second decade of mechanical flight.

Category C controls vary the most widely between different types of airplanes. These types of controls have also evolved most radically over the history of mechanical flight. A small, low-cost training plane connects the pilot's control to the aerodynamic controls with cables or push rods; hydraulic lines and associated motors perform the same function in high-speed commercial airliners. Electrical conducting wire or even fiber optic lines might be used to carry the control signal from the cockpit to an electrical motor at the surfaces in other commercial airplanes or high performance military airplanes.

The airplane responds to the movement of the primary category A cockpit controls in a number of ways. The ele-

Location of Airplane Controls

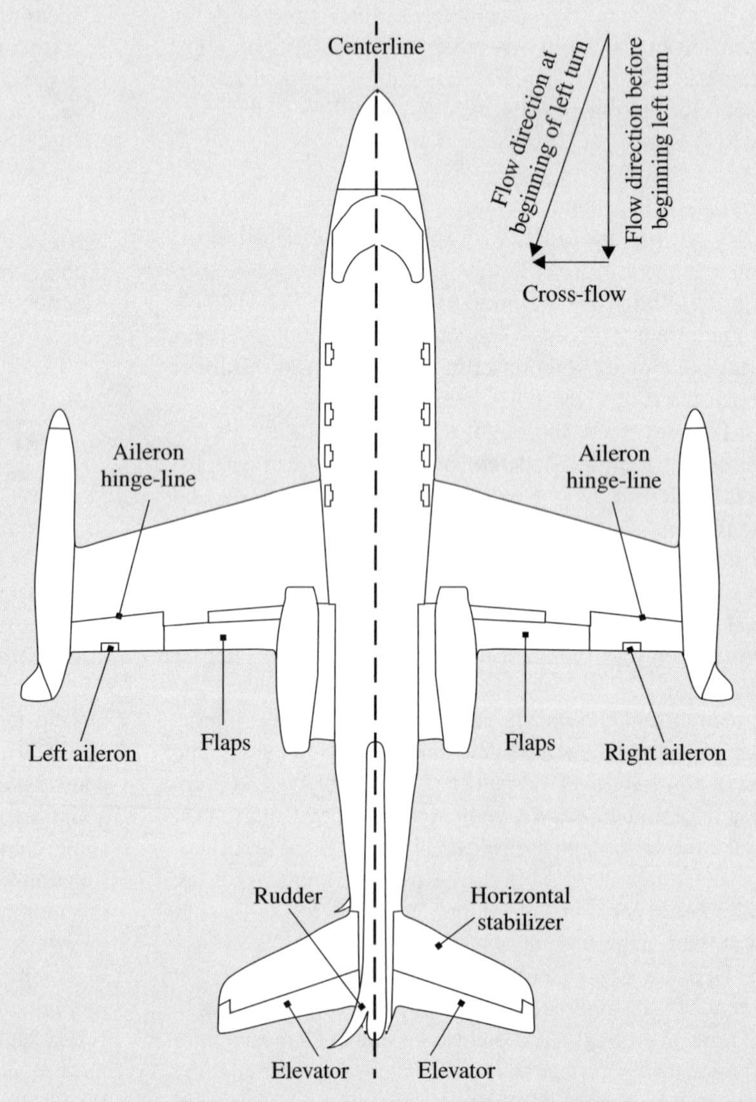

vator is deflected to change the pitch angle of the airplane: When the trailing edge of the elevator is moved upward, a down force is generated on the horizontal stabilizer. The result is that the nose of the airplane pitches upward. The airplane will pitch in the opposite direction if the trailing edge of the elevator is moved downward.

When the rudder is moved to the left side of the airplane, from the point of view of the pilot, a side force to the right is applied to the vertical stabilizer. This force swings, or yaws, the nose of the airplane to the left. Reversing the direction of the rudder movement will reverse the yaw direction.

Finally, movement of the ailerons causes the airplane to roll. The ailerons move differentially; when one aileron moves upward, the other moves downward. On the wing with the downward aileron, there is a slight increase in the lift. On the wing with the upward aileron, there is a slight decrease in lift. The unbalance in lift between the two wings causes the airplane to roll.

Cockpit controls are connected to the airplane's external controls. Moving the stick back brings up the elevator trailing edge, placing a down force on the horizontal stabilizer. The horizontal stabilizer and tail goes down while the nose goes up. Reversing the direction of the stick movement reverses the motion of the pitch of the airplane.

The rudder is moved by pushing on the rudder pedals located on the floor of the airplane. Pressing the left pedal causes the trailing edge of the elevator to move to the left, resulting in the application of a side force to the right on the vertical stabilizer. The tail of the airplane moves to the right, and the nose moves to the left. Moving the nose left or right is called yawing the airplane left or right.

Finally, the ailerons are moved by either sideways motion of the stick or rotation of the control wheel. To roll the airplane to the right, for example, the stick is moved to the right, lowering the left aileron and raising the right aileron.

Secondary Controls

The secondary aerodynamic controls are the tabs and the flaps, both of which can be operated by the pilot from the cockpit. The tab is a small elevator hinged to the trailing edge of the elevator. To hold the nose up for a prolonged period of time, the pilot must continually apply a backward force on the stick to keep the elevator in the up position. By moving the tab downward, in this case, a small force through leverage balances the much larger force on the elevator, with the result that there is no or little stick force required of the pilot to keep the elevator trailing edge upward. Tabs are found also on the rudder and aileron. On an airplane with two wing-mounted engines, a rudder tab is nearly essential in helping the pilot set and hold the extreme rudder deflection required for single-engine flight.

Flaps deflect in unison, unlike ailerons, which move differentially. Flaps help maintain lift, especially during low-speed flight. The pilot can control the deflection angle of the flap. Flaps are deflected at maximum deflection for landing and at a small angle for takeoff.

There are three basic flap designs. The split flap, the simplest but least effective, consists of a small plate that comes down from the lower surface of the wing. Because this flap does not change the contour of the wing, it primarily produces drag. The plain flap changes the shape of the wing and therefore produces lift as well as drag. The slotted flap is derived from the plain flap with special attention given to the junction of the flap and the wing. The design of this junction is crucial to the flap's effectiveness. The Fowler flap is the most effective and the most mechanically complicated flap. When deflected, it changes not only the shape but also the area of the wing.

There are two types of control used on airplanes: primary and secondary. Primary controls are the elevator, rudder and ailerons, and the primary cockpit controls are the stick and rudder, located in the cockpit. The secondary controls are tabs and flaps. The flaps allow the airplane to fly at lower speeds than would otherwise be possible. The tab allows the pilot to remove any forces required to hold control deflections. Tabs are usually located on the elevator and can also be found on the rudder and ailerons.

Frank J. Regan

Bibliography

Hubin, W. N. *The Science of Flight: Pilot-Oriented Aerodynamics*. Ames: Iowa State University Press, 1992. A book requiring some familiarity with high school algebra, but there are many sections that are entirely descriptive. There is much basic information about the control of airplane flight.

Illamn, Paul E. *The Pilot's Hand Book of Aeronautical Knowledge*. Rev. ed. Blue Ridge Summit, Pa.: TAB Books, 1991. A description of airplane control mostly from a pilot's point of view, requiring only basic arithmetic.

Raymer, Daniel P. *Aircraft Design: A Conceptual Approach*. 3d ed. Reston, Va.: American Institute of Aeronautics and Astronautics, 1999. A highly recommended, comprehensive, and up-to-date book on airplane design, directed at the engineering student, but featuring many sections requiring little more than high school algebra.

Stinton, Darrel. *The Design of the Airplane*. New York: Van Nostrand-Reinhold, 1985. An excellent introduction to airplane design.

Taylor, John W. R. *The Lore of Flight*. New York: Crescent Books, 1974. A massive, well-illustrated, oversized book featuring nontechnical descriptions of airplanes and spacecraft, and covering controls and cockpit instruments.

See also: Ailerons and flaps; Airplanes; Cockpit; Landing procedures; Rudders; Stabilizers; Tail designs; Takeoff procedures; Wing designs

Flight plans

Definition: Documents used to track the progress of aircraft in flight.

Significance: The three types of flight plans, visual, instrument, and defense visual, enable air traffic controllers and flight service specialists on the ground to more accurately sequence aircraft into the nation's air traffic flow. Flight plans can also assist rescuers should an aircraft go down unexpectedly.

Flight Plan Information and Procedures

A flight plan contains information such as the aircraft's type, color, speed, special navigational equipment, and amount of fuel carried on board. It also contains information on the intended route of the flight, the expected cruising altitude, the destination airport and any potential alternate airports, the number of passengers aboard, and contact information for the pilot. It is important for the pilot to keep the various flight service stations informed about the progress of a flight should there be any unexpected delay during the flight. Flight service stations will begin to attempt to locate an aircraft if its flight plan is not expressly closed by the pilot within thirty minutes after the original expected time of arrival listed on the plan. It is precisely for this reason that many pilots are encouraged to file flight plans, as a source of insurance, especially if no one other than the pilot is aware of the pilot's intentions for a particular flight. Should something unexpected happen during the flight that forces it down away from the intended airport, the aircraft's whereabouts will be sought within a relatively short span of time. In the event of such an occurrence, the specialist in charge of the flight plan will begin attempting to locate the aircraft by telephone after the flight is thirty minutes overdue. Calls are initially made to the intended destination airport, to airports surrounding the intended destination, and even to the airport of departure in an attempt to locate the aircraft on the ground. Most such searches end with these telephone calls, because most often pilots merely forget to close their plans upon arrival at their destination. If such attempts are unsuccessful, however, more elaborate search procedures, involving local law enforcement, the Civil Air Patrol, and ultimately the United States Armed Services, are initiated.

There are three main types of flight plans: visual flight rules (VFR) flight plans, instrument flight rules (IFR) flight plans, and defense VFR (DVFR) flight plans.

Visual Flight Plans

VFR flight plans are most typically used by pilots of small, privately owned and operated aircraft, who operate aircraft by using outside visual references to the earth's surface. VFR flight plans are managed by a network of federal flight service stations across the United States. There is approximately one flight service station per state in the United States. These stations are primarily responsible for the gathering and dissemination of weather and other critical flight information to pilots for use in planning and flying a particular flight. The pilot places a VFR flight plan on file with a flight service station, either over the telephone or with an Internet-capable computer, shortly before departing on a flight. Most often, pilots place flight plans on file for cross-country flights to another destination, generally 50 nautical miles or more away. This flight plan is generally filled out in paper form and relayed over the telephone to a flight service specialist, who then copies the relevant data and keeps it on file for activation by the pilot. The specialist then enters the flight plan into an appropriate computer for dissemination to other flight service facilities as necessary.

Instrument Flight Plans

The second type of flight plan used in the United States is the instrument flight plan for pilots engaging in flight under IFR. These flight plans are reserved for civil aircraft using more complex forms of navigation than are used in small aircraft. All scheduled airlines and most other large aircraft operate under instrument flight rules. Although the routing and alternate airport requirements vary slightly on IFR flight plans, they are otherwise identical to VFR flight plans. IFR flight plans may be filed by the pilot, as are VFR flight plans, or they may be filed by a company dispatcher in the case of scheduled airline operations. Whereas VFR flight plans are optional at the discretion of the pilot, IFR flight plans are a requirement for flight under instrument flight rules, because they are used by air traffic controllers for the scheduling and coordination of air traffic. IFR flight plans are closed automatically by an air traffic controller upon the aircraft's landing at an airport, as long as that airport has an operating control tower. Otherwise, the pilot is responsible for closing the IFR flight plan, as with the VFR plan. Search-and-rescue procedures are the same for an IFR flight as for a VFR flight. However, an overdue aircraft operating under IFR will be missed sooner than an overdue aircraft operating under VFR, because IFR aircraft are either under radar surveillance or are required to make periodic position reports to air traffic control.

Defense VFR Flight Plans

The third type of flight plan is known as a DVFR flight plan. This type of a plan is filed when a VFR flight is entering the United States from another country or U.S. territory and will be penetrating the U.S. Air Defense Identification Zone (ADIZ). This is an area located just off shore that the U.S. Department of Defense uses for positively identifying all aircraft entering the United States. This type of flight plan is required of all VFR pilots entering the United States. It is otherwise the same as the previous two types of flight plans mentioned, except that a pilot must list the exact point of intended entry into the United States.

R. Kurt Barnhart

Bibliography

Federal Aviation Administration. "Air Traffic Procedures." In *Aeronautical Information Manual*. Washington D.C.: U.S. Government Printing Office, 2001. An annually updated aeronautical manual for pilots, containing information on how to operate in the air traffic environment.

Jeppesen Sanderson. "Communication and Flight Information." In *Private Pilot Manual*, Englewood Colo.: Jeppesen Sanderson, 1998. A good introductory explanation of the flight information services available to pilots, with illustrations and accompanying videos that help the novice pilot gain a greater understanding of flight planning concepts.

Welch, John F., ed. "Air Traffic Control." In *Van Sickle's Modern Airmanship*. 6th ed. Blue Ridge Summit, Pa.: TAB Books, 1990. A more advanced look at air traffic procedures, designed for the advanced student in aeronautical studies.

See also: Air traffic control; Airports; Federal Aviation Administration; Landing procedures; Pilots and copilots; Takeoff procedures; Safety issues; Training and education

Flight recorder

Also known as: Black box

Definition: An instrument that records the performance and condition of an aircraft in flight.

Significance: The data retrieved from a flight data recorder can be used to generate a computer-animated video reconstruction of the flight of an aircraft, making possible the investigation and analysis of aircraft accidents or other unusual occurrences.

Flight Data Recorders

An aircraft flight recorder records many different operating conditions of a flight and provides information that may be difficult or impossible to obtain by any other means. By regulation in most countries in the world, newly manufactured aircraft must monitor at least twenty-eight important parameters. These include time, altitude, airspeed, heading, vertical acceleration, and aircraft pitch. Some recorders can record the status of more than three hundred additional in-flight characteristics that can aid in an accident investigation. Some of these include flap position, autopilot mode, and even smoke alarms. To ensure that a large amount of information is recorded, a flight recorder is able to record for at least twenty-five hours.

Computer programs have been written to take flight recorder data and reconstruct animated videos of aircraft flight. The animation allows the investigation team to view the last moments of a flight prior to an accident. In the event of an accident, investigators can visualize the instrument readings, power settings, airplane's attitude, and other important characteristics of a given flight.

Cockpit Voice Recorders

A cockpit voice recorder records the flight crew's voices, as well as other sounds within the cockpit. Communications with air traffic control, automated radio weather briefings, and conversation between the pilots and ground or cabin crew are recorded. Sounds of interest to an investigation board, including engine noise, stall warnings, landing gear extension and retraction, and any clicking or popping noises, are typically recorded. Based on these sounds, important flight parameters, such as speed, system failures, and the timing of certain events can often be determined.

In the event of an accident, an investigation committee creates a written transcript of the cockpit recorder tape. Local standard times associated with the accident sequence are determined for every event on the transcript.

The cockpit voice recorder (left) and the flight data recorder (held by engineer) help investigators reconstruct the events leading up to crashes and other midair incidents. (AP/Wide World Photos)

This transcript contains all the pertinent portions of the cockpit recording. Due to the highly sensitive nature of the verbal communications inside the cockpit, a high degree of security is provided for the cockpit recorder tape and its transcript. The timing of release and the content of the written transcript are strictly regulated.

History

The idea of a device to record both the voices and the instrument readings in the cockpit of an aircraft was originally conceived by Dr. David Warren at the Aeronautical Research Laboratory in Melbourne, Australia, in the 1950's. A demonstration unit was constructed in 1957. Although Australian aviation authorities did not initially approve the device, it was taken to Great Britain and the United States for further development.

On June 10, 1960, a Trans-Australian Airlines Fokker F-27 crashed while landing at an airport in Queensland, Australia, killing all twenty-nine people on board. The subsequent board of inquiry was unable to arrive at any definite conclusions as to the factors underlying the accident. The board recommended that all airliners be fitted with flight recorders. In 1961, Australia became one of the first countries to make flight recorders mandatory in aircraft. Any craft with a takeoff weight greater than 12,568 pounds must carry both a cockpit voice recorder and a flight data recorder.

Flight recorders and cockpit voice recorders, also known as black boxes, are actually painted bright orange to aid in their recovery following an accident. They have provided critical clues in solving the mysteries associated with many of the world's air disasters and have also been invaluable in helping to prevent future accidents.

Specifications

Flight recorders and cockpit voice recorders are housed in titanium boxes that are lined with many layers of insulating material. This design protects the recorders against impacts that produce accelerations up to 3,400 times the acceleration of gravity, against fires of up to 2,000 degrees

Fahrenheit (1,093 degrees Celsius), and against pressures at water depths of up to 20,000 feet. The recording devices are protected against contact with seawater and inadvertent erasure of recorded information. These specifications preserve the devices in the most serious accidents and in extreme climatic conditions. In addition, the boxes are fitted with battery-powered ultrasonic beacons that aid with underwater recovery. The beacons can transmit pulses from water depths of up to 14,000 feet for at least thirty days over a range of 2 miles.

Because they are more reliable and require minimal maintenance, computer memory chips have replaced most magnetic tapes as the recording media. Flight recorders are connected to a flight data acquisition unit that processes, digitizes, and formats the data for recording on the memory chips. Both the flight recorder and the cockpit voice recorder are carried in the tail of an aircraft. Flight data recorders reveal what happened in an accident, whereas cockpit data recorders reveal why it happened.

Even before a crash occurs, it is possible to monitor the safety of flights by using a quick access recorder. This device records even more parameters than a typical flight recorder and samples the data at higher rates for a longer duration of time. The data are stored on an optical disk and can be studied to identify problems before they become fatal. A ground-based computer analyzes the data and determines what is going wrong, rather than what went wrong.

Alvin K. Benson

Bibliography

Launius, Roger D., ed. *Innovation and the Development of Flight*. College Station: Texas A&M University Press, 1999. Excellent description of the history of the technological innovations in aeronautics.

Trujillo, Anna C. *Effects of Historical and Predictive Information on the Ability of Transport Pilot to Predict and Alert*. Hampton, Va.: National Aeronautics and Space Administration, 1994. Useful technical discussion about the operation and uses of flight recorders.

United States Federal Aviation Administration. *Airworthiness and Operational Approval of Digital Flight Data Recorder Systems*. Washington, D.C.: U.S. Federal Aviation Administration, 1999. Description of flight recording systems and necessary specifications to meet certification.

Veatch, D. W., and R. K. Bogue. *Analogue Signal Condition for Flight Test Instrumentation*. Neuilly-sur-Seine, France: North Atlantic Treaty Organization, 1986. Discusses the operation and mechanics of flight recorders.

See also: Accident investigation; Airplanes; Cockpit; Communication; Emergency procedures; National Transportation Safety Board; Pilots and copilots; Safety issues

Flight schools

Definition: An institution that provides the education and training necessary for a student to learn to pilot an aircraft.

Significance: Flight schools educate pilots and prepare them for certification. Flight schools teach pilots who intend to fly for their personal enjoyment, who intend to pilot commercial aircraft, and who intend to pilot military aircraft.

History

In the early days of aviation, there were no government regulations to control the certification of pilots. Learning to fly was largely a matter of experimentation, observation of others who knew how to fly, and trial and error. As the field of aviation evolved, the need for more formal methods of training pilots became apparent. Flight schools first began to appear in the late 1920's. Parks College was the first flight school to be awarded a Transport and Limited Commercial Ground and Flying School Certificate, granted in 1929 by the U.S. government. During the Great Depression years of the 1930's, the few flight schools in existence were fortunate if they were able to stay in business, and significant growth in flight training did not occur until the early 1940's.

The outbreak of World War II generated a need for a number of pilots, each of whom needed to be trained to a certain standard in a relatively short amount of time. In 1939, the U.S. Congress appropriated four million dollars to create the Civilian Pilot Training Program. The flight training done under this program was conducted at more than 400 colleges nationwide. After World War II, there continued to be a strong interest in aviation, particularly by the returning veterans. The G.I. Bill (1944) provided funding for veterans to obtain flight training, and thousands of students took advantage of this program. This source of income provided a foundation for flight schools to continue to grow and prosper.

Pilot training today is regulated by the Federal Aviation Administration (FAA), an agency of the federal government. The FAA issues Federal Aviation Regulations (FARs), which are the rules that govern aviation in the

United States. These rules include the certification of pilots and aircraft and the governance of flight operations. Flight schools train prospective pilots to meet the certification requirements specified by the FARs for various levels of pilot certificates and ratings.

Types of Schools

There are a number of different types of institutions that provide flight training and education in the United States. These include fixed-base operators (FBOs), collegiate aviation programs, proprietary professional aviation academies, and military programs. The type of flight school best suited to a particular student depends on that student's goals and intentions in aviation.

FBOs are businesses that operate at airports. They often provide a variety of services to the aviation community, including aircraft rental, maintenance, refueling, and the sale of aviation equipment, in addition to pilot training. Pilot training at this type of facility is typically tailored to an individual's schedule and personal goals. This type of flight school is usually attended by students who are interested in flying for pleasure or for personal business transportation.

Collegiate aviation programs, available at both two-year and four-year institutions, are designed for those students who wish to pursue a career as a pilot. Both types of institutions typically provide flight training through at least the Commercial Pilot Certificate, and usually the Certified Flight Instructor Certificate. Graduates of two-year programs receive an associate of science degree, whereas graduates of four-year programs receive a bachelor of science degree. In addition to completing the required ground and flight training for a Commercial Pilot Certificate, students at these institutions complete course work in a variety of areas important to understanding aviation. These may include maintenance, weather, aerodynamics, and aviation management courses. There are more than one hundred colleges and universities, large and small, that offer flight training as part of the curriculum for a degree. The Council on Aviation Accreditation is the accrediting body for collegiate aviation programs, and most reputable college programs have received accreditation by this organization.

Proprietary professional aviation academies are also designed for those students who wish to enter the aviation profession as a pilot. These schools typically provide training through at least the Commercial Pilot Certificate, and often through the Certified Flight Instructor Certificate. Enrollment in this type of school is most often a full-time endeavor. Since the late 1980's, the educational requirement for career advancement to a position as a pilot for a major airline has been a four-year college degree, so a number of proprietary aviation academies are also associated with a collegiate institution.

Military flight schools are utilized to train those personnel accepted into a branch of the U.S. Armed Forces for a pilot position. These programs provide high-quality initial training in basic piloting skills, followed by training in the specific type of aircraft and operation to which the person is to be assigned. The training period for both the initial course and the advanced course is typically one year each. The U.S. Army, Air Force, Navy, Marine Corps, and Coast Guard each have personnel assigned to pilot positions. The training for these personnel are conducted at various military bases throughout the country.

Whatever the type of institution, a flight school will provide a fleet of aircraft in which to conduct training, and a staff of certified flight instructors (CFIs) to provide flight training. Both the size of the aircraft fleet and the size of the CFI staff may vary from one to more than one hundred.

Types of Training Offered

At any of the civilian institutions described above, flight training may be conducted under either FAR Part 61 or FAR Part 141. Part 141 specifically describes minimum requirements regarding training facilities, personnel, course syllabi, and student performance rates for FAA-approved flight schools. Programs conducted under Part 141 are subject to continuing oversight and approval by the FAA. Collegiate and proprietary aviation academies are typically certified under FAR Part 141, although a number of FBOs also have Part 141 certification. FAR Part 61 specifically governs the certification of aircraft and pilots, and flight training can also be conducted under this part. Often, training for students who are interested in flying for their personal benefit or enjoyment is conducted under Part 61 at a local airport FBO, whereas training for students who desire a career as a pilot is conducted at a Part 141 school. Part 141 schools tend to be more structured and formalized, and Part 61 schools tend to be tailored more toward the individual requirements of the person receiving training. For example, a businessperson who wants to obtain a pilot certificate for transportation purposes may desire to participate in flight training only twice a week and at a different time each week. This type of schedule is often best accommodated at a local airport FBO under Part 61. A person interested in a career as a pilot would most likely desire to pursue this goal in a full-time capacity, and many Part 141 schools can accommodate this arrangement.

Types of Certificates

There are a number of types of pilot certificates issued by the FAA. These include the Recreational Pilot Certificate, the Private Pilot Certificate, the Commercial Pilot Certificate, the Certified Flight Instructor Certificate and the Airline Transport Pilot (ATP) Certificate. Training to obtain any one of these certificates involves a specified minimum of both flight and ground training, often called ground school. Both the Recreational and Private Pilot Certificates are designed for individuals who wish to fly for their own personal enjoyment. The Recreational Pilot Certificate has a number of limitations, such as the requirement that recereational pilots remain within 50 nautical miles of the departure airport, carrying no more than one passenger, and that a recreational pilot not fly an aircraft with more than four seats. The Private Pilot Certificate allows more freedom, with no limit on passengers or distance from the departure airport.

The Commercial Pilot Certificate is required in order for a pilot to be paid for flying an aircraft. The CFI Certificate is required to be able to instruct others in flight training, and the ATP Certificate is required to be a captain (or pilot in command) of an aircraft operated by a commercial air carrier. In addition to these certificates, an important rating that can be added to the Private and Commercial Certificates is the instrument rating. The ATP Certificate essentially includes an instrument rating as part of its privileges and limitations. The instrument rating allows pilots to fly in bad weather, called instrument meteorological conditions, which include such things as clouds or low visibilities. Before obtaining an instrument rating, pilots are restricted to visual meteorological conditions, which means they must maintain certain minimum visibilities and distances from clouds. If a person intends to use aviation as a dependable and regular means of personal transportation, obtainment of an instrument rating is essential. If a pilot is to fly for hire, an instrument rating is likewise required.

One additional rating that must be obtained before flying an airplane that has more than one engine is a multiengine rating. Since most flight students first learn to fly in a single-engine aircraft, this rating is usually added to an existing Private or Commercial Single-Engine Certificate. CFI and ATP Certificates also specify whether the pilot has single-engine privileges, multiengine privileges, or both.

Ground Training

Training conducted at flight schools, while often termed flight training, in reality consists of both ground training and training in an actual aircraft. Ground training may be conducted in a formal classroom setting, with a number of students receiving instruction from a teacher, or it may be conducted informally by a student's flight instructor before or after a flight. Typically, Part 61 flight schools tend to use more informal methods, whereas larger Part 141 flight schools and college programs tend to use traditional classroom settings for ground school. Again, the best method depends on the interests and background of the flight student.

Ground school covers a variety of topics, including applicable FARs, aircraft systems and performance, aerodynamics, weather, flight planning, and navigation. Often, flight schools own one or more flight-training devices in addition to their fleet of aircraft. These flight-training devices are more simplified versions of what are commonly known as flight simulators. Most often, they have a cockpit mock-up and a rudimentary visual display. However, there is no movement of the device in response to aircraft control movements. These devices are used most heavily during training for the Instrument Rating. Students working on this rating receive training in these flight-training devices in addition to conventional ground school and flight training in an aircraft.

Obtaining a Pilot Certificate

To obtain any level of pilot certificate or rating, an applicant must do a number of things. First, the ground and flight instruction specified by the FARs must be obtained from and certified by a CFI. A knowledge test, administered in a computer-based testing format, must be taken and passed with a minimum score of 70 percent. An appropriate medical certificate must be obtained from an aviation medical examiner for the level of certificate desired. For example, for a Private Pilot Certificate, a third-class medical certificate is required. For a Commercial Pilot Certificate, a second-class medical certificate is required, and for an Airline Transport Pilot Certificate, a first-class medical certificate is required. Finally, a practical test is conducted by a pilot examiner. This test consists of both an oral exam and a flight exam. During the oral exam, the examiner will cover items such as aerodynamics, weather, aircraft systems, aircraft performance, and flight planning. During the flight, a series of maneuvers will be evaluated to determine whether the applicant meets the minimum standards specified for the certificate for which he or she is applying. If the check ride is satisfactory, the student will be issued the certificate for which he or she applied.

Selecting a Flight School

A flight school is best selected by considering the needs of an individual. Such items as the location of the school and the schedule of lessons are key issues, as are the types of training typically conducted and the structure of the school, for instance, whether it is geared toward those interested in aviation as a profession or toward those interested in learning to fly for fun. Other things to consider are the size and availability of the training aircraft fleet and the availability of instructional staff. The school's safety record, how long the school has been in operation, and its reputation are also important. In addition, maintenance of the training fleet should be examined. Many schools offer an introductory flight lesson, during which a CFI will allow a prospective student to manipulate the controls of the airplane in flight. This provides an opportunity for the prospective student to examine the flight environment firsthand, as well as a chance to experience a representative training aircraft and instructor.

One aspect of the decision regarding a flight school selection involves whether to select a FAR Part 141-approved flight school or a FAR Part 61 flight school. To obtain a Private Pilot Certificate, thirty-five hours of flight training are required under Part 141, whereas forty hours of flight training are required under Part 61. However, the national average of flight hours to obtain a Private Pilot Certificate ranges from sixty-five to seventy hours, so it would be an error to base a decision to use a Part 141 school instead of a Part 61 facility solely on the flight-time requirement for a Private Pilot Certificate. If a student is interested in pursuing a Commercial Pilot Certificate, there is a flight-time benefit in utilizing a Part 141 flight school. The flight time required for a Commercial Pilot Certificate is 250 hours under Part 61 and 190 hours under Part 141.

Cost of Flight Training

The cost of flight training varies widely depending on the area of the United States in which a student resides and the type of flight school attended. Many flight schools offer package deals for flight instruction, but it is important to understand what items are included in the package. During flight training in an aircraft, both an airplane rental fee and a flight instructor's hourly fee are charged. Aircraft used for instruction usually have a digital recording clock, called a Hobbs meter, which records the amount of flight time for a given flight by subtracting the Hobbs meter reading at the beginning of the flight from the Hobbs meter reading at the end of a flight. Preflight and postflight briefing time, which is conducted by a student's CFI and which is necessary for effective flight training, is also billed.

The most common type of package offer includes the cost of these items up to a certain number of hours, with excess hours becoming the student's responsibility if they are required. Other packages may guarantee obtainment of a certificate, with no maximum number of hours specified, although there are often many other stipulations in this kind of package. The minimum time required by the FARs to obtain a Private Pilot Certificate under FAR Part 61 is forty hours: twenty hours with an instructor, called dual instruction, and twenty hours of solo flight time. For flight-school package offer-comparison purposes, however, an average student usually requires from sixty-five to seventy hours to obtain a Private Pilot Certificate, with forty to forty-five flight hours of dual instruction and twenty-five hours of solo flight time.

The total cost of a university education, including the obtainment of Commercial Pilot and Certified Flight Instructor Certificates as well as a four-year degree at a private university, can equal more than $100,000. However, much of this cost would also be incurred in the course of obtaining a bachelor's degree from a private university in a field other than aviation. The cost is typically less at state-supported universities and less still at junior colleges or community colleges. Often, two-year program graduates can continue their studies at a four-year university to complete a bachelor of science degree.

The cost to attend a proprietary professional academy, usually resulting in the obtainment of Commercial Pilot and Certified Flight Instructor Certificates, can range from approximately $50,000 to $85,000. This type of program is often selected by individuals who have already obtained a four-year college degree and who are interested in changing careers. The sole focus on flight training allows such individuals to accelerate their training so they can begin to pursue their new career path. In addition, there are students who choose to enroll in this type of program right after high school, and then serve as certified flight instructors while earning their college degrees.

Other Types of Flight Schools

The preponderance of flight schools in existence in the United States are for airplane pilots. However, in addition to training for pilot certificates for airplanes, there are also flight schools that conduct specialized training in other types of aircraft or operations. For example, helicopter pilots, glider pilots, pilots involved in agricultural operations, and seaplane pilots are required to receive appropriate ground and flight training for the type of operation and aircraft they pilot. Some large flight schools conduct these types of training in addition to more traditional airplane pilot training, whereas other schools choose to specialize in a niche market.

Wendy S. Beckman

Bibliography

Phillips, Wayne. "A Wealth of Options: Choosing Your Educational Opportunities." *AOPA Flight Training Magazine* (December, 2000). An article examining the flight school options available to those who are interested in pursuing a career as a pilot. In addition, this magazine is a source of continuing information regarding flight schools and flight training.

University Aviation Association. *Collegiate Aviation Guide*. Auburn, Ala.: University Aviation Association, 1999. This publication is a directory of 119 institutions offering degree programs in aviation.

Willits, Pat, ed. "Discovering Aviation." In *Private Pilot Manual*. Englewood, Colo.: Jeppesen Sanderson, 2000. This chapter provides basic information regarding the role of a flight school in obtaining a Private Pilot Certificate.

See also: Federal Aviation Administration; Military flight; Pilots and copilots; Training and education

Flight simulators

Definition: Devices which are used to enable a person to experience flight situations and/or movements without actually flying in an aircraft or spacecraft.

Significance: Flight simulators can save time and money in the training of pilots. They are also used in flight testing to investigate the stability, control characteristics, and behavior of aircraft and spacecraft, allowing detailed and realistic simulations of all aspects of flight with no risk to either vehicle or pilot.

Flight simulation involves the use of a ground-based device to enable a pilot, student pilot, or an aerospace engineer to experience or evaluate the behavior of an aircraft or spacecraft in flight. The inside of the simulator looks like the cockpit of an airplane or spacecraft.

In the course of the first century of human aviation, flight simulators have evolved from crude devices consisting of little more than a chair and a set of imitation controls mounted on a wood platform that can be pitched and rolled by training personnel, to multimillion-dollar computer-controlled aircraft or spacecraft cockpits that can duplicate every conceivable motion and reaction of the real vehicle.

Every child who has placed a chair in a large cardboard box and used anything from a broomstick to a baseball bat to pretend to control a make-believe airplane has experienced flight simulation at a very basic level. Flight simulators allow people to "pretend" to fly.

Teaching Tools

An important use of flight simulators is to help teach pilots how to fly an airplane with only their instruments to tell them the position, attitude, and direction of flight of their airplane. An important aspect of such training is teaching pilots that they cannot rely on the body's natural senses of sight and balance to fly under "instrument flight conditions"; they learn to fly using the information provided by the flight instruments alone. A simple desktop computer screen and a set of airplanelike controls can be used with any of many flight simulator computer codes to accomplish this task. Older pilots will recall training in simulators that were made to resemble little airplanes with small wings and tails and that were mounted on mechanically or hydraulically powered platforms designed to move the small cockpit like an airplane as the pilot "flew" the trainer, using an array of instruments identical to those on a real instrument panel.

Pilots of craft, from fighters to general aviation craft to space shuttles, train in sophisticated flight simulators in which the pilot can see realistic in-flight images of sky, terrain, and airports and learn to fly the vehicle using both its instruments and the simulated view from the cockpit. The simulator can, with the flick of a switch or the turning of a knob, subject pilots to the conditions they would face with the loss of an engine, severe turbulence and weather, loss of part of the control system, or almost any other emergency imaginable.

There is continued debate about whether training is more or less effective when the simulator moves to replicate the body forces which pilots might experience in training maneuvers. Both moving and nonmoving simulators are used in teaching pilots how to react to almost any situation that may be encountered, ranging from an ordinary flight to a severe emergency.

Research and Development Tools

These same simulators are used to study ways to improve the control systems of airplanes and spacecraft. Engineers can write "control law" equations that will alter the way the vehicle behaves in flight, simulating everything from a shift in payload weight, to the loss of a rudder in combat, to a complete redesign of the airplane wing or tail. Simulators are used to investigate such changes and events without risk of loss of life or vehicle in a flight test. If there is any question of control system failure or problems in an aircraft accident, simulators are used to determine the effect of that loss on the performance and handling of the plane and to compare the test results to the facts known about the accident. Using these control laws, every newly designed aircraft is "flown" for hundreds of hours in the simulator before a test airplane ever leaves the ground; it has become commonplace for the test pilot to report after the first flight that the plane flew just like it did in the simulator.

Some of the world's most sophisticated flight simulators are used in the design and development of military aircraft such as fighters. Several government facilities have twin simulators in which two fighter pilots can fly simulated dog-fights against each other with the "enemy" simulator programmed to handle like real enemy aircraft. The simulators are coupled in such a way that the two pilots can see the opponent aircraft projected onto huge screens surrounding their multimillion-dollar full-motion flight simulators. This type of simulation allows the military to determine the best maneuvers for use in aerial combat and to design or redesign their aircraft and control systems to give them the edge in a fight.

Dozens of very sophisticated flight simulator programs and games now on the market allow anyone with a home computer to experience flight simulation. Many of these programs provide excellent simulations of actual airplane motion and control effectiveness, rivaling that of real flight training simulators. Some of the best such programs have been developed using the control laws of real aircraft, both modern or historic, and can give users an outstanding feel for the thrill of flight in their airplane of choice.

James F. Marchman III

Bibliography

Boyne, Walter J. *Flying: An Introduction to Flight, Airplanes, and Aviation Careers*. Englewood Cliffs, N.J.: Prentice-Hall, 1980. A guide for anyone who is interested in getting into the world of flying airplanes.

Dickinson, B. *Aircraft Stability and Control for Pilots and Engineers*. London: Pitman Press, 1968. An older college-level text written with test pilots in mind.

Nelson, Robert C. *Flight Stability and Automatic Control*. New York: McGraw-Hill, 1989. A college-level text which includes a discussion of flight simulation and the operation of simulators.

Rolfe, J. M., and K. J. Staples, eds. *Flight Simulation*. Reprint. New York: Cambridge University Press, 1988. Twelve essays cover the basic principles and uses of flight simulators.

See also: Accident investigation; Airplanes; Cockpits; Fighter pilots; Flight control systems; Instrumentation; Pilots and copilots; Space shuttle; Spaceflight; Testing; Training and education

Flying Fortress

Also known as: B-17, Flying Fort, Fortress

Date: First prototype built in 1935; production ended in 1945

Definition: A four-engined heavy bomber of World War II; one of the most important bombers of any kind in that war, it was legendary for the amount of battle damage it could absorb and still fly back to its base.

Significance: Considered one of the greatest military aircraft of World War II, the Boeing B-17 was a mainstay of the U.S. daylight strategic bomber fleet that helped defeat Germany. The B-17 also confirmed belief in the efficacy of strategic bombing, which continued as U.S. military policy for the next three decades.

The B-17 was born in 1937. The Boeing company had privately designed and made the Model 299 prototype when the U.S. Air Corps needed a heavy bomber to replace the smaller, slower Martin B-10. The early B-17's, B through D models, had no tail turret, no forward armament, and no ball turret; they had .30-caliber machine guns, which proved too light for defense against German fighters. After some combat experience with the early B through D models, the B-17E was designed with increased armament, entering service in 1941. This became the general type on which all later models were based. Over twelve thousand B-17's of all types were built through 1945. Most B-17's saw service in Europe, although they were also used in the Pacific and other theaters.

Specifications

The B-17 was a midwing monoplane constructed of aluminum skin over a steel and aluminum framework. The wingspan was just over 103 feet; the plane was just over 74 feet long. The four engines, two in each wing of the aircraft, were turbocharged 1,000-horsepower Wright radial engines, powering the plane to a maximum speed of more than 300 miles per hour, with a service ceiling of between 34,000 and 38,000 feet. The aircraft was unpressurized; aircrew breathed from an oxygen system and were thickly dressed against the thin air and subzero temperatures of the high bombing altitudes. Armor in the plane shielded each position. The B-17 was capable of carrying up to 17,600

The B-17 Flying Fortress was legendary for the amount of damage it could sustain on a bombing mission and still return its crew home. (AP/Wide World Photos)

pounds of bombs, depending on the length of the flight, the types of bombs carried, the target, and the amount of fuel needed. Without extra tanks in the bomb bay, the maximum amount of fuel that could be carried was 780 gallons.

The B-17 carried a crew of ten men in its fuselage: the pilot and copilot in the cockpit, bombardier and navigator in front of them in the nose of the plane; the flight engineer (who also was top turret gunner) standing in a motorized turret directly behind and slightly above them; and the radio operator (who also operated a machine gun) behind the flight engineer. Beyond the bomb bay were two waist gunners, each with a .50-caliber machine gun; at their feet was the revolving Sperry ball turret, which contained one gunner, curled up in the ball turret and operating two machine guns. The tail of the plane contained two more .50-caliber machine guns, operated by the tail gunner. There were various configurations—some standard, some individually rigged by crews after hard experience—of .50-caliber machine guns in the nose, operated by the navigator and bombardier, until the development of the G model, which had a chin turret containing two .50-caliber machine guns under the nose.

B-17's were modified in several different configurations. Some B-17's dropped lifeboats mounted under the belly of the plane; others were fitted with radar and used for search and antisubmarine warfare. The YB-40, conceived as a bomber escort, carried no bombs, but had extra machine guns and ammunition. It proved to be too slow and was not built in numbers.

Combat and After

One major reason for the fame of the Flying Fortress was the amount of battle damage it could absorb and still make it back to base after a bombing run. Flying Fortresses could return to base with only two of the four engines operating; with tail, wing, and nose surfaces sheared off; and with holes made by bullets and cannon shells peppering the fuselage. Many wounded airmen credited the B-17 with saving their lives. The aircraft was not invulnerable, however—four thousand B-17's were destroyed in combat during the war. Attrition was highest before Allied fighters with auxiliary fuel tanks began escorting the bombers on missions in 1943. B-17's usually flew in a combat formation that maximized the defensive firepower of the aircraft as well as the aircraft surrounding it. Missions could last up to eight hours from takeoff to the return of the planes to their bases.

B-17's were superseded in the Pacific by the larger, faster B-29's. Many B-17's were converted and used as aerial spraying planes, transports, water-bombing planes for fighting forest fires, and for other uses. Of the over twelve thousand Flying Fortresses built, most were scrapped after the war, but many B-17's remain intact, retained or reconverted to military status by collectors and historic aircraft foundations. Many serve as static displays or "gate guards" at military (especially Air Force) bases and museums throughout the United States and Europe. Many others are still flying in the United States and Great Britain as part of historic aircraft collections.

Robert Whipple, Jr.

Bibliography

Caidin, Martin. *Flying Forts*. New York: Bantam, 1990. A detailed account of the history of the B-17 and its missions around the world during World War II, by one of America's best aviation writers.

Dorr, Robert E. *U.S. Bombers of World War II*. London: Arms and Armour, 1989. Contains a critical appraisal of the B-17 and its effectiveness in the air war over Europe.

Jablonski, Edward. *Flying Fortress: The Illustrated Biography of the B-17's and the Men Who Flew Them*. Garden City, NY: Doubleday, 1965. One of the definitive accounts of the history of the B-17 and its operations in World War II.

See also: Air Force, U.S.; Bombers; Dresden, Germany, bombing; Firefighting aircraft; Military flight; World War II

Flying Tigers

Also known as: American Volunteer Group

Date: From April 15, 1941, to July 4, 1942

Definition: A group of American civilians flying fighter planes for the Chinese against the Japanese during World War II.

Significance: The Flying Tigers helped keep supply lines open to the Chinese Nationalist government in southern China during World War II.

Background

In 1941, war between the United States and Japan seemed imminent. The Japanese had been bombing Chinese cities since 1937 and had virtually destroyed China's Air Force. The idea for a group of American volunteer pilots to assist the Chinese in their struggle, similar to the Lafayette Escadrille of American volunteers who flew for France during World War I, had been brewing for some time.

Claire Lee Chennault and the Volunteers

Claire Lee Chennault, a retired Army Air Corps captain and air advisor to China, was authorized by Nationalist leader Chiang Kai-Shek to form a volunteer group consisting entirely of American airmen to protect China's skies and to help train Chinese aviators. The idea was sold to President Franklin D. Roosevelt, who, on April 15, 1941, signed an executive order authorizing the formation of the American Volunteer Group (AVG), as the Flying Tigers were officially known.

The order permitted members of the U.S. Navy, U.S. Marine Corps, and U.S. Army Air Corps to resign from their branches of service with the assurance that they would be reinstated to their former rank or grade upon completion of their contract. The Flying Tigers were to defend the Burma Road, China's lifeline to Burma and Indian Ocean ports. Because the United States technically remained at peace with Japan in April of 1941, the plan required some subterfuge. Central Aircraft Manufacturing Company (CAMCO) was chosen as a cover. CAMCO, owned by William Pawley, had an aircraft factory at Loiwing, China, supplying parts and planes to Chiang's air force. The volunteers signed one-year contracts with CAMCO to perform certain services not technically relating to combat.

Training and Combat

In September, 1941, Chennault and the volunteers gathered in Toungoo, Burma (now Myanmar), about 170 miles north of Rangoon. At Toungoo, Chennault taught and trained his pilots in the intricacies of the P-40 Tomahawk, the volunteers' fighter plane, and taught them how to use the P-40 against the Japanese fighter pilots. Chennault emphasized that although the Japanese fighter planes had superior maneuverability and rate of climb, the P-40 had superior armor, firepower, and diving speed.

The Flying Tigers first clearly demonstrated their abilities on December 20, 1941, when they attacked a formation of ten Japanese bombers on its way to Kunming (K'un-ming), the capital of Yunnan (Yün-nan) Province, China. Only one Japanese bomber returned safely to its base. During January and February, 1942, the Flying Tigers began compiling the extraordinary record of victories that placed them firmly in history.

Flying into History

Over the skies of Myanmar and China in January and February, 1942, the Flying Tigers destroyed at least 217 enemy aircraft in thirty-one encounters and lost only six pilots. It was during these two months that the Chinese dubbed the AVG the Flying Tigers. AVG personnel painted the sharp-toothed mouths of sharks on the noses of their P-40's.

The men of the AVG did not always fly P-40's. In March and April, 1942, the AVG obtained P-40E Kittyhawks as replacements for lost P-40's. The P-40E's had all the instrumentation that the original Tomahawks never had as well as six free-firing .50-caliber machine guns in the wings and bomb racks. When Japanese planes were hit by a P-40E's machine guns, the Japanese planes would often disintegrate in the face of a single well-aimed burst.

Disbandment

After the Flying Tigers disbanded, members who wished were absorbed on July 4, 1942, into the United States Tenth Air Force, which became the nucleus of the China Air Task Force and was reorganized in March, 1943, as the Fourteenth Air Force. This group remained under the command of Chennault, who was promoted to brigadier general. During their six and one-half months of aerial combat, the Flying Tigers destroyed 297 enemy planes, with another 153 probably destroyed. Twenty-two AVG personnel lost their lives. The Flying Tigers were so important to the war effort that Winston Churchill lauded them on the floor of Parliament in London. Starting with the Japanese attack on Pearl Harbor, the United States suffered a number of disasters. As a result, Americans saw the AVG in Myanmar as one early bright spot in the war against Japan.

A great many Flying Tigers later became airline captains. One even started his own airline, with the help of several members of the group. Some became test pi-

The Flying Tigers painted shark teeth on their P-40's to symbolize their ferocity in keeping Chinese supply lines open in the face of Japanese aggression in the early days of World War II. (Digital Stock)

lots. Others went on to successful military and business careers.

Dana P. McDermott

Bibliography

Baisden, Chuck. *Flying Tiger to Air Commando.* Atglen, Pa.: Schiffer, 1999. An enlisted man's story of over twenty years of service to his country, including the Army Air Corps and the American Volunteer Group, better known as the Flying Tigers.

Bond, Charles R., and Terry H. Anderson. *A Flying Tiger's Diary.* College Station: Texas A & M University Press, 1993. The wartime diary of General Charles Bond, who flew as a Flying Tiger in southern China during World War II.

Ford, Daniel. *Flying Tigers: Claire Chennault and the American Volunteer Group.* Washington, D.C.: Smithsonian Institution Press, 1995. An account of General Claire Chennault and the volunteers who fought for China against the Japanese both before and after Pearl Harbor.

Losonsky, Frank S., and Terry M. Losonsky. *Flying Tiger: A Crew Chief's Story.* Atglen, Pa.: Schiffer, 1996. The war diary of a Flying Tiger crew chief from the Third Pursuit Squadron, describing much of the unit's history from the pilot's viewpoint.

See also: Air Force, U.S.; Fighter pilots; Pearl Harbor, Hawaii, bombing; World War II

Flying wing

Also known as: All-wing and *nurflügel*

Definition: The American term given to airplanes that are predominantly the lifting component, the wing.

Significance: The all-wing design was among the earliest aerodynamic ideas for reducing drag and costs through high efficiency. Its simplicity has challenged generations of designers, teaching them many aerodynamic fine points.

Early Development

Before the invention of the airplane, English physicist Sir George Cayley, who in 1853 built the first manned glider, suggested that flying machines would be most efficient if they were only a wing. After the airplane became a reality in the early twentieth century, most successful airplane designs were linear. Their noses sported vertical or horizontal stabilizers, or perhaps an engine, and worked backward toward the cockpit, wings, rudimentary fuselage, and vertical or horizontal stabilizers, or both.

Airplanes are engineered to suit mathematical logic and economic reality. Pilots seek aerodynamic poise, while passengers seek comfort and amenities. Operators measure an aircraft's reliability, and accountants measure its economy. Most people also judge airplanes for their inspiring beauty. One design, the flying wing, exhibited grace, economy, and performance. Inspired by Cayley's belief in eliminating the drag and weight of fuselages and tails, the flying wing has long been aviation's Holy Grail. Because all-wing airplanes need fewer parts and construction steps than do conventional designs, they are more energy-efficient both to build and to operate. Still, flying wings are rare.

During and after World War I, the development of aircraft engines quickly overpowered the aerodynamic drag produced by early airplanes. This development spelled doom for the flying wing design. Because airplane designs were still new, people had little preconception of how airplanes should appear. By the 1940's, airplanes had proved viable, and society's view of airplanes included a fuselage and tail. The economic boom and low energy costs had made conventional designs inefficiencies tolerable.

The earliest flying wings, sporting vertical stabilizers, were not purely wing-only designs. In 1907, British airpline designer John William Dunne showed that conventional tails were unnecessary. His balanced aerodynamics have infused tailless and all-wing airplanes ever since. By 1930, Germany's Walter and Reimar Horten first flew their model all-wing airplane, called a *nurflügel* in Germany. In 1933, the Hortens flew a manned all-wing glider model called the HO-1. Knowledge gained from the HO-1 inspired the HO-5, a twin-engine machine potentially leading to an all-wing fighter. German general Ernst Udet, long appreciating the Hortens' *nurflügel* ideas, succumbed to political blame for other project failures and in November, 1941, took his own life. Germany's all-wing idea had lost a patron, and the proposed all-wing fighter languished until 1945, when advancing American soldiers discovered one nearly completed twin-jet HO-9 fighter. Other Horten all-wings flew as developmental projects. However, the most ambitious project, the HO-18 long-range heavy bomber, remained only an idea.

Northrop's All-Wing Designs

In the United States, John Knudsen Northrop dreamed of flying wings. His single-seat 1933 design, the Model 1, had thin, tubular twin booms supporting a conventional horizontal stabilizer with twin rudders. Northrop began testing a true *nurflügel* in July, 1940, when the N-1M first flew. It was a flying laboratory, designed to change configurations between flights. Changes in wingtip droop, sweepback (the taper of the wing's leading edge), and wing dihedral (angle of the wingtips) could provide a bank of information. The Americans considered building a medium bomber based on the N-1M. Encouraged, Northrop continued testing predominantly all-wing airplanes, including four N-9M engineering test airplanes. Although one N-9M crashed, the remaining crafts saw duty as trainers, giving pilots firsthand experience with the flight characteristics of all-wing airplanes.

Following World War II, Northrop built the prototype XB-35 all-wing bomber, a six-engine, propeller-driven pusher design that was a true wing-only machine. Jet engine technology was making propellers obsolete for combat airplanes, so Northrop soon rebuilt the XB-35 into the YB-49, an eight-engine all-jet bomber. Although the YB-49 project did suffer airplane loss, the design was sound, and officials who were influential in procurement favored the U.S. Air Force's adoption of the YB-49. However, the project was mired in political intrigue, dooming America's flying wing. So sour were the feelings of those involved that Air Force secretary Stuart Symington ordered not only the flying wing's cancellation in 1950 but also the destruction of all XB airframes. None remained even for museum display.

The Stealth Bomber

Four decades later, on November 22, 1988, the Air Force unveiled the B-2 stealth bomber, designed by the Northrop Corporation, to the American public. Several years earlier, Northrop company officials had secretly revealed the airplane to the eighty-five-year-old Northrop, who died the following year, knowing that his dream would finally fly. In 1996, the McDonnell Douglas Corporation revealed its idea of a BWB-1 (blended-wing body) airliner to seat 800 passengers. Efficient design would permit the three-engine jet to operate at about two-thirds the cost of a conventional airplane of the same capacity. In 2001, Northrop Grumman used company funds to construct an all-wing uncrewed combat air vehicle (UCAV) for aircraft carrier use. With leading edges swept back 55 degrees and trailing edges sweeping forward 30 degrees, the inherently stealthy design was a logical step in airplane development toward autonomous, or "smart" aircraft. To meet the military's need for extended endurance, the Northrop Grumman Corporation began to look at extensions on each wingtip, transforming the UCAV's aggressive arrowhead shape into the more elegant, traditional all-wing design.

David R. Wilkerson

Bibliography

Campbell, J. M., and G. R. Pape. *Northrop Flying Wings: A History of Jack Northrop's Visionary Aircraft*. Atglen, Pa.: Schiffer, 1995. A comprehensive examination of Northrop Aircraft's triumphs and tragedies and of Jack Northrop's impact on America's ultimate all-wing bomber, includes a valuable selection of lists, drawings, and color and black-and-white photographs.

Kohn, Leo. *The Flying Wings of Northrop*. Milwaukee, Wis.: Aviation Publications, 1974. A brief but well-researched text, with black-and-white photos and a good reproduction of Northrop Aircraft's pilot manual for the YB-49 bomber.

Myhra, David. *The Horten Brothers and Their All-Wing Aircraft*. Atglen, Pa.: Schiffer, 1998. A look at the personality-driven aviation industry, from the perspective of the all-wing's chief proponents and opponents, with an excellent selection of black-and-white photographs and line drawings.

Norris, G., and M. Wagner. *Giant Jetliners*. Osceola, Wis.: MBI, 1997. A well-researched and easily read discussion of heavy airliners, profusely illustrated with color photographs.

Wall, R., and D. A. Fulghum. "New Demonstrator Spurs Navy UCAV Development." *Aviation Week and Space Technology* 154 (February 19, 2001).

See also: Aerodynamics; Airplanes; Bombers; Sir George Cayley; Experimental aircraft; Heavier-than-air craft; Stabilizers; Stealth bomber; Ultralight aircraft; Uninhabited aerial vehicles; Wing designs

Fokker aircraft

Definition: Aircraft designed and produced by Anthony Fokker or by companies under his ownership or direction or that bear his name.

Significance: Fokker aircraft have played a significant role in the history of aviation. Innovative designs and construction techniques, combined with foresight into the needs of both military and civilian aviation, kept Fokker companies at the forefront of aircraft design and manufacture for nearly ninety years.

Anthony Fokker was born in Kediri, Java, in 1890. After his family returned to the Netherlands, Fokker began a lifelong commitment to aviation. When he was twenty-one, he started an aviation company in Wiesbaden, Germany. Fokker's first two attempts to build viable aircraft ended in crashes; the Spin I hit a tree in 1910, and the Spin II crashed in 1911. In 1913, however, Fokker's Spin III model tested successfully and was purchased by the German military. Prior to the outbreak of World War I, Fokker made overtures to both the British and Dutch governments concerning purchase of his aircraft for military purposes. He was rejected by both, and so turned his attentions to designing exclusively for the German military authority.

World War I

The first true fighter aircraft to appear in World War I were Fokkers. Fokker produced 7,600 aircraft for Germany during World War I. Of these, his most famous designs include the Fokker Eindecker series, the Fokker Dr-I triplane, the Fokker D-VII, and the Fokker E-V/D-VIII.

The Fokker Eindecker monoplanes caused a revolution in concepts of employing aircraft as weapons. Fokker produced about 450 Eindeckers in four versions, E-I to E-IV, with the E-III produced in the greatest numbers. The Eindecker was the first aircraft to effectively employ a fixed, forward-firing machine gun that was synchronized with the engine to fire bullets through the propeller arc, an innovation credited to Anthony Fokker. The machine gun was aimed by pointing the entire plane at the target. The results achieved with these machine-gun-equipped Fokkers were so spectacular that during 1915, when they reigned over the Western Front, the era is referred to as the "Fokker Scourge," and Allied aircraft referred to as "Fokker Fodder."

The Fokker Dr-I was the result of a triwing design concept first built by the British Sopwith Company in 1917. No less than thirty-four prototypes were tested by the German military to counter the Sopwith. Of the planes tested, only the Fokker Dr-I triplane was produced. The plane was small, light, and exceptionally agile. The Fokker design was unique in that it had no wire bracing between the wings, only a single strut connecting the lifting surfaces near the tips. It was the first aircraft to employ the Göttingen 298 airfoil with a 13 percent thickness ratio, a feature adopted on almost all subsequent Fokker designs. This airfoil gave the Dr-I one of the lowest zero-lift drag coefficients of all World War I fighter aircraft. The Dr-I was issued to elite fighter squadrons and used in combat for less than a year. The Dr-I is one of the most recognizable of all aircraft ever manufactured, inexorably linked to its most famous pilot, Manfred von Richthofen, the "Red Baron."

In 1917, Fokker and Reinhold Platz designed a new aircraft using input from Manfred von Richthofen. The result was the Fokker D-VII. The plane had a squarish airframe equipped with an in-line engine and an air-cooled radiator. The most advanced feature of the D-VII was its internally braced cantilever wings with thick airfoil sections and a wooden structure. The first of these planes reached the front in April, 1918, and by October, eight hundred were in active service. Popular with German pilots, the D-VII was strong and fast, and it performed superbly at high altitudes. Most aviation historians view the D-VII as the most advanced and outstanding fighter plane of World War I. The quality of the Fokker D-VII was acknowledged by the terms of the Treaty of Versailles. Article IV stated that all Fokker D-VII planes had to be handed over to the Allies, the only aircraft to be specifically targeted by the armi-

Events in Fokker History

1910: Aviation pioneer Anthony H. G. Fokker builds his first aircraft, named Spin (Dutch for "spider")

1912: Fokker establishes an airplane factory at Johanneshal, Germany, where he develops the Dr-I triplane flown by Manfred von Richthofen, the Red Baron, during World War I.

1914-1918: Fokker develops German pursuit planes during World War I and invents a timing mechanism for the shooting of forward-mounted machine guns through an airplane's propeller blades.

1919: Fokker builds a factory in Amsterdam, the Netherlands.

1920: Fokker designs the F.II, one of the first passenger transport planes.

1922: Fokker moves to the United States, where he eventually builds three more aircraft factories.

1939-1945: Fokker's company designs several successful military aircraft used during World War II, including the Fokker G-1.

1955: The Fokker F-27 Friendship turboprop aircraft makes its first flight.

1964: The Fokker F-28 Fellowship jet makes its first flight.

1983: Two new Fokker aircraft are launched to replace the F-27 and the F-28: the Fokker 50 and the Fokker 100.

1996: Fokker Aircraft declares bankruptcy and is reorganized as Fokker Aviation, which is acquired by Stork.

1999: Fokker Aviation is renamed Stork Aerospace Group.

stice. After the war ended, Fokker managed to smuggle two hundred dismantled aircraft, five hundred engines, and other machine parts to the Netherlands, where he started his own factory at Sciphol outside of Amsterdam. During the 1920's, the Fokker D-VII became the mainstay of the Dutch Air Force.

Fokker Between the Wars

In 1918, the German Air Force sponsored a fighter design competition. Twenty-five prototypes were submitted; five were Fokker-designed monoplanes. The Fokker D-VIII parasol monoplane was the winner. It entered production too late to affect the war's outcome, but its design concepts were a significant change in aircraft theory. Unlike earlier aircraft, the D-VIII had a wing that was tapered in both platform and thickness ratio, and it was covered entirely in plywood, giving it great strength and rigidity. The tapered wing reduced wing weight and stress, while increasing aerodynamic efficiency and strength, giving the plane a higher rate of roll.

In July, 1919, N.V. Nederlandsche Vliegtuigenfabriek was incorporated in Amsterdam. Although Anthony Fokker was its managing director, his name was not included in the company name because people had not forgotten that during the war, Fokker had designed some of the most effective German military aircraft at his Fokker Flugzeug-Werke GmbH factory in Germany. Often accused of choosing the wrong side during the war, Fokker always pointed to the fact that before the outbreak of hostilities, both Great Britain and Holland had turned down the aircraft he had offered them. Because of his notoriety, however, it was not until much later that the name Fokker was included in the corporate title. A number of well-known civilian and military aircraft were produced by Fokker between the World Wars.

In October, 1919, another aviation company was incorporated in the Netherlands, N.V. Koninklijke Maatschappij (KLM). Fokker became KLM's main supplier of aircraft and remained so for years. Due to contracts with KLM, orders for Fokker civilian aircraft increased worldwide. Fokker set up factories in the United States and by the late 1920's had become the largest aircraft manufacturer in the world. Numerous aircraft were built under license, and Fokker planes were used by airlines the world over.

The success of postwar Fokker aircraft was linked to a simple construction technique in which the fuselage and the tail section were made of welded steel pipe. In 1933, Douglas Aircraft Company began marketing a modern, streamlined, all-metal aircraft with a retractable undercarriage, and Fokker realized too late that he had stuck with his cheap and simplistic design theory for too long. The DC-2 and DC-3 forced Fokker from the airliner market, when KLM made Douglas their main supplier. It was not until 1958 that Fokker placed a new passenger airliner on the market.

World War II and After

During World War II, production of Fokker aircraft came almost to a standstill. Between 1940 and 1945, when the Netherlands was occupied by Nazi Germany, the Fokker factory was used for the repair and construction of German military aircraft. By the war's end, Allied bombing had reduced the Fokker factory to ruins, and salvageable tools and machines had been plundered by the retreating Germans. After the war, the Dutch government decided that aircraft production in the Netherlands should resume, and the government consolidated its aircraft industries into one company, Fokker. Reconstruction of the Fokker Company had to start from scratch, forcing the company to lag behind its competitors. In countries such as the United States, Great Britain, and Germany, the war had given the aviation industry a great boost, with the outcome being new designs, technologies, and engines, especially the jet engine. Initially, all Fokker was able to do was provide services to refit DC-3's and convert military aircraft into civilian passenger aircraft. Later, however, the postwar Fokker Company developed a number of successful small business aircraft and military trainers, including the S-11 and S-14 models. The company also began the assembly and licensed production of military aircraft designed by others, including the Sea Fury, the Gloster Meteor fighter jets, the Hawker Hunter, the F-104 Starfighter, and the F-5, and later participated in coproduction of the F-16 Fighting Falcon. Nonetheless, these contracts did not make Fokker an independent manufacturer of passenger aircraft. In 1949, on its thirtieth anniversary, the company changed its name to N.V. Koninklijke Nederlandse Vliegtuienfabriek Fokker.

In 1955, Fokker developed the F-27 Friendship, the successor to the popular DC-3, and later, the F-28 Fellowship. Market research indicated there was a demand for a replacement for the famous DC-3. The new aircraft needed a capacity of forty to fifty passengers. While they never became stars in the aircraft marketplace, both the F-27 and F-28 were dependable workhorses and continued to be used in the air fleets of many countries through the beginning of the twenty-first century. Between 1958 and 1986, 786 F-27's were sold, making it the most successful turboprop aircraft in the Western market. The F-28 was the first passenger jet developed by Fokker. It was less successful than the F-27 because of competition from the Boeing 737 and the DC-9. Fokker delivered 241 F-28's between 1968 and 1986. The successors to the F-27 and F-28 were the F-50 and F-100 airliners, introduced in the late 1980's. The largest export order in Dutch history was the sale of seventy-five F-100's to American Airlines. During this period of production, the Fokker Company employed more people than did the civil aircraft division of McDonnell Douglas. However, Fokker was deeply in debt from the cost of the simultaneous development of these two new aircraft, as well as from losses resulting from an all-time low in currency exchange rates and an unexpected drop in aircraft demand.

In an attempt to save the company, Fokker became one of the first major corporations to implement cross-border corporate integration by merging with the German VFW Company. When this merger ended in failure, Fokker attempted a second merger with DASA in the 1990's. Unfortunately, the Fokker Company went bankrupt despite the fact that its product line was well liked and respected, and orders for Fokker products were backlogged. In 1997, the last airliner to bear the Fokker name was assembled at Amsterdam Airport Schiphol, marking the end of a tradition whose origins traced to the earliest days of aviation. In the late 1990's, the Stork Company bought what remained of Fokker and converted it to the specialized manufacture of major components, electric and power distribution systems, and advanced aerospace materials and maintenance. The Fokker Aircraft Group still exists and is in partnership with several global aircraft projects. Following a 1997 decision by the Netherlands government, it participates in projects with Airbus.

Randall L. Milstein

Bibliography

Angelucci, Enzo. *The Rand McNally Encyclopedia of Military Aircraft: 1914-1980*. New York: Rand McNally, 1980. This richly illustrated book is an excellent reference and does a fine job in outlining the history of military aircraft.

Fokker, A. H. G., and Bruce Gould. *Flying Dutchman: The Life of Anthony Fokker.* New York: Arno Press, 1931. A somewhat self-promotional autobiography, but one that does well in outlining the early developments of Fokker aircraft and military innovations such as the synchronized forward-firing aircraft mounted machine gun.

Loftin, L. K. *Quest for Performance: The Evolution of Modern Aircraft. NASA SP-468*. Washington, D.C.: NASA Scientific and Technical Information Branch, 1985. A very easy-to-read and thorough history of aviation and aircraft design. While this book is out of print, it can be found in the United States government collections in most research libraries. It is also available over the Internet at the NASA Web site.

See also: Airbus; Fighter pilots; Luftwaffe; Manufacturers; Military flight; Manfred von Richthofen; Triplanes; World War I; World War II

Food service

Definition: Meals and beverages served by airlines on board an aircraft in flight.

Significance: Airline food quality among the biggest airlines in the United States has never been remarkable, but some regional carriers have made a point of serving tasty meals that regularly garner rave reviews from frequent fliers. Several international airlines, notably those in Asia and Europe, have made food service a priority, especially in first and business classes.

Airline Food

Although airline food quality has often been the butt of jokes, most airlines continue to strive for quality in the meals they provide to passengers. The prevailing view of U.S. domestic airline food is hardly complimentary, but two airlines, Hawaiian Airlines and Midwest Express, are noted for their cuisine. Singapore Airlines, Cathay Pacific, and Japan Air Lines are also known for quality food in all classes of service. First Nations Air, serving the Inuits' Nunavut territory in Canada, also has become recognized for its meal service, notably its salmon dinners.

On Midwest Express, a regional carrier with hubs in Milwaukee and Omaha, meals are served on china with linen napkins. Every meal ends with complimentary wine or champagne. On some flights, chocolate chip cookies are baked on board. Midwest Express spends more than twice the per-meal industry average on food services, and the extra effort is evident.

Between 1995 and 2000, the amount spent on food service by an average airline declined by 15 to 20 percent. An economy-class meal cost an average of $4.81 in the year 2000. Midwest Express spent an average of $10.05. Amounts spent on food service among other U.S. carriers ranged from $8.44 per passenger by American Airlines to $.26 per passenger on Southwest Airlines, which serves nothing but light snacks. Food service for business- and first-class passengers can be considerably more expensive, with a meal, including beverages, sometimes costing more than $30. Singapore Air spends up to $45 per passenger on its first-class food and beverage service.

Midwest Express

Midwest Express was ranked during 1998 and 1999 as the number-one U.S. passenger air carrier in the *Zagat Airline Survey*, which compiled the opinions of 31,500 frequent fliers. The *Zagat Airline Survey* rates seventy airlines for their comfort, service, and food on its signature thirty-point scale. In the 1999 *Zagat Airline Survey*, Midwest Express ranked first in all three categories. Midwest Express was given high marks for its food and service, with the only complaint being that the airline did not fly to enough destinations. Midwest Express not only offers excellent meal service but also provides first-class seat size to all passengers at coach-seat fares.

Midwest Express has been expanding, a fact illustrated by its announcement, in April of 2000, of a one-half-billion-dollar purchase order for new 717-series jets from Boeing. The deal represented twenty new aircraft, with options on twenty to thirty more. This single deal almost doubled the size of the Midwest Express fleet. By 2001, Midwest Express and its commuter partner, Skyway Airlines, were responsible for 37 percent of the passenger traffic at Milwaukee's Mitchell International Airport.

International Airlines

Food quality has long been a concern on several airlines outside the United States. Israel's El Al, for example, hired special consultants to improve its food, adopting a regional Mediterranean style. The airline maintains thirteen kitchens along its routes, which serve roughly 25,000 portions per day. A typical new-style El Al menu consists of Mediterranean appetizers with smoked salmon, roasted oregano-scented spring chicken with couscous, or lasagna Bolognaise with olives and basil, and fresh fruit pie.

Major airlines serve a wide variety of special-order meals, including kosher, vegetarian, Hindu, Muslim, low-fat, gluten-free, peanut-free and lactose-free options. Many airlines also offer so-called bland meals, containing no spices or seasonings, which are designed for people with allergies. Swissair offers a children's meal served in a small briefcase that includes games and snacks.

Caterers

Most airline food is provided not by the carriers themselves, but by independent contractors, such as LSG Skychefs, Host Marriott Services, and Dobbs International. In 1999, LSG Skychefs, which operates 210 kitchens, generated $2.7 billion in sales as it prepared food for 260 airlines. Marriott served 200 carriers and generated $1.2 billion in sales, and Dobbs served 100 airlines and generated $890 million.

Because food service is so important to passengers and because airlines serve 200,000 meals and snacks daily, airlines and airline food caterers give a great deal of consideration, evaluation, and planning into the meals they serve. Airlines receive more comments about the food on flights than about any other aspect of air travel, and the major complaints come when no meal at all is served on a flight.

In 1999, American Airlines launched its Flagship Service menu on all first- and business-class transcontinental flights, with menu items created by four noted chefs from across the United States. At about the same time, Swissair introduced its natural gourmet menu, created from all-natural, fresh foods that are easy to digest. Swissair uses only organically grown foods, and its meats and poultry are fed on natural grains and are free from chemicals or additives. The idea of adopting organic foods came from the airline's international passengers, who are extremely health-conscious and aware of their fat and protein intake. Also contributing to Swissair's move was an increasing trend toward more natural, environmentally friendly choices in food.

Passengers who sample Swissair's natural-gourmet menu might dine on salmon trout with wild garlic sauce, mixed rice with spelt, or sautéed veal with chanterelles and chives. Because Swissair has always emphasized Swiss national products in its cuisine, there are, of course, Swiss chocolates. Swissair's food service subsidiary, Gate Gourmet, supplies meals not only for the Swiss carrier but also for other airlines around the world, including Qantas in Australia. Gate Gourmet produces 48,000 meals per day for seventy carriers. Its bakery in Zurich makes 75,000 buns, croissants, pies, cakes, and other baked goods daily.

In December, 2000, Northwest Airlines and *Food & Wine* magazine initiated a partnership to collaborate on a selection of wine and champagne for Northwest's domestic first class and international business class. Northwest also expanded its on-board wine selection, acquiring its own small wine cellar offering eighteen varieties of wine.

Northwest Airlines, perhaps taking a cue from Midwest Express, also upgraded its food service. Recent new entrées introduced in Northwest's World Business Class include beef tenderloin with smoked pepper sauce, cheese-potato soufflé and squash medley, and oven-roasted chicken with wild mushroom sauce and blue cheese tortellini. North-

Early airlines offered their passengers simple snacks that could be served without much preparation, since early commercial planes were not able to offer facilities to heat and cool food. (Hulton Archive)

west hired some of the best-known chefs in restaurants along its routes to create the new entrées.

In 2001, Las Vegas-based National Airlines became the world's first airline to use a revolutionary in-flight beverage cart that provides faster customer service while reducing the company's costs. The beverage cart, developed by Sterling Beverage Systems, uses the postmix technology commonplace in restaurants, but which had not been adopted in the airline industry. The Sterling cart was also expected to reduce airline costs, because less fuel is burned due to decreased cabin weight.

A Professorship of Airline Food

"Airline food is often a subject of ridicule, but meals in the sky are no laughing matter," according to Peter Jones, who, during 2001, was appointed as the world's first professor of airline food. Jones holds the appointment, which was funded by the International Flight Catering Association (IFCA), a group linking the several organizations that are involved in the airline-food business, at the University of Surrey in Guildford, Great Britain.

The appointment demonstrates the highly sophisticated business that airline catering has become. The daily delivery of hundreds of thousands of meals to airline passengers requires both culinary and logistical skill. The IFCA's intent is to train people who will continue to improve standards.

Despite these improvements, however, passengers should not have unrealistic expectations of airline food. Passengers may not always appreciate how much effort goes into the making of an airline meal. In the high-altitude environment of airline flight, even the most delicately prepared food is affected in taste, texture, and consistency. Meals must be prepared hours in advance, transported to the aircraft, and served in confined conditions with a minimum of equipment. Food served on airline flights has been kept at precisely controlled temperatures for hours and subjected to a barrage of bacteriological and quality control checks.

Adding to the complexity, airline crew also has its own food, often with different dishes for the captain and flight attendants. On several airlines, most notably the Asian carriers, crew meals vary according to the rank of each person eating them.

Computers are increasingly allowing airlines to personalize service. In a few years, passengers will be able to reserve their meals from a menu, as they do in restaurants.

Bruce E. Johansen

Bibliography

Bridges, Linda. "No Pie in the Sky." *National Review* 46 (June 13, 1994): 72. A wry review of airline food from a passenger's point of view.

Cooke, Kieran. "The Appointment of the World's First Professor of Airline Food." *Financial Times*, January 23, 2001, 19. An article describing the airline food industry's establishment of an academic chair and an institute for airline cuisine at an English university.

Dulen, Jacqueline. "Flights of Fancy." *Restaurants and Institutions* 109, no. 20 (August 1, 1999): 14. An article detailing the logistical problems that airlines face in serving many thousands of meals daily on complex schedules.

Holcomb, Henry J. "Midwest Express Tops National Airline Survey." *Philadelphia Inquirer*, March 21, 2001, n.p. An account of Midwest Express's number-one rating for food quality and service in the Zagat survey.

"Midwest Express Still Number One with Frequent Fliers." *Milwaukee Journal-Sentinel*, March 21, 2001, p. D6. A view of Midwest Express's business prospects from Milwaukee, its primary hub.

Sheridan, Margaret. "Institutional Food Service." *Restaurants and Institutions* 109, no. 25 (September 15, 1999): 100. A description of the scope and complexity of airline food services.

See also: Air carriers; Airline industry, U.S.; American Airlines; Flight attendants; Northwest Airlines; Qantas; Singapore Airlines; Swissair

Forces of flight

Definition: The so-called four forces—gravity, drag, lift, and thrust—that act upon an airplane in straight-and-level unaccelerated flight.

Significance: Weight and drag are forces of nature inherent of any object lifted from the ground and moved through the air. The forces of lift and thrust are artificially caused to overcome the forces of weight and drag and enable an airplane to fly.

Humans' first attempts to fly, inspired by birds, were limited until humans realized they could not fly like birds. Birds, with their very light weight, great strength, and complex biological design, can use their wings to create both lift and thrust to overcome the natural forces of weight and drag, and to maintain control. Humans, in contrast, had to invent a different approach to meet any success in aviation. The functions of lift and thrust had to be separated. For that, wings and engines were introduced. While wings produce lift, engines produce thrust.

Following the first flights made by Orville and Wilbur Wright in December, 1903, the pace of aeronautical development accelerated, and the progress made in overcoming the natural forces in the aviation industry in following decades was dramatic. The understanding of natural forces is thus as important for an airplane's aerodynamics as the creation of artificial forces to counterbalance these natural forces. The engine and propeller combination is designed to produce thrust to overcome drag. The wing is designed to produce lift to overcome weight, or gravity. In unaccelerated, straight-and-level flight, which is coordinated flight at a constant altitude and heading, lift equals weight and thrust equals drag. Nevertheless, lift and weight will not equal thrust and drag. In everyday vocabulary, the upward forces balance the downward forces, and forward forces balance the rearward forces. This statement is true whether or not the contributions due to weight, drag, lift, and thrust are calculated separately. Any inequality between lift and weight will result in the airplane entering a climb or descent. Any inequality between thrust and drag while maintaining straight-and-level flight will result in acceleration or retardation until the two forces become balanced. However, there are a couple of paradoxes surrounding this information. The first paradox is that in a low-speed, high-power climb, the amount of lift is less than the amount of weight. In this situation, thrust is supporting part of the weight. The second paradox is that in a low-power, high-speed descent, the amount of lift is again less than the amount of weight. In this situation, the drag is supporting part of the weight. In light aircraft, the amount of lift ordinarily is approximately ten times the amount of drag.

The motion of an aircraft through the air depends on the size of these four forces. The weight of an airplane is determined by the size and material used in the airplane's construction and on the payload and fuel that the airplane carries. The lift and drag are aerodynamical forces that depend on the shape and the size of the aircraft, air conditions, and the flight speed and direction relative to the air velocity. The thrust is determined by the size and type of the propulsion system used in the airplane and on the throttle setting selected during the flight.

The relative wind velocity acting on the airplane contributes a certain amount of force, called total aerodynamic force. This force can be resolved into two components perpendicular to each other along the directions of lift and drag. Lift is the component of aerodynamic force directly perpendicular to the relative wind velocity. Drag is the component of aerodynamic force acting parallel to the relative motion of the wind. Weight is the force directed always downward toward the center of the earth. It is equal to the mass of the airplane multiplied by the acceleration due to the gravity, or the strength of the gravitational field. Thrust is the force produced by the engine and is usually more or less parallel to the long axis of the airplane.

Weight

Weight, or gravity, is the force which always acts downward, toward the center of the earth. It is the total sum of the masses of all its components and contents multiplied by the strength of the gravity, commonly referred to as the number of g's. The weight may be considered to act as a single force, representing all its components and contents, through a single point called the center of gravity.

Weight is the most reliable force, which always acts in the same direction and gradually decreases as airplane fuel is used. The center of gravity shifts as the weight is redistributed.

Although the terms "mass" and "weight" are often confused with each other, it is important to distinguish between them. Mass is a property of a body itself and measures a body's quantity of matter. Weight, in contrast, is a force representing the force of gravity acting on a body. It is also loosely called gravity.

To illustrate the difference, one could describe an object that is taken to the Moon, where the force of gravity is weaker, about one-sixth that on Earth. On the Moon, the object will weigh only about one-sixth as much as it did on Earth. The mass of the object will be the same on the Moon or anywhere else. In other words, it will continue to have the same amount of matter.

Drag

When an object moves relative to a fluid, either a gas or a liquid, the fluid exerts a frictional force on the object. This force which is referred to as a drag force, is due to the viscosity, or stickiness, of the fluid and also, at high speeds, to the turbulence behind and around the object. To characterize the motion of an object at different speeds relative to the fluid and to understand the associated drag, it is useful to understand Reynolds numbers.

The Reynolds number depends on the properties, such as length and velocity, of the fluid and the object relative to the fluid. In case of an airplane, which flies through air, the Reynolds number for air is smaller than that for water because of the lower density of the air. For example, an object of one millimeter long moving with a speed of 1 millimeter per second through water has the same Reynolds number as an object 2 millimeters long moving at a rate of 7 millimeters per second in the air. The drag manifests itself differently for different Reynolds numbers associated to it.

When the Reynolds number is less than 1, as in the case of fairly small objects, such as raindrops, the viscous force is directly proportional to the speed of the object. For large Reynolds numbers, usually above a value between about 1 and 10, there will be turbulence behind the body, known as wake, and hence, the drag force will be larger and it increases as the square of the velocity instead of its linear dependence on the velocity. When the Reynolds number approaches a value of around 1,000,000, the drag force increases abruptly. For above this value, turbulence exits in the layer of fluid lying next to the body all along its sides. For streamlined objects, however, there will be less turbulence and, hence, less drag. The flow is said to be streamlined of laminar flow if the flow is smooth, such that neighboring layers of the fluid slide by each other smoothly.

There are several types of drag, subdivided and classified according to their action on an airplane. Pressure drag is the force pushing a horizontally moving object against the front vertical surface of the object. Friction drag is produced on a horizontally moving object by applying a force along the surface of the object. Friction drag is proportional to the viscosity of the fluid. Fortunately, air has rather low viscosity, so in most situations the amount of friction drag is small compared to that of pressure drag. In contrast, pressure drag does not depend very strongly on viscosity. Instead, it depends on the density of the air.

Both friction drag and pressure drag create a force proportional to the area involved and the square of the airspeed. Part of the pressure drag that a wing produces depends on the amount of lift it is producing. This part of the drag is called induced drag. The rest of the drag is called parasite drag. The part of the parasite drag that is not due to friction is called form drag, because it is extremely sensitive to the detailed form and shape of the airplane.

A streamlined object can have ten times less form drag than a nonstreamlined object of comparable frontal area. The peak pressure in front of the two shapes will be the same. However, the streamlined shape causes the air to accelerate, so the region of highest pressure is smaller, and more importantly, the streamlined shape cultivates high pressure behind the object that pushes it forward, thus canceling most of the pressure drag. This situation is called pressure recovery. An object moving through the air has a high-pressure region in front, but a properly streamlined object will have a high-pressure region in back as well. However, streamlining is never perfect; there is always at least some net pressure drag.

Induced drag also contributes to pressure drag whenever lift is being produced, even for perfectly streamlined objects in the absence of separation. The flow pattern near a nonstreamlined object is not symmetric fore and aft because the streamlines separate from the object as they go around the sharp corners of the plate. Except in the cases of very small objects or very low speeds, pressure drag is larger than friction drag, even for well-streamlined objects.

The pressure drag of a nonstreamlined object is much larger still. For this reason even the smallest parts of high-performance aircraft, such as fuel-cap handles, are precisely aligned with the airflow. An inevitable exception involves the air that has to flow through the engine compartment to cool the engine. A lot of the air has to flow through narrow channels. The resulting friction drag, called cooling drag, amounts to 30 percent of the total drag in some airplanes.

Unlike pressure drag, friction drag cannot possibly be canceled. It can, however, be minimized. The way to minimize friction drag is to minimize the total area, called wetted area, that has high-speed air flowing along it. The way to reduce form drag is to minimize separation by streamlining all parts.

It is often convenient to express the drag force as a dimensionless quantity by the coefficient of drag. In that case, the drag force is proportional to the coefficient of drag, the density of the air, the square of the true airspeed, and the relevant area, which is typically taken to be the wing area excluding the surface area of the fuselage.

In the mushing regime, most of the drag is induced drag. As the airplane goes more slowly, induced drag increases dramatically, and parasite drag becomes almost neglible. At high airspeeds, parasite drag is dominant, and induced drag becomes almost negligible. In a high-speed regime that includes normal cruise, the power required increases rapidly with increasing airspeed.

Parasite drag is the dominant contribution to the coefficient of drag, and it is more or less independent of airspeed. Induced drag decreases as the airspeed increases, but this is a relatively minor contribution in this regime. Ways of reducing induced drag include wing tapering, wingtip modification, and employing washout and a high aspect ratio. The aspect ratio is defined as the ratio between the span and the mean chord. The mean chord, in turn, is the ratio between the wing area and the wingspan.

Lift

Airplane wings and other airfoils are designed to deflect the air so that, although streamline flow is largely maintained, the streamlines are crowded together above the wing. Just as the flow lines are crowded together in a pipe constriction where the velocity is high, so the crowded streamlines above the wing indicate that the airspeed is greater than below the wing. Hence, according to Bernoulli's principle which states that velocity increases as pressure decreases, the air pressure above the wing is less than that below the wing, and there is a net upward force, which is called dynamic lift, or lift.

In fact, Bernoulli's principle is only one aspect of the lift on a wing. Wings are usually tilted slightly upward so that air striking the bottom surface is deflected downward. The change in momentum, a product of mass and velocity, of the rebounding air molecules results in an additional upward force on the wing. As the air passes over the wing, it is bent down. The bending of the air is the action; the reaction is the lift on the wing. To generate sufficient lift, a wing must divert air down. To increase the lift, either or both the diverted air and downward velocity must be incremented.

The downward velocity behind the wing is called downwash. The vertical downward airspeed varies as the angle of attack. The angle of attack is the angle of the chord line. The direction of the relative airflow on the wing, along the chord line, or chord length, is the distance from the loading edge of the wing to the trailing edge. As the wing moves along while the air is diverted at the rear end of the wing, it is pulled up at the leading edge, also giving rise to upwash. This upwash contributes negatively to the lift. Turbulence also plays an important role in contributing to the lift.

Like drag, lift can also be expressed in a dimensionless quantity in terms of the coefficient of lift. In that way, the lift force is proportional to the coefficient of lift and the density of the air, the square of true airspeed and relevant area. The coefficient of lift is a ratio that basically measures how effectively the wing turns the available dynamic pressure into a useful average suction over the wing. The dynamic pressure is the product of the air density and the square of the velocity. This is the difference between total pressure and static pressure. The total pressure is the pressure in air that has been brought to rest from the free stream, and the static pressure is the ambient pressure at the same level as the aircraft. In actual flight, pilots are not free to make any amount of lift they want. The lift is nearly always equal to the weight multiplied by the load factor; the coefficient of lift depends directly on the load factor, and inversely, on the square of the airspeed. Because of the airspeed squared, the airplane must fly at a very high coefficient of lift in order to support its weight at low airspeeds.

As there is a center of gravity, there is also a center of pressure, which is a point through which the resultant lift acts. The center of pressure changes with change of wing shape. A number, called the lift-drag ratio, is considered best when it produces the most efficient speed for maximum range with minimum drag.

Thrust

A force pushing an airplane, or any object, forward is called thrust. The thrust is produced by the engines of the airplane or by the flapping of a bird's wings. The engines push fast-moving air out behind the plane, by either propeller or jet. The fast-moving air causes the plane to move forward, countering drag.

Since the Wright brothers first flew in 1903, aeronautical engineers have created a multitude of airplane types, every one of which has dealt with the same four forces of weight, drag, lift, and thrust. All people have to deal with the challenges of stability with respect to these forces. Flying faster than the speed of sound has its own special demands, but the underlying forces of weight, drag, lift, and thrust remain the same.

In some sense, it is easier to fly in space, which is devoid of air, than it is to fly in air. However, spaceflight has its own special challenges. In space, one must deal with only two forces, weight and thrust. Thrust provides the force to lift a rocket into space. Once in orbit, a spacecraft no longer needs propulsion. Short bursts from smaller rockets are used to maneuver the spacecraft. To change its orientation, a spacecraft applies torque, a twisting force, by firing small rockets called thrusters or by spinning internal reaction wheels.

M. A. K. Lodhi

Bibliography

Barnard, R. H., and D. R. Philpott. *Aircraft Flight*. 2d ed. Essex, England: Addison-Wesley Longman, 1995. An excellent, nonmathematical text on aeronautics, in which illustrations and physical descriptions, rather than equations, are used to explain virtually all aspects of airplane flight.

Craig, Gail. *Stop Assuming Bernoulli! How Airplanes Really Fly*. Anderson, Ind., Regenerative Press, 1997. A vivid description of airplane flight that clarifies some misconceptions about the forces of flight.

Giancoli, D. C. "Fluids." In *Physics with Application*. 3d ed. Englewood Cliffs, N.J.: Prentice Hall, 1991. A brief description of underlying physical principles of forces of flight with simple equations and good illustrations.

Wegener, Peter P. *What Makes Airplanes Fly? History, Science, and Applications of Aerodynamics*. New York: Springer-Verlag, 1991. A well-written and well-illustrated but slightly technical review of the historical development of aerodynamics and airplanes.

See also: Aerodynamics; Aeronautical engineering; Airplanes; Gravity; History of human flight; Microgravity; Orbiting; Propulsion; Roll and pitch; Spaceflight; Wind tunnels

Steve Fossett

Date: Born on April 22, 1944, in Garden Grove, California

Definition: Prominent balloonist, aviator, and yachtsman who has set many world records for flight.

Significance: Ever since his twenties, Fossett engaged in daredevil adventures, sailing marathons, flying jet planes, climbing mountains, and racing automobiles. His most publicized and spectacular feats have been in ballooning.

Steve Fossett is a millionaire stockbroker who engages in adventurous hobbies. As a youngster, he started rock climbing. As a college student at Stanford University, he climbed mountains and swam the Hellespont in Turkey—a classical test of strength and endurance that he swam both ways.

After earning his master of business administration degree at Washington University in St. Louis, Missouri, Fossett moved to Chicago, where he made his fortune. However, he continued to engage in dangerous adventurous sports, continuing to try, despite many failures, until he succeeded. "I always thrive under pressure," he claimed. In the late 1990's, he concentrated on sailing and ballooning, sports in which he believed he could set world records. In

January, 1997, he attempted to be the first person to circumnavigate the world in a hot-air balloon. He competed against three teams, the British Virgin Group of Richard Branson, an Australian team, and a Swiss team. Compared to the millions that his competitors were spending, Fossett's bare-bones *Solo Spirit* balloon was relatively inexpensive, at $300,000.

Fossett's 1997 flight failed when he had to land in northern India after six days, but he set a record for endurance and distance. His balloon traveled at heights of 18,000 to over 28,000 feet. With a broken heater, temperatures in the balloon were 15 degrees Fahrenheit. Fossett failed in a second attempt, in August, 1998, when he had to land in Russia. Later that year, he began his third attempt in North Africa, but he was forced to land in the Pacific after China refused permission to fly over their airspace. A rival team of Bertrand Piccard and Brian Jones, however, accomplished the feat shortly afterward. Fossett then sought to set sailing records in his Play Station Maxi Catamaran. In February, 2000, along with copilots Darrin Adkins and Alex Tai, Fossett set the around-the-world record for medium-weight airplanes in his Citation X two-engine business jet. The trip took 41 hours, 13 minutes, and 11 seconds, about 6 hours less than the previous record. His average speed was 559 miles per hour. The same year, he set the U.S. coast-to-coast records for private planes in both directions in his Citation.

His August, 1998, balloon flight set the record for the longest solo aircraft flight and the second longest balloon flight. He also holds the record for several other distance and speed flights in his private plane.

Steve Fossett's sixth attempt at an around-the-world balloon flight ended on August 17, 2001, when bad weather forced him to land in Brazil after successfully flying from Western Australia over the Pacific Ocean to South America. (AP/Wide World Photos)

In August of 2001, Fossett made a fifth attempt at a transglobal balloon flight, but bad weather forced him to land in southern Brazil, just one day after he had reached the halfway point of his trip. Despite the curtailed effort, however, Fossett still managed to set a new record for the longest solo balloon flight, with a trip lasting 12 days and 13 hours. With Fossett's failed attempt, the solo transglobal balloon record remained unclaimed.

Frederick B. Chary

Bibliography

Conniff, Richard. "Racing with the Wind." *National Geographic* 192, no. 3 (September, 1997). A good article describing the 1997 balloon competition. Includes pictures, diagrams, and maps.

Gannon, Robert. "The Great Balloon Race." *Popular Science* 248, no. 5 (May, 1996). A description of the preparations by Fossett and other racers for the competition to circumnavigate the globe, with illustrations and graphic designs of the Solo Challenger and other balloons.

Hogan, David. "Up, Up, and Away." *Current Science* 83, no. 6 (November 14, 1997). Describes Fossett's 1997 balloon flight with illustrations and diagrams.

See also: Balloons; Richard Branson; Lighter-than-air craft; Record flights; Transglobal flight

Franco-Prussian War

Date: From 1870 to 1871

Definition: A war fought between France and Prussia over the issue of Spanish succession.

Significance: The Franco-Prussian War was one of the first uses of air power in warfare, though pigeons played a more important role than balloons.

Background

On July 19, 1870, France declared war on Prussia after unsuccessful negotiations between the two nations concerning who would be the next Spanish king. The Prussians were backing a candidate in Spain friendly to them, but the French, fearful of being surrounded by a Prussian-Spanish alliance, called for negotiations. After the talks failed, the Prussian premier, Otto von Bismarck, deliberately angered the French by insulting their government in the famous Ems telegram. French emperor Napoleon III had been assured that France's army was so powerful it could never be defeated. Early in the war, however, the French suffered massive defeats at the hands of the Prussians, and on September 1, 1870, the French army surrendered. Napoleon III, who had become ill while fighting with his troops, was taken prisoner.

Hot-air Balloons

In Paris, the French set up a provisional government of national defense that took away Napoleon's power and established the Third Republic. The Prussian army then surrounded and laid siege to Paris, totally blockading the new government from the rest of France. The leaders of the Third Republic needed a way to communicate with its armies outside of Paris. On September 22, a solution was provided from above. A number of large hot-air balloons, which had been built for an international exposition in 1867, were found in a sad state of disrepair. One balloon, the *Neptune*, was patched together and piloted out of the city. The astonished Prussians watched, powerless to bring the balloon down. After three hours in the air, the balloon's pilot landed in friendly territory with messages for the commanders of the remaining French army.

Over the next several days, four more balloons took off safely and reached their destinations without being shot out of the sky. A means of sending messages to the countryside had been established, and the minister of the post office officially established a regular "Balloon Post."

Carrier Pigeons

It soon became apparent that balloons, at the mercy of the wind, could not be effectively steered in a desired direction. Although dirigibles, motor-powered balloons, were being tested at the time of the siege, they were not yet available for service. Balloons flew out of Paris, but, because of the wind patterns, they could not return.

Postal officials tried various solutions to the problem. They once sent five sheepdogs out of the city by balloon with the intent that the dogs could be sent back with mail tied to their backs, but none of the dogs were ever seen again. They floated hollow metal balls with messages inside down the River Seine, but none was ever recovered.

A Parisian carrier pigeon owner's club, L'Esperance, contacted the government to suggest that urban pigeons could be sent away with messages bound to their legs and could then be sent back to Paris with new messages. Government officials initially laughed at the idea. However, the club secretary eventually found a willing listener at the Central Telegraph Offices, the wires to which had been cut by the Prussians early in the siege. A pigeon loft was built on the roof of the Telegraph Office on September 4, 1870. Before the system could work, however, pigeons would have to be taken out of Paris and trained to return to their new loft.

On September 10, the first pigeons were taken out of Paris by balloon and flown to the city of Tours in southern France. Members of L'Esperance began rounding up the limited supply of available carrier pigeons, most of which were untrained. The principal supplier of pigeons was the club's president, Mr. Cassier, who had fifty-two birds in his loft. Of all fifty-two birds that saw service in the war, only two survived. During the remaining five months of the siege, hundreds of carrier pigeons were taken out of Paris by balloon.

Once the pigeons arrived in army headquarters at Tours, they were fed, rested, and prepared for the 130-mile return flight to Paris. Before they were released, the pigeons had messages placed in small metal containers attached to their legs. More than three hundred pigeons left Paris by balloon during the siege, but only fifty-nine successfully made the return flight to Paris. The others fell into Prussian hands or were blown off course by severe weather, particularly in the winter months of January and February, when only six of sixty-five pigeons safely made the journey back to Paris. The weather and the Prussians were not the pigeons' only enemies, however. Hawks killed some, as did human hunters with shotguns seeking food for their families. The pigeon service was ended on February 1, 1871, after the Prussians lifted their siege.

The pigeons had an important impact on the morale of the Parisians, even though France lost the war and signed a humiliating peace treaty. The fifty-nine pigeons that successfully made the return flight carried more than 95,000 messages to Paris. The citizens of Paris even built them a monument after the war through private contributions.

Microfilm Messages

One interesting result of these pigeon flights was the development of the first microfilm messages. A Parisian photographer developed a method of taking pictures of official messages and reducing them in size by using his camera. These microfilms made it easier for pigeons to carry large numbers of documents that would have been too heavy on paper. The microfilm was placed in tubes attached to one of the bird's legs. With this method, pigeons could carry twenty or more messages at a time rather than one or two. To improve their chances of getting through Prussian lines, the same messages were sent by several pigeons at the same time. Carrier pigeon flights thus played a major role in the Franco-Prussian War and aided in the survival of the people of Paris.

Leslie V. Tischauser

Bibliography

Hayhurst, John D. *The Pigeon Post into Paris, 1870-1871.* London: privately printed, 1970. A detailed analysis of this first use of air power that includes many statistics and illustrations.

Horne, Alistair. *The Fall of Paris.* New York: St. Martin's Press, 1965. Describes the war and the introduction of balloons and pigeons in detail.

Milner, John. *Art, War, and Revolution in France, 1870-1871.* New Haven: Yale University Press, 2000. Includes several interesting illustrations of air power during the war.

See also: Balloons; Birds; Buoyant aircraft; Lighter-than-air craft; Reconnaissance

Frequent flier miles

Definition: Marketing tools used by airlines to keep and maintain loyal customers.

Significance: Frequent flier miles and other frequent traveler programs (FTPs) are a vital part of an airline's strategic marketing plan to gain the repeat business of an airline's most profitable customers.

Airline Reward Programs

Programs designed to reward loyal customers for their patronage are not new, having their roots in trading stamps and coupon books issued by grocery stores and gas stations to create brand loyalty and product differentiation.

Prior to the Airline Deregulation Act of 1978 that eliminated the U.S. government's control over airline operations, the U.S. government established airline fares and routes, and airlines emphasized service and prestige to gain potential customers. However, after the Airline Deregulation Act was passed, airline routes were opened to competition, and airlines could adjust their fares to meet market conditions.

Deregulation caused an immediate and tremendous flurry of activity among start-up carriers. New airlines marketed their seats based solely on price. Consequently, many large carriers began to lose market share among their most profitable customers, frequent business travelers who paid the highest fares.

In response to this loss, American Airlines became the first airline to offer a frequent flier program in 1981. This program, AAdvantage Travel Awards, offered customers "miles," or points, based on the distances they flew. Customers could trade in accumulated miles for free flights or upgrades on the airline. The program, initially targeted as a promotion to a few top customers, grew rapidly in both popularity and profitability and proved successful in capturing repeat business. Other domestic U.S. airlines quickly followed American Airlines' example, as did non-U.S. airlines, and soon nearly all airlines had a frequent traveler program or were tied into one.

Role of Technology

Two major technological innovations made FTPs possible: the computer reservations system (CRS) and the sophisticated computer technology that allows the CRS system to exist. Early CRSs were used by airlines to track internal airline bookings. Although all major airlines had automated their systems by the 1970's, travel agents could not access information on all available flights between destinations without great difficulty, because these systems were not linked.

The first CRS used by travel agents that expanded flight availability information from that of a single airline to that of all airlines was American Airline's Semi-Automated Business Research Environment (SABRE) system, developed in conjunction with IBM in the 1960's and marketed to travel agenices beginning in 1975. Other airlines developed similar programs that competed with SABRE for acceptance by travel agents. The substantial increase in the number of existing flights made possible by airline deregulation accelerated the acceptance of computer reservation technology.

The airlines' heavy reliance on computer technology to keep track of seats and fare distributions propelled the evolution of CRS technology. In the early 1980's, tremendous leaps in computing technology enabled the increasing sophistication of CRSs. Passengers' travel histories could be tracked with greater detail and easier access than ever before. The most frequent customers could be easily identified, and marketing programs could be geared directly to them.

The increase in air travel fueled expansion in other service industries tied to air travel. Hotel, rental car, and credit card companies and others tied into the CRS programs offered superior service to the frequent customer. The CRS system has become a multidimensional tool that enables travelers to easily plan, price, and book almost any aspect of travel worldwide.

CRS as an FTP Tool

With information about ticket buyers gained through CRS bookings, the airlines could track their most valued customers, those who traveled frequently on high-yield markets. Initially, the airlines were able to search their databases to track bookings and correlate tickets sold with telephone numbers and customers' names. These became the first members of the airlines' frequent flier programs. As FTPs became more sophisticated, customers could be identified by means of rating systems that varied between airlines. The airlines define a set of guidelines by which a customer is rated, and the highly valued core customers are tracked to determine how they are generating revenue for the airline.

Expansion of FTPs

Ever innovative, American Airlines enlisted hotels and rental car companies as partners in its mileage program, giving additional points to customers who used the associated company's services. This arrangement enabled customers to accumulate their awards more quickly. By the end of 1981, most major U.S. airlines had frequent flier programs with other service partners. The battle to win the business of the lucrative frequent traveler had begun in earnest. A few airlines, hotels, and rental car companies were unenthusiastic about the mileage awards programs and dismissed them as a short-lived gimmick. However, their quick acceptance by consumers and the fierce competition of those airlines vying for the business traveler's business soon proved to be disastrous to the bottom line of those airlines that did not participate.

Airlines integrated their FTPs with companies as diverse as credit card companies, telephone companies, hotels, cruise lines, and many types of retailers. Airline marketers continue to pursue different avenues of mileage programs to keep their FTPs dynamic and diverse.

Cost of FTPs

FTPs generate both administration costs and direct costs. Administrative costs include the tracking and marketing of the program to customers. Direct costs are incurred when a customer seeks to redeem the award for travel. These costs include variable passenger expenses, or the incremental costs—such as meals, beverages, fuel, reservations, and handling—of carrying an additional passenger. They also include the additional cost incurred if passengers were to redeem travel on the airlines' FTP partner. Displacement costs are those incurred by the elimination of a seat from an otherwise fare-paying passenger. These costs are sometimes partially offset by the revenue of a traveling companion. Diversion, or dilution, costs are incurred when the redeeming passenger would have purchased a ticket on the airline, regardless of having an award. Many of these costs are controlled by limiting the number of seats per flight available for reward programs.

FTP Marketing Benefits

Marketing to proven loyal customers is a cost-effective use of a business's limited marketing dollars. FTP awards have become so widely accepted that they now form a core part of airlines' products and are expected by customers.

However, airline rewards continue to evolve. Airline officials realize that the number of miles a customer accumulates over time is not the single best indicator of that customer's value to the company. A customer who infrequently travels long distances may generate more miles on the airline than another customer who travels more frequently on short routes, but the short-haul customer may generate more income to the airline's bottom line. More frequent, short-haul travelers are more valuable to the airline not only because they buy more tickets but also because the short-haul, business-route tickets they buy tend to have a higher margin of profit for the airline.

As a response, the airline industry has shifted from simply selling seats toward a more customer- or relationship-based marketing strategy, realizing the customer's value over the long term. Loyal customers are rewarded for their value to the airline, further increasing their satisfaction. The types of awards offered to frequent travelers have also evolved. Initially, awards were granted in the form of free flights or upgrades from a coach seat to a business-class or first-class seat. Research into frequent traveler's preferences has shown that special recognition, preferred service, and privileges such as exclusive lounges or other perks are highly valued. As technology advances, the ability to track an individual's purchasing behavior and preferences has enabled airlines to individually market different types of award.

Frequent traveler programs are an integral part of an airline's strategic marketing. Customers expect rewards, and the airlines compete fiercely for their loyalty. As the travel industry expands and the need for travel services rises, FTPs will continue to evolve to meet the customers' demands.

Veronica T. Cote

Bibliography

Ellis, John M. "The New Role of Frequent Flier Programs." In *Handbook of Airline Marketing*, edited by G. F. Butler and M. R. Keller. Washington, D.C.: Aviation Week Group, 1998. A comprehensive collection of articles written by experts in the field, addressing critical issues and subjects of airline marketing.

Garvett, Donald S. "Frequent Traveler Programs: Moving Targets." In *Handbook of Airline Marketing*, edited by G. F. Butler and M. R. Keller. Washington, D.C.: Aviation Week Group, 1998. A comprehensive collection of articles written by experts in the field, addressing critical issues and subjects of airline marketing.

Wells, Alexander T. *Air Transportation: A Management Perspective*. 4th ed. Belmont, Calif.: Wadsworth, 1999. A textbook covering all major topic areas in the air transportation field.

Zakreski, Eugene. "Beyond Frequent Fliers: Knowing Customers as a Foundation for Airline Growth." In *Handbook of Airline Marketing*, edited by G. F. Butler and M. R. Keller. Washington, D.C.: Aviation Week Group, 1998. A comprehensive collection of articles addressing critical issues and subjects of airline marketing written by experts in the field.

See also: Air carriers; Airline Deregulation Act; Airline industry, U.S.; American Airlines; Mergers; Ticketing

G

Yuri Gagarin

Date: Born on March 9, 1934, in Klushino, near Gzhatsk, Smolensk Oblast, Soviet Union; died on March 27, 1968, near Moscow, Soviet Union

Definition: Russian cosmonaut whose 108-minute Earth orbital flight on April 12, 1961, represented humankind's first space travel.

Significance: As pilot of the Soviet orbital mission Vostok 1, Gagarin ushered in the space age by proving that a human being could endure the rigors of liftoff, reentry, and weightlessness and still perform the manual operations essential to space flight.

After a primarily vocational education, Russian cosmonaut Yuri Alekseyevich Gagarin entered pilot training at the First Chkalov Orenburg Military School for Pilots. In the autumn of 1957, he graduated with high honors from Orenburg and joined the Soviet Air Force as a junior lieutenant. From late 1957 until the spring of 1960, he served as a military fighter pilot in the Arctic. In 1960, he was selected as a member of the first group of Soviet cosmonauts.

On the morning of April 12, 1961, Gagarin literally flew into history on board the spaceship Vostok 1, which launched at 9:07 A.M. Moscow time. The flight was automated for fear that the weightlessness of space might disable the pilot. A key was available in a sealed envelope in case it became necessary to take control in an emergency. In a preflight speech, Gagarin commented that he had always waited for this moment and that he was glad to "meet nature face to face, in an unprecedented encounter." The rocket accelerated to a peak of 5 g's, indicating that Gagarin felt five times his normal weight. Fourteen minutes after liftoff, Gagarin reported that the capsule had achieved Earth orbit. He then tested his food and water samples and reported no side effects to the weightlessness. During the 108-minute flight, he made one elliptical Earth orbit, the apogee of which was about 203 miles above sea level. The orbital speed was approximately 17,000 miles per hour. The payload included life-support equipment as

Russian Yuri Gagarin's 108-minute flight on April 12, 1961, marked the first human voyage into space. (Hulton Archive)

well as communications equipment that relayed information on Gagarin's condition. As planned, at about 20,000 feet, Gagarin ejected and descended under his own parachute and landed southwest of the Saratov region, near Smelovka, Saratskaya.

Following his historic flight, Gagarin received many honors in recognition of his Vostok mission. He was named a hero of the Soviet Union and was awarded the Order of Lenin and the K. E. Tsiolkovsky Gold Medal of the U.S.S.R. Academy of Sciences. Later, a crater on the far side of the Moon was named after him. On March 27, 1968, Gagarin was killed in an accident while test piloting a MiG-15 aircraft near Moscow. The event caused a great deal of shock and spawned numerous conspiracy theories and rumors within the Soviet Union, whereas Western powers alleged that Gagarin had been drunk at the time. Two years after his death, he was posthumously inducted to the International Aerospace Hall of Fame.

Monish R. Chatterjee

Bibliography

Cole, Michael D. *Vostok 1: First Human in Space*. Springfield, N.J.: Enslow, 1995. A book presenting six milestones in space exploration in brief yet dramatic narratives containing a wealth of interesting detail.

Harpole, Tom. "Saint Yuri." *Smithsonian Air & Space Magazine* (December, 1998/January, 1999). An interesting study of a brave adventurer and explorer.

Kennedy, Gregory P. *The First Men in Space*. New York: Chelsea House, 1991. An account of the pioneering roles of the first Soviet and American men in space.

See also: Astronauts and cosmonauts; History of human flight; Russan space program; Spaceflight

Roland Garros

Date: Born on October 6, 1888, in Saint Denis, France; died on October 5, 1918, near Vouziers, France

Definition: French pilot and aviation pioneer who was the first person to fly an airplane across the Mediterranean and is credited with being the first fighter pilot.

Significance: Garros helped to devise an apparatus that allowed a machine gun to fire effectively between the blades of an airplane's moving propeller, and he successfully tested the device against German aircraft in World War I.

Roland Garros was born in 1888 and was an avid sportsman who developed a passion for aviation. He entered air races, placing second in the 1911 Paris-to-Rome competition and in April, 1913, winning the International Air Rally of Monaco. He also set several world altitude records, and on September 23, 1913, made history by being the first person to fly cross the Mediterranean. With the outbreak of war in 1914, he joined the French Air Corps as a lieutenant.

At the beginning of World War I, military planes mainly flew observation missions. Then pilots began to use rifles or revolvers to fire at enemy aircraft and ground troops. When a machine gun was mounted to the plane, a second man was needed and he could fire the gun only from the rear, so that the bullets would not hit the blades of the moving propeller. Pilots and aircraft designers alike quickly realized the urgent need to develop a forward-firing machine gun that enabled the pilot to aim his aircraft and gun in the same direction. Frenchman Raymond Saulnier, of the Morane-Saulnier aircraft company, had designed steel plates that would fit on the propeller blade to deflect most of the mounted gun's bullets. Garros calculated, however, that 7 percent of the bullets could still hit the propeller. Determined to make the deflector shields more effective, Garros worked to improve on Saulnier's invention by adding small steel wedges to the propeller blades. Garros then had his plane fitted with the new deflection plates and with a Hotchkiss machine gun with its trigger connected by a wire to the cockpit. On April 1, 1915, Garros was ready to test the new device in combat. He encountered a German reconnaissance plane, aimed his machine gun and shot down the Albatros B-II. During the next two weeks, Garros shot down four more German planes.

The Germans were unable to explain how Garros managed to fire his machine gun successfully through the propeller blades until April 18, 1915, when Garros was forced to land his aircraft behind German lines. The established practice of the time was that a downed pilot would burn his aircraft as quickly as possible to prevent its falling into enemy hands. Garros's plane, however, was too damp and did not burn. After they captured Garros's plane, the Germans studied it carefully and soon determined the specific workings of the propeller deflectors. Under the direction of Dutch aircraft designer Anthony Fokker, the Germans fitted their planes with similar yet slightly improved devices and thus launched the era of aerial combat.

The Germans treated Garros respectfully and placed him in an elite prisoners' camp. He remained a prisoner of war until 1918, when he managed to escape to Holland. He then traveled back to France and returned to aerial combat.

Garros was shot down and killed on October 5, 1918, at Vouziers in the Ardennes region, where he was buried. The French commemorated the aviation pioneer in 1928 by naming their new international tennis stadium in Paris the Roland Garros Stadium.

Ellen Elghobashi

Bibliography

Bowen, Robert Sidney. *They Flew to Glory: The Story of the Lafayette Flying Corps*. New York: Lothrop, Lee & Shepard, 1965. The story of the American pilots who volunteered to fly for the French Air Corps in World War I; the first chapter summarizes the development of the airplane's role in combat.

Clark, Alan. *Aces High*. New York: G. P. Putnam's Sons, 1973. A detailed study of the aircraft and ace pilots of World War I, with excellent period photographs.

Franks, Norman L. R., and Frank W. Bailey. *Over the Front*. London: Grub Street, 1992. A complete listing of the ace pilots of the U.S. and French air services during World War I.

Robertson, Bruce, ed. *Air Aces of the 1914-1918 War.* Letchworth, England: Harleyford, 1959. Discusses the important aces from Britain, America, Italy, Belgium, France, Germany, Russia, and Austro-Hungary, with numerous detailed appendices.

See also: Airplanes; Fighter pilots; Military flight; Record flights; World War I

Gemini Program

Date: From December 7, 1961, to November 15, 1966

Definition: America's crewed spaceflight program that placed humans into Earth orbit.

Significance: The Gemini Program placed humans into Earth orbit and taught astronauts how to track, maneuver, and control orbiting spacecraft; dock with other orbiting vehicles; and reenter Earth's atmosphere and land at specified locations, all necessary to the execution of an Apollo mission.

Evolution of the Gemini Program

Prior to the formation of the National Aeronautics and Space Administration (NASA), a number of crewed space concepts had been investigated within the military. After the Soviet Union launched the world's first artificial satellite, Sputnik 1, on October 4, 1957, the U.S. reaction could easily be described as one of panic, with fear centering on the suspicion that the Soviet Union would assume technological leadership over the free world.

The United States' first attempt to send a satellite into space failed miserably. Vanguard 1 blew up on the launch pad in December, 1957, before the eyes of the world. On January 31, 1958, an Army-based group including Wernher von Braun successfully placed Explorer 1 into orbit. Within months, President Dwight D. Eisenhower, with Congressional approval, created NASA as a civilian space agency. Its first major endeavor was the Mercury project, the goal of which was to send an astronaut into orbit before the Russians. However, the Soviets scored another major first when, on April 12, 1961, they launched cosmonaut Yuri Gagarin into space. He completed one Earth orbit before safely returning to Earth, landing within the Soviet Union. The Soviets led the space race thanks to their proficiency with heavy-lift boosters, a strength they continued to exploit for many years. This strength played a major role in Gagarin's achieving orbit before NASA's Mercury astronauts.

On May 5, 1961, Alan Shepard became the first American to enter space. Shepard launched atop a Redstone rocket, which did not have sufficient thrust to lift his Mercury capsule into orbit. Shepard flew a fifteen-minute-long suborbital profile, arcing up to 115 miles altitude and splashing down in the Atlantic Ocean off the coast of Cape Canaveral. Just three weeks later, President John F. Kennedy committed NASA to sending a man to the Moon and back before the year 1970.

To achieve this goal, NASA initiated the Apollo Program. However, the proposed three-man Apollo vehicle was far too big a step over the existing primitive single-astronaut Mercury spacecraft that could orbit the earth for only a brief period. Further, the Mercury spacecraft was incapable of orbital maneuvers of the type necessary for achieving Apollo's goal using what was termed a lunar orbit rendezvous (LOR) technique. LOR involved having the Apollo spacecraft separate into two portions, one that remained in lunar orbit and another that took two astronauts wearing protective pressure suits down to the lunar surface. This separation necessitated a rendezvous after lunar exploration, so that all three astronauts could reunite for the journey back to Earth. Apollo missions would last between eight and fourteen days and would require very precise reentry maneuvers in order to bring the crew safely through a narrow corridor in the earth's atmosphere where the spacecraft would survive reentry heating.

The Gemini Program was therefore developed as an interim means whereby all of the techniques necessary for Apollo missions could be assessed and refined in low-Earth orbit. Gemini astronauts would build up experience with orbital maneuvering, living for prolonged periods of time in weightlessness, rendezvous and docking separately launched target vehicles, and controlled reentries involving splashing down near recovery forces. Whereas the Mercury spacecraft carried one astronaut and the Apollo spacecraft would carry three, Gemini was designed to build on the Mercury experience with a modular spacecraft capable of flying two astronauts for up to two weeks in duration.

Astronaut Gordon Cooper is recovered after the splashdown of the Gemini V space capsule on August 29, 1965. (NASA)

Gemini Flight Operations

The first two Gemini missions were uncrewed test flights to rate the Titan II launch vehicle and to qualify critical Gemini spacecraft systems. Gemini 1 was launched successfully from Cape Kennedy on April 8, 1964. A Gemini spacecraft mock-up was placed in orbit as a result of proper Titan II booster performance. The spacecraft was meant neither to be separated from the booster's second stage nor returned to Earth.

Gemini 2, which included the first fully operational test spacecraft, was launched on January 19, 1965. It followed a suborbital profile designed to stress the Gemini spacecraft heat shield's ability to manage reentry heating. The spacecraft was safely recovered from the Atlantic Ocean.

Gemini 3 launched on March 23, 1965, with astronauts Virgil "Gus" Grissom and John W. Young aboard. Over the course of three orbits, the astronauts performed three different maneuvers to change the spacecraft's orbit, the first time that orbital maneuvers were executed on a crewed spacecraft. Russian cosmonauts had flown aboard Vostok and Voskhod capsules that had been largely automated in nature.

Just prior to the Gemini 3 mission, the Russians had scored a major advance when cosmonaut Alexei Leonov departed his Voskhod 2 spacecraft for a brief walk in space. The next Gemini mission, Gemini IV, attempted the first extravehicular activity (EVA) performed from a NASA spacecraft. Gemini IV launched on June 3, 1965, with astronauts James A. McDivitt and Edward White aboard. McDivitt attempted to fly close to the spent Titan II booster's second stage after spacecraft separation, consuming a great deal of fuel in the process. Because the mission was meant to last four days, a NASA first, it was decided to halt that maneuver to save fuel. On the mission's third orbit, White opened his hatch and proceeded to exit from the spacecraft using a small gas-powered thruster gun to move about while remaining tethered to Gemini IV by a life-support umbilical. White quickly depleted his gas supply, but he spent a total of twenty-three minutes floating about before returning to Gemini IV's cabin. After four days, McDivitt flew a manually controlled rolling reentry, and Gemini IV splashed down within range of recovery forces in the Atlantic Ocean. Gemini IV began NASA's evolutionary buildup toward a two-week-long mission, and its astronauts spent as much total time in space as had that astronauts of all previous NASA crewed flights combined.

Gemini V launched on August 21, 1965, with astronauts L. Gordon Cooper and Charles "Pete" Conrad on board. This mission marked the first use of fuel cells utilizing liquid oxygen and hydrogen to produce electrical power. Problems with systems associated with the fuel cells surfaced early in flight, forcing the cancellation of a rendezvous exercise and a powering-down of the spacecraft. Those pressure problems diminished later in the flight, and Cooper and Conrad were able to remain aloft in Gemini V for nearly eight full days before returning to Earth, splashing down in the Atlantic Ocean with no major physiological problems encountered during their record-setting flight.

The next Gemini mission was a planned rendezvous and docking with an uncrewed Atlas-Agena docking target. The original flight plan called for the liftoff of the Agena docking target on top of an Atlas rocket about an hour before that of the Gemini VI spacecraft. Once the Agena had reached the correct orbit, Gemini VI would be launched. Over a period of time, Gemini VI would catch, rendezvous, and join or dock with the Agena. Both of these vehicles would be launched from Cape Kennedy, Florida, from separate launch pads.

The first launch attempt of Gemini VI was made on October 22, 1965. Unfortunately, the Agena suffered a failure shortly after it separated from the Atlas booster and was lost. Astronauts Walter Schirra and Thomas Stafford were already in their spacecraft, but, with no Agena target in space, Gemini VI was scrubbed in favor of proceeding with Gemini VII, a two-week-long flight of astronauts Frank Borman and Jim Lovell in December, 1965. Gemini Program managers decided to alter Gemini VI's mission, renaming it Gemini VI-A, and to use Gemini VII as a target with which to rendezvous if Cape Kennedy personnel could refurbish the launchpad sufficiently quickly following Gemini VII's launch to permit a second Gemini liftoff within a two-week period.

Gemini VII was launched on December 4, 1965, and entered stable orbit. An attempt to launch Gemini VI-A on December 12 resulted in an engine shutdown on the pad, but the crew was safe. Then, on December 15 with the booster refurbished, Gemini VI-A lifted off and began a four-orbit chase, closing to within a foot of Gemini VII's nose. Even without docking, this dual flight verified the capability of astronauts to execute the maneuvers needed for Apollo. Also, the Gemini VII crew proved that astronauts could survive weightlessness during the longest Apollo flights. Gemini VI-A returned to Earth, splashing down in the Atlantic Ocean on December 16. Gemini VII executed its reentry on December 18, landing close to the same recovery vessel that had recovered Gemini VI-A.

Gemini VIII launched on March 16, 1966, with astronauts Neil Armstrong and David Scott on board. Their liftoff came one orbit after an Atlas booster delivered an Agena to orbit. Armstrong and Scott executed a rendezvous over the course of four orbits and docked to their Agena vehicle. Within one half-hour, Gemini VIII and its Agena entered a rolling motion that threatened structural stability. Armstrong undocked and backed away from the Agena. As the problem involved a Gemini VIII thruster firing uncontrollably, the roll rate increased, forcing Armstrong to regain control by firing other thrusters dedicated for reentry. Gemini VIII had to be terminated early, and Armstrong and Scott splashed down in a backup recovery zone in the Pacific Ocean after only seven orbits.

After Gemini IX's Agena target, launched on May 17, 1966, failed to reach orbit, the mission was postponed. An alternate target called an Augmented Target Docking Adapter (ATDA) was launched on June 1, 1966, but the crewed Gemini flight, now renamed IX-A, could not follow. Two days later, astronauts Stafford and Eugene Cernan launched from Pad 19. When Gemini IX-A approached the ATDA after a three-orbit rendezvous, the astronauts found the ATDA's forward shroud had not cleanly separated from the docking mechanism. They used the ATDA for several different rendezvous exercises, but no docking was possible. Cernan attempted a spacewalk meant to last one full orbit, but he ran into difficulties working on a jet backpack he intended to test-fly up to 100 feet away from Gemini IX-A. His visor fogged over, and he had to terminate the EVA and return to the cabin. Gemini IX-A splashed down in the Atlantic Ocean after 45 orbits.

Gemini X included a pair of rendezvous exercises involving the Agena VIII target vehicle and an Agena launched one orbit before astronauts John W. Young and Michael Collins. Gemini X and Agena X were launched on July 18, 1966. After docking Gemini X and Agena X together, a rocket firing of Agena X's main engine propelled the docked complex toward rendezvous with Agena VIII. Collins performed a tethered EVA in the proximity of Agena VIII and also performed another spacewalk while standing up in his seat to perform astronomical observations. This marked the first time that all major objectives of the Gemini Program were demonstrated in one single mission. Gemini X splashed down in the Atlantic Ocean on July 21.

Gemini XI and its Agena were launched on September 9, 1966. Astronauts Conrad and Richard F. Gordon completed a rendezvous on their first orbit and docked to Agena XI. Gordon performed a stand-up spacewalk and an umbilical EVA, the latter requiring early termination, after Gordon overstressed his life-support chest pack's ability to keep him cool. Using Agena XI's propulsion system, Conrad and Gordon were able to temporarily boost their spacecraft up to a record 850-mile altitude. They splashed down in the Atlantic Ocean after executing the first computer-controlled reentry.

Gemini XII and its Agena were launched on November 11, 1966. Astronauts Lovell and Edwin "Buzz" Aldrin performed several rendezvous and docking exercises, expanding NASA's experience base. Perhaps the most important aspect of the final Gemini mission involved Aldrin's three periods of spacewalking, two of the stand-up variety and one umbilical. He spent a total of almost five and one-half hours outside the spacecraft and demonstrated methods that overcame problems encountered by earlier Gemini spacewalkers. Gemini XII landed on November 15 in the Atlantic Ocean.

Historical Context

From March, 1965, to November, 1966, Gemini astronauts flew ten crewed missions, greatly expanding NASA's crewed space flight experience beyond that of the original seven Mercury astronauts. During that period, not a single Russian cosmonaut flew in space, and NASA finally overcame the early Soviet lead in space technology. Gemini flights investigated virtually all aspects of an Apollo mission and laid the foundation for the successful achievement of a crewed lunar landing in July, 1969.

David G. Fisher

Bibliography

Collins, Michael. *Liftoff.* New York: Grove Press, 1988. Provides an astronaut's perspective of the Gemini and Apollo Programs.

Hacker, Barton C., and James M. Grimwood. *On the Shoulders of Titans: A History of Project Gemini.* Washington, D.C.: Government Printing Office, 1977. Provides a thorough historical chronicle of program engineering and management evolution.

Schirra, Walter M., Jr., with Richard N. Billings. *Schirra's Space.* Boston: Quinlan Press, 1988. Provides an astronaut's perspective of NASA's crewed space flight programs from Mercury through Apollo.

See also: Aerospace industry, U.S.; Apollo Program; Neil Armstrong; Astronauts and cosmonauts; Mercury project; National Advisory Committee for Aeronautics; National Aeronautics and Space Administration; Russian space program; Alan Shepard; Spaceflight

John Glenn

Date: Born on July 18, 1921, in Cambridge, Ohio

Definition: The first U.S. astronaut to orbit the earth (1962) and the world's oldest astronaut (1998).

Significance: Glenn is a symbol of the evolution of the American space program. His first space mission restored American pride in the space race with the Soviet Union, while his last space mission demonstrated that the elderly can make important contributions to society.

Early Life

Reared in New Concord, Ohio, John Herschel Glenn, Jr., developed a great love and respect for his parents, who taught him that he had unlimited possibilities, and that, with hard work, he could achieve whatever goals he set for himself. His mother, an elementary school teacher, taught Glenn to love reading and learning. When Glenn was eight years old, he accompanied his father, a plumbing contractor, on a job to Cambridge, Ohio. During this trip, Glenn's father arranged for his son's first flight on an airplane, after which Glenn was hooked on flying. Model airplanes became his favorite hobby, and he dreamed of someday becoming a pilot.

In high school, Glenn participated in football, basketball, and tennis; played the trumpet in orchestra; and served as a school newspaper reporter and student body officer. After graduating from high school in 1939, he enrolled in Muskingum College to study chemical engineering. He also entered a civilian pilot training program and earned his flying license in 1941. Upon the U.S. entry into World War II (1939-1945), Glenn decided it was his patriotic duty to enlist for naval aviation training. After graduation, Glenn received a commission in the U.S. Marine Corps Reserve. By March, 1943, he had earned his wings and was promoted to a Marine second lieutenant. He married his childhood sweetheart, Annie Castor, on April 6, 1943.

War Experience

Assigned to Marine Fighter Squadron 155, Glenn spent a year flying F-4U Corsair fighters on a variety of bombing and reconnaissance missions against Japanese garrisons in the Marshall Islands. He flew fifty-nine combat missions and was hit by enemy fire five different times. After returning to the United States, his principal duties were as a flight instructor. He was promoted to the rank of captain in July, 1945. In December, 1946, he was assigned as a member of Marine Fighter Squadron 218 to patrol North China in support of General George C. Marshall's World War II peace terms. From June, 1948, until December, 1950, he served as an instructor in advanced flight training in Corpus Christi, Texas.

During the Korean War (1950-1953), Glenn flew jets in ground-support missions for the Marines and in air-to-air combat as an exchange pilot in the new Air Force F-86 Sabre jets, completing a total of ninety missions between February and September, 1953. Glenn had many close calls that often caused him to return to base with a seemingly unflyable aircraft. In the last nine days of fighting in Korea, Glenn downed three Soviet-built MiG-15's in fierce combat along the Yalu River.

For his military service during World War II and the Korean War, Glenn received four Distinguished Flying Crosses and eighteen Air Medals. He rose steadily through the ranks, becoming a captain in 1945, a major in 1952, and a colonel in 1959. In 1954, he was assigned to the Navy's test pilot school in Patuxent River, Maryland. Upon graduation, he served as a project officer on a number of aircraft. On July 16, 1957, he set a record for the first coast-to-coast, nonstop, supersonic flight in an F-8U Crusader jet fighter, flying from Los Angeles to New York in three hours and twenty-three minutes. For this event, Glenn received his fifth Distinguished Flying Cross.

Space Flight and Politics

Spurred by the success of the Russian satellite Sputnik, the United States established Project Mercury in 1958. Glenn was named as one of the seven Mercury astronauts in April, 1959. Motivated by his deep religious faith, hard work ethic, and tenacious devotion to duty, he helped win the widespread public support that the space program needed.

Glenn was selected to serve as backup pilot for the suborbital flights of Alan Shepard and Virgil "Gus" Grissom in 1961. He was then chosen as the first American to orbit the earth, orbiting three times in the *Friendship 7* on February 20, 1962. The mission restored American pride in the space race with the Soviet Union.

After convalescing from a severe inner-ear injury caused by a fall in February, 1964, Glenn retired from the Marines in January, 1965. He was elected to four consecutive terms as a U.S. senator from Ohio, beginning in 1974. He made an unsuccessful bid for the Democratic presidential nomination in 1984. On the thirty-fifth anniversary of his historic flight (February 20, 1997), Glenn announced that he would retire from the Senate at the end of his fourth term in 1998.

The Oldest Astronaut

While Glenn sought additional funding for the National Aeronautics and Space Administration (NASA) in 1995, he reviewed some documents on the physical changes that happen to astronauts in orbit. He was amazed at the similarities between the effects of zero gravity on the body and the natural aging process on Earth. Consequently, he began petitioning NASA for the opportunity to go back into space and study the effects of weightlessness on older Americans. After much perseverance, on January 15, 1998, he was granted his wish of going back into space.

After a thirty-seven-year hiatus from space flight, Glenn spent months of training, experimenting, baseline medical tests to become the oldest person to travel into space. As a member of the nine-day space shuttle *Discovery* mission from October 29 to November 7, 1998, the seventy-seven-year-old Glenn conducted numerous experiments that focused on osteoporosis and the immune system's adjustments to the aging process. Glenn's contributions demonstrated that the elderly can still make important contributions to society. Glenn stands out as a symbol of courage, honor, and lifelong devotion and service to his family and his country.

Alvin K. Benson

Bibliography

Bredeson, Carmen. *John Glenn Returns to Orbit: Life on the Space Shuttle*. Berkeley Heights, N.J.: Enslow, 2000. Excellent account of the details associated with Glenn's last space mission.

Glenn, John, and Nick Taylor. *John Glenn: A Memoir*. New York: Bantam, 1999. Glenn's account of his life and career, from astronaut to U.S. senator and back to astronaut.

Streissguth, Thomas. *John Glenn*. Minneapolis, Minn.: Lerner, 1999. A detailed biography.

Vogt, Gregory L. *John Glenn's Return to Space*. Brookfield, Conn.: Millbrook Press, 2000. An inspiring story of Glenn's accomplishments as an astronaut, with behind-the-scenes details of the space race, including the challenges and technological developments.

See also: Apollo Program; Astronauts and cosmonauts; Korean War; Marine pilots, U.S.; Mercury project; Model airplanes; National Aeronautics and Space Administration; Alan Shepard; Space shuttle; Spaceflight; Supersonic aircraft; Test pilots; World War II

Gliders

Definition: Any one of a number of types of winged, heavier-than-air craft, having no motive power other than gravity. Sailplanes are a specific type of glider than can ascend as well as descend.

Significance: Gliders were the predecessors to motorized flight. Information gained from experiments with gliders made motorized flight possible.

A glider is launched from a raised elevation and is capable only of forward movement through air while at the same time losing altitude. The relation between forward momentum and loss of altitude is a glider's sink rate, and gliding is the motion of the craft's controlled descent. The history of glider development is essentially the process of experimentation to minimize a glider's sink rate, while giving the glider pilot increasing control over the movement or flight of the glider while airborne. Eventually, after centuries of experimentation, aviation technology developed to the point where gliders could be constructed and flown in ways that permitted the glider pilot to slow and even reverse the rate of descent. The process of flying a glider using the energy from thermal air currents to regain altitude lost by the downward force of gravity is called soaring. The type of glider capable of being flown in such a way is termed a sailplane. It is basically a high-performance glider designed specifically for soaring. Post-World-War-II gliders are more correctly called sailplanes to distinguish them from earlier gliders, regardless of size and precise configuration, that were not capable of regaining lost altitude in a controlled manner after they had been launched.

Earliest History

The process of experimentation with gliders that led to modern sailplanes took place over the course of centuries. As long as humans have watched birds in flight, humans have wanted to imitate them. Many of the earliest attempts at human flight are thinly documented or are mythological. One of the earliest stories of human flight is the account of Daedalus and his son Icarus. As related in *Metamorphoses* (c. 8 C.E.; Eng. trans. 1567), a collection of tales by the Roman writer Ovid, Daedalus was imprisoned by the Cretan king Minos. While watching sea gulls in flight, Daedalus got the idea to fashion wings from discarded gull feathers held together with candle wax. Using these birdlike wings, Daedalus and Icarus escaped. Daedalus wisely kept to a course midway between earth and heaven, but Icarus flew too close to the sun. The wax holding his wings together melted and he plummeted to his death. This cautionary tale of Daedalus and Icarus set the stage for much later thinking about human flight. Most people were of the opinion that humans had no business trying to fly, but there was a small group of adventurers and inventors who disregarded this opinion.

Medieval Attempts at Flight

There are numerous undocumented passages in medieval historical sources stating that humans achieved flight aboard or attached to gigantic kites, perhaps similar to present-day hang gliders. The Italian mathematician Giovanni Danti is reported to have tried to fly over Lake Trasimeno in Italy in the late 1500's. John Damian, another Italian, reportedly constructed a pair of wings and jumped off the wall of a castle belonging to King James IV of Scotland. He plummeted to Earth, breaking his leg. Leonardo da Vinci, a fifteenth century Italian artist, scientist, and inventor, seriously examined the possibility of human flight. Using comparative zoology and architectural and mathematical studies, da Vinci concluded that humans were too heavy to be kept aloft by feathered wings modeled on the wings of birds. Da Vinci thought that batlike wings in which the skin is stretched over a lightweight skeleton was more likely to sustain the weight of a human in flight. Da Vinci also designed rudimentary parachutes and a type of ornithopter or bird-imitating flapping machine that is considered an early prototype to the modern helicopter. Although da Vinci's flying inventions are theoretically possible, it was almost three hundred years before they were actually built, tested, modified, and put into practice.

Nineteenth Century

The Englishman Sir George Cayley systematically examined the problems associated with human flight. In 1809, he published the results of his experiments with small, uncrewed glider models, each of them with V-shaped wings and a tail stabilizer. Using his horses to supply the forward momentum, Cayley performed a brief, barely controlled glider flight in 1853. William Henson tried to develop Cayley's experiments further by adding a steam-powered motor to the air craft. Such an engine made the aircraft far too heavy to get off the ground, but Henson improved Cayley's glider designs, eventually designing a fixed, single-wing glider with a bird-tail-shaped tail, a rudder, and landing gear. Henson's friend John Stringfellow built a small model glider with a small steam engine that could fly under specific circumstances, but he did not build a model big enough to carry the weight of a human. F. H. Wenham, another Englishman, also studied birds to investigate possibilities for human flight. Wenham concluded that a slightly arched wing set at an angle, rather than a flat wing surface, could lift more weight. He also thought that a connected series of shorter, arched wings rather than one set of long, flat wings might sustain a person in flight, if only a means could be found to lift the craft off the ground initially.

Frenchmen Jean-Marie Le Bris and Felix Du Temple both built uncrewed, motorless gliders. Le Bris fashioned his glider in the shape of an albatross and Du Temple constructed the first propeller-driven aircraft to lift off from the ground under its own power. Neither craft could stay aloft for more than a few seconds nor could their flight path be controlled. In the late nineteenth century, the German Otto Lilienthal built numerous single-winged gliders, each with a fixed tail for stability. The pilot stood in the center of the glider with the glider frame attached around his waist. By making over two thousand flights off a small hill, Lilienthal learned how to move his weight to steer the glider. His longest flight was approximately 200 feet. On August 9, 1896, Lilienthal attached a small motor to his glider and launched himself off the hill. The wind shifted and he crashed, suffering fatal injuries. Percy Pilcher, a Scotsman who had known Lilienthal, modified his own triplane glider based on Lilienthal's experiments. Pilcher conducted numerous glider flights, one as long as 750 feet, before being killed in a glider accident in September, 1899. A naturalized American, Octave Chanute, was also influenced by Lilienthal's experiments. He designed numerous gliders and tested them on the beach at Lake Michigan near Chicago, Illinois. He had two-, three-, and five-winged models with rear stabilizers, each controlled in flight by shifts in the pilot's weight. Chanute kept careful records of his experiments with equilibrium while aloft, information he shared with the Wright brothers.

Early Twentieth Century

Wilbur Wright and his brother Orville Wright grew up primarily in Dayton, Ohio. They initially made their living repairing bicycles while pursuing aeronautical experiments as a hobby. Beginning with a series of kites, the Wright brothers developed a system of wing-warping that greatly increased the pilot's ability to control the flight of an aircraft. The Wright brothers spent part of each year from 1900 to 1905 at Kitty Hawk, North Carolina, testing gliders they had designed in Dayton. The 1900 glider weighed 52 pounds and had 18- by 5-foot wings. It was not substantial enough to lift a pilot in a controlled flight. In 1901, the glider was much bigger, having 22- by 7-foot wings. The longest piloted flight, by Wilbur, was 400 feet. During the winter of 1901, the Wright brothers reworked information from Chanute and Lilienthal in order to solve problems with both lift and control. Using this new information, the 1902 biplane glider weighed 116 pounds and had a 32-foot wingspan. It incorporated various design changes to provide more lift, including a forward monoplane elevator, as well as a fixed rudder linked to the wing-warping or shaping system that allowed the pilot to control the glider's flight. The longest flight of the 1902 testing session was 622 feet, lasting 26 seconds. The original patent issued to the Wright brothers covered the modifications included in the 1902 glider design. Returning to design experiments, the Wrights constructed a glider that could carry a 12-horsepower engine and have two propellers. On December 17, 1903, Wilbur Wright flew 852 feet, staying aloft for 59 seconds, the first documented pilot-controlled motorized flight in history. The Wrights continued to refine their aircraft designs in 1904 and 1905, gradually increasing both the length of and control over motorized flights. In 1911, Orville returned briefly to gliders, setting a glider flight record of 9 minutes, 45 seconds.

World War II

Once motorized flight had been demonstrated, gliders seemed rather primitive. All the major powers in World War I used motorized airplanes, not gliders. After the end of World War I, however, attention returned to gliders. The Treaty of Versailles ending the war prohibited the Germans from building new planes with engines. The treaty did not mention the building of motorless gliders. Thus, throughout the 1920's and early 1930's, thousands of young German men learned to fly as glider pilots. They formed the core of the Nazi Luftwaffe in World War II. In 1930, the three Schweizer brothers—Bill, Paul, and Ernest—began to build gliders for sale to enthusiasts in the United States. In 1932, the Soaring Society of America was founded to regulate the small but growing hobby in America. *Soaring Magazine*, still in publication, debuted in 1937.

The Germans were the first to recognize the potential military applications of gliders. The first military glider capable of carrying troops and equipment was the DFS-230. On May 11, 1940, ten DFS-230's carrying seventy-eight glider troops attacked and captured Eben Emael in Belgium, due in large measure to the element of surprise. Other countries quickly took notice. The United States produced thousands of small TG-2 and TG-3 gliders, as well as jumbo gliders such as the Laister-Kauffman CG-10A Trojan. The British also built large numbers of various types of gliders to use in aerial observation, as well as in troop and equipment transport.

The idea of parachute troops or airborne infantry was in its infancy in early World War II. Rather than trying to coordinate hundreds of individual soldiers in parachute drops, the conventional wisdom of the time thought it made more sense to airlift troops in platoons in gliders. Unfortunately, glider pilots and troops suffered very high casualty rates, in excess of 50 percent. The Germans tried an unsuccessful glider assault on Crete. Many gliders were blown off course, some crashed, some landed intact but far from the designated landing zone. On July 9, 1943, the Allies tried a joint American-British glider assault on Sicily. Of the 144 troop gliders involved in the assault, 69 landed in the ocean rather than on land, 10 were apparently shot down, only 12 were able to land intact, and only 4 of those landed within the designated landing zone. The Allies also tried glider assaults in Burma, with similar disastrous results.

Post-World War II

After Word War II, many military pilots turned to gliding as a recreational pursuit. Inexpensive military surplus gliders were readily available. By the mid-1950's, there was a large enough recreational market to spur further refinements in glider design. Invented in 1928, the variometer, a piece of equipment that allowed the pilot to measure even small differences in altitude, became standard on every glider, which became technically sailplanes, able to both ascend and descend. National and international championships are held annually for different design classifications of sailplanes, with various contests for speed, altitude, duration of flight, distance covered and accuracy in landing at a designated spot. All rules and standards concerning sailplane construction and classification, as well as sailplane pilot training requirements in the United States, are regulated by the Federal Aviation Administration.

Victoria Erhart

Bibliography

Editors of *Flying Magazine. America's Soaring Book.* New York: Charles Scribner's Sons, 1974. Covers the modern sport of soaring.

Gannon, Robert. *Half Mile Up Without an Engine.* Englewood Cliffs, N.J.: Prentice-Hall, 1982. Particularly informative on the Wright brothers and later developments.

Josephy, Alvin M., Jr., ed. *The American Heritage History of Flight.* New York: American Heritage, 1962. A comprehensive source for all periods of the history of aviation.

See also: Aerodynamics; Sir George Cayley; Octave Chanute; Forces of flight; Hang gliding and parasailing; Heavier-than-air craft; Otto Lilienthal; Luftwaffe; Military flight; World War II; Wright brothers

Robert H. Goddard

Dates: Born October 5, 1882, in Worcester, Massachusetts; died August 10, 1945, in Baltimore, Maryland
Definition: Inventor of rocket components.
Significance: Goddard pioneered rocket technology in the early twentieth century.

Robert Hutchings Goddard was born on October 5, 1882, in Worcester, a middle-class suburb of Boston, Massachusetts. He graduated from Worcester Polytechnic Institute in 1908, and continued his graduate education at nearby Clark University, where he earned his M.A. in 1910 and his Ph.D. in 1911.

By 1914, Goddard had applied for and won two patents; one for a liquid fuel rocket, the second for a multistage rocket. These awards gave him a standing in the scientific community and eventually led to some financial support from the Smithsonian Institution.

His report *A Method of Reaching Extreme Altitudes* was published by the Smithsonian in early 1920. The publi-

Robert Goddard conducted rocket experiments in his own backyard, gaining a reputation as something of a mad scientist, until his theories were proved viable and he was recognized as one of the founders of modern spaceflight. (NASA)

cation drew attention beyond scientific circles due to Goddard's suggestion that jet propulsion could be the technology to achieve escape velocity and fly to the Moon. He was mocked for this idea and became something of an embarrassment to his family. This and other proposals earned him the nickname of "Moon Man" in the popular press.

His backyard experiments became well known and feared. However, before pressure from the authorities, the public, and his family prevailed on him to relocate, on March 16, 1926, he successfully launched the first liquid fuel rocket at Auburn, Massachusetts. It traveled 184 feet in 2.5 seconds. In the same year, Fritz Lang produced the motion picture *The Woman in the Moon*, based on the efforts of Goddard and a German rocket hobbyist. Interestingly, the film foreshadowed the German V-2 series. This movie was seen by virtually all scientists who became the backbone of the German and later the U.S. space program.

In 1929, Goddard and his small team of assistants moved their testing operations to Roswell, New Mexico. Here, isolated from the world and after many failures, he achieved success in 1930 when he and his team fired an 11-foot, liquid-fueled rocket to an altitude of 2,000 feet at a velocity of over 500 miles per hour. In 1932, a critical step came with a gyroscopically controlled rocket. Goddard's list of rocket achievements is impressive, including: components of a ramjet engine via a rocket fuel pump, regenerative cooling of combustion chambers, instrument payloads and recovery systems, guidance vanes, and gimbaled engines.

By March, 1935, Goddard was launching gyroscopically controlled rockets to 7,800 feet at over 700 miles per hour. His paper "Liquid Propellant Rocket Development," a primer on rocket technology, was published by the Smithsonian in 1936. With the threat of war looming in Europe, Lieutenant John Sessums visited Goddard in New Mexico to assess the military value of Goddard's rocket work. He concluded that it was of little or no value. In 1940, Goddard offered his research, patents, and facilities for use by the military, an offer that was ignored.

Goddard died at the end of the war on August 10, 1945. He survived long enough to see his dream realized in the German V-2 effort. The technical foundation he established was impressive, and he registered more than two hundred patents. On May 1, 1959, he was honored when the National Aeronautical and Space Administration named the space center at Greenbelt, Maryland, for him.

Richard C. Jones

Bibliography

Braun, Wernher von, and Frederick I. Ordway, III. *History of Rocketry and Space Travel*. 4th rev. ed. New York: Harper & Row, 1985. An excellent overview of the subject, aimed at the interested reader or amateur, with good coverage of the early days of the rocket clubs in the United States and Germany.

Coil, Suzanne M. *Robert Hutchings Goddard: Pioneer of Rocketry and Space Flight*. New York: Facts on File, 1992. A biography aimed at young adult readers.

Goddard, Esther C., and Edward C. Pendray, eds. *The Papers of Robert H. Goddard.* 3 vols. New York: Dover, 1970. A collection of Goddard's scientific papers.

Lehman, Milton. *This High Man: The Life of Robert H. Goddard.* New York: Farrar, Straus, 1963. An authorized biography written for the general reader, showing the human side of rocket technology and giving insights into Goddard's life.

See also: Engine designs; Propulsion; Ramjets; Rocket propulsion; Rockets; National Aeronautics and Space Administration

Goodyear blimp

Also known as: Lighter-than-air craft

Date: First flight on May 24, 1917

Definition: A lighter-than-air craft that achieves lift by filling the balloon-type structure with 6 million cubic feet of helium.

Significance: Early airships provided the U.S. Navy with reconnaissance and aircraft carrier capabilities while modern blimps are used primarily for television coverage and commercial advertising. Capable of staying afloat for up to eleven days at a time, the Goodyear blimps could fly from the United States to Europe and back again without refueling.

History

Goodyear established an Aeronautics Department in 1910 for the purpose of manufacturing and marketing rubber-impregnated fabrics and coatings for airplanes. By 1912, the company had constructed its first balloon, and in 1916 it purchased 720 acres of land southeast of Akron, Ohio, for the construction of the Wingfoot Lake Airship Base. After the United States entered World War I in 1917, the U.S. Navy ordered sixteen B-type airships, nine of which Goodyear manufactured. The first airship flew on May 24, 1917. The following year the Navy placed an order for ten C-type airships. The Navy took over the Wingfoot Lake facility from 1918 to 1921 for the training of pilots and further experiments and testing of the aircraft.

Between the wars, Goodyear built two airships, the *Wingfoot Express* and the *Pony*, which used hydrogen instead of helium. In 1925, Goodyear entered the commercial market. A single-engine helium-inflated *Pilgrim* preceded the 86,000-cubic-foot, twin-engine, TZ-type blimp. Each new airship resulted in the production of larger envelopes. During the 1930's, the U.S. Navy ordered two giant rigid airships from Goodyear. The aircraft, measuring over 200 yards long and requiring 6.5 million cubic feet of helium to ascend, relied on an internal metal frame to maintain its shape. The USS *Akron* and the USS *Macon*, used as aerial aircraft carriers equipped with small planes that could be deployed and retrieved, operated for two years before being destroyed during storms.

The next generation of blimps, built in the 1940's and 1950's, functioned as surveillance airships along coastal areas after the attack on Pearl Harbor. In the post-World War II period, Goodyear purchased several airships back from the Navy and outfitted them with neon night sign panels equipped with a grid for a running sign. Goodyear continued to manufacture airships for the Navy, constructing the largest nonrigid airship in 1960. Modifications to the airships since 1960 include major car and power plant changes and a 147,300-cubic-foot envelope. In 1966, the Goodyear blimp at the Indy 500 auto race included four running night signs in color. In 1972, the company moved

Goodyear Blimp Technical Statistics

Stars and Stripes and *Spirit of Goodyear*

Length: 192 feet
Width: 50 feet
Height: 59.5 feet
Volume: 202,700 cubic feet
Maximum Gross Weight: 12,840 pounds
Maximum Speed: 50 miles per hour
Cruise Speed: 30 miles per hour
Power Plant: Two 210-horsepower fuel-injected, air-cooled aircraft piston engines
Propellers: Two-blade constant speed, 78 inches
Passengers: 5 plus pilot
Operating Altitude: 1,000-3,000 feet, 10,000 feet maximum
Car: Aluminum and welded steel tube
Maximum Car Length: 22.75 feet
Fins, Rudders, and Elevators: Polyester fabric over aluminum and welded steel
Empennage: "+" configuration
Envelope: Neoprene-impregnated polyester fabric, two-ply
Night Sign Lights: Over 165,000 LEDs with over 256 colors
Landing Gear: Fixed

its facility from Akron to Spring, Texas. In 1986, additional alterations to the blimp included the use of twin vectorable turbine engines with ducted propellers, "X" fins, and a 247,800-cubic-foot envelope. Between 1917 and 1996, Goodyear produced 347 airships. By 2000, Goodyear maintained five airships worldwide: the *Eagle* in the City of Carson, California; the *Spirit of Goodyear* in Akron, Ohio; the *Stars and Stripes* in Pompano Beach, Florida; the *Spirit of Europe*; and the *Spirit of the Americas*, based in Saõ Paulo, Brazil.

Goodyear Blimp Technical Statistics

Spirit of Europe and *Spirit of the Americas*

Length: 130 feet
Width: 34 feet
Height: 44 feet
Volume: 70,000 cubic feet
Gondola Length: 14 feet
Gondola Width: 5 feet
Seating: 5 maximum
Engines: Two 68-horsepower Limbach L-2000
Fuel Capacity: 60 gallons
Maximum Speed: 53 miles per hour
Maximum Flight Capacity: 15 hours
Turning Radius: 750 feet diameter

Construction

Goodyear has manufactured three types of airships. The rigid airship has an internal frame of aluminum alloy that supports the balloon, but the weight of the frame requires the construction of long structures to maintain a proper weight-to-volume ratio. The semirigid airship incorporates a rigid lower keel and a pressurized envelope above. The nonrigid airship, the most advanced of the three types, uses the internal pressure of the gases to maintain the shape of the envelope and has no internal framework.

The anatomy of the blimp includes nose cone battens that stiffen the nose of the airship, helping to distribute the weight when the craft is moored and preventing damage to the nose of the ship. Behind the nose is the forward ballonet, an airbag within the envelope, which releases air through valves during ascent and lets air out through the scoops during descent. The air scoops take air from the props to fill the ballonets when additional air is required. When the airship is not flying and the engines are idle, the air scoops receive air from an electric blower. Four air valves control the release of air from the ballonets. The helium valve, located in the Goodyear logo on the ship, acts as a safety valve for the helium gas within the main envelope. Two inside envelopes, called catenary curtains, each 30 degrees off center, are attached by suspension cables and are sewn into the main envelope. The aft ballonet works in conjunction with the forward ballonet to achieve a nose-up or nose-down position. The ship is controlled by rudders (vertical fins) used for steering and elevators (horizontal fins) used to control the ascent and descent of the craft. Attached underneath the envelope is a car-passenger compartment measuring approximately 23 to 35 feet long and capable of holding a pilot and six passengers.

Propulsion is achieved by one of two different types of engines. Two 6-cylinder, gasoline-powered airplane engines that generate 210 horsepower can reach a top speed of 50 miles per hour. The second system uses two turboprop engines that generate a combined total of 840 horsepower and can reach speeds of 65 miles per hour. The average rate of speed maintained by both types usually averages 30 to 40 miles per hour.

The exterior of the Goodyear blimps are covered by 3,780 light boards with red, blue, and green light-emitting diodes capable of altering intensity to produce a total of 256 colors. Although one pilot operates the aircraft, the ground crew consists of fifteen individuals who work as aircraft mechanics, electronic technicians, or riggers. The ground crew follows the blimp across the country on a bus which functions as a traveling command and control center.

Cynthia Clark Northrup

Bibliography

Payne, Lee. *Lighter Than Air: An Illustrated History of the Airship*. New York: Orion Books, 1991. Payne concentrates on current and future airship technology as well as information about the role that airships have played in commercial and military history.

Sullivan, George E. *Famous Blimps and Airships*. New York: Dodd, Meade, 1988. Easy-to-understand reference work that describes the difference between rigid and nonrigid airships. The author also provides a description, history, and construction information for several different airships, including the Goodyear blimp.

Topping, A. Dale. *When Giants Roamed the Sky: Karl Arnstein and the Rise of Airships from Zeppelin to Goodyear*. Akron, Ohio: University of Akron Press, 2000. Although this work chronicles the life of Karl

Arnstein, the designer of the Zeppelin, the author also provides information concerning the development of the airship industry in the United States during the first few decades of the twentieth century.

See also: Blimps; Buoyant aircraft; Dirigibles; Lighter-than-air craft; Manufacturers; Military flight; Reconnaissance

Gravity

Definition: The force that all objects in the universe exert on all other objects as a result of their mass.

Significance: The origin of weight and the cause of the downward acceleration ("falling") of unsupported objects, gravity must be overcome by lift in order to sustain aerial flight, and must be properly exploited during spaceflight to successfully achieve orbit.

Gravitational Force

Physicists identify four fundamental forces that account for all known physical phenomena: gravity, electromagnetism, the strong nuclear force, and the weak nuclear force. Gravity is the weakest of the four, despite its overwhelming influence in everyday life, and it has a cosmic role in controlling the structure and evolution of the universe. Gravitation is dominant on a cosmic scale because it is long range, extending to infinity. The strong and weak nuclear forces, while much stronger than gravity, are of very short range and confined to the interior of the atomic nucleus. Electromagnetism also extends to infinity, but electric charges, which are the source of electromagnetic forces, come in both positive and negative forms that by and large cancel each other out, leaving only a relatively tiny net effect. Gravity, by contrast, is always attractive, therefore always additive and reinforcing. With a sufficient amount of mass, gravity can be made arbitrarily large. It is only because of the tremendous mass of the earth that gravity becomes the dominant force in everyday life.

Sir Isaac Newton in 1684 recognized that the gravitational force between two widely separated bodies must be proportional to the mass of each and weaken as the square of the distance between them. The gravitational force of the earth on an object is called its weight, and Newton's law states that one object twice the mass of another will weigh twice as much. It is on this basis that the mass of an object can be measured using devices such as balances and scales that actually determine weight. The law also specifies that weight will diminish with distance, so that objects at high altitudes will weigh less than they do at sea level. This loss of weight is real and easily measured with modern instruments. It must be stressed, however, that there is no corresponding loss of mass.

Because gravity extends to infinity, weight never vanishes completely and there is no such thing as true weightlessness. What is typically thought of as "weight" is actually the counterforce of the ground that supports objects and prevents them from falling due to the gravitational force. The "weightlessness" experienced by astronauts in orbit is actually free fall. In the 1580's, Simon Stevin experimentally discovered that all objects fall in a gravitational field at exactly the same rate, a result whose importance was first recognized and widely disseminated by Galileo Galilei in 1638. Astronauts in orbit are continuously falling toward Earth, but the spacecraft enclosing them is falling in exactly the same direction at exactly the same rate. As there is no relative motion, the astronauts float in the cabin as though gravity has gone away.

Orbital Motion

Orbital motion is a combination of free fall with a large velocity at right angles to the direction of fall. Absent the gravitational force, a satellite would move away from Earth in a straight line that would eventually carry it off toward infinity. The gravitational force pulls the satellite from this straight-line motion onto a path that curves around Earth and closes onto itself repetitively. This is an orbit.

Newton's law of gravitation explains Johannes Kepler's three laws of orbital motion: satellites travel in ellipses with the gravitational source (the primary) at one focus of the ellipse; a line joining the primary to the satellite sweeps out equal areas in equal times as the satellite moves around the orbit; and the cube of the average distance from the primary to the satellite is proportional to the square of the orbital period.

Moving objects possess energy of motion called kinetic energy, equal to one-half of their mass multiplied by the square of their velocity. Satellites in orbit are continuously speeding up as they fall toward the primary, thereby gaining kinetic energy, and slowing down as they coast away from it, losing kinetic energy. Since the total amount of energy in a system, kinetic plus potential, can never increase or decrease, the gain or loss of kinetic energy must be balanced by a gain or loss from another source called gravitational potential energy. The gravitational potential energy of two objects mutually attracted by a gravitational force is proportional to the product of their masses divided by the distance between them.

An object in free fall decelerates as it coasts upward, eventually coming to a stop when all of its kinetic energy has been converted to potential energy. It then starts to fall downward, converting potential energy back to kinetic energy and accelerating as it does so. Because the gravitational force weakens with distance, the amount of additional kinetic energy needed to reach ever-greater heights is limited. Objects moving fast enough to have kinetic energies that exceed this limit will never stop rising and will coast away from Earth forever. The velocity associated with this energy is referred to as the local escape velocity. Escape velocity at Earth's surface is approximately 7 miles per second (11 kilometers per second).

The total energy of a satellite determines the size of its orbit, its orbital speed, and its orbital period through Kepler's second and third laws. Satellites in low-Earth orbit travel at slightly less than 5 miles per second (7.7 kilometers per second). The atmosphere at that altitude is extremely thin but still capable of exerting significant drag on objects traveling at such high velocities. Drag is a dissipative force which converts kinetic energy to heat and ordinarily slows objects down, but satellites under the influence of drag drop closer to the earth, converting potential energy to kinetic energy as they do so, and surprisingly end up traveling faster. When the total energy is no longer sufficient to maintain orbit, the satellite reenters the atmosphere.

In order for the satellite to reach the ground at rest, all of its orbital kinetic and potential energy must be converted into heat. Temperatures become so great that the air around the reentering satellite becomes hot enough to glow. In uncontrolled reentry, too much of the heat builds up within the satellite and the satellite vaporizes, a fate common to small meteors. Crewed spacecraft control reentry and survive by discharging the heat overboard.

Gravitational Effects

The gravitational pull of the Moon is felt daily in the rising of the tides. Additionally, as Earth rotates underneath the tides, it pulls the bulge of water from west to east, working against the pull of the Moon. This produces a small tug on the Moon in the direction of its orbital motion and slightly increases the Moon's total energy. As a consequence, the Moon's orbit increases in size a small but measurable amount. The orbital period of the Moon increases as a result, and the month gets slightly longer. Correspondingly, the drag of the tides on the ocean floor slows down Earth, increasing the length of the day.

Although Earth and the Moon appear to be made of hard, rigid rock, each is flexible enough to bend in response to their mutual gravitational attraction. This allows tides to rise in the rock itself. Rock tides on Earth contribute to the braking effect of the ocean tides, but are very small in comparison. Earth's gravity also raises rock tides on the Moon, which have, over billions of years, slowed the Moon's rotation down to the point that the length of the lunar day exactly equals the orbital period: one month. As a consequence, the Moon always keeps one face toward Earth, and humankind is only privileged to see the other side of the Moon through photographs taken from lunar orbit.

This curious circumstance is called tidal locking and it is not at all rare. A majority of the natural moons in the solar system are tidally locked to their parent planet. Tidal locking is the inevitable result of the gravitational interaction of one flexible body orbiting another. When deliberately used by satellite designers to keep one end of an oblong satellite pointed toward Earth, it is referred to as gravity-gradient stabilization. (Space shuttle pilots put the shuttle into gravity-gradient stabilization during sleep periods so that noisy thruster firings to maintain attitude can be avoided.)

General Relativity

Although gravity was the first force to be mathematically described by physicists, it remains the least understood. Stevin's and Galileo's observation that all objects fall at exactly the same rate in response to the gravitational force inspired Albert Einstein in 1915 to go beyond Newton's law of gravitation to propose the theory of general relativity. Based on Einstein's theory of special relativity, which unites space and time into a four-dimensional universe, general relativity describes gravity as the result of localized space-time curvature in this four-dimensional universe.

General relativity predicts that clocks at high altitudes will run faster than identical clocks at low altitudes. This prediction has been verified and this phenomenon had to be included in the design of the Global Positioning System (GPS) in order to achieve required accuracy and precision. Very accurate and stable atomic clocks flown on GPS satellites consistently run faster than identical clocks on the ground.

General relativity also explains the cosmological expansion of the universe and the bizarre properties of black holes. The expansion of the universe was discovered by

Edwin Hubble in 1925 through measurement of the frequency shifts of light emitted by distant galaxies. Almost all proved to be moving radially away from the Milky Way, with a speed of recession proportional to distance away: A galaxy twice as far away as another recedes from Earth twice as fast. This shocking phenomenon proved to be a direct and natural expectation of the general theory of relativity.

Apparently, the universe originated billions of years ago in a big bang that flung matter outward in all directions. Over the course of billions of years, gravity pulled the matter into clumps out of which galaxies, stars, and planets formed. Because the galaxies attract each other gravitationally, the expansion should slow as time goes by. If the universe does not contain enough matter to make the local escape velocity of the galaxies everywhere greater than the current recession velocity, then the expansion will go on forever. If the universe does contain enough matter, then the expansion will eventually slow to a halt and the universe will contract back into a single mass, possibly to explode again and expand into a brand new and different universe.

Black holes are objects whose surface gravity is so strong that in regions inside what is called the escape horizon, the local escape velocity is greater than the speed of light. As a basic tenet of special relativity is that nothing can travel faster than light, nothing that ever falls through the escape horizon can ever get out again. A second consequence is that anything that falls through the escape horizon continues to fall all the way to the center of the black hole where it and all other infalling matter are crushed to zero volume and infinite density. It appears that the laws of physics themselves cease to hold under these conditions.

Gravity and Unified Field Theory

Certain aspects of general relativity have not been reconciled with quantum theory, the branch of physics that explains the behavior of objects at atomic and sub-atomic levels. Physicists have succeeded in uniting the theory of electromagnetism and the theory of the weak nuclear force into one theory of electroweak interactions. They are confident that eventually the theory of electroweak interactions and the theory of the strong nuclear force will be united into a grand unified theory. The ultimate quest of theoretical physics is a single theory uniting this eventual grand unified theory with general relativity, capable of explaining all four fundamental forces, and by extension, everything in the universe.

Billy R. Smith, Jr.

Bibliography

Layzer, D. *Constructing the Universe.* New York: Scientific American Library, 1984. A history of astronomy's changing view of the structure of the universe. Illuminates the basic properties of gravity and delightfully illustrates the primary role of gravity in cosmology.

Misner, Charles W. *Gravitation.* New York: W. H. Freeman, 1973. A very popular textbook, still in print after more than a quarter of a century, explaining gravitational physics in depth.

Schwinger, J. *Einstein's Legacy: The Unity of Space and Time.* New York: Scientific American Library, 1986. This book requires some knowledge of algebra to be fully understood, but even without the mathematics contains a wealth of information on both special and general relativity accessible to the thoughtful and careful reader.

See also: Forces of flight; Microgravity; Orbiting; Satellites; Spaceflight

Guernica, Spain, bombing

Date: April 26, 1937

Definition: The bombing of a small town in northern Spain by units of the German Condor Legion during the Spanish Civil War (1936-1939).

Significance: An operational test of the German Luftwaffe's strategy of Blitzkrieg, the bombing of Guernica created an international outcry and was also a portent of the mass bombings of civilians during World War II.

On April 26, 1937, Guernica, a Basque town in northern Spain with a population of about 7,000 people, was almost totally devastated. Fire and explosions destroyed most of the town's wooden houses, its two hospitals, and its surrounding farmhouses and village areas. Many civilians were burned to death in their houses, while survivors who ran into the streets were machine-gunned to death.

Among the few structures that survived unscathed was Casa de Juntes, the repository of a valuable historical archive. The church of Santa Maria was largely untouched, as was the famous Guernica oak tree, where the kings of Spain had traditionally taken an oath to respect the rights of their subjects, who in turn pledged their allegiance.

Events Leading to the Guernica Bombing

Mid-July, 1936: The Spanish Civil War begins.

Late July, 1936: The first units of the German Condor Legion arrive in Spain.

March 31, 1937: Nationalist General Emilio Mola begins a military campaign against Republican armies in northern Spain.

April 25, 1937: Condor Legion chief of staff Wolfram von Richthofen formulates plan for attack on Guernica.

April 26, 1937, 3:45 P.M.: Condor Legion planes leave airports in Vitoria and Burgos for the attack on Guernica, with the Rentería Bridge as the purported target.

Political Background

The destruction of Guernica became one of the most famous events of the Spanish Civil War, in which a Republican government consisting of parties on the Spanish Left was challenged by the conservative Nationalist armies of General Francisco Franco. The civil war continued the century-old struggle between monarchists and republicans in Spain. King Alfonso XIII had left the country in 1931, and in the 1936 elections, a Popular Front of socialists and other leftist parties had taken parliamentary control. There were fears within the Spanish military that the Popular Front was a communist-supported political device that might introduce communism into Spain. In response to the anticlerical traditions of the Spanish Left, much of the hierarchy of the Spanish Catholic Church also opposed the Spanish Republic.

In July, 1936, General Franco, commander of Spanish troops in Morocco, assembled a Nationalist force to oppose the Republic. This began a civil war in which much of the Spanish army, landowners, businesspeople, and the Church opposed the Republic, while agricultural workers, urban workers, and portions of the middle class supported it. Also siding with the Republican government were many Basques who, although one of the most devoutly Catholic segments of the population, had historically sought independence.

The war, seen by some as one of fascism versus communism, drew assistance from a number of other countries. The Republicans drew volunteer fighters from a variety of nations, including the United States and the Soviet Union. Nazi Germany and Fascist Italy supported the Nationalists, although Franco discouraged discussion of this fact.

Germany's Condor Legion

In military terms, the most significant aid from Nazi Germany was some 50,000 troops of the Condor Legion, a unit of the German Luftwaffe, or air force, assigned to fight in the civil war. Hermann Göring, chief of the Luftwaffe, sent the Condor Legion on the condition that it would remain under German command.

Although Nationalist leaders denied it, it was at their request that Nazi Germany had sent the Condor Legion. Although the Legion included tanks and antiaircraft batteries, its main value in the Spanish Civil War lay in its air power, consisting of four bomber squadrons of twelve bombers each, plus four fighter squadrons. It was the most powerful air arm ever assembled, exceeding the firepower of the combined air forces of World War I (1914-1918).

The role of the Condor Legion in the destruction of the Republican-controlled Guernica was not immediately clear. When Nationalist forces occupied the town several days after the disaster, they blamed the Republican forces for the devastation, claiming to find evidence that Republican forces had used explosives and arson to cause the damage. The German minister of war repeatedly cabled the Condor Legion asking who was responsible for the attack, and the Legion's radio operator in Spain was ordered to reply, "Not the Germans."

Even the international press took sides. British newspapers gave wide coverage to the incident, but some European newspapers were slow to report it. Others virtually ignored the incident, and in countries such as France, where there were fears that the war might spread into other parts of Europe, press reports sometimes reflected the editors' political inclinations. Some accepted the Nationalist claims that Guernica had been vandalized by Republican forces. In the United States, press reports of the destruction appeared quickly, even in those newspapers, such as those of the Hearst chain, that had editorially thrown their support to Franco.

Von Richthofen's Role

In their book *Guernica: The Crucible of World War II* (1975), Gordon Thomas and Max Morgan-Witts reported on the results of examination of archival files in Freiburg, Germany, and of interviews with more than forty survivors of the Guernica disaster and some dozen surviving members of the Condor Legion. Although many records were destroyed in World War II, the family of Wolfram von Richthofen, a cousin of Manfred von Richthofen, the famous "Red Baron" of World War I, and one of the commanding officers of the Condor Legion, allowed Thomas and Morgan-Witts to examine von Richthofen's papers. The authors concluded that German planes had indeed attacked Guernica and that some of the key decisions leading to the attack were made by von Richthofen.

Although he was not an admirer of Göring, von Richthofen had accepted Göring's offer of a planning position in the new German Air Ministry during the 1930's. Responding to a personal appeal from Franco, German chancellor Adolf Hitler had pledged to aid the Nationalist cause. Shortly after the Condor Legion arrived in Spain in late 1936, von Richthofen was appointed chief of staff. He was especially interested in the idea of sudden, overwhelming air attacks delivered with speed and precision, attacks that would later be known as Blitzkrieg. The Spanish Civil War became an opportunity for the Germans to test such theories of aerial warfare.

In March, 1937, General Emilio Mola, commander of the Nationalist armies in northern Spain, began a campaign against the Basque strongholds in that part of the country. Some 50,000 Nationalist troops participated in the campaign against Republican forces greatly weakened by an inability to buy arms and ammunition abroad. As Republican forces retreated, northern towns such as Guernica became filled with refugees and retreating soldiers. The normal population of Guernica swelled by several thousand.

Von Richthofen, responding to reports of Republican troops retreating from Vizcaya toward Bilbao, identified as a possible target a bridge that the troops would have to cross, the Rentería Bridge in Guernica. Meeting with Spanish field commanders and representatives of the Italian air force in Spain, von Richthofen emphasized that the retreating Republican forces had been slowed, if not halted, in the narrow canyons approaching Guernica, creating a bottleneck that provided bombing opportunities. Von Richthofen's immediate superior, General Hugo Sperrle, who had criticized the effectiveness of Nationalist forces, viewed bombing as a way to compensate for the deficiencies of his Nationalist allies.

The degree of Nationalist complicity in, or advance knowledge of, the bombing of Guernica remains unknown. Before planning the attack in detail, von Richthofen met with Juan Vignon, General Mola's chief of staff. Although there appears to be no detailed record of the conversation, von Richthofen is quoted as saying "Anything that moves on that bridge or those roads can be assumed to be unfriendly." When the town of Durango had been bombed by Legion planes on March 12, 1937, it had been on Sperrle's orders. At that time, von Richthofen had posted a memorandum noting that while targets were always military, bombing might be done "without regard for the civilian population." Sperrle, however, expressed unhappiness at the degree of civilian casualities at Durango, including fourteen nuns killed in their convent.

Details of the Bombing

For the three-hour attack on Guernica, a force of 43 bombers and fighters was assembled on airfields at Vitoria, some 50 miles away, and at Burgos, more than 120 miles away. Together, these planes would carry 100,000 pounds of incendiary, shrapnel, and high-explosive bombs. Among the planes selected for the mission was the Junkers Ju-52 bomber, considered less accurate because of its outdated bomb sights and because its spherical bombadier's chamber had to be lowered from the floor of the airplane for bombing runs, contributing to instability.

Although six Heinkel He-51 fighters participated in a diversionary attack on the town of Munditibar, the primary target of the day was Guernica. An experimental squadron of new Heinkel He-111 bombers, more maneuverable than the Ju-52 bombers and able to deliver incendiaries at 200 miles per hour, was regarded as too valuable to be used as the main force in the bombing. The Heinkels acted instead as pathfinders. The commander of the experimental squadron, Rudolf von Moreau, flew over Guernica in a new Heinkel bomber first, dropping his load of bombs on the Rentería Bridge after determining that there were no antiaircraft defenses. After this run, he joined the remainder of the experimental squadron for a bombing run over the town. They were protected by six Messerschmitt Bf-109 fighters.

A squadron of Messerschmitt fighters also provided protection for the Ju-52 bombers, the central wave of the attack, forming a protective umbrella above them. The Junkers bombers attacked in waves with a one-mile gap between them. They bombed in chains of three aircraft and flew in "V" formations at an altitude of about 6,000 feet, a height that would likely cause a large number of misses. To maintain the element of surprise, the bombers approached the bridge from the side, rather than straight on.

This mode of attack convinced many on the ground that the civilian casualties were not accidental. In fact, on the Republican side, much attention was given to a photograph of the first run of Junkers bombers approaching Guernica, taken by Father Eusebio Arronategui of Guernica. The photograph, which shows Junkers bombers approaching three abreast, was regarded as evidence that civilians, and not the narrow Rentería Bridge, were the main target. After passes by the bombers, the Messerschmitt fighters returned and attacked Guernica. One eyewitness said he saw the Messerschmitts flying north to south through the town and "firing all the time." A squadron of He-51's carried out low-level attacks using machine guns and dropping smaller bombs.

When a member of the Condor Legion complained about the use of incendiary bombs, he was told that von Richthofen wanted the mission to proceed, and quickly. The choice of the relatively inaccurate Junkers bombers, the bombing heights utilized, and the high amount of explosive power sent on the mission, 400 pounds for every square yard of the target bridge, later raised questions about the lack of concern shown for damage to civilian centers. Interviewed thirty-seven years later, some members of the Condor Legion insisted that they did not know that Republican troops were present in Guernica and that their bombs were sent off target by unexpectedly high winds. Many survivors disagreed, insisting the lack of wind allowed them to contain some of the fires.

It is known today that a valid military target, the Unceta Munitions Factory, was operating within Guernica, but the management of the plant was so pro-Franco that Republican soldiers had been posted throughout the plant to keep an eye on production. There appears to have been no discussions about the plant among the German pilots.

Aftermath

Upon their return to Germany in May, 1939, the troops of the Condor Legion were greeted personally by Göring, who announced that the "volunteers" would receive medals. In June, the entire force, some 15,000 strong, paraded through Berlin, led by von Richthofen and Sperrle. Hitler, in a welcoming speech, hailed the "heroes of Spain" for "teaching a lesson to our enemies." Von Richthofen, who participated in German air attacks against Poland in 1939, became, at 47, the youngest field marshal in the German air force.

Captured by Allied troops in southern Germany in 1945, Sperrle was among the Nazis included in the war crimes trials at Nürnberg in 1948. When a former Basque minister of justice, Jesús Leizaola, asked that the bombing of Guernica be added to the list of charges, the tribunal refused, insisting that all charges be confined to activities during World War II. Sperrle eventually was acquitted of all charges against him. At the trials, Göring commented that the Spanish Civil War had been an opportunity for him to "try out" his new air force.

The bombing of Guernica became the single-most famous event of the Spanish Civil War. Although other Spanish towns were bombed during the war, Guernica created a special controversy. The apparently systematic destruction of a town that was seen mainly as a communications center served as a precursor of World War II bombing policies and treatment of civilians. Because the town itself was of questionable military importance, the event came to be viewed as an attempt to terrorize civilians, particularly the Basques. Although the incident underlined the extent to which Franco's German and Italian allies were involved in the war, the Spanish general reportedly would not allow the subject to be discussed in his presence, at least not in public.

When, in 1937, the Spanish artist Pablo Picasso was asked by the Spanish government to contribute a work for the country's pavilion at the Paris Exhibition, he entered *Guernica*, a melange of distorted and grotesque faces, bodies, and animals that was viewed as a condemnation of both the bombing and war in general. Picasso's work guaranteed that the name "Guernica" would not fade into memory.

Niles R. Holt

Bibliography

Southworth, Herbert Rutledge. *Guernica! Guernica! A Study of Journalism, Diplomacy, Propaganda, and History.* Berkeley: University of California Press, 1977. A detailed and thoroughly documented volume focusing on the continuing debate over the truth about Guernica. Includes much information on the newspaper coverage of the event.

Thomas, Gordon, and Max Morgan-Witts. *Guernica: The Crucible of World War II.* New York: Stein and Day, 1975. A highly readable narrative that mixes the personal memories of many Guernica survivors with accounts of military leaders over strategy. Its conclusions, that the Condor Legion attacked Guernica partly as a test of bombing tactics, are based heavily on eyewitness accounts and interviews.

Thomas, Hugh. *The Spanish Civil War.* New York: Harper & Row, 1977. The standard work concerning the events before and during the Spanish Civil War and a balanced and dispassionate account. The author believes that Guernica was attacked mainly because of its value as a communications center for Republican armies. He also discusses the efforts of the German government to cover up their involvement after an international outcry over the bombing.

See also: Bombers; Luftwaffe; Messerschmitt aircraft; Military flight; Spanish Civil War; World War II

ENCYCLOPEDIA OF
FLIGHT

Alphabetical Index of Entries

Categorized Index of Entries

Aircraft Design

Animal Flight

Aviation Careers

Instruments and Controls

Manufacturers

Military Flight

Organizations, Programs, and Agencies